Microwave Engineering

Annapurna Das
School of Electronics and Communication Engineering
College of Engineering
Anna University
Chennai, India

Sisir K Das
SAMEER-Centre for Electromagnetics
Ministry of Information Technology
Chennai, India

Higher Education

Boston   Burr Ridge, IL   Dubuque, IA   New York
San Francisco   St. Louis   Bangkok   Bogotá   Caracas   Kuala Lumpur
Lisbon   London   Madrid   Mexico City   Milan   Montreal   New Delhi
Santiago   Seoul   Singapore   Sydney   Taipei   Toronto

Higher Education

MICROWAVE ENGINEERING

Published by McGraw-Hill, a business unit of The McGraw-Hill Companies, Inc., 1221 Avenue of the Americas, New York, NY 10020. Copyright © 2008 by The McGraw-Hill Companies, Inc. All rights reserved. No part of this publication may be reproduced or distributed in any form or by any means, or stored in a database or retrieval system, without the prior written consent of The McGraw-Hill Companies, Inc., including, but not limited to, in any network or other electronic storage or transmission, or broadcast for distance learning.

Some ancillaries, including electronic and print components, may not be available to customers outside the United States.

This book is printed on acid-free paper.

1 2 3 4 5 6 7 8 9 0 BKM/BKM 0 9 8 7 6

ISBN 978–0–07–352950–9
MHID 0–07–352950–8

Senior Sponsoring Editor: *Michael S. Hackett*
Developmental Editor: *Rebecca Olson*
Executive Marketing Manager: *Michael Weitz*
Project Coordinator: *Melissa M. Leick*
Senior Production Supervisor: *Kara Kudronowicz*
Associate Media Producer: *Christina Nelson*
Senior Designer: *David W. Hash*
Cover Designer: *Rokusek Design*
(USE) Cover Image: *©PhotoDisc*
Compositor: *Techbooks*
Printer: *Book-mart Press*

This book was previously published by Tata McGraw-Hill Publishing Company Limited, New Delhi, India, copyright © 2000.

Library of Congress Cataloging-in-Publication Data

Das, Annapurna.
    Microwave engineering / Annapurna Das, Sisir K. Das. — 1st ed.
        p.   cm.
    Includes index.
    ISBN 978–0–07–352950–9 — ISBN 0–07–352950–8 (alk. paper)
    1. Microwave devices. I. Das, Sisir K. II. Title.

TK7876.D36   2008
621.381'3—dc22
                                        2006041924
                                        CIP

www.mhhe.com

Series Foreword

The McGraw-Hill Companies has been a leader in providing trusted information and analysis for well over a century. From the Industrial Revolution to the Internet Revolution, The McGraw-Hill Companies has filled a critical need for information and insight by helping individuals and businesses in the field of Engineering.

As early as 1910, the McGraw-Hill book company was making a difference on college campuses with the publication of its first series, Electrical Engineering Texts, outlined and edited by Professor Harry E. Clifford of Harvard University. McGraw-Hill's Electrical Engineering textbooks have shaped engineering curricula worldwide. I am thrilled that I have been invited to be the Global Series Editor in Electrical Engineering, helping to shape how the next generation of electrical engineering students around the globe will learn.

As advances in networking and communications bring the global academic community even closer together, it is essential that textbooks recognize and respond to this shift. It is in this spirit that we will publish textbooks in the McGraw-Hill Core Concepts in Electrical Engineering Series. The series will offer textbooks for the global electrical engineering curriculum that are reasonably priced, innovative, dynamic, and will cover fundamental subject areas studied by Electrical and Computer Engineering students. Written with a global perspective and presenting the latest in technological advances, these books will give students of all backgrounds a solid foundation in key engineering subjects.

Microwave Engineering by Annapurna Das of Anna University and Sisir K. Das of the Centre for Electromagnetics, Chennai, is an introductory textbook for undergraduate and graduate students. It serves as a fundamental introduction to microwave engineering. The book covers microwave vacuum tube devices, antennae, and radar and industrial applications of microwave systems. Furthermore, microwave measurements and radiation hazards are discussed in two separate chapters. Numerous examples and exercises are provided for the student.

This book has been reviewed and assessed for use in engineering classrooms at all levels. Like *Microwave Engineering,* each book in the Core Concepts series presents a comprehensive, straightforward, and *accurate* treatment of an important subject in Electrical & Computer Engineering. With their clear approach, contemporary technology, and international perspective, Core Concepts books are an unmistakable choice for professors wanting understandable, concise engineering textbooks that adhere to the standards of The McGraw-Hill Companies.

—Richard C. Dorf,
University of California, Davis
Series Editor, The Core Concepts Series in Electrical and Computer Engineering

About the authors

Annapurna Das, born in India, received her B.Sc. (Hons.) and M.Sc. degrees in Physics from Calcutta University, India, M.Tech. degree in Microwave Electronics and Ph.D. degree in Electrical Engineering from University of Delhi, India.

As a senior research fellow of CSIR, New Delhi, she was engaged in doctoral research and teaching undergraduate courses in the Electrical Engineering Department of Delhi College of Engineering, during 1980 to 1982 and doctoral research at the Indian Institute of Technology, Chennai, during 1983 to mid 1985.

She joined the School of Electronics and Communication Engineering, Anna University in September 1985 where she is presently an Assistant Professor. She is in charge of the microwave laboratory and is engaged in a number of training, research and teaching programmes of the Anna University in the areas of microwaves, EMI/EMC, Electromagnetics and antenna.

She is a life member of Society of EMC Engineers (India) and Indian Society of Technical Educations (ISTE). Her current and major areas of research interest are microwaves, EMI/EMC, Micro-strip Antennas and Electromagnetics.

Sisir K. Das, born in India, obtained his B.Sc. (Hons) Physics and B.Tech. in Radio Physics and Electronics degrees from Calcutta University, M.Tech. and Ph.D. degrees in Electronics and Electrical Communication Engineering from the Indian Institute of Technology, Kharagpur and Anna University respectively. After a brief spell of teaching and research at the University of Delhi, he joined the Department of Electronics where he undertook the responsibility of setting up the Centre for Electromagentics at Chennai.

Dr. Das now heads the EMC Division at SAMEER Centre for Electromagnetics, Chennai. He is engaged in a number of training, research and consultancy assignments in the areas of EMI/EMC, electromagnetic scattering and antennas. He is associated with several Indo-U.S. collaborative research projects on EMC with NIST, USA.

Dr. Das is an Associate Editor of the IEEE Trans. on EMC, senior member of IEEE, life member of the Society of EMC Engineers (India) and also the Editor of the Journal of the *Society of EMC Engineers (India)*. His current areas of interest are Microwave measurements, EMC measurements and mitigation techniques.

Preface

In recent years microwaves have been used extensively in radars, transmission of television programmes, astronomic research, radio spectroscopy, domestic ovens and many other things. This rapid progress in microwave electronics has created an increasing demand for trained microwave engineering personnel.

This book is intended for the undergraduate and postgraduate students specializing in electronics. It will also serve as reference material for engineers employed in the industry. The fundamental concepts and principles behind microwave engineering are explained in a simple, easy-to-understand manner. Each chapter contains a large number of solved problems which will help the students in problem solving and design.

Chapter 1 introduces the subject with its background and applications. Chapter 2 explains the basic transmission line theory introducing the concepts of impedance characteristics, mismatch effects, loss characteristics and graphical solution techniques using Smith Chart, which will help in the design and analysis of microwave components and circuits. Chapter 3 is devoted to basic EM theory and methods of solution in different coordinate systems to help in understanding microwave propagation in free space and in transmission guides.

Since the microwave circuits involve different configurations of signal guiding structures, Chapter 4 deals with the characteristics of microwave transmission lines, viz. rectangular and circular waveguides, coaxial lines, strip lines, microstrip lines and slot lines. The excitation principles of different modes in these structures are also described. Concepts of impedance transformation and broad band matching for maximum power transfer and its design are covered in Chapter 5.

Quantities such as voltage, current and impedance cannot be measured directly at microwave frequencies where the signal is propagated as electromagnetic waves. The directly measurable quantities are the amplitude and phase of a wave reflected from any discontinuity relative to the incident wave amplitude and phase, respectively. These can be described in terms of scattering parameters through which the field equations are linearly related. The properties of scattering matrix, analysis of various passive microwave circuits and devices using S-parameters are described in Chapter 6.

Chapter 7 describes various resonating structures at microwave frequencies. Analysis and design of microwave filters are covered in Chapter 8. Chapters 9 and 10 are devoted to microwave active devices. Since a large number of such devices have been designed during the last few years, only a selected number of them, which have important applications in many common microwave facilities are described and analyzed.

Chapter 11 on microwave applications will help the reader understand many practical aspects of microwave engineering. Topics covered in this chapter include microwave propagation equations, antennas, radar and communication systems, and industrial applications of microwaves.

Chapter 12 is written on microwave radiation hazards and protection techniques. Beginners may not be aware of these hazards which can be potentially dangerous in high power applications. This chapter, therefore, provides knowledge on safety measures.

Finally, Chapter 13 on microwave measurements links theory with practice. This will help in establishing the experimental set-up in the laboratories and for precision measurements in research institutions and laboratories.

The authors are indebted to all the textbooks that they have encountered as students, and as teachers or scientists. They sincerely thank the reviewers especially Prof. Ramesh Garg, Department of Electronics and Electrical Communication Engineering, IIT Kharagpur, who reviewed the manuscript of this text at its various stages and made valuable suggestions. Constructive criticism and suggestions from the readers of the book will be gratefully accepted.

ANNAPURNA DAS
SISIR K DAS

Contents

Chapter 1

INTRODUCTION

1.1 INTRODUCTION

Microwave is a region in the electromagnetic (EM) wave spectrum in the frequency range from about 1 GHz (= 10^9 Hz) to 100 GHz (= 10^{11} Hz). This corresponds to a range of wavelengths from 30 cm to 0.3 mm in free space. The free space is characterised by the electrical medium parameters-permittivity $\varepsilon_o = 10^{-9}/36\pi$ farad/m, permeability $\mu_o = 4\pi \times 10^{-7}$ henry/m and conductivity $\sigma_o = 10^{-14}$ mho/m. During World War II, microwave engineering became a very essential consideration for the development of high resolution radars capable of detecting and locating enemy planes and ships through a narrow beam of EM energy. Such a beam could be achieved by means of a paraboloid antenna of large diameter compared to the wavelength of radiation. Therefore, size of the antenna to be carried by an airplane on board could be reduced by using such short wavelength of the microwaves.

The microwave band is capable of making economic transmission of a large number of communication channels or TV programs by modulating all these channels or programs into a single microwave carrier and transmitting them over one communication link. This is possible because the modulation side-bands are a few per cent of the microwave carrier frequency.

Because of very short wavelengths, microwaves are capable of almost freely propagating through the ionized layers in the atmosphere. This facilitates radio astronomic research of space, and communication between the ground stations and space vehicles.

According to the quantum theory, the quantum of energy at microwave frequencies becomes comparable to the difference in energies between adjacent energy levels of atoms and molecules. Moreover, molecular, atomic and nuclear systems exhibit resonances in the microwave range. These lead microwaves to become a powerful tool in microwave radio spectroscopy for material analysis.

In more recent years microwaves are widely used in domestic microwave ovens for rapid cooking and also in industrial and medical uses for microwave heating.

However, microwave engineering has a marked difference from the conventional electronics engineering because of the short wavelengths involved. Conventional low-frequency circuit analysis based on Kirchhoff's laws and voltage-current concepts of the distributed transmission line theory no longer apply. This is because the propagation time of electrical effects from one point in a circuit to another point at microwave frequencies is comparable with the period of the oscillation of currents and charges. Moreover, conventional circuits or lines radiate out the microwave energy as electromagnetic waves resulting in high loss in signal transmission. Thus microwave transmission involves propagation of EM waves consisting of changing electric and magnetic fields in a medium.

These specific properties of microwaves steered engineers to develop techniques for microwave guided structures such as waveguides, coaxial lines, strip lines, microstrip lines, slot lines and microwave sources like klystrons, magnetrons, travelling wave tubes, backward wave oscillators, microwave solid state devices and other microwave circuit components like, attenuators, phase-shifters, isolators, circulators, directional couplers, detectors, mixers, etc.

Microwave frequencies are grouped into several smaller bands which are designated as listed in Table 1.1.

Table 1.1 *Microwave bands*

New Designation	Frequency (GHz)	Old Designation	Frequency (GHz)
C	0.5–1.0	VHF	0.5–1.0
D	1.0–2.0	L	1.0–2.0
E	2.0–3.0	S	2.0–4.0
F	3.0–4.0	C	4.0–8.0
G	4.0–6.0	X	8.0–12.4
H	6.0–8.0	Ku	12.4–18.0
I	8.0–10.0	K	18.0–26.5
J	10.0–20.0	Ka	26.5–40.0
K	20.0–40.0	v	40.0–75.0
L	40.0–60.0		
M	60.0–100.0		

1.2 HISTORICAL RESUME

There are some pioneers who have laid the foundations of microwave engineering. Some of them are worth mentioning and described as follows.

James C. Maxwell, the founder of the electromagnetic theory of radiation, presented to the Royal Society in 1864 a paper titled "A Dynamical Theory of the Electromagnetic Field" which described the properties of electromagnetic fields in terms of 20 equations. These equations constitute the well known *Maxwell's equations*. He had predicted theoretically the existence of electric and magnetic fields associated with electromagnetic wave propagation.

In 1893, *Heinrich Hertz* first conducted an experiment to show that a parabolic antenna fed by a dipole, on excitation by a spark discharge, sends a signal by wave motion to a similar receiving arrangement at a distance. He gave a strong experimental support for the theoretical conclusions drawn by Maxwell for electromagnetic fields.

William Thompson (1893) developed the waveguide theory for propagation of microwaves in a guided structure. Later, *Lodge* (1897-1899) established the mode properties of propagation of EM waves in free space and in a hollow metallic tube known as the waveguide.

Sir J.C.Bose (1895, 1897, 1898) generated millimeter waves using a circuit developed in his laboratory and used these waves for communication, and developed microwave spectrometers, diffraction gratings, polarimeters and detectors for conducting microwave experiments. He also developed microwave horn antennas which are still considered to be useful feeds for reflector antennas.

In 1937, the microwave vacuum tube Klystron was developed by the *Russel and Varian Bross. J.D. Kraus* (1938) developed corner reflector antenna for electromagnetic wave transmission. *Kompfner* (1944) developed the microwave travelling wave tube. *Percy Spencer* (1946) built the microwave oven for domestic cooking.

Some of the modern devices were developed after 1950. *Deschamps* (1953) developed the microstrip antenna. J.B.Gunn (1963) developed the Gunn diode for microwave generation using solid state materials such as GaAs.

There are numerous events in the field of microwaves all of which are not mentioned here. Interested readers may refer text books, journals and scientific magazines for further information.

REFERENCES

1.1 "Historical Perspectives of Microwave Technology", Special Centennial Issue, *IEEE Trans.*, Vol. MTT-32, September, 1984.

1.2 Montgomery, C G, *Technique of Microwave Measurements*, McGraw-Hill Book Company, New York, 1947.

1.3 Ginzton, E L, *Microwave Measurements*, McGraw-Hill Book Company, New York, 1957.

1.4 Bailey A E, (Ed.), *Microwave Power Engineering*, Academic Press, New York, 1961.

Chapter 2

BASIC TRANSMISSION LINE THEORY

2.1 INTRODUCTION

When a microwave signal is transmitted through two open conductor transmission lines of length larger than the wavelength, (energy is lost due to radiation). This loss is reduced or eliminated by making transmission of microwave signals through wave guiding structures. The wave guiding structure may consist of two co-axial conductors or two parallel plates or it may be a single hollow conductor called *waveguide*. Many other forms are also possible which will be described later. The two conductor lines can be analysed in terms of voltage, current, and impedance by the distributed circuit theory. Waveguides can be analysed from the solution of Maxwell's field equations. Since the general characteristics of electromagnetic wave propagation in all these lines are the same, the basic transmission line theory concerning voltage standing waves, reflection and impedance is applicable to problems in all these guiding structures when excited by a single mode.

2.2 TRANSMISSION LINE EQUATIONS

At microwave frequencies the line parameters are distributed along the whole length of the line in the z-direction. A small length of the line can be represented by an equivalent symmetrical T-network with constant parameters R, L, G, and C per unit length as shown in Fig. 2.1, where R is the resistance per unit length which takes into account the ohmic loss in the line conductor, L is the inductance per unit length which takes into account the magnetic energy storage around the conductor, G is the conductance per unit length which takes into account the dielectric loss between the line conductors and C is the capacitance per unit length which appears due to two conductors at different potential and represents the electric energy storage. In this section it is assumed that the line is excited by a matched generator ($Z_g = Z_0$).

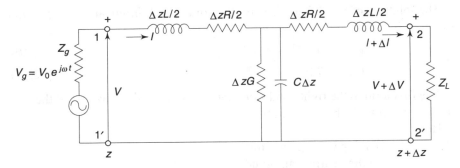

Fig. 2.1 *Equivalent circuit of a small section of a uniform transmission line*

The line voltage and current per unit length decrease as z increases and are expressed by

$$\frac{dV}{dz} = -(R + j\omega L)I = -ZI \tag{2.1}$$

$$\frac{dI}{dz} = -(G + j\omega C)V = -YV \tag{2.2}$$

where

$$Z = R + j\omega L \tag{2.3}$$

$$Y = G + j\omega C \tag{2.4}$$

are called the *series impedance* and *shunt admittance* per unit length of the line. Eliminating I from Eq. 2.1 and V from Eq. 2.2, the voltage and current wave equations are, respectively,

$$\frac{d^2V}{dz^2} = \gamma^2 V \tag{2.5}$$

$$\frac{d^2I}{dz^2} = \gamma^2 I \tag{2.6}$$

Here

$$\gamma = \sqrt{(ZY)} = \sqrt{[(R + j\omega L)(G + j\omega C)]}$$

$$= \alpha + j\beta \tag{2.7}$$

and is called the *propagation constant*. α is called the *attenuation constant* and β, the *phase constant*. At high frequencies when the losses in the line are small, Eq. 2.7 gives

$$\alpha = 1/2\left(R\sqrt{(C/L)} + G\sqrt{(L/C)}\right) \tag{2.8}$$

and

$$\beta = \omega\sqrt{(LC)} \tag{2.9}$$

The solutions of Eqs 2.5 and 2.6 give the voltage and current at any point z and are given by

$$V(z) = V_s e^{-\gamma z} + V_r e^{+\gamma z} \tag{2.10}$$

$$I(z) = I_s e^{-\gamma z} + I_r e^{+\gamma z} \tag{2.11}$$

The first term in the right hand side represents the *incident wave* and the second term represents the *reflected wave*.

Here

V_s = sending voltage amplitude
I_s = sending current amplitude
V_r = reflected voltage amplitude
I_r = reflected current amplitude

The phase velocity of the wave is given by

$$V_p = \omega/\beta = 1/\sqrt{(LC)} \tag{2.12}$$

2.3 CHARACTERISTIC AND INPUT IMPEDANCES

When the reflected wave in the line is zero, the ratio $V(z)/I(z)$ is called the *characteristic impedance* of the line and is defined by

$$Z_0 = V_s/I_s = \sqrt{\left[\frac{R + j\omega L}{G + j\omega C}\right]} \tag{2.13}$$

When $R \ll \omega L$ and $G \ll \omega C$ for low loss lines and also at microwave frequencies, $Z_0 \approx (L/C)$. The impedance of the line at z looking towards the load is called the *input impedance* of the line:

$$Z_{in} = \frac{V(z)}{I(z)} = Z_0 \frac{V_s e^{-\gamma z} + V_r e^{\gamma z}}{V_s e^{-\gamma z} - V_r e^{\gamma z}} \tag{2.14}$$

When the line of length l is terminated by a load Z_L, as shown in Fig. 2.2, the input impedance becomes

$$Z_{in} = \frac{(Z_L \cosh \gamma l + Z_0 \sinh \gamma l)}{(Z_0 \cosh \gamma l + Z_L \sinh \gamma l)} Z_0 \tag{2.15}$$

(i)   If the line is short-circuited, $Z_L = 0$ and $V_L = 0$, therefore,

$$Z_{inSC} = Z_0 \tanh \gamma l \tag{2.16}$$

(ii) For an open-circuited line, $Z_L = \infty$, $I_L = 0$, and

$$Z_{inOC} = Z_0 \coth \gamma l \tag{2.17}$$

(iii) For a loss-less line $\alpha = 0$, $\gamma = j\beta$ and Eqs 2.15–2.17 reduce to

$$Z_{in} = Z_0 \frac{Z_L + jZ_0 \tan \beta l}{Z_0 + jZ_L \tan \beta l} \tag{2.18}$$

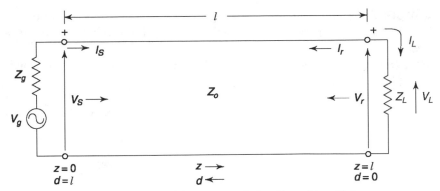

Fig. 2.2 *A uniform transmission line of length l*

$$Z_{inSC} = j Z_0 \tan \beta l \qquad (2.19)$$

$$Z_{inOC} = -j Z_0 \cot \beta l \qquad (2.20)$$

Example 2.1 A telephone line has $R = 6$ ohms/km, $L = 2.2$ mh/km, $C = 0.005$ µf/km, and $G = 0.05$ µmho/km. Determine Z_0, α, β at 1 kHz. If the line length is 100 km, determine the attenuation and phase shift of the signal. Calculate the phase velocity of the signal.

Solution

$$\omega = 2\pi \times 1000 = 6280 \text{ rad/sec}$$

$$Z_0 = \sqrt{\left[\frac{R + j\omega L}{G + j\omega C} \right]}$$

$$= \sqrt{\frac{6 + j\,6280 \times 2.2 \times 10^{-3}}{0.05 \times 10^{-6} + j\,6280 \times 0.005.10^{-6}}}$$

$$= 692.5 - j\,11.52 \text{ ohm}$$

$$\tau = \sqrt{((R + j\omega L)(G + j\omega C))}$$

$$= \sqrt{(6 + j6280 \times 2.2 \times 10^{-3})(0.05 \times 10^{-6} + j\,6280 \times 0.005 \times 10^{-6})}$$

$$= 0.0045 + j\,0.0213 = \alpha + j\beta$$

Therefore, $\alpha = 0.0045$ Np/km and

$$\beta = 0.0213 \text{ rad/km}$$

For 100 km length,

attenuation $= 0.45$ Np $= 8.686 \times 0.45 = 3.91$ dB

phase shift $= 2.13$ rad

$$\text{Phase velocity} = \frac{\omega}{\beta} = \frac{6280 \text{ (rad/s)}}{0.0213 \text{ (rad/km)}} = 294.84 \times 10^3 \text{ km/s}$$

Example 2.2 In Example 2.1 if a signal of 1 V supplies power to the line terminated in its characteristic impedance, find the power delivered at the receiving end.

Solution Sending end voltage $V_s = 1$ V , $l = 100$ km, $R_e(Z_0) = 692.5$ ohm
Load impedance $Z_L = Z_0$, $Z_{in} = Z_0$, $\gamma = 0.0045 + j\,0.0213$

Sending end current $I_s = \dfrac{V_s}{Z_{in}} = \dfrac{1.0}{692.5 - j\,11.52} = 1.444 \times 10^{-3} \underline{/0.9531}\,°$amp

Receiving end current $I_L = I_s\,e^{-\gamma l}$

$$= [1.444 \times 10^{-3}\underline{/+ 0.9531}\quad e^{-(0.0045\,+\,j0.0213)l}\quad \text{amp}$$

Therefore,

$$|I_L| = 1.438 \times 10^{-3}\text{ amp}$$

Power delivered at load $= |I_L^2| \times \text{Real }(Z_L)$

$$= |I_L^2| \times \text{Real }(Z_0)$$
$$= (1.438 \times 10^{-3})^2 \times 692.5$$
$$= 14.31 \times 10^{-4}\text{ Watts}$$

Example 2.3 A lossless transmission line with characteristic impedance 500 ohms is excited by a signal of voltage $10\ \angle 0°$ volts at 1.2 MHz. If the line is terminated by Z_L at a distance 1 km, calculate (a) input impedance of the line for $Z_L = \infty$ and 0, (b) the voltage at the mid point of the line for $Z_L = Z_0$.

Solution

$$f = 1.2 \text{ MHz}, \lambda = c/f = \frac{300}{1.2} = 250 \text{ m}$$

$$\beta l = \frac{2\pi l}{\lambda} = \frac{2\pi \times 1000}{250} = 8\ \pi\,\text{rad}$$

(a)   For $Z_L = \infty$, $Z_{in} = -j\,Z_0 \cot \beta l$

$$= -j\,500 \cot (\,8\pi\,) = -j\,\infty \text{ ohm}$$

(The line is essentially an O.C. because $l = 4\ \lambda$.)

For $Z_L = 0$,

$$Z_{in} = j\,Z_0 \tan \beta l$$
$$= j\,500 \tan (8\pi) = 0 \text{ ohm}$$

(b)   For $Z_L = Z_0$,

$$V(z) = V_s\,e^{-\gamma z}$$
$$\alpha = 0$$

or $V(l/2) = V_s\,e^{-j\beta l/2}$

$$= 10e^{-j4\pi} \text{ Volt}$$

2.4 REFLECTION AND TRANSMISSION COEFFICIENTS

Due to the presence of a reflected wave, the voltage reflection coefficient at any point z on the line is defined by

$$\Gamma(z) = \frac{V_r\,e^{\gamma z}}{V_s\,e^{-\gamma z}} = \frac{V_r\,e^{\gamma(l-d)}}{V_s\,e^{-\gamma(l-d)}} = \frac{V_r\,e^{\gamma l}\,e^{-2\gamma d}}{V_s\,e^{-\gamma l}}$$

$$= \Gamma_L\,e^{-2\gamma d}\,;\,z = l - d \tag{2.21}$$

where

$$\Gamma_L = \frac{V_r\,e^{\gamma l}}{V_s\,e^{-\gamma l}} \tag{2.22}$$

is called the *load end voltage reflection coefficient* and d is the distance meas-ured from the load end. Since current reverses its direction at the load end, the current reflection coefficient

$$\Gamma_c = \frac{I_r\,e^{\gamma l}}{I_s\,e^{-\gamma l}}$$

$$= \frac{-V_r\,e^{\gamma l}}{V_s\,e^{-\gamma l}}$$

$$= -\Gamma(z) \tag{2.23}$$

When a line is terminated by a matched load, maximum power will be trans-ferred to the load without any reflection and the line is said to be matched termi-nated. Under this condition

$$Z_{in} = Z_L = Z_0\,,\text{ and } \Gamma = 0 \tag{2.24}$$

It is to be noted that since

$$\Gamma_{SC} = -1,\,\Gamma_{OC} = 1,\text{ and } \Gamma_{matched} = 0$$

therefore, $$-1 < \Gamma(z) < +1 \tag{2.25}$$

From Eqn. 2.21

$$\Gamma(z) = \Gamma_L\,e^{-2\alpha d}\,e^{-j2\beta d}$$

$$= |\Gamma_L|\,e^{j\theta}{}_L\,e^{-2\alpha d}\,e^{-j2\beta d} \tag{2.26}$$

where $$Z_L = |Z_L|\,e^{j\theta}{}_L$$

Thus for a line with losses as we move from load end towards the generator (d increases), the complex reflection coefficient changes in an inward spiral way as shown in Fig. 2.3. For a lossless line ($\alpha = 0$), the complex reflection coefficient changes only in phase and describes a circle as shown in Fig. 2.4.

It must be noted that the reflected wave from the mismatched load will again be reflected from the source end if the source impedance is not matched to the line ($Z_g \neq Z_0$).

From Eqs 2.15, 2.22 and 2.26

$$Z_{in} = Z_0\,\frac{1 + \Gamma_L\,e^{-2\gamma d}}{1 - \Gamma_L\,e^{-2\gamma d}}$$

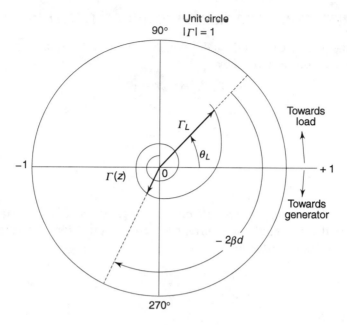

Fig. 2.3   *Locus of reflection coefficient for a uniform line with losses*

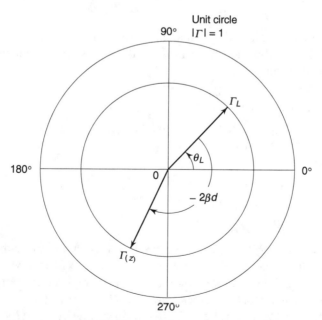

Fig. 2.4   *Locus of reflection coefficient for a uniform lossless line*

$$= Z_0 \; \frac{1 + \Gamma(z)}{1 - \Gamma(z)} \qquad\qquad (2.27)$$

From Eqs 2.10 and 2.11 the transmitted voltage and current at the load end are

$$V_L = V_s e^{-\gamma l} + V_r e^{\gamma l} \tag{2.28}$$

$$I_L = \frac{V_L}{Z_L} = \frac{V_s e^{-\gamma l}}{Z_0} - \frac{V_r e^{\gamma l}}{Z_0} \tag{2.29}$$

Solving Eqns 2.28 and 2.29, the voltage reflection and transmission coefficients are obtained as

$$\Gamma_L = \frac{V_r e^{\gamma l}}{V_s e^{-\gamma l}} = \frac{Z_L - Z_0}{Z_L + Z_0}; \tag{2.30}$$

$$T = \frac{V_L}{V_s e^{-\gamma l}} = \frac{2Z_L}{Z_L + Z_0} = 1 + \Gamma_L; \tag{2.31}$$

$$= \text{Transmission coefficient for current}$$

The power P_L transmitted to a mismatched load Z_L is given by

$$P_L = \text{incident power} - \text{reflected power}$$

or, $$\frac{\text{power transmitted to the load}}{\text{incident power}} = 1 - \frac{\text{reflected power}}{\text{incident power}}$$

or, $$\frac{|V_L|^2/Z_L}{|V_s e^{-\gamma l}|^2/Z_0} = \frac{Z_0}{Z_L}|T|^2$$

$$= 1 - |\Gamma_L|^2 \tag{2.32}$$

2.5 STANDING WAVE

In a mismatched terminated line, the incident and reflected signals interfere to produce a standing wave pattern along the line. From Eqns 2.10, 2.11, 2.21 and 2.23, the voltage and current at any point z on the line $(0 < z < l)$ are given by

$$V = V_s (e^{-\gamma z} + \Gamma_L e^{\gamma z}) = V_s e^{-\gamma(1-d)} (1 + \Gamma_L e^{-2\gamma d}) \tag{2.33}$$

$$I = \frac{V_s}{Z_0} (e^{-\gamma z} - \Gamma_L e^{\gamma z}) = \frac{V_s e^{-\gamma(1-d)}}{Z_0} (1 - \Gamma_L e^{-2\gamma d}) \tag{2.34}$$

For a complex load Z_L, $\Gamma_L = |\Gamma_L| e^{j\theta_L}$; then

$$|V| = |V_s e^{-\alpha(1-d)}| |1 + \Gamma_L e^{-2\gamma d}|$$

$$= |V_s e^{-\gamma(1-d)}| \left[(1 + |\Gamma_L| e^{-2\alpha d})^2 - 4 |\Gamma_L| e^{-2\alpha d} \right.$$

$$\left. \sin^2 (\beta d - \theta_L/2) \right]^{1/2} \tag{2.35}$$

For a lossless line $\alpha = 0$, then

$$|V| = |V_s| \left[(1 + |\Gamma_L|)^2 - 4 |\Gamma_L|^2 \sin^2(\beta d - \theta_L/2) \right]^{1/2} \tag{2.36}$$

Eqns 2.35 and 2.36 show that,

(1) magnitude of the line voltage varies between maximum values of

$$|V_s \, e^{-\alpha(1-d)}| \, (1 + |\Gamma_L| \, e^{-2\alpha d}), \text{ when } \beta d - \theta_L/2 = n\pi \qquad (2.37)$$

and,

(2) minimum values of

$$|V_s \, e^{-\alpha(1-d)}| \, (1 - |\Gamma_L| \, e^{-2\alpha d}), \text{ when } \beta d - \theta_L/2 = n\pi + \pi/2 \qquad (2.38)$$

where n is an integer.

In the lossless case, these maximum or minimum values are same throughout their locations on the line. In a line with losses Γ decreases exponentially by a factor of $e^{-2\alpha d}$ and a standing wave pattern appears as shown in Fig. 2.5. Successive maxima and minima are spaced a distance $z = \pi/\beta = \lambda/2$.

Since the current reflection coefficient is $-\Gamma_L$, the current maxima corresponds to voltage minima and vice-versa. Therefore, 'Voltage Standing Wave' (VSW) and 'Current Standing Wave' (CSW) are 90 degrees out of phase along the line. Other than the perfect short or open circuits the amplitude of the incident and reflected waves are not equal, so that the minimum of the voltage and current standing waves are not zero. Figure 2.6 shows the nature of standing waves along the line near the load end when terminated with different resistive and reactive loads.

The magnitude of the standing waves is measured in terms of 'Standing Wave Ratio' (SWR) defined by

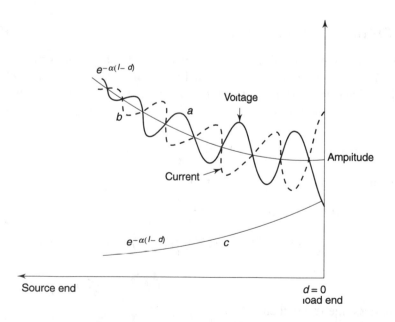

Fig. 2.5 *(a) Standing wave pattern in a lossy line (b) Envelope of incident-wave amplitude (c) Envelope of reflected-wave amplitude*

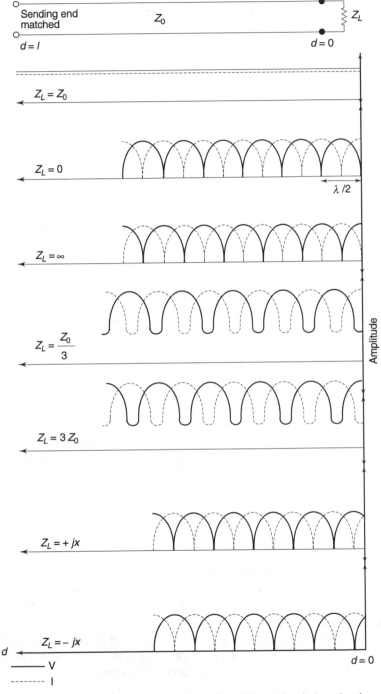

Fig. 2.6 *Voltage standing wave pattern for different loads in a lossless line*

$$\text{SWR} = \frac{\text{max. voltage or current}}{\text{min. voltage or current}} = S$$

or,

$$S = \frac{1+|\Gamma_L|\,e^{-2\alpha d}}{1-|\Gamma_L|\,e^{-2\alpha d}} \; ; \quad \alpha \neq 0 \tag{2.39}$$

or,

$$S = \frac{1+|\Gamma_L|}{1-|\Gamma_L|}; \quad \alpha = 0 \tag{2.40}$$

It can be written that,

$$|\Gamma_L| = \frac{S-1}{S+1} \tag{2.41}$$

Table 2.1 shows the value of VSWR and reflection coefficients for different load values. From Eqs 2.30 and 2.40 for resistive load in a lossless line,

$$S = \frac{1+\dfrac{R_L \sim R_0}{R_L + R_0}}{1-\dfrac{R_L \sim R_0}{R_L + R_0}} = \frac{R_L + R_0 + R_L - R_0}{R_L + R_0 - R_L - R_0}$$

$$= \frac{R_L}{R_0} \geq 1, \text{ for } R_L \geq R_0 \tag{2.41a}$$

$$= \frac{R_0}{R_L} \geq 1, \text{ for } R_L \leq R_0 \tag{2.41b}$$

Table 2.1 *Reflection coefficient and VSWR*

Load	Voltage reflection coefficient	s
0	-1	∞
∞	1	∞
Z_0	0	1
$+jx$	$\exp[j\,2\tan^{-1}(X/Z_0)]$	∞
$-jx$	$\exp[-j2\tan^{-1}(X/Z_0)]$	∞

These relations show that $-1 \leq \Gamma_L < +1$ and $1 \leq S < \infty$. It can be shown that for a lossless line with real characteristic impedance Z_0, the line impedances at voltage max. and min. points are purely real, and also have max. and min. values, respectively. At voltage maximum or current minimum,

$$Z_{\text{in}} = Z_{\text{max}} = \frac{V_{\text{max}}}{I_{\text{min}}} = \frac{E_i\,(1+|\Gamma_L|)}{I_i\,(1-|\Gamma_L|)} = Z_0 S \tag{2.42}$$

and at voltage minimum or current maximum

$$Z_{in} = Z_{min} = \frac{V_{min}}{I_{max}} = \frac{E_i \, (1 - |\Gamma_L|)}{I_i \, (1 + |\Gamma_L|)} = Z_0/S \qquad (2.42a)$$

Example 2.4 A 50 ohm lossless line connects a signal of 100 kHz to a load of 100 ohm. The load power is 100 mW. Calculate the
(a) voltage reflection coefficient
(b) VSWR
(c) position of the first V_{min} and V_{max}.
(d) impedance at V_{min} and V_{max} and values of V_{max} and V_{min}.
Solution

$$Z_0 = 50 \text{ ohm}, f = 100 \text{ kHz}, Z_L = 100 \text{ ohm}$$

$$P_L = 100 \text{ mW}, \lambda = \frac{c}{f} = \frac{3 \times 10^8}{100 \times 10^3} = 3000 \text{ m}$$

(a)
$$\Gamma_L = \frac{Z_L - Z_0}{Z_L + Z_0} = \frac{100 - 50}{100 + 50} = \frac{1}{3}$$

(b)
$$\text{VSWR} = \frac{1 + \Gamma_L}{1 - \Gamma_L} = \frac{1 + 1/3}{1 - 1/3} = 2$$

(c) For real $Z_L > Z_0$, first V_{max} is located at the load and first V_{min} is located at a distance $\lambda/4 = 750$ m from the load end.

(d) Load power $P_L = \dfrac{V^2_{max}}{Z_L} = 100 \text{ mW}$

or,
$$V_{max} = \sqrt{(100 \times 100 \times 10^{-3})} = \sqrt{10} \text{ volts}$$

$$V_{min} = V_{max}/S = \sqrt{10}/2 \text{ volts}$$

At V_{max}, $Z_{in} = SZ_0 = 2 \times 50 = 100$ ohm

At V_{min}, $Z_{in} = Z_0/S = 50/2 = 25$ ohm

2.6 MISMATCH LOSSES IN TRANSMISSION LINES

Due to mismatch between the input and output terminations of a lossy transmission line, four losses are often defined in microwave circuits. These are attenuation loss, insertion loss, reflection loss, and return loss. They are also interrelated by expressions involving reflection coefficient, transmission coefficient, VSWR and attenuation constant as given below. The attenuation loss is a measure of the power loss due to signal absorption in the device:

$$\begin{array}{l}\textbf{Attenuation loss (dB)}\\ \text{(in the line or device)}\end{array} = 10 \log \frac{\text{Input energy} - \text{Reflected energy at the input}}{\text{Transmitted energy to the load}}$$

$$= 10 \log \frac{|V_s|^2 - |V_r|^2}{\left(|V_s|^2 - |V_r|^2\right) e^{-2\alpha l}}$$

$$= 8.686 \, \alpha l \qquad (2.43)$$

The reflection loss is a measure of power loss during transmission due to the reflection of the signal as a result of impedance mismatch.

$$\begin{aligned} \textbf{Reflection loss (dB)} \\ \textbf{(at a plane)} \end{aligned} = 10 \log \frac{\text{Input energy}}{\text{Input energy} - \text{Reflected energy}}$$

$$= 10 \log \frac{1}{1 - |\Gamma|^2}$$

$$= 10 \log \frac{(S + 1)^2}{4S} \qquad (2.44)$$

The transmission loss is a measure of loss of power due to transmission through the line or device.

$$\begin{aligned} \textbf{Transmission loss (dB)} \\ \textbf{(due to the line)} \end{aligned} = 10 \log \frac{\text{Input energy}}{\text{Transmitted energy}}$$

$$= 10 \log \frac{|V_s|^2}{\left(|V_s|^2 - |V_r|^2\right) e^{-2\alpha l}}$$

$$= 10 \log \frac{e^{2\alpha l}}{1 - |\Gamma|^2}$$

$$= 8.686 \, \alpha l + 10 \log \frac{1}{1 - |\Gamma|^2}$$

$$= \text{Attenuation loss(dB)} + \text{Reflection loss (dB)} \qquad (2.45)$$

The return loss is a measure of the power reflected by a line or network or device.

$$\textbf{Return loss (dB)} = 10 \log \frac{\text{Input energy to the device}}{\text{Reflected energy at the input of the device}}$$

$$= - 20 \log |\Gamma| \qquad (2.46)$$

The insertion loss is a measure of the loss of energy in transmission through a line or device compared to direct delivery of energy without the line or device. let P_1 be the power received by the load when connected directly to source without the line or device, and P_2 the power received by the load when the line or the device is inserted between the source and the load, while the input power is held constant. Then

$$\textbf{Insertion loss (dB)} = 10 \log \frac{P_1}{P_2} \qquad (2.47)$$

The insertion loss is contributed by

1. Mismatch loss at the input
2. Attenuation loss through the device
3. Mismatch loss at the output

Example 2.5 A lossless transmission line with characteristic impedance of 300 ohm is fed by a generator of voltage $\angle 0°$ volts and impedance 100 ohm. The line is 100 m long and is terminated by a resistive load of 200 ohm. Calculate the reflection loss, the transmission loss and the return loss.

Solution

Given $Z_g = 100$ ohm, $Z_0 = 300$ ohm, $l = 100$ m, $Z_L = 200$ ohm.

Load reflection coefficient $\Gamma_L = \dfrac{Z_L - Z_0}{Z_L + Z_0} = \dfrac{200 - 300}{200 + 300} = -1/5$

$$\text{Reflection loss} = 10 \log \frac{1}{1 - |\Gamma|^2}$$

$$= 10 \log \frac{1}{1 - (1/25)}$$

$$= 10 \log (25/24)$$

$$= 0.1773 \text{ dB}$$

Transmission loss = Attenuation loss(dB) + Reflection loss (dB)

$$= 10 \log 25/24$$

$$= 0.1773 \text{ dB; Attenuation loss} = 0 \text{ dB}$$

Return loss $= -20 \log |\Gamma| = -20 \log (1/5) = 13.98$ dB

2.7 SMITH CHART

In order to avoid the tedious solution methods used above to solve complicated transmission line problems, Phillip H. Smith devised a simple graphic tool, named after him as the *Smith chart*, where the normalised impedance, or admittance or reflection coefficient is plotted to read the magnitude in the radial direction and phase in the angular direction directly. The chart is constructed from the equations for the input and load impedances normalised with respect to the characteristic impedance of the line in the following manner:

$$\text{Normalised input impedance} = \frac{Z_{\text{in}}}{Z_0} = \frac{1 + \Gamma_L\, e^{-2\gamma d}}{1 - \Gamma_L\, e^{-2\gamma d}} \tag{2.48}$$

$$\text{Normalised load impedance} = \frac{Z_L}{Z_0} = \frac{1 + \Gamma_L}{1 - \Gamma_L}$$

$$= r + j\,x \tag{2.49}$$

where r and x are the real and imaginary parts of the normalised load impedance. Separating the complex load reflection coefficient into real and imaginary parts, Γ_r and Γ_i, respectively, we can write

$$\Gamma_L = |\Gamma_L|\, e^{j\theta_L} \;=\; \frac{Z_L/Z_0 - 1}{Z_L/Z_0 + 1}$$

$$= \Gamma_r + j\Gamma_i \tag{2.50}$$

Reflection coefficient at $z = l - d$ can be expressed as

$$\Gamma_d = |\Gamma_L|\, e^{-2\alpha d}\, e^{j(\theta_L - 2\beta d)} \tag{2.51}$$

From Eqs 2.49 and 2.50

$$(\Gamma_r - 1)^2 + (\Gamma_i - 1/x)^2 = (1/x)^2 \tag{2.52}$$

$$\left(\Gamma_r - \frac{r}{1+r}\right)^2 + \Gamma_i^2 = \left(\frac{1}{1+r}\right)^2 \tag{2.53}$$

Equations 2.52 and 2.53 represent two families of circles in the complex reflection coefficient plane ($|\Gamma_L|$, θ_L) as shown in Fig. 2.7 which is called the *Smith chart*. The characteristics of the Smith chart are described below in terms of the normalised impedance $Z_L/Z_0 = r + jx$, the normalised admittance $Y_L/Y_0 = g + jb$, and the normalised length l/λ.

1. Eq. 2.52 represents the constant reactive (x) circles with radius $1/x$ and centres at 1, $1/x$, $-\infty \le x \le \infty$.

2. Eq. 2.53 represents the constant resistance (r) circles with radius $1/(1+r)$ and centres at $\left[\dfrac{1}{1+r}, 0\right]$ on the Γ_r axis, $0 \le r \le \infty$.

3. Since the line impedance and the reflection coefficient change as we move towards the generator or towards the load, the distance along the line in terms of wavelength is given along the circumference of the chart, ($|\Gamma| = 1$), clockwise towards the generator or anticlockwise towards the load, respectively. One complete rotation covers a distance $\lambda/2$ along the line; the impedance and the reflection coefficient repeat themselves at these intervals.

4. The upper half of the chart represents inductive reactance jx and lower half capacitive reactance $-jx$.

5. Since admittance is reciprocal of the impedance, the Smith chart can also be used for normalised admittance where the resistance scale reads the conductance and the inductive reactance scale reads capacitive reactance and vice-versa.

6. At a point of maximum voltage, line impedance $(Z_L/Z_0)_{max} = S$ and at minimum voltage, $(Z_L/Z_0)_{min} = 1/S$.

7. The centre O of the chart ($S = 1$) represents matched impedance, extreme right of the horizontal radius represents an open circuit ($S = \infty$, $Z_L/Z_0 = \infty$, $\Gamma = 1$) and extreme left represents short circuit ($S = \infty$, $Z_L/Z_0 = 0$, $\Gamma = -1$).

8. The distances along the line are normalised with respect to wavelength and are measured toward the generator and also toward the load along the periphery or unit circle ($|\Gamma| = 1$).

9. Circle passing through O ($s = 1$) is called the unit VSWR circle.

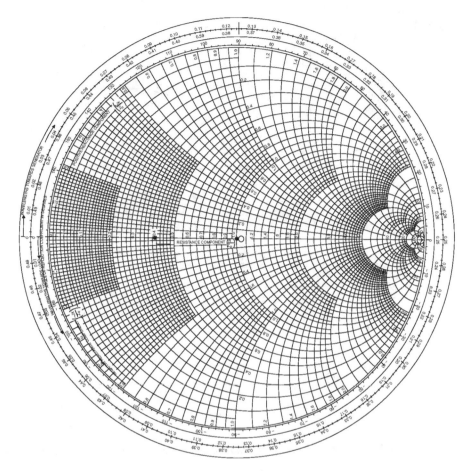

Fig. 2.7 *Constant r and x circles in the reflection coefficient plane—Smith chart.*

2.7.1   Applications of the Smith Chart

The Smith chart is conveniently used to determine the input and load impedances in a lossless transmission line from the values of the voltage standing wave ratio and the position of the voltage minimum on the line. The Smith chart also simplifies the procedure of finding stub positions and its length for impedance matching. These are described as follows.

2.7.1.1   *To determine input impedance of the lossless transmission line of length l and terminated by a load Z_L*

The graphical solution involves the following steps using the Smith chart as shown in Fig. 2.8.

1. Locate the normalised load impedance on the Smith chart at, say A.
2. Join the centre point O (1, 0) and A. Extend OA up to the periphery to cut it at B.

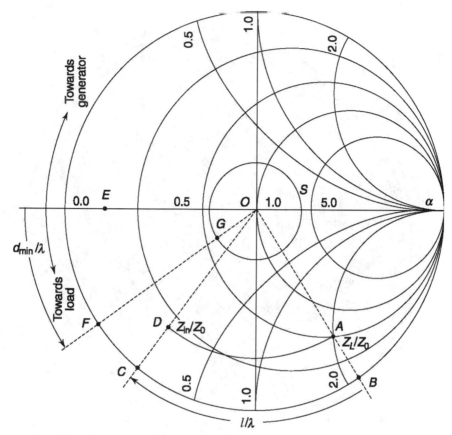

Fig. 2.8 *Determination of impedance of lossless line using Smith chart*

3. A rotation towards the generator by l/λ from the load point A gives a point C on the unit circle which corresponds to input point of the line.
4. In movement along the line and the corresponding rotation on the Smith chart, reflection coefficient and VSWR both remain constant in lossless line.
5. A circular arc is drawn with O as the centre that passes through A and intersects the line OC at D.
6. The point of intersection D gives the required normalised input impedance of the line.

Example 2.6 A transmission line of length 100 m and characteristic imped-ance of 100 ohm is terminated by a load $Z_L = 100 - j200$ ohm. Using the Smith chart determine the line impedance and admittance, at 25 m from the load end at a frequency of 10 MHz.

Solution

Given f = 10 MHz, Z_0 = 100 ohm, Z_L = 100 − j200 ohm,

l = 100 m, d = 25 m.

$$\lambda = c/f = 300/10 = 30 \text{ m}$$

$$d/\lambda = 25/30 = .833$$

Normalised load impedance $Z_L / Z_0 = (100 - j200)/100$
$$= 1 - j2$$

The normalised load impedance is located at A on the Smith chart (Fig.2.9). With origin O as the centre , a circle is drawn with radius OA. From the load point A, a normalised distance $d/\lambda = 0.833$ is moved along the circumference towards the generator to reach the point B.

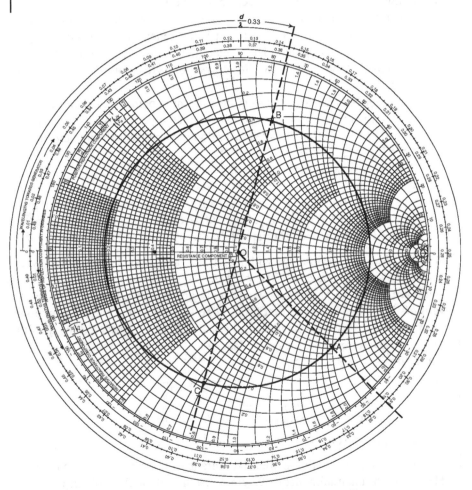

Fig. 2.9 *Calculation of line impedance using the Smith chart*

The normalised impedance at B is the $0.45 + j\,1.2$. Therefore, line impedance at 25 m distance from the load is

$$Z_{in} = (0.45 + j\,1.2) \times 100 = 45 + j\,120 \text{ ohm.}$$

Draw a line from point B through O to cut the circle at point C. This point C gives normalised admittance $0.27 - j\,0.73$ of the line at a distance 25 m from the load end. Therefore, admittance at 25 m distance from the load is

$$Y_{in} = (0.27 - j\,0.73)/100 = 0.0027 - j\,0.0073 \text{ mho}$$

2.7.1.2 To find the load impedance from the VSWR and voltage minimum position

In a laboratory where the VSWR S and position of the first voltage minimum d_{min} can be measured, the Smith chart in Fig. 2.8 gives a very simplified method of solution for load impedance in the following manner:

1. Draw a VSWR circle S with centre at O $(1 + jO)$.
2. When the line is shorted, the first voltage minimum occurs at the place of the load. This minimum shifts towards the generator by an amount d_{min}/λ when the line is loaded. The impedance at voltage minimum is pure resistance of magnitude $1/S$. This corresponds to a point E on the left half of the real axis.
3. Move above distance d_{min}/λ from the minimum point E along the periphery toward the load and locate this position on the periphery as F.
4. Join O and F and find the intersection point G between the line OF and the VSWR circle.
5. Point G represents the normalised load impedance.

There are many other uses of the Smith chart in transmission line theory and microwave engineering, viz. design of stubs for impedance matching, design of microwave filters, to determine the effects of discontinuities in a line and so on. These will be discussed at appropriate places.

Example 2.7 A lossless line with a characteristic impedance 50 ohm is terminated by an impedance Z_L. The voltage standing wave maximum and minimum are found 2.5 V and 1 V respectively and distance between successive minima is 5 cm. The line is first terminated by a short and then the unknown load, so that a shift in voltage minimum of 1.25 cm is observed towards the generator. Determine the load impedance.

Solution

$$Z_0 = 50 \text{ ohm}, V_{max} = 2.5 \text{ V}, V_{min} = 1 \text{ V}$$
$$\text{VSWR } S = V_{max}/V_{min} = 2.5$$

Distance between successive minima is $\lambda/2 = 5$ cm

Therefore, $\lambda = 10$ cm

Under short circuit condition, the first voltage minimum occurs at load point. Hence the shift in minimum of 1.25 cm = distance of the first voltage minimum d_{min} from the load.

$$d_{min}/\lambda = 1.25/10 = 0.125$$

Draw a constant S-circle for $S = 2.5$ with O as the centre and mark a V_{min} point A on the circle along the real axis on Smith chart (Fig. 2.10).

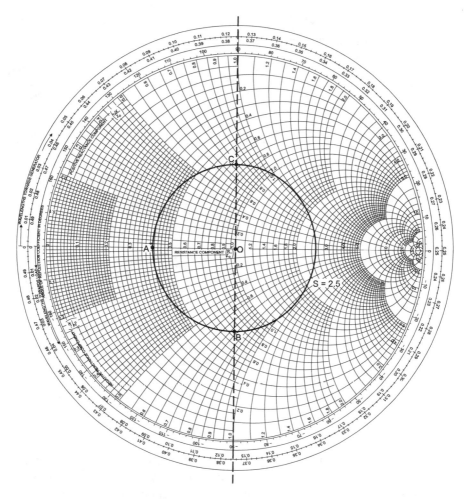

Fig. 2.10 *Calculation of load impedance using the Smith chart*

Travel $d_{min}/\lambda = 0.125$ from point A towards the load to reach point B, where normalised impedance is $0.69 - j\,0.72$. Diametrically opposite point C on the circle is the load admittance point.

Therefore, load impedance at B is $Z_L = (0.69 - j\,0.72) \times 50$

$$= 34.5 - j\,36 \text{ ohm}$$

Load admittance at C is $Y_L = 1/Z_L = 0.02006 \angle 46.23° \text{mho}$

2.8 HIGH FREQUENCY LINES

At considerably high frequencies i.e. in the microwave region where the guiding structures are made of very high conductivity materials and low loss dielectric materials, $\omega L \gg R$, and $\omega C \gg G$. Therefore from Eq. 2.13

$$Z_0 = \sqrt{(L/C)} \tag{2.54}$$

is real and other parameters are

$$\gamma = \alpha + j\beta = j\omega\sqrt{LC} \tag{2.55}$$

$$\alpha = 0 \tag{2.56}$$

$$\beta = \omega\sqrt{(L/C)} \tag{2.57}$$

$$v_p = \omega/\beta = 1/\sqrt{LC} \tag{2.58}$$

Therefore, high frequency lines are called low-loss or lossless lines. All the theories described in the previous paragraphs for $\alpha = 0$ also apply to the high frequency lines.

EXERCISES

2.1 Distinguish between the characteristic impedance and the input impedance of a line. Under what conditions will they be the same? In a lossless line how can the input impedance be made purely inductive, capacitive, infinite, and zero?

2.2 Sketch the standing wave patterns for voltage and current along a line when the termination is (a) open circuit, (b) short circuit, (c) purely inductive load, (d) purely capacitive load, (e) matched load Z_0, (f) resistive impedance $> Z_0$ and $< Z_0$.

2.3 A lossless line with a characteristic impedance 50 ohm is terminated by a load Z_L of VSWR 1.2 and fed by a matched generator of power 100 mW at the input. Calculate the minimum and maximum values of voltages along the line.

2.4 A lossless transmission line with the characteristic impedance 100 ohms produces a voltage minimum at a distance of 20 cms from the load end when short circuited. When the short is replaced by a load, the same voltage minimum is shifted by 9 cms towards the load. If the VSWR of the load is 3.0, using the Smith chart, find the value of the load.

2.5 A transmission line of length 10 m with a characteristic impedance of 50 ohms, $C = 65$ pF/m, and $\alpha = 0.06$ dB/m is fed by a matched generator and terminated by a matched load. If ac source voltage is $10/\angle 0$ at 100 MHz, find the voltage at the load.

2.6 A lossless transmission line with a characteristic impedance of 50 ohms is terminated by a load of $40 + j\,30$ ohms. If the magnitude of the incident wave is 10 V, find the maximum and minimum values of voltage on the line and VSWR at the load.

2.7 Calculate the incident power for the transmission line of Example 2.6 and also the net power delivered to the load.

2.8 A lossless transmission line has $C = 95$ pF/m, $Z_0 = 100$ ohms. If the operating frequency is 3 GHz, Calculate

(a) the phase constant
(b) the phase velocity V_p
(c) the wavelength on the line.

REFERENCES

2.1   Johnson, W C, *Transmission Line and Networks*, McGraw-Hill, New York, 1950.

2.2   Potter, J L, and S J Fich, *Theory of Networks and Lines*, Prentice-Hall of India (Pvt.) Ltd., New Delhi, 1965.

2.3   Dworsky, L N, *Modern Transmission Line Theory and Applications*, John Wiley & Sons, New York, 1979.

2.4   Sinnema, W, *Electronic Transmission Technology: Lines, Waves and Applications*, Prentice-Hall, Inc., Englewood Cliffs, N.J., 1979.

2.5   Collins, R E, *Foundations for Microwave Engineering*, McGraw-Hill Book Co., New York, 1992.

2.6   Ramo, S, J R Whinnery, and T Van Duzer, *Fields and Waves in Communication Electronics*, John Wiley & Sons, Inc., New York, 1985.

2.7   Sander, K F, and G A L Reed, *Transmission and Propagation of Electromagnetic Waves*, Cambridge University Press, Cambridge, 1978.

2.8   Slater, J C, *Microwave Transmission*, McGraw-Hill Book Company, Inc., New York, 1942.

2.9   Smith, P H, *Electronic Applications of the Smith Chart*, McGraw-Hill, New York, 1969.

2.10 Wheeler, H A, "Reflection Charts Relating to Impedance Matching," *IEEE Trans. on Microwave Theory and Technique*, Vol. MTT-32, no. 9, Sept. 1984.

Chapter 3

PROPAGATION OF ELECTROMAGNETIC WAVES

3.1 INTRODUCTION

In an electromagnetic wave propagation in a medium, a changing magnetic field will induce an electric field and vice-versa. The waves propagate with two oscillatory fields, one electric (E) and other magnetic (H). The waves are produced by radiation from time varying current sources or accelerated charges. The regions close to the radiating sources are most likely to have high intensity of fields having both longitudinal and transverse components with respect to the direction of propagation. In general, such locations are characterised by complicated field structure, including reactive (stored) and real (propagated) energies, irregular phase surfaces, and unknown field polarisation.

The propagation characteristics of electromagnetic waves in general depend on the electrical parameters (σ, ε, μ) of the medium and the presence of the boundaries or interfaces between two media. In free space, at sufficiently large distances from the source, wave propagation takes place in "Transverse Electromagnetic Mode" (TEM) from one point to another, where E and H fields are orthogonal to each other and both are perpendicular to the direction of propagation as shown in Fig. 3.1 for sinusoidally time-varying vector fields.

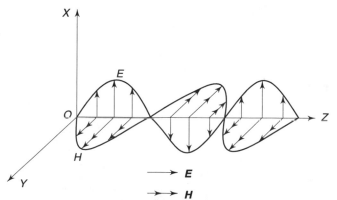

Fig. 3.1 *TEM wave propagation*

3.2 WAVE IMPEDANCE

The two fields, **E** and **H**, are related by an impedance, called the wave impedance, given by

$$\eta = |\mathbf{E}_t|/|\mathbf{H}_t| \qquad (3.1)$$

Here $\mathbf{E}_t$ and $\mathbf{H}_t$ represent transverse components with respect to the direction of
propagation.

One experiences variable wave impedance as one moves away from the source
of the electromagnetic radiation. It is seen from Eq. 3.1, that a very high impedance is encountered in the region where E-field is predominant. This happens near
a high impedance voltage driven source like a straight wire. On the other hand if
the waves are generated by a low impedance current driven source like a loop or
coil, the wave impedance is very low in the vicinity of the source where the
H-field predominates. In these predominant field regions most of the energy is
stored in the dominant field components E or H. As one moves away from the E or
H field source, correspondingly, other fields H or E are induced more and more.
Finally after a distance r_0 the fields are stabilised to give a constant wave impedance in the region called the far field region characterised by the TEM mode of
propagation. Figure 3.2 shows the variation of wave impedance with distance
from the source.

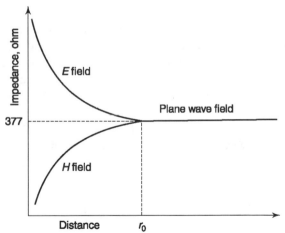

Fig. 3.2 *Variation of wave impedance with distance*

3.3 WAVE PROPAGATION

The electromagnetic signal may be propagated from the transmitter to a distant
receiver as a direct wave, ground reflected wave, surface wave or as a sky wave
depending on the frequencies. These propagation paths are explained in Fig. 3.3.

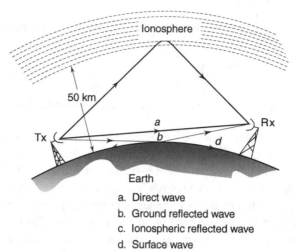

a. Direct wave
b. Ground reflected wave
c. Ionospheric reflected wave
d. Surface wave

Fig. 3.3 *Electromagnetic wave propagation above the earth*

The ionospheric region (50 km above the earth) contains gases which are ionized due to solar radiation. All the frequencies which are lower than a critical frequency f_c, given by

$$f_c = 9 \, (N_{max})^{1/2} \text{ Hz} \tag{3.2}$$

are reflected back to the receiver, on the earth, where N_{max} is the maximum electron density in the ionosphere. Since microwave frequencies are greater than f_c, microwave signals are not reflected back to the earth, but penetrate the ionosphere to travel in space. The component of the microwaves, diffracted around the surface of the earth, is guided by the ground-air interface as a surface wave which gets attenuated very fast. Therefore, only propagation of surface-to-surface microwave takes place through direct and ground reflected waves in the atmosphere. Because of the curvature of earth and the presence of other obstacles like buildings and trees, microwave signals are attenuated and the range of microwave direct transmission is limited to about 50 kms between a transmitter and a receiver.

3.4 ELECTROMAGNETIC WAVE EQUATION

The electromagnetic field in wave propagation in a medium can be solved from the Maxwell's equations:

$$\nabla \times \mathbf{E} = -\partial \mathbf{B}/\partial t \tag{3.3}$$

$$\nabla \times \mathbf{H} = \mathbf{J} + (\partial \mathbf{D}/\partial t) \tag{3.4}$$

$$\nabla \cdot \mathbf{D} = \rho \tag{3.5}$$

$$\nabla \cdot \mathbf{B} = 0 \tag{3.6}$$

$$\nabla \cdot \mathbf{J} = -\partial \rho/\partial t \tag{3.7}$$

The electrical properties of the medium (σ, ε, μ) offer constitutive relations.

$$\mathbf{D} = (\varepsilon' - j\varepsilon'')\,\mathbf{E} = \text{Electric displacement vector} \tag{3.8}$$

$$\mathbf{B} = \mu\mathbf{H} = (\mu' - j\mu')\,\mathbf{H} = \text{Magnetic induction vector} \tag{3.9}$$

$$\mathbf{J} = \sigma\mathbf{E} + \mathbf{J}_i = \text{Electric current density} \tag{3.10}$$

$$\varepsilon = \varepsilon_r \varepsilon_0 = \text{Electric permittivity} \tag{3.11}$$

$$\mu = \mu_r \mu_0 = \text{Magnetic permeability} \tag{3.12}$$

Here $\mathbf{J}_i$ is the externally impressed current density producing the electromagnetic fields. σ, ε'' and μ'' account for the joule heating loss, dielectric loss and magnetic loss in the medium, respectively. The loss tangent of the medium is given by

$$\tan\delta = \frac{\omega\,\varepsilon'' + \sigma}{\omega\,\varepsilon'} \tag{3.13}$$

For time harmonic fields $\left(e^{j\omega t}\right)$ the above equations reduce to

$$\nabla \times \mathbf{E} = -j\omega\mu\mathbf{H} \tag{3.14}$$

$$\nabla \times \mathbf{H} = (\sigma + j\omega\varepsilon)\,\mathbf{H} \tag{3.15}$$

$$\nabla \cdot \mathbf{D} = \rho \tag{3.16}$$

$$\nabla \cdot \mathbf{B} = 0 \tag{3.17}$$

$$\nabla \cdot \mathbf{J} = -j\omega\rho \tag{3.18}$$

Since both E and H in these equations are unknown, solution for electric or magnetic fields can be obtained from wave equations involving either the E or H fields derived from Eqs 3.14-3.18.

$$\nabla^2\mathbf{E} + K^2\mathbf{E} = j\omega\mu\mathbf{J} - \frac{\nabla\nabla\cdot\mathbf{J}}{j\omega\,\varepsilon}; \quad \nabla^2\mathbf{H} + K^2\mathbf{H} = -\nabla\mathrm{x}\mathbf{J} \tag{3.19}$$

where

$$K = \sqrt{-j\omega\mu(\sigma + j\omega\varepsilon)} = \sqrt{-j\omega\mu\sigma + \omega^2\mu\varepsilon} = \alpha + j\beta \tag{3.20}$$

Here K is called the *propagation constant* of the medium. In a medium with finite conductivity σ, a conduction current density $\mathbf{J} = \sigma\mathbf{E}$ will exist to give attenuation α because of joule heating. In a non-conducting lossless medium, $\sigma = 0$ and $K = k = \omega\,(\mu\,\varepsilon)^{1/2}$. At frequencies below the optical region the propagation constant in a good conductor ($\sigma \ll w\varepsilon$) becomes

$$K = (1+j)\,\sqrt{\pi f \mu\sigma} = \alpha + j\beta \tag{3.21}$$

and in free space where ($\sigma \simeq 0$), $k = \omega\,(\mu_0\varepsilon_0)^{1/2}$

Microwave signal can penetrate a conductor with $1/e$ times the attenuation at a depth called 'skin depths' which is given by

$$\delta = \frac{1}{\sqrt{\pi f \mu\sigma}} = \frac{1}{\alpha} \tag{3.22}$$

The propagating wave characterised by the wave impedance in free space is

$$\eta_0 = \sqrt{\frac{\mu_0}{\varepsilon_0}} = 120 \, \pi \text{ ohm} \tag{3.23}$$

and that in lossy and conducting media are

$$\eta = \sqrt{\frac{j\omega\mu}{\sigma + j\omega\varepsilon}}$$

and

$$\eta = \sqrt{\frac{\omega\mu}{2\sigma}} (1 + j) = (1 + j) \, R_s \tag{3.24}$$

where R_s is the *conductor surface resistance* given by

$$R_s = \sqrt{\frac{\omega\mu}{2\sigma}} \tag{3.25}$$

3.5 ELECTROMAGNETIC ENERGY AND POWER FLOW

When a potential source supplies time varying currents in a conductor it emits electromagnetic energy. The energy supplied by the source is partly stored in the electric and magnetic fields inside a closed volume in the immediate vicinity of the conductor and partly propagated away as an electromagnetic wave in a lossless transmission medium, such as free space, coaxial line, waveguides, etc.

For the steady state time harmonic excitation $\left(e^{j\omega t}\right)$, the time-average energies stored in the electric and magnetic fields in a volume are given by

$$W_e = R_e \frac{1}{4} \int_V \mathbf{E} \cdot \mathbf{D}^* dV = \varepsilon'/4 \int_V E^2 \, dV \tag{3.26}$$

$$W_m = R_e \frac{1}{4} \int_V \mathbf{H} \cdot \mathbf{B} \cdot dV = \mu'/4 \int_V H^2 \, dV \tag{3.27}$$

where peak values are used for the complex fields $\mathbf{E}$ and $\mathbf{H}$.

For electromagnetic wave propagation, the time average power transmitted through a unit area is called the *Poynting vector*, $\mathbf{P}$.

$$\mathbf{P} = 1/2 \, (\mathbf{E} \times \mathbf{H}^*) \tag{3.28}$$

The time average complex power flow across a surface S is, therefore

$$\mathbf{P} = 1/2 \int_s (\mathbf{E} \times \mathbf{H}^*) \cdot d\mathbf{B} \tag{3.29}$$

3.6 POYNTING THEOREM

The *Poynting theorem* states that the total complex power impressed by a current source within a region is equal to the sum of the time average power dissipated as heat inside the region, the complex power transmitted from the region and $2\omega j$ times the difference between time average magnetic and electric energies stored within the region.

The total power flow through a closed surface S enclosing a volume V can be written as

$$\frac{1}{2}\oint_S \mathbf{E} \times \mathbf{H}^* \cdot \mathbf{dS} \tag{3.30}$$

Using Gauss's theorem

$$\frac{1}{2}\oint_S \mathbf{E} \times \mathbf{H}^* \cdot \mathbf{dS} = \frac{1}{2}\int_V \nabla \cdot (\mathbf{E} \times \mathbf{H}^*) \, dV \tag{3.31}$$

Using vector identity and Maxwell's equations with complex conjugate field H we can write from Eqs 3.14 and 3.15

$$\nabla \cdot (\mathbf{E} \times \mathbf{H}^*) = \mathbf{H}^* \cdot \nabla \times \mathbf{E} - \mathbf{E} \cdot \nabla \times \mathbf{H}^* = j\omega \mathbf{H} \cdot \mathbf{H}^* - j\omega \mathbf{D}^* \cdot \mathbf{E} - \mathbf{E} \cdot \mathbf{J}^* \tag{3.32}$$

Substituting Eq. 3.32 in 3.31

$$\frac{1}{2}\oint_S \mathbf{E} \times \mathbf{H}^* \cdot \mathbf{dS} = -j\omega/2\int_V (\mathbf{B} \cdot \mathbf{H}^* + \mathbf{E} \cdot \mathbf{D}^*) \, dV - 1/2\int_V \mathbf{E} \cdot \mathbf{J}^* \, dV \tag{3.33}$$

For a medium with electrical parameters $\mu = \mu' - j\mu''$, $\varepsilon = \varepsilon' - j\varepsilon''$ and σ, Eqs 3.8 – 3.10, 3.26, 3.27 and 3.33 yield

$$R_e \frac{1}{2}\oint_S \mathbf{E} \times \mathbf{H}^* \cdot \mathbf{dS} + \omega/2\int_V \left(\mu'' H^2 + \varepsilon'' E^2\right) dV + 1/2\int_V \sigma E^2 dV$$

$$= -1/2\int_V \mathbf{E} \cdot \mathbf{J}_i dV \tag{3.34}$$

and $\quad$ Im $\dfrac{1}{2}\oint_S \mathbf{E} \times \mathbf{H}^* \cdot \mathbf{dS} = -2\omega (W_m - W_e)$ $\hfill$ (3.35)

Here the externally impressed current density $\mathbf{J}_i$ is assumed real. Eq. 3.34 shows that the real power P_i impressed by the current source within the volume V represented by the right hand side term is the sum of real electromagnetic power flow P_r through the closed surface S from volume V represented by the first term in the left hand side, the power loss due to electric and magnetic polarisation damping forces represented by the second term in the left hand side, and joule heating loss P_L represented by the third term in the left hand side. Eq. 3.35 states that the imaginary part of the complex power flow from V is equal to the $2w$ times the total reactive energy stored in the magnetic and electric fields in V. In general, when the losses are small ($\varepsilon'' \ll \varepsilon''$ and $\mu'' \ll \mu'$),

$$P_i = P_r + j2\omega (W_m - W_e) + P_L \tag{3.36}$$

3.7   EQUIVALENT CIRCUIT PARAMETERS OF PROPAGATION LINE

The distributed electrical circuit parameters R, L, G and C of a microwave transmission line which account for the joule heating loss, magnetic energy stored, the loss in the medium and the electric energy stored, respectively, can be expressed by

$$R = \frac{2P_L}{II^*} = \frac{\int_V \sigma E^2 \, dV}{I^2} \tag{3.37}$$

$$L = \frac{4W_m}{II^*} = \frac{\mu' \int_V E^2 \, dV}{I^2} \tag{3.38}$$

$$G = \sigma_0 \, C/\varepsilon' \tag{3.39}$$

$$C = \frac{II^*}{4\omega^2 \, W_e} = \frac{I^2}{\omega^2 \varepsilon' \int_V E^2 \, dV} \tag{3.40}$$

where σ_0 is the electrical conductivity of the medium inside the line. The characteristic impedance of the line is given by

$$Z_0 = \sqrt{\frac{R + j\omega L}{G + j\omega C}} \approx \sqrt{\frac{L}{C}} \tag{3.41}$$

at microwave frequencies ($\omega L \gg R$, $\omega C \gg G$) where reactive parts predominate the loss parameters R and G.

3.8 BOUNDARY CONDITIONS

Solution to Maxwell's equations in a region V is not unique unless the behavior of the field on the boundary of V is known or at the boundary, of material bodies with different electrical properties is known. These boundaries may be the interface between two different dielectrics or between a dielectric and a conducting media.

Consider a boundary surface S, containing a surface charge density ρ_s, separating the regions 1 and 2 characterised by (σ_1, ε_1, μ_1) and (σ_2, ε_2, μ_2), respectively, as shown in Fig. 3.4.

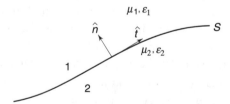

Fig. 3.4 *Boundary surface between two media*

The following four basic boundary conditions must be satisfied by the fields for a specific solution.

1. The tangential components of electric field intensity are continuous across the boundary

$$E_{1t} = E_{2t} \tag{3.42}$$

2. The normal components of magnetic flux density are continuous across the boundary

$$B_{1n} = B_{2n} \tag{3.43}$$

3. The normal components of electric flux density are discontinuous at the boundary by the surface charge density ρ_s

$$D_{1n} - D_{2n} = \rho_s \tag{3.44}$$

4. The tangential components of magnetic field are discontinuous at the boundary by the surface current density J_s

$$H_{1t} - H_{2t} = J_s \tag{3.45}$$

If one of the media is a perfect conductor ($\sigma = \infty$, $\varepsilon_r = 1$, $\mu_r = 1$) and other one is a perfect dielectric ($\sigma = 0$, $\varepsilon_r = 1$, $\mu_r = 1$), then

$$\hat{n} \times \mathbf{E} = \mathbf{E}_t = 0 \tag{3.46}$$

$$\hat{n} \cdot \mathbf{B} = \mathbf{B}_n = 0 \tag{3.47}$$

$$\hat{n} \cdot \mathbf{D} = \mathbf{D}_n = \rho_s \tag{3.48}$$

$$\hat{n} \times \mathbf{H} = \mathbf{H}_t = J_s \tag{3.49}$$

Where $\hat{n}$ represents unit normal vector at the interface.

Example 3.1 Let a region 1 ($y < 0$) be free space and region 2 ($y > 0$) be a dielectric with no surface charge on the plane interface (xy plane at $y = 0$). In region 1

$$\mathbf{D}_1 = \hat{x}\ 10\,x^2 + \hat{y}\ (6 + y) + \hat{z}\ 10\,z$$

Find $\mathbf{D}_2$ and $\mathbf{E}_2$ in the region 2 at the boundary ($y = 0$)

Solution

Since the normal component of $\mathbf{D}$ is continuous at $y = 0$, $\mathbf{D}_{2y} = \mathbf{D}_{1y} = 6$, at $y = 0$.

Therefore, $E_{2y} = 6/\varepsilon_2$

Now, $\mathbf{E}_1 = \mathbf{D}_1/\varepsilon_1 = 1/\varepsilon_1 \cdot [\ \hat{x}\ 10\,x + \hat{y}\ (6 + y) + \hat{z}\ 10\,z]$

Since tangential components of $\mathbf{E}$ are continuous at $y = 0$

$$E_{2x} = E_{1x} = \frac{10x^2}{\varepsilon_1} \quad \text{and} \quad E_{2z} = E_{1z} = \frac{10z}{\varepsilon_1}$$

Therefore, at the boundary ($y = 0$)

$$\mathbf{E}_2 = \hat{x}\ E_{2x} + \hat{y}\ E_{2y} + \hat{z}\ E_{2z}$$

$$= \hat{x}\ \frac{10x^2}{\varepsilon_1} + \hat{y}\ \frac{6}{\varepsilon_2} + \hat{z}\ \frac{10z}{\varepsilon_1}$$

and $$\mathbf{D}_2 = \varepsilon_2 \mathbf{E}_2 = \hat{x}\ \frac{\varepsilon_2}{\varepsilon_1}\ 10\,x^2 + \hat{y}\ 6 + \hat{z}\ \frac{\varepsilon_2}{\varepsilon_1}\ 10\,z$$

3.9 POLARISATION OF WAVES

An electromagnetic wave is said to be polarised in the direction of **E** vector. The polarisation of waves is named according to the locus that the tip of **E** describes while propagation in a medium. The polarisation that the wave is called linear or circular or elliptical when the tip of the **E** vector describes a straight line, or a circle or an ellipse, respectively, as shown in Fig. 3.5.

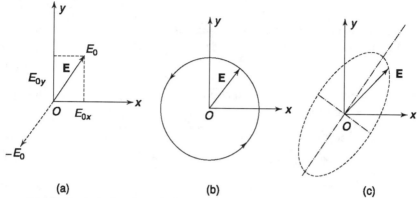

(a)	(b)	(c)

Fig. 3.5 *Polarisation of electromagnetic waves (a) linear polarisation (b) circular polarisation (c) elliptical polarisation*

Elliptical polarisation

For the propagation of TEM waves in the z-direction, the **E** vector, in general has two orthogonal components at any point on the $z = $ constant plane,

$$E_x = E_{0x} \sin \omega t \tag{3.50}$$

$$E_x = E_{0x} \sin (\omega t + \theta) \tag{3.51}$$

The resultant field

$$E = (E_x^2 + E_y^2)^{1/2}$$

or,

$$E^2 = E_{0x}^2 \sin^2 \omega t + E_{0y}^2 \sin^2 (\omega t + \theta)$$

or,

$$\frac{E_{0x}^2}{E^2} \sin^2 \omega t + \frac{E_{0y}^2}{E^2} \sin^2 (\omega t + \theta) = 1 \tag{3.52}$$

This Eq. 3.52 represents an ellipse with its semi-major and semi-minor axes inclined to the x and y axes. Thus the **E** vector constantly changes both its magnitude and direction describing an ellipse. The rotation rate of **E** around the origin is ω radian per second. Such a wave is said to be an elliptically polarised plane wave.

Circular polarisation

When $\theta = \pi/2$ with $E_{0x} = E_{0y} = E_0$, Eq. 3.52 reduces to a circle

$$E_x^2 + E_y^2 = E_0^2 \tag{3.53}$$

Here the tip of **E** describes a circle of radius E_0. This wave is called a *circularly polarised plane wave.*

Linear polarisation

In a linearly polarised wave with sinusoidal time variation $\theta = 0$, and the components of $\mathbf{E}$ are $E_x = E_{0x} \sin \omega t$, $E_y = E_{0y} \sin \omega t$, at any point on a $z = $ constant plane. Then

$$|\mathbf{E}| = \sqrt{E_x^2 + E_y^2} = E_0 \sin \omega t \tag{3.54}$$

Here the tip of $\mathbf{E}$ describes a plane surface and is constant in direction.

3.10 PLANE WAVES IN UNBOUNDED MEDIUM

When microwaves propagate with phase remaining constant over a set of planes, it is called a *plane wave*. In addition, if the magnitude also remains constant, the wave is called a uniform plane wave having the following properties.

(a) The electric and magnetic fields are mutually perpendicular to each other and to the direction of propagation, i.e. the wave is transverse electromagnetic (TEM).

(b) $\mathbf{E}$ and $\mathbf{H}$ fields are always in time phase.

(c) The magnitudes of the two fields are always constant.

(d) The stored energies are equally divided between the $\mathbf{E}$ and $\mathbf{H}$ fields.

(e) The power transmitted by the two fields is in the direction of propagation.

The mathematical solution of plane wave in an unbounded source free medium is obtained from the wave Eq. 3.19 in rectangular coordinates (x, y, z), with the right hand side of the equation equated to zero and can be written as

$$\mathbf{E}\,(r) = \mathbf{E}_0\, e^{-\mathbf{K} \cdot \mathbf{r}} \tag{3.55}$$

where $\mathbf{E}_0$ is the amplitude of the wave propagating in the direction specified by the propagation vector

$$\mathbf{K} = \hat{x}\,K_x + \hat{y}\,K_y + \hat{z}\,K_z \tag{3.56}$$

The Expression for $\mathbf{K}$ is given by Eq. (3.20) $\mathbf{r}$ is the position vector of the point of observation, as explained in Fig. 3.6.

The uniform plane wave solution (3.55) gives plane phase surfaces $\mathbf{K} \cdot \mathbf{r} = $ constant. $\mathbf{E}$ remains constant on a constant-phase plane.

The solution for $\mathbf{H}$ is obtained from the Maxwell's equation,

$$\mathbf{H} = \mathbf{K} \times \mathbf{E}/\omega\mu \tag{3.57}$$

The propagating wave is characterised by the wave impedance given by Eqs 3.23 and 3.24.

3.11 WAVE VELOCITIES

In a lossless medium $\sigma = 0$, $\alpha = 0$, and $K = j\beta$. If waves propagate along the z-axis, the uniform plane wave solution can be written as

$$E_x = E_0 \sin (\omega t - \beta z) \tag{3.58}$$

$$H_y = E_0/\eta \sin (\omega t - \beta z) \tag{3.59}$$

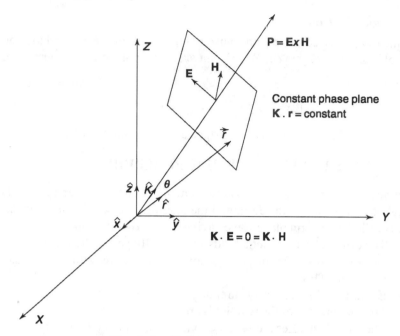

Fig. 3.6 *Plane wave propagation*

The argument of the sine functions of the above equations is the phase ϕ of the fields.

$$\phi = \omega t - \beta z \qquad (3.60)$$

with frequency $\quad f = \dfrac{\omega}{2\pi} = \dfrac{1}{T} \qquad (3.61)$

where T is the time period. The phase constant is defined by $\partial \phi / \partial z = \beta$ for a fixed value of time .

The wavelength λ is the distance, along the z-direction, in which the phase changes by 2π radians for a fixed value of time: $\lambda = 2\pi/\beta$.

The velocity of the wave propagation is the velocity of a constant phase and is called *phase velocity*.

$$\partial z / \partial t = V_p = \omega/\beta = \lambda f \qquad (3.62)$$

The velocity at which the energy is propagated is expressed by

$$V_g = d\omega/d\beta \qquad (3.63)$$

and is called the *group velocity*.

Example 3.2   The electric field of a uniform plane wave is given by

$$\mathbf{E} = \hat{x}\, 20 \sin (3\pi 10^8 t + (\pi z) + \hat{y} 20 \cos (3\pi\, 10^8 t - (\pi z)\ \text{V/m}.$$

Find (a) phase velocity, (b) the corresponding magnetic field $\mathbf{H}$, and (c) the polarisation of the wave.

Solution

(a) $\omega = 3\pi 10^8$ rad/s , $\beta = \pi$ rad/m , $V_p = \omega/\beta = 3 \times 10^8$ m/s

(b) Given $\mathbf{E} = \hat{x}\, E_x + \hat{y}\, E_y$

$$\mathbf{H} = \frac{\hat{z} \times \mathbf{E}}{\eta} = \frac{\hat{z}x\,(\hat{x}\, Ex + \hat{y}\, Ey)}{\eta}$$

$$= \hat{y}\, \frac{20}{377} \sin(3\pi10^8\, t - \pi z) - \hat{x}\, \frac{20}{377} \cos(3\pi10^8\, t - \pi z);\ \text{A/m}$$

$$= 0.0531\,[-\hat{x}\,\cos(3\pi10^8\, t - \pi z) + \hat{y}\, \sin(3\pi10^8\, t - \pi z)];\ \text{A/m}$$

(c) **E** field has two orthogonal components with equal amplitude and out of phase by 90°. Thus the wave is circularly polarised.

Now at $\qquad z = 0, t = 0, \mathbf{E} = \hat{y}\, 20$ or $E_y = 20$ V/m

At $\qquad z = 0, t = T/4 = 2\pi/4\omega$ or $\omega t = \pi/2$

$\qquad \mathbf{E} = \hat{x}\, 20 \sin \pi/2 + \hat{y}\, 20 \cos \pi/2$

$\qquad = \hat{x}\, 20$

or, $\qquad E_x = 20$ V/m

$\qquad z = 0, t = T/8 = 2\pi/8\omega$ or $\omega t = \pi/4$

$\qquad \mathbf{E} = \hat{x}\, 20 \sin \pi/4 + \hat{y}\, 20 \cos \pi/4$

$\qquad = (\hat{x} + \hat{y})\, 20/\sqrt{2}$

or, $\qquad \mathbf{E} = 20$ V/m

Thus as time progresses, the tip of **E** rotates from y axis to x axis in the anti-clockwise direction when looked in $+z$ direction. Therefore, the wave is left circularly polarised.

Example 3.3 The electric field of a plane wave is represented by $\mathbf{E} = 10\, \hat{y}\, \cos(10^9\, t + 30\, z)$ V/m. Determine the (a) magnetic field **H**, (b) phase velocity, and (c) dielectric constant of the medium, where $\mu = \mu_0$

Solution

E field amplitude $E_0 = 10$ V/m

$$\omega = 10^9\ \text{rad/sec}$$

$$\beta = 30\ \text{rad/m}$$

(a) $\nabla \times \mathbf{H} = \dfrac{\partial \mathbf{D}}{\partial t} = \varepsilon\, \dfrac{\partial \mathbf{E}}{\partial t} = -\hat{y}\varepsilon 10 \sin(10^9\, t + 30z)\, 10^9$

or, $\hat{y}\left[\dfrac{\partial \mathbf{H}_z}{\partial \hat{x}} - \dfrac{\partial \mathbf{H}_x}{\partial \hat{z}}\right] = -\hat{y}\, 10^{10}\varepsilon \sin(10^9\, t + 30z)$

For plane wave with E_y, in z-direction, $H_z = 0$. Therefore,

$$-\hat{y}\, \dfrac{\partial \mathbf{H}_x}{\partial \hat{z}} = -\hat{y}\, 10^{10}\, \varepsilon \sin(10^9\, t + 30z)$$

or, $\dfrac{\partial \mathbf{H}_x}{\partial \hat{z}} = \varepsilon 10^{10} \sin \left(10^9 t + 30z\right)$

Therefore, $\mathbf{H}_x = -\varepsilon 10^{10} \dfrac{\cos \left(10^9 t + 30z\right)}{30}$

$$= -\varepsilon 10^9 \dfrac{\cos \left(10^9 t + 30z\right)}{3} \text{ A/m}$$

or, $\mathbf{H} = -\hat{x}\, \varepsilon \dfrac{10^9}{3} \cos \left(10^9 t + 30z\right) \text{ A/m}$

(b) $\mathbf{V}_p = \dfrac{\omega}{\beta} = \dfrac{10^9}{30} = \dfrac{10^8}{3} \text{ m/s}$

(c) $\mathbf{V}_p = \dfrac{1}{\sqrt{\mu \varepsilon}}$

or, $\mu \varepsilon = \dfrac{1}{V_p^2} = \left(\dfrac{3}{10^8}\right)^2 = 9 \times 10^{-16}$

Since $\mu = \mu_0,\, \varepsilon = \dfrac{9 \times 10^{-16}}{\mu_0} = \dfrac{9 \times 10^{-16}}{4\pi \times 10^{-7}}$

Dielectric constant of the medium

$$\varepsilon_r = \dfrac{\varepsilon}{\varepsilon_0} = \dfrac{9 \times 10^{-16} \times 36\pi}{4\pi \times 10^{-7} \times 10^{-9}} = 81$$

3.12 PLANE WAVES IN LOSSY DIELECTRIC

In a lossy dielectric $\varepsilon = \varepsilon' - j\varepsilon''$ and $\sigma \ll \omega \varepsilon$. The intrinsic impedance of the dielectric is given by

$$\eta = \sqrt{\dfrac{j\omega\mu}{\sigma + j\omega\varepsilon}} = \sqrt{\dfrac{\mu}{\varepsilon}} \left(1 - j\dfrac{\sigma}{\omega\varepsilon}\right)^{-1/2}$$

$$\simeq \sqrt{\dfrac{\mu}{\varepsilon}} \left[1 + \dfrac{j\sigma}{2\omega\varepsilon}\right] \tag{3.64}$$

where $\sigma/\omega\varepsilon$ is the *loss tangent*. For low loss case $(\sigma/\omega\varepsilon \ll 1)$ and

$$K = jk \left(1 - j\sigma/\omega\varepsilon\right)^{1/2}$$
$$= jk \left(1 - j\sigma/2\omega\varepsilon\right)$$
$$= k\sigma/2\omega\varepsilon + jk$$
$$= \alpha + j\beta \tag{3.65}$$

where $k = \omega(\mu \varepsilon)^{1/2}$ is the propagation constant in dielectric (μ, ε).
Therefore, the attenuation and phase constants are given by

$$\alpha = \frac{\sigma}{2}\sqrt{\frac{\mu}{\varepsilon}} \qquad (3.66)$$

$$\beta = \omega(\mu\varepsilon)^{1/2} = k \qquad (3.67)$$

3.13   PLANE WAVES IN LOSSLESS DIELECTRIC

In the case of lossless dielectric medium, $\varepsilon = \varepsilon'$, $\varepsilon'' = 0$, $\sigma = 0$. Therefore, the
intrinsic impedance and the propagation constant in lossless dielectric are given
by

$$\eta = (\mu/\varepsilon)^{1/2} \qquad (3.68)$$

$$\beta = \omega(\mu\varepsilon)^{1/2} \qquad (3.69)$$

For the free space (ε_0, μ_0), $\eta = \sqrt{\dfrac{\mu_0}{\varepsilon_0}} = 120\,\pi\,\text{ohm}$.

3.14   PLANE WAVES IN GOOD CONDUCTOR

In a good conductor the conduction current is normally much greater than the
displacement current even at microwave frequencies (below the optical region). It
is useful to consider the approximation $\sigma \gg \omega\varepsilon$. The propagation constant can be
written as

$$K = (j\sigma\omega\mu)^{1/2} = (1+j)\sqrt{\pi f \mu\sigma} = \alpha + j\beta \qquad (3.70)$$

Here

$$\alpha = \beta = (\pi f\sigma\mu)^{1/2} \qquad (3.71)$$

The fields inside the conductor are then

$$\mathbf{E} = \mathbf{E}_0 e^{-\alpha\mathbf{K}\cdot\mathbf{r}} e^{-j\beta\mathbf{K}\cdot\mathbf{r}} \qquad (3.72)$$

$$\mathbf{H} = \frac{1+j}{\sqrt{2}}\sqrt{\frac{\sigma}{2\pi f\mu}}(\mathbf{K}\times\mathbf{E}) \qquad (3.73)$$

Both fields decrease with penetration, falling to $1/e$ of their surface values in a
distance equal skin depth.

$$\delta = \sqrt{\frac{1}{\pi f\mu\sigma}} = \frac{1}{\alpha} = \frac{1}{\beta} \qquad (3.74)$$

$\delta \to 0$ as $\sigma \to \infty$ and is smaller for higher frequencies for finite conductivity.
The intrinsic impedance of a good conductor is given by

$$\eta = \sqrt{\frac{j\omega\mu}{\sigma}} = (1+j)\sqrt{\frac{\omega\mu}{2\sigma}} = (1+j)R_s \qquad (3.75)$$

where $\qquad R_s = (\omega\mu/2\sigma)^{1/2}$ (3.76)

The time average of the electrical energy density and the magnetic energy density inside the conductor are

$$W_e = \varepsilon E^2/2 \qquad (3.77)$$

$$W_m = \sigma E^2/2\omega \qquad (3.78)$$

The ratio of these energies is $\omega\varepsilon/\sigma$, which is a very small number. Thus within the conductor the electrical energy density is negligible in comparison to the magnetic energy, and good conductors are essentially not penetrated by electric fields but penetrated by magnetic fields, especially at lower frequencies. The phase velocity inside the conductor is

$$V_p = \omega\delta \qquad (3.79)$$

Example 3.4 A microwave signal at 2.45 GHz is transmitted through a medium having $\sigma = 2.17$ S/m, $\varepsilon = 47\ \varepsilon_0$ and $\mu = \mu_0$. Find K, λ, V_p, R_s and η.
Solution

$$K = jk\,(1 - j\sigma/\varepsilon\omega)^{1/2}$$

$$k = \omega\,(\mu\varepsilon)^{1/2} = 2\pi f\,(\mu_0\varepsilon_0\varepsilon_r)^{1/2}$$

$$= 2 \times 3.14 \times 2.45 \times 10^9 \times (4\pi \times 10^{-7} \times 47 \times 10^{-9}/36\pi)^{1/2}$$

$$= 351.78$$

$$\sigma/\omega\varepsilon = \frac{2.17 \times 36\pi}{2\pi \times 2.45 \times 10^9 \times 47 \times 10^{-9}} = 0.33921$$

$$K = j\,351.78\,\sqrt{(1 - j0.33921)} = 361.478\ \angle 80.63°$$

$$= 58.85 + j\,356.65 = \alpha + j\beta$$

Therefore, $\alpha = 58.85$ NP/m

$\beta = 356.65$ rad/m

$$\lambda = \frac{2\pi}{\beta} = 0.0176\ \text{m}$$

$$V_p = \frac{\omega}{\beta} = \frac{2\pi f}{\beta} = \frac{2\pi \times 2.45 \times 10^9}{356.65}$$

$$= 0.04316 \times 10^9\ \text{m/s}$$

$$R_s = \sqrt{\frac{\omega\mu_0}{2\sigma}} = \sqrt{\frac{2\pi \times 2.45 \times 10^9 \times 4\pi \times 10^{-7}}{2 \times 2.17}}$$

$$= 6.676 \times 10 = 66.76\ \text{ohm}$$

$$\eta = \sqrt{\frac{j\omega\mu_0}{\sigma + j\omega\varepsilon}} \cong \sqrt{\frac{\mu_0}{\varepsilon_r\varepsilon_0}}\left(1 - j\frac{\sigma}{\omega\varepsilon}\right)^{-1/2}$$

$$= \frac{377}{(47)^{1/2}} \frac{1}{[1 - j0 \cdot 33921]^{1/2}}$$

$$= 52.1 \angle 18.7° \text{ ohm}$$

Example 3.5 The electric field of a uniform plane wave in a medium $(\sigma = 4\text{S/m}, \varepsilon_r = 80, \mu_0)$ propagating in $+z$ direction is given by

$$\mathbf{E} = \hat{x}\, 10 \cos 5 \times 10^4\, \pi t, \text{ V/m}$$

at $z = 0$. Find the instantaneous and the time-average power flow per unit area along the z direction.

Solution

$$\omega = 5 \times 10^4 \pi \text{ rad/s}$$

$$f = \frac{\omega}{2\pi} = 2.5 \times 10^4 \text{ Hz} = 25 \text{ kHz}$$

$$K = j\omega\sqrt{(\mu_0 \varepsilon)} \sqrt{(1 - j\sigma/\varepsilon\omega)}$$

$$= j2\pi \times 2.5 \times 10^4 \sqrt{\frac{4\pi \times 10^{-7} \times 80 \times 10^{-9}}{36\pi}}$$

$$\sqrt{1 - j\frac{4 \times 36\pi}{2\pi \times 2.5 \times 10^4 \times 80 \times 10^{-9}}}$$

$$= \alpha + j\beta$$

Therefore, $= \alpha \cong j\beta = 0.63$

$$\eta = \sqrt{\frac{j\omega\mu_0}{\sigma + j\omega\varepsilon}} = 0.222 \angle 45°$$

Therefore, $\mathbf{E}\,(z, t) = \hat{x}\, 10 \cos (5 \times 10^4 \pi t - 0.63\, z)\, e^{-0.63\,z}$ V/m

$$\mathbf{H}\,(z, t) = \hat{y}\, \frac{10}{0.222 \angle 45°} \cos(5 \times 10^4 \pi t - 0.63\, z)\, e^{-0.63z} \text{ A/m}$$

$$= \hat{y}\, 45.02 \cos[(5 \times 10^4 \pi t - 0.63\, z - \pi/4]\, e^{-0.63z} \text{ A/m}$$

The instantaneous power flow per unit area is the Poynting vector $\mathbf{P} = \mathbf{E} \times \mathbf{H} = \hat{z}\, 450.2 \cos (5 \times 10^4 \pi t - 0.63z) \cos (5 \times 10^4 \pi t - \pi/4 - 0.63z)\, e^{-1.26z}$ W/m^2.
The time-average power flow per unit area along z direction

$$\mathbf{P}_{av} = |\mathbf{E} \times \mathbf{H}^*|\, dt = 1/2\, EH \text{ W/m}^2$$

$$= 225.1\, e^{-1.26z} \cos \pi/4$$

$$= 159.2\, e^{-1.26z} \text{ W/m}^2$$

3.15 PLANE WAVES AT THE INTERFACE OF TWO MEDIA

When a plane wave incidents on a plane interface xy between two lossless dielectric media 1 and 2, the wave is partially reflected back to the incident medium 1 and partially transmitted to the second medium as shown in Fig. 3.7. Let us assume that θ_i, θ_r and θ_t are the angles of incidence, reflection and transmission, respectively. Any arbitrary polarisation of the field can be decomposed into two components, one in the plane of incidence (parallel polarisation) and another perpendicular to the plane of incidence (perpendicular polarisation). In both the cases, the reflection and refraction follows Snell's law: $\theta_i = \theta_r$

$$\frac{\sin \theta_t}{\sin \theta_i} = \frac{n_1}{n_2} = \sqrt{\frac{\mu_1 \, \varepsilon_1}{\mu_2 \, \varepsilon_2}} \tag{3.80}$$

where n_1, n_2 are the indices of refraction of the media 1 and 2, respectively.

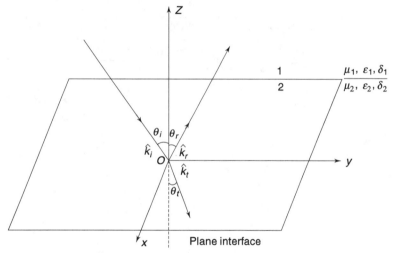

Fig. 3.7 *Reflection and transmission at a plane interface*

For the parallel polarisation of incident wave, the reflection and transmission coefficients are given by

$$\Gamma_{11} = \frac{\eta_2 \cos \theta_t - \eta_1 \cos \theta_i}{\eta_2 \cos \theta_t + \eta_1 \cos \theta_i} \tag{3.81}$$

$$T_{11} = \frac{2 \, \eta_2 \cos \theta_t}{\eta_2 \cos \theta_t + \eta_1 \cos \theta_i} \tag{3.82}$$

where η_1, η_2 are the intrinsic impedance of media 1 and 2, respectively,

$$\eta_1 = \sqrt{\frac{\mu_1}{\varepsilon_1}} = \frac{\eta_0}{\sqrt{\varepsilon_{r_1}}}, \quad \eta_2 = \sqrt{\frac{\mu_2}{\varepsilon_2}} = \frac{\eta_0}{\sqrt{\varepsilon_{r_2}}} \tag{3.83}$$

The z-directed wave impedances in media 1 and 2 are given, respectively, by

$$Z_1 = \frac{E_x^{(1)}}{H_y^{(1)}}\bigg|_{z=0} = \eta_1 \cos\theta_i, \; Z_2 = \eta_2 \cos\theta_t \qquad (3.84)$$

For the perpendicular polarisation of incident waves

$$\Gamma_\perp = \frac{\eta_2 \sec\theta_t - \eta_1 \sec\theta_i}{\eta_2 \sec\theta_t + \eta_1 \sec\theta_i} \qquad (3.85)$$

$$\Gamma_\perp = \frac{2\eta_2 \sec\theta_t}{\eta_2 \sec\theta_t + \eta_1 \sec\theta_i} \qquad (3.86)$$

The z-directed wave-impedances in media 1 and 2 are given by

$$Z_1 = \frac{\eta_1}{\cos\theta_i}, \; Z_2 = \frac{\eta_2}{\cos\theta_t} \qquad (3.87)$$

Figure 3.8 shows the nature of variation of $\Gamma_{||}$ and $\Gamma_\perp$ with incidence angle θ_i.

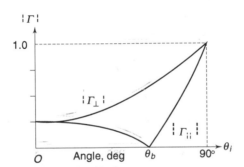

Fig. 3.8 *Variation $\Gamma_{||}$ and $\Gamma_\perp$*

3.15.1 Total Transmission

If the angle of incidence is varied, there will exist an angle $\theta_i = \theta_b$ say, called the *Brewster angle*, when $\Gamma_{||} = 0$ and total transmission of parallel polarised waves occurs at the interface. Under this condition from Eq. 3.81.

$$\theta_i = \theta_b = \sin^{-1}\sqrt{\frac{\varepsilon_2/\varepsilon_1 - \mu_2/\mu_1}{\varepsilon_2/\varepsilon_1 - \varepsilon_1/\varepsilon_2}} = \sin^{-1}\sqrt{\frac{\varepsilon_2}{\varepsilon_1 + \varepsilon_2}} \; \text{ for } \mu_1 = \mu_2 = \mu_0$$

$$(3.88)$$

$\Gamma_\perp$ never becomes zero so long as two media are different ($\eta_1 \neq \eta_2$). Hence Brewster angle does not exist for perpendicular polarisation. Any arbitrarily polarised wave incident at θ_b will be reflected with **E** polarised parallel to the interface and other component of **E** is totally transmitted.

3.15.2 Total Reflection

The incident wave is totally reflected back in the same medium for $|\varGamma_{||}| = 1$. It is seen from Eqns 3.84 and 3.88 that this cannot occur for real values of θ_i and θ_t. However, if $\sin \theta_t > 1$ for both polarisations (θ_t imaginary), $\sin \theta_i > \sqrt{\mu_2 \varepsilon_2 / \mu_1 \varepsilon_1}$ as seen from Eq. 3.80. Therefore, there exists a critical angle θ_c given by $\sin \theta_c = \theta_c \sqrt{\mu_2 \varepsilon_2 / \mu_1 \varepsilon_1}$, such that a wave incident on a plane interface at an angle equal to or greater than this angle will be totally reflected.

3.16 REFLECTION FROM AND TRANSMISSION THROUGH A PLANE INTERFACE BETWEEN A CONDUCTING AND DIELECTRIC MEDIA

The behaviour of a microwave signal incident at a plane interface between a conducting medium and a dielectric medium can be analysed assuming a TEM wave propagation for the incident wave in a lossless dielectric medium 1 as shown in Fig. 3.7, where medium 2 is a conductor. In the incident (dielectric) medium and the transmitted medium (conductor) the fields satisfy the wave equation. The propagation contents of the incident and reflected wave are $K = \omega\sqrt{\mu\varepsilon}$ and that of the transmitted waves in the conducting medium is

$$K = \omega \sqrt{\mu\varepsilon} \sqrt{(1 - j\sigma/\varepsilon\omega}$$

$$\cong (1 + j) \sqrt{\mu f \sigma \pi} \quad \alpha + j\beta \text{ for } \sigma \gg \varepsilon\omega \qquad (3.89)$$

Here the attenuation constant α and phase constants β are equal:

$$\alpha = \beta = \sqrt{\mu f \sigma \pi} \qquad (3.90)$$

The incident fields are represented by

$$\mathbf{E}_i = \mathbf{E}_{0i} e^{-j\mathbf{k}_i \cdot \mathbf{r}} \qquad (3.91)$$

$$\mathbf{H}_i = k\hat{k}_i \times \mathbf{E}_i / \mu\omega \qquad (3.92)$$

The reflected fields are

$$\mathbf{E}_r = \mathbf{E}_{0r} e^{-j\mathbf{k}_{ri} \cdot \mathbf{r}} \qquad (3.93)$$

$$\mathbf{H}_r = k\hat{k}_r \times \mathbf{E}_r / \mu\omega \qquad (3.94)$$

The transmitted fields inside the conductor are, therefore,

$$\mathbf{E}_t = \mathbf{E}_{0t} e^{-j\mathbf{k} \cdot \mathbf{r}} = \mathbf{E}_{0t} e^{-j\alpha \mathbf{k}_t \cdot \mathbf{r}} \cdot e^{-j\beta \mathbf{k}_t \cdot \mathbf{r}} \qquad (3.95)$$

$$\mathbf{H}_t = \frac{KK \times \mathbf{E}_t}{\omega\mu} = (\alpha + j\beta)\frac{\hat{k}_t \times \mathbf{E}_t}{\omega\mu} = \frac{1+j}{\sqrt{2}} \sqrt{\frac{\sigma}{\omega\varepsilon}} (\hat{k}_t \times \mathbf{E}_t) \qquad (3.96)$$

Two features of Eqs 3.98 and 3.99 are immediately evident and are given as follows.

1. The conductivity gives rise to exponential damping of the wave in metal.
2. The E and H fields are no longer in phase.

Both E and H fields decrease with penetration, falling to $1/e$ of their surface values in a distance equal to skin-depth $\delta = \sqrt{\dfrac{1}{\pi f \mu \sigma}}$. From Eqs 3.95 and 3.96, the attenuation coefficient in the conductor in surface normal direction $\sqrt{\pi f \mu \sigma}$ cos θ_t. As seen from Eqns. 3.80 and 3.81 the ratio of electric and magnetic energy densities inside the conductor is $\varepsilon \omega/\sigma \ll 1$ at frequencies below the optical region. Therefore, electric field is highly attenuated inside the conductor whereas the magnetic field is considerably transmitted through.

The laws of reflection and Snell's law are given by

$$\theta_i = \theta_{r,} = \frac{\sin \theta_i}{\sin \theta_t} = \sqrt{\frac{\sigma}{2\omega \varepsilon}} \tag{3.97}$$

We can write

$$K \cos \theta_t = \frac{K}{\beta} (\beta^2 - k_i^2 \sin^2 \theta_i)^{1/2} \approx K \tag{3.98}$$

when $\sigma \gg \omega \varepsilon$ or $\beta \gg k_i$. It shows that to a very high degree of approximation, the propagation constant normal to the interface in the conductor is nearly K and is independent θ_i since cos $\theta_t \approx 1$, for all $\theta_i < 90°$.

The wave impedance in the conductor in the direction normal to the interface is

$$\eta = \frac{E_t}{H_t} \cos \theta_t \approx (1+j) \sqrt{\frac{\omega \mu}{2\sigma}} \tag{3.99}$$

For a nonmagnetic conductor ($\mu = \mu_0$), the ratio of the reflected energy to the incident energy for normal incidence is

$$|\Gamma|^2 \approx 1 - 2 \sqrt{\frac{2\omega \varepsilon}{\sigma}} \tag{3.100}$$

Therefore, reflection loss is given by

$$R(dB) = 10 \log_{10} \sqrt{\frac{\sigma}{8\omega \varepsilon}} \tag{3.101}$$

The ratio of the transmitted energy at the interface to the incident energy for normal incidence is

$$|\Gamma|^2 \approx \frac{4\omega \varepsilon}{\sigma} \tag{3.102}$$

Therefore, the absorption loss for transmission up to a depth d is

$$A(dB) = 8.686 \, \alpha d = 8.686 \, d \, (\pi f \mu \sigma)^{1/2} \tag{3.103}$$

3.17 PROPAGATION OF MICROWAVES IN FERRITE

The propagation characteristics of microwaves in a ferrite magnetic material are utilised to develop many microwave non-reciprocal devices. Ferrites are complex solids represented by $M^{+2}O.Fe_2O_3$, where M^{+2} is the ion of a divalent metal (Cobalt, Nickel, Zinc, Magnesium, Cadmium, Iron, Manganese, Chromium, Copper, etc. or a mixture of these). The specific resistance of ferrites is very high (of the order of $10^7 - 10^8$ ohm m), which is 10^{14} times as high as those of metals, relative permittivity is of the order of $10 - 15$, and loss tangent $\tan \delta = 10^{-4}$ (low loss at microwave frequencies). The permeability of the ferrite is an asymmetric tensor and can be represented by

$$\tilde{\mu} = \mu_0 (1 + \tilde{\chi}_m) \tag{3.104}$$

where $\tilde{\chi}_m$ is the *tensor magnetic susceptibility* given by

$$\tilde{\chi}_m = \begin{bmatrix} \chi_m & jK & 0 \\ jK & \chi_m & 0 \\ 0 & 0 & 0 \end{bmatrix} \tag{3.105}$$

The relative permeabilities of ferrites are of the order of several thousand. Thus, ferrites are good dielectrics, but exhibit magnetic anisotropy. It has nonreciprocal electrical properties i.e., (1) the transmission coefficient for microwave propagation through ferrite is not the same for different directions, (2) nonreciprocal rotation of the plane of polarisation.

The magnetic properties of ferrites result mainly from the magnetic dipole moment **m** associated with the electron spin. In absence of any source the net angular momentum of a full shell is zero. Only electron spins in the incomplete shell are effective to produce a net dipole moment per atom. Thus a ferrite can be regarded as a collection of N effective spinning electrons per unit volume giving rise to total magnetic dipole moment M = **Nm**. In the absence of an external magnetic field, the dipole moments are oriented at random and the combined effect is very small.

In a steady external magnetic field H_0 in the direction other than that of the **m** of an electron, **m** precess gyromagnetically around H_0 due to the torque $\tau = \mathbf{m} \times \mu$ H_0 and tends to align the electron spin with H_0. The frequency of precessions is given by

$$f_0 \text{ (MHz)} \cong 2.8 \, \mathbf{H}_0 \text{ oersted} \tag{3.106}$$

The direction of precession is determined by H_0 and is clockwise looking along H_0 as shown in Fig. 3.9. The angle of precession $\psi = \angle \mathbf{H}_0$, **m** decreases due to friction and the ferrite is magnetised with magnetisation momentum **M** when the electron spins are aligned with H_0.

In addition to external steady magnetic field, if a *rf* circularly polarised magnetic field **H** at frequency $f \cong f_0$ is applied perpendicular to H_0, the angle of precession ψ of **M** and also the amplitude of induced procession will tend to increase at $f \cong f_0$ when the direction of rotation of **H** and **M** coincide. Due to magnetic frictional loss, the amplitude and the angle of **M** attain a steady state, describing a conical surface around $\mathbf{H_0}$. The energy continuously supplied by the *rf* field is dissipated as heat in the ferrite. When the interaction between the microwave field and the electrons is reduced, the ferrites show lower losses. When **H** rotates opposite to **M**, the ferrite dissipates no time average power and exhibit low loss.

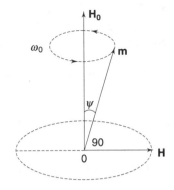

Fig. 3.9 *Precession of spinning electron in a steady magnetic field*

Therefore, microwave propagation in ferrites shows a gyromagnetic resonance with a peak of loss for clockwise polarisation of **H** which coincides with that of **M** and a flat low loss for opposite polarisation as shown in Fig. 3.10 where the horizontal scale represents $\mathbf{H_0}$ or frequency f_0 since both are linearly related.

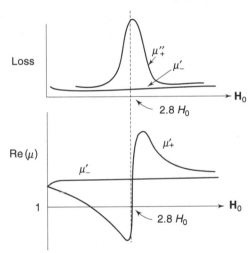

Fig. 3.10 *Gyromagnetic resonance in ferrite*

The clockwise and anticlockwise polarisations of propagating waves produce total ac fields

$$B_t^+ = \mu_0 M^+ + B^+ = \mu_+ H^+ \tag{3.107}$$

$$B_t^- = \mu_0 M^- + B^- = \mu_- H^- \tag{3.108}$$

respectively, and thus introduce two complex permeabilities denoted by plus and minus subscripts, as given by

$$\mu_+ = \mu'_+ - j\mu''_+ \tag{3.109}$$

$$\mu_- = \mu'_- - j\mu''_- \tag{3.110}$$

From Fig. 3.10 it is seen that the real and imaginary parts of the relative permeability μ_- are independent of the externally applied steady magnetic field. But μ_+ shows a resonant behaviour at a value of $\mathbf{H}_0$ for a given frequency. Since a linearly polarised plane wave can be considered to be composed of two circularly polarised waves, above analysis explains the non-reciprocal behaviour of ferrite materials to microwave propagation.

3.18 FARADAY ROTATION IN FERRITES

A plane circularly polarised wave propagating in $\mathbf{H}_0$ direction will have two different propagation constants given by

$$\beta^+ = \omega\sqrt{\varepsilon\,\mu_+} = \frac{2\pi}{\lambda^+} \quad \text{Clockwise} \tag{3.111}$$

$$\beta^- = \omega\sqrt{\varepsilon\,\mu_-} = \frac{2\pi}{\lambda^-} \quad \text{Anticlockwise} \tag{3.112}$$

Therefore, for a linearly polarised wave propagating along $\mathbf{H}_0$ the plane of polarisation rotates. This phenomenon is a non-reciprocal one. The rotation of the electric field of a linearly polarised wave passing through a magnetised ferrite is known as *Faraday rotation* as explained below in Fig. 3.11.

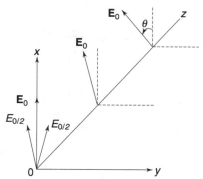

Fig. 3.11 *Faraday rotation in ferrite*

Let a linearly polarised TEM wave propagate in ferrite along the z-axis with $\mathbf{E}$ $= \hat{\mathbf{x}}\,Ex$ at $z = 0$. The linearly polarised wave may be decomposed into sum of clock and anti-clockwise circularly polarised waves. The component waves propagate with different phase constants β^+ and β^-, respectively, and the two corresponding electric field vectors rotate at different rates. Over a distance λ the resultant linearly polarised wave will undergo a phase delay of

$$(\beta^+ + \beta^-)\,\lambda = 4\pi \tag{3.113}$$

Therefore,

$$\lambda = 4\pi/(\beta^+ + \beta^-) = 2\,\lambda^-\lambda^+/(\lambda^+ + \lambda^-) \tag{3.114}$$

and corresponding rotation angle

$$(\beta^+ - \beta^-)\frac{\lambda}{2} = \left(\frac{\beta^+ - \beta^-}{\beta^+ + \beta^-}\right) 2\pi \tag{3.115}$$

The rotation per unit distance is $\theta = (\beta^+ - \beta^-)/2$. This property is utilised in designing ferrite isolator and circulators which will be described in Chapter 6 .

EXERCISES

3.1 The magnitude of an E-field for a plane wave traveling in free space is 200 volts per meter. Find the average power flow per unit area in the direction of propagation. What is the power flow through a circular surface of radius 1m that is perpendicular to the direction of propagation?

3.2 The velocity of an electromagnetic wave traveling in a dielectric material is 2.5×10^8 m/s. What is the relative permittivity of the dielectric and the intrinsic impedance for the medium?

3.3 For a plane wave in free space, the magnitude of the H field is 0.05 A/m. Find the average power flow per unit area. Determine the maximum power flow through an area of 4 cm^2.

3.4 A microwave oven is operated at 2.5 GHz with a field of magnitude 100 V/m. A cup of water has a volume of 100 cm^3 and relative permittivity ε_r'' = 9. Calculate the power being supplied to the water.

3.5 The half-space $z > 0$ is filled with a material with relative permeability 10 and dielectric constant 2.5. When a x-polarised plane wave is incident normally on this material from air, calculate the reflection and transmission coefficients.

3.6 A uniform plane wave of frequency 100 kHz is incident on the plane interface between free space ($z < 0$) and sea water ($z > 0$, $\sigma = 4$ S/m, ε_r = 80 and $\mu = \mu_0$). Determine (a) the amplitude of the incident wave electric field for which the amplitude of the transmitted wave electric field at $z = $ 1m is 1 mV/m ; and (b) the depth at which the transmitted wave electric field is $1/e$ times the incident wave field.

3.7 Calculate the total electric energy at an instant in a plane wave contained in a volume one wavelength long in the direction of propagation with cross-sectional area 1 m^2. Find the Poynting vector.

3.8 A plane wave is $\mathbf{E} = \hat{\mathbf{x}} E_0 \exp[-j(k_y y + k_z z)]$ is obliquely incident at a conducting plane xz at $y = 0$ at an angle θ_i with respect to the normal. Show that the $\mathbf{H}$ associated with the incident and reflected plane waves is

$$\mathbf{H} = -2E_0 \sqrt{\frac{\varepsilon_0}{\mu_0}} \, (\cos\theta_i \, \cos k_y y \, \hat{z} + j \sin\theta_i \, \sin k_y \sin k_y y \, \hat{y}) \exp(-jk_z z)$$

3.9 A plane wave propagates at a frequency 10 GHz in teflon ($\sigma = 0$) and carries a total power of 10 watts. Calculate the total power lost in a 10 m^2 path length and 1m cross-section in the material.

3.10 Calculate the phase velocity and attenuation constant for glass of dielectric constant 5 and conductivity 10^{-12} mhos/m.

REFERENCES

3.1 Ghose, RN, *Microwave Circuit Theory and Analysis,* McGraw-Hill Book Co. Inc., 1963.

3.2 Collin, RE, *Foundation for Microwave Engineering*, McGraw-Hill, Inc., International Editions, 1992.

3.3 Harrington, RF, *Time Harmonic Electromagnetic Fields*, McGraw-Hill Book Co., 1961.

3.4 Clarricoats, PJB, *Microwave Ferrites*, Chapman and Hall Ltd., London, 1961.

3.5 Fox, AG, Miller and Weiss, MT, Behaviour and Application of Ferrites in the Microwave Region, *Bell System. Tech. J.*, Vol. 34, 1955.

3.6 Aulock, V, H Wilhelm and EF Cliffored, *Linear Ferrite Devices for Microwave Applications*, Academic Press, New York, 1968.

3.7 Marcuvitz, N, *Waveguide Handbook* (M.I.T. Radiation Laboratory Series, Vol. 10), McGraw-Hill, New York, 1951.

3.8 Ramo, S, JR Whinnery and T Van Duzer, *Fields and Waves in Communication Electronics*, 2nd ed., John Wiley, New York, 1984.

3.9 Kraus, JD, *Electromagnetics*, 3rd ed., McGraw-Hill Book Company, New York, 1984.

3.10 Hayt, WH, Jr. *Engineering Electromagnetics*, 5th ed., McGraw-Hill Book Company, New York, 1989.

3.11 Johnk, CTA , *Engineering Electromagnetic Fields and Waves*, 2nd Ed., John Wiley and Sons, Inc, New York, 1988.

3.12 Wait, JR, *Electromagnetic Wave Theory*, Harper & Row Publishers, Inc., New York, 1985.

3.13 Stratton, JA, *Electromagnetic Theory*, McGraw-Hill Book Company, New York, 1941.

3.14 Shen, LC, and JA Kong, *Applied Electromagnetism*, Brooks-Cole, Calif., 1983.

Chapter 4

MICROWAVE TRANSMISSION LINES

4.1 INTRODUCTION

The microwave circuits and devices constitute a section or sections of microwave transmission lines or waveguides. Microwave signals are propagated through these lines as electromagnetic waves and scattered from the associated junction of these lines to travel in well-defined direction or ports. This chapter attempts to bridge the gap between electromagnetic theory and microwave circuits by highlighting some salient concepts for different configurations of these lines or guides. Conventional open wire lines are unsuitable for microwave transmission because of the high radiation losses that occur when the wavelength becomes smaller than the physical lengths of these lines at high frequencies. The structures considered here are (1) multi conductor lines, viz. coaxial lines, strip lines, microstrip lines, slot lines, and coplanar lines, (2) single conductor lines, such as rectangular waveguides, circular waveguides, ridge waveguides and (3) open-boundary structures, viz. dielectric rods. In the first category of lines, the mode of transmission is a TEM or quasi-TEM wave. In the second category of lines, the modes are either TE or TM waves or both. The third category of lines support, in general, a combination of TE and TM waves called the hybrid HE modes, except possible axi-symmetric modes which are either purely TE or TM waves. The relations for input impedance, reflection coefficient, transmission coefficient, characteristic impedance, and so forth, given in Chapters 2 and 3 are applicable to microwave guides operating in a single mode.

4.2 IDEAL COAXIAL LINE

An ideal coaxial line consists of an inner circular perfect conductor and an outer circular perfect conductor as shown in Fig. 4.1. The space between the two conductors is filled ideally with a uniform lossless homogeneous dielectric having a dielectric constant ε_r. The two conductors are at two different potentials. The dominant mode of propagation in a symmetric coaxial line is TEM wave. The field configuration of mode is shown in Fig. 4.1. Since TEM mode does not have

a cut-off frequency, a coaxial line is a broadband device. For propagation in the
+ z direction, the fields inside the line cross-section are expressed by

$$\mathbf{E} = \mathbf{E}_t = \mathbf{E}_0(\rho,\phi)e^{-jkz} = \frac{V_0}{\ln(b/a)}\frac{\hat{\rho}}{\rho}e^{-jkz} \tag{4.1}$$

$$\mathbf{H} = \mathbf{H}_t = \pm\frac{\hat{z}\times\mathbf{E}_0(\rho,\phi)}{\eta}e^{-jkz} = \frac{V_0}{\eta\ln(b/a)}\frac{\hat{\phi}}{\rho}e^{-jkz} \tag{4.2}$$

where $k = \omega\sqrt{\mu\varepsilon}$ and $\eta = \sqrt{\mu/\varepsilon}$, the wave impedance for TEM waves. Here a
and b are the outer and inner radii of the inner and outer conductors, respectively.

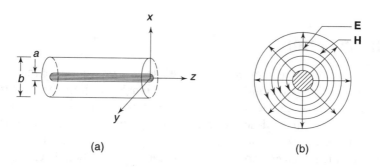

(a) (b)

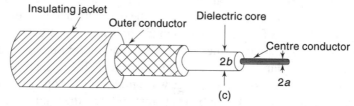

(c)

Fig. 4.1 *Coaxial lines and TEM wave fields — E-field lines*
— H-field lines

The current density on the outer surface of the inner conductor is given by

$$\mathbf{J_s} = \hat{n}\times\mathbf{H} = \hat{\rho}\times\mathbf{H}_t = \hat{z}\frac{V_0\,e^{-jkz}}{\eta a\,\ln(b/a)} \tag{4.3}$$

The power flow through the coaxial line is expressed by

$$P = \frac{1}{2}\,\mathrm{Re}\int_a^b\int_0^{2\pi}\mathbf{E}\times\mathbf{H}^*\cdot\mathbf{dS} = \frac{\pi V_0^2}{\eta\ln(b/a)} \tag{4.4}$$

where $\mathbf{dS} = \rho\,\mathrm{d}\rho\,\mathrm{d}\phi$. The characteristic impedance of the symmetric coaxial line
for non magnetic dielectric is given by

$$Z_0 = \frac{1}{2\pi}\sqrt{\frac{\mu}{\varepsilon}}\,\ln\left(\frac{b}{a}\right) = \frac{60}{\sqrt{\varepsilon_r}}\,\ln\frac{b}{a}\ \text{ohm} \tag{4.5}$$

where $\varepsilon_r = \varepsilon\,\varepsilon_0$, the dielectric constant of the dielectric medium between the conductors.

4.2.1 Coaxial Line with Small Losses

The power loss in the coaxial line is caused by the finite conductivity σ of the conductors and the dielectric loss in the medium between the conductors. In a lossy dielectric, $\varepsilon = \varepsilon' - j\varepsilon''$ and $\sigma_d/w\,\varepsilon \ll 1$, where σ_d is the conductivity of the dielectric. Under such condition, the propagation constant is complex, given by, $K = \alpha + j\beta \approx jk\,(1 - j\sigma_d/2\omega\,\varepsilon)$. Thus

$$\alpha_d = \frac{\sigma_d}{2}\sqrt{\frac{\mu}{\varepsilon}} \tag{4.6}$$

$$\beta = \omega\sqrt{\mu\varepsilon} = k \tag{4.7}$$

where α_d and β are the attenuation constant in the dielectric and the phase constant, respectively.

Due to the finite conductivity σ of the conductors, the conductors exhibit a non-zero surface impedance given by,

$$\eta_c = (1 + j)\sqrt{\frac{w\mu}{2\sigma}} = \frac{1 + j}{\sigma\delta} \tag{4.8}$$

and a surface current density $\mathbf{J}_s = \hat{n} \times \mathbf{H}$. Here $\delta = \dfrac{1}{\sqrt{\pi f \mu\sigma}}$ is the skin depth.

Therefore , the electric field must have a non-zero tangential component given by $\eta_c\,\mathbf{J}_s$ at the surface. Since the surface current is axial, an axial component of electric field must be present, resulting in a non-TEM wave mode of propagation. This axial component of **E**-field gives rise to a component of the Poynting vector directed into the conductor, resulting in power loss in the conductor, given per unit length, by

$$P_c = \frac{1}{2}\,\mathrm{Re}\,\eta_c \oint_{s_1 + s_2} \mathbf{J}_s \cdot \mathbf{J}_s^* \cdot dl$$

$$= \frac{R_s}{2} \oint_{s_1 + s_2} |\mathbf{H}|^2\, dl \tag{4.9}$$

where integration is taken around the periphery $s_1 + s_2$ of the two conductors. Due to this power loss, the power propagated along the line decreases according to a factor $(e^{\alpha_c z})^2 = e^{-2\alpha_c z}$. The rate of decrease of power with distance is

$$-\frac{\partial P}{\partial z} = 2\alpha_c P$$

$$= \text{Power loss per unit length} \tag{4.10}$$

$$= P_l$$

Therefore, the attenuation constant due to conductor loss is given by

$$\alpha_c = \frac{P_l}{2P} = \frac{R_s \int\limits_{s_1 + s_2} |\mathbf{H}|^2 \, dl}{\sqrt{\dfrac{\mu}{\varepsilon}} \int\limits_{s} |\mathbf{H}|^2 \, ds} \tag{4.11}$$

The total attenuation constant of the line becomes $\alpha = \alpha_c + \alpha_d$.

The distributed line parameters of symmetric coaxial transmission lines can be computed from magneto static and electrostatic field considerations. These parameters are given by the following relations

$$R = \frac{R_s}{2\pi} \left(\frac{1}{a} + \frac{1}{b} \right) \tag{4.12}$$

$$L = \sqrt{\mu_0 \, \varepsilon'} \, Z_0 \approx \frac{\mu_0}{2\pi} \ln \frac{b}{a} \tag{4.13}$$

$$G = \frac{w \, \varepsilon'' \, C}{\varepsilon'} \tag{4.14}$$

$$C = \frac{\sqrt{\mu_0 \, \varepsilon'}}{Z_0} \cong \frac{2\pi \, \varepsilon'}{\ln \dfrac{b}{a}} \tag{4.15}$$

For low-loss lines

$$\alpha_d = \frac{GZ_0}{2} \quad \alpha_c = \frac{R}{2Z_0} \tag{4.16}$$

$$\alpha = \frac{1}{2} \left(R \sqrt{\frac{C}{L}} + G \sqrt{\frac{L}{C}} \right) \tag{4.17}$$

and

$$Z_0 \cong \sqrt{\frac{L}{C}} \tag{4.18}$$

For power transmission, coaxial lines are used up to about 3GHz and for small signal transmission, up to about 18GHz. For cables with $G \rightarrow 0$, the attenuation is minimum for $b/a = 3.6$. For an air dielectric $\varepsilon_r = 1$ and $b/a = 3.6$ gives $Z_0 = 76.86$ ohm. With a solid polyethylene dielectric, $\varepsilon_r = 2.3$ and $Z_0 = 50.67$ ohm for minimum attenuation. Since the attenuation coefficient does not vary rapidly with b/a, 50 ohm lines are very commonly used for minimum attenuation.

Example 4.1 An air-filled coaxial transmission line has outer and inner conductor radii equal to 6 cm and 3 cm, respectively. Calculate the values of (a) inductance per unit length, (b) capacitance per unit length and (c) characteristic impedance of the line.

Solution

Given $b = 6$ cm, $a = 3$ cm

(a) $L = \mu_0/2\pi \ln b/a = 4\pi \times 10^{-7}/2\pi \ln 6/3 = 1.386 \times 10^{-7}$ h/m

(b) $C = 2\pi\varepsilon_0/\ln b/a = 2\pi \times 10^{-9}/(36\pi \ln 6/3) = 0.08 \times 10^{-9}$ f/m

(c) $Z_c = \sqrt{(L/C)}$

$$= \sqrt{[\mu_0/\varepsilon_0 \, ((\ln b/a)/2\pi)^2]}$$

$$= 120\pi \times \ln (b/a)/2\pi = 41.59 \text{ ohm}$$

4.2.2 Higher Order Modes

At sufficiently high frequencies, higher order modes are generated in a coaxial line. The lower cut-off frequencies of the higher order modes in a coaxial line are very high. Thus, higher order mode interference is not encountered at the normal operating (lower) frequencies in a conventional coaxial lines. The lowest order higher order modes in coaxial lines are TE_{11} and TM_{01}. The cut-off wavelength of these modes are

$$\lambda_C \, (TE_{11}) \cong \pi(a + b) \qquad \lambda_C \, (TM_{01}) \cong 2(b - a) \qquad (4.19)$$

Therefore, the average circumference of the inner and outer conductors in the propagating cross-section of the coaxial line should be less than the operating wavelength in order to prevent higher order mode interference. The field configuration of these higher order modes are shown in Fig. 4.2.

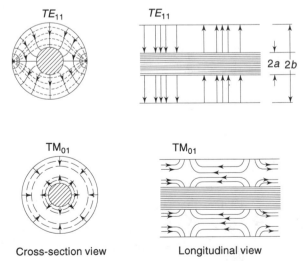

| TE₁₁ | TE₁₁ |
| TM₀₁ | TM₀₁ |

Cross-section view   Longitudinal view

Fig. 4.2 *Higher order modes in coaxial lines — E-field lines $\cdots$ H-field lines*

4.2.3 Unbalance Characteristics of Coaxial Lines

A coaxial line is said to be an unbalanced line because the two conductors are not at equal and opposite potentials with respect to the ground. In that, the outer conductor is commonly grounded and the propagation path on this side ideally has the ground potential. Whereas, the other one has different potential with respect to ground. Any interconnected equipment between the lines introduces unequal impedances in both lines and would present unequal impedances to ground. Therefore, when a coaxial cable is used to feed balanced circuits, such as dipole antennas or ordinary parallel-wire lines, impedance mismatch occurs.

Example 4.2   The primary constants for a coaxial cable at 1 GHz, are $L =$ 250 nh/m, $C = 95$ pf/m, $R = 0.06$ ohm/m, and $G = 0$. (a) Determine the attenuation coefficient α, (b) the phase constant β, (c) the phase velocity μ_p, (d) relative permittivity ε_r, and (e) power loss for a length of 10 m, when the input power is 500 watts.

Solution

$$\omega L = 2\pi \times 1 \times 10^9 \times 250 \times 10^{-9} = 1570.8 \text{ ohm/m}$$

$$\omega C = 2\pi \times 1 \times 10^9 \times 95 \times 10^{-12} = 596.6 \times 10^{-3} \text{ mho/m}$$

Therefore,

$$\omega L \gg R, \ \omega C \gg G = 0$$

Hence,

$$Z_0 \simeq (L/C)^{1/2} = [(250 \times 10^{-9})/(95 \times 10^{-12})]^{1/2}$$
$$= 51.3 \text{ ohm}$$

(a) $$\alpha = R/2Z_0 = 0.06/(2 \times 51.3) = 5.85 \times 10^{-4} \text{ Np/m}$$
$$= 5.08 \times 10^{-3} \text{ dB/m}$$

(b) $$\beta = \omega (\mu\varepsilon)^{1/2} = \omega(LC)^{1/2}$$
$$= 2\pi \times 1 \times 10^9 \times (250 \times 95 \times 10^{-9} \times 10^{-12})^{1/2}$$
$$= 968.3 \times 10^9 \times 10^{-10} \ (10^{-1})^{1/2}$$
$$= 96.83/(10)^{1/2}$$
$$= 30.6 \text{ rad/m}$$

(c) $$u_p = 1/(LC)^{1/2}$$

$$= \frac{1}{(250 \times 95 \times (10^{-9} \times 10^{-12})^{1/2}}$$
$$= 2.05 \times 10^8 \text{ m/s}$$

(d) $$c = \sqrt{\varepsilon_r u_p}$$

or, $$\varepsilon_r = (c/u_p)^2$$
$$= [(3 \times 10^8)/(2.05 \times 10^8)]^2 = 2.14$$

(e) $\quad P_{\text{loss}} = P_{\text{in}} \times 2\alpha \times 1$

$\qquad\qquad = 500 \times 2 \times 5.85 \times 10^{-4} \times 10$

$\qquad\qquad = 5.85$ watts

4.3 PLANAR TRANSMISSION LINES

Miniaturisation of microwave circuits has taken place through the development of planar transmission lines which are flat two or multi conductor transmission lines having low profile and light weight. This geometry allows control of the characteristic impedance of the line by defining the line dimensions in a single plane and is therefore suitable for microwave integrated circuits. The complete transmission line circuit can be fabricated in one step by *thin film technology* and *photolithography techniques*. Among the several configurations, there are four basic forms which are widely used in microwave integrated circuits. These are (1) strip lines (2) microstrip lines, (3) slot lines, and (4) coplanar strip lines as shown in Fig. 4.3.

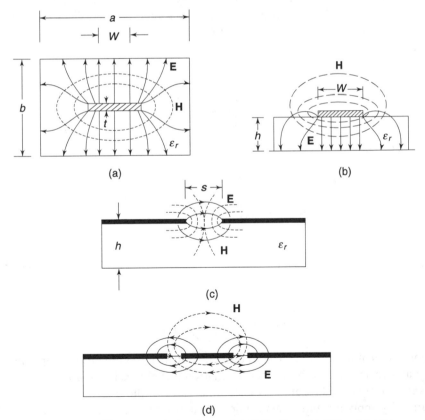

Fig. 4.3 *Planar transmission lines (a) Strip line (b) Microstrip line (c) Slot line (d) Coplanar strip line*

4.3.1 Strip Lines

Strip lines consist of a central thin conducting strip of width $W \gg$ its thickness t placed inside a low-loss dielectric substrate of thickness b between two wide ground plates as shown in Fig. 4.3(a). The propagation characteristics in such a line is nearly TEM mode where most of the electric field lines are perpendicular to the centre and its parallel ground conductors and are concentrated over the width of the centre conductor. There are also fringing field lines at the edges of the centre strip as shown in Fig. 4.3(a). The fringing fields are extended up to a certain distance from the edges of the centre strip, beyond which the fields are practically zero. When $b < \lambda/2$ the field cannot propagate in the transverse direction and decrease exponentially. The energy is confined inside the line cross-section provided the width a of the ground plane is at least five times greater than the spacing b between the plates. Therefore, no vertical side walls are necessary at the two transverse ends. The commonly used dielectrics were Teflon, polyolefine, polystyrene, etc. These lines are used over the frequency range from 100 MHz to 30 GHz. For a symmetric homogeneous strip line having centre strip thickness $t \to 0$, the characteristic impedance and the field distributions can be analysed accurately by the conformal transformation technique. The characteristic impedance of the symmetric stripline having zero strip thickness can be expressed as follows:

(1)   For $W/b \leq 0.5$

$$Z_0 = \frac{30}{\sqrt{\varepsilon_r}} \ln \left[\frac{2\left(1 + \sqrt{k}\right)}{\left(1 - \sqrt{k}\right)} \right] \text{ ohm} \qquad (4.20)$$

(2)   For $W/b > 0.5$

$$Z_0 = \frac{30\pi^2}{\sqrt{\varepsilon_r} \ln \left(\frac{2\left(1 + \sqrt{k'}\right)}{\left(1 - \sqrt{k'}\right)} \right)} \text{ ohm} \qquad (4.21)$$

where
$$k = \text{sech} \left(\frac{\pi W}{2b} \right) \qquad (4.22)$$

$$k' = \sqrt{1 - k^2} = \tanh \left(\frac{\pi W}{2b} \right) \qquad (4.23)$$

For a finite non-zero thickness of the centre strip of a symmetrical strip line, an exact analysis using conformal transformation techniques is possible but this results in a complicated formula. An accurate value of the impedance can also be obtained by applying numerical techniques, such as the method of moments. For a wide centre strip having non-zero thickness of a symmetric stripline ($W/b \gg 0.35$), the characteristic impedance is given by

$$Z_0 = \frac{94.15}{\sqrt{\varepsilon_r}} \left[\frac{WK}{b} + \frac{C_f}{8.854\,\varepsilon_r} \right]^{-1} \text{ohm} \qquad (4.24)$$

where

$$K = \frac{1}{1 - t/b} \qquad (4.25)$$

$C_f =$ fringing capacitance in pF/m, which arises due to fringing electric field at the edges of the strip.

$$= \frac{8.854\,\varepsilon_r}{\pi} [2K \ln (K + 1) - (K - 1) \ln (K^2 - 1)] \text{PF/m} \quad (4.26)$$

The characteristic impedance Z_0 for a partially shielded strip line vs W/b with t/b as a parameter is plotted in Fig. 4.4. It is seen that the value of Z_0 decreases with the increase of W/b and also increase with t/b. In practice, MICs use thickness t of the order of 1.5 to 3 mils and error introduced for the zero thickness assumption is negligible.

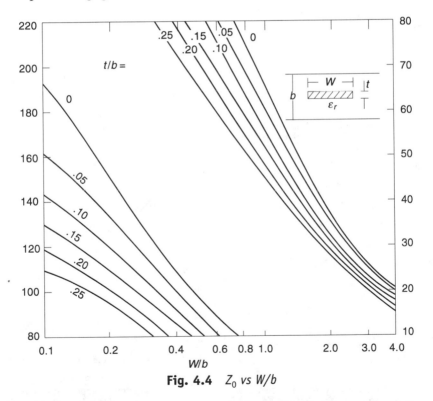

Fig. 4.4 Z_0 vs W/b

For large W/b, the strip line can be completely shielded to avoid radiation. A shielded strip line consists of a very thin strip centre conductor ($t/W \ll 1$) embedded in a low loss or lossless dielectric medium enclosed in a rectangular metallic box as shown in Fig. 4.3(a). For a completely closed strip line with large W, there is more uniform TEM field region between the centre conductor and the horizontal plates. But due to proximity between the edges of the centre plate and the side walls, fringing fields increase at the edges of the strip and some field lines terminate on the side walls. Here fringe capacitance increases and the characteristic impedance decreases considerably. Since W is large, it may be necessary that the thickness of the strip is not negligible for requirement of mechanical strength to support the central conductor when there is air-dielectric. The variations of Z_0 with different dimensional changes are shown in Fig. 4.5 for a wide strip enclosed in a shielded box.

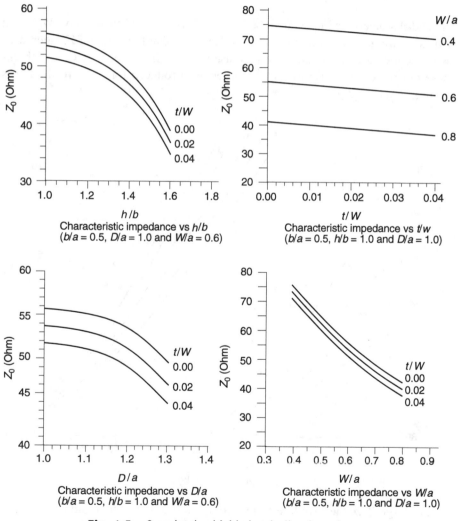

Fig. 4.5 *Completely shielded strip line impedances*

Figure 4.6 shows electric field distributions in the cross-section of a typical shielded strip line where the fields are nearly uniform within the range of the centre conductor. Since the propagation mode in a strip line is nearly TEM mode, the guide wavelength in the line is $\lambda_0 / \sqrt{\varepsilon_r}$.

4.3.1.1 *Higher order modes in strip lines*

The operating band width of a strip line is considerably large for TEM mode of propagation. The upper frequency limit is set by the presence of nearest higher order TE_{10} and TM_{11} modes. The approximate expressions of cut-off wavelength for these modes are

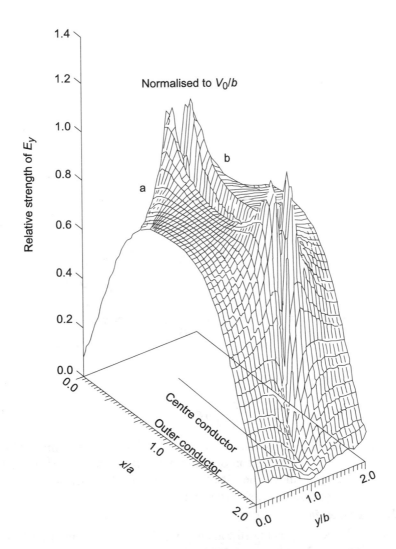

Fig 4.6 *(Contd)*

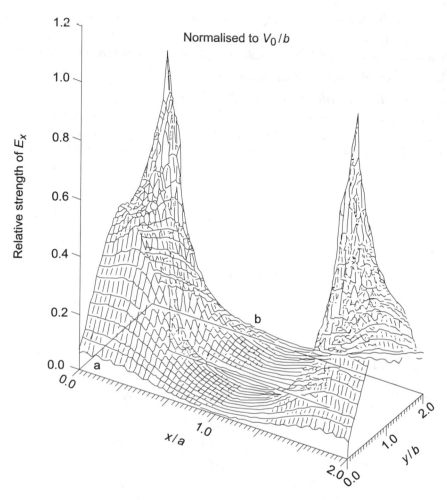

Fig 4.6 *Field distribution in the cross section of a shielded strip line (h/b = 0.5, W/a = 0.6)*

$$\lambda_0 \ (TM_{11}) = 2b\sqrt{\varepsilon_r} \qquad\qquad (4.27)$$

$$\lambda_0 \ (TE_{10}) = \left(2W + \frac{\pi b}{2}\right)\sqrt{\varepsilon_r} \qquad\qquad (4.28)$$

For a considerably wide strip, the cut-off conditions can be estimated by analysing the cross-section using conformal transformation or numerical methods, viz., finite element method. The cut-off wavelengths of higher order modes for typical shielded strip lines are shown in Fig. 4.7.

4.3.1.2 Losses in strip lines

For low loss dielectric substrate, the attenuation factor in the strip line arises from conductor losses and can be expressed by

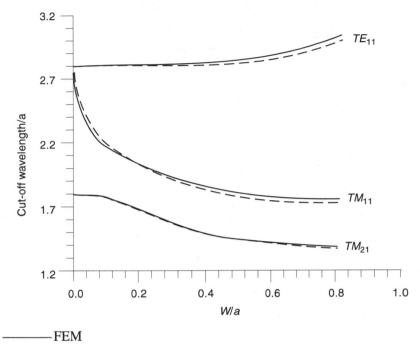

Fig. 4.7 *Cut-off wavelength vs centre conductor width for the higher order TE and TM modes (b/a=1.0, $\varepsilon_r = 1$)*

$$\alpha_c = \frac{R_s}{Z_0 b} \frac{\pi W/b + \ln(4b/\pi t)}{\ln 2 + \pi W/2b} \quad \text{neper/unit length; } W \geq 2b, t \leq \frac{b}{10} \quad (4.29)$$

where $R_s = \sqrt{\dfrac{\pi f \mu}{\sigma}}$

4.3.1.3 Excitation of strip lines

Strip lines are excited by a coaxial line which interfaces the strip lines by means of special launcher or connector. This connector consists of thin flat small centre conductor which forms the centre conductor of the co-axial line and two rectangular outer conductors joined together to the ground planes as shown in Fig. 4.8(a). Strip lines are suitable for the design of passive circuits but inconvenient for mounting active component.

4.3.2 Microstrip Lines

A microstrip line consists of a single ground plane and a thin strip conductor on a low loss dielectric substrate above the ground plate. Since the size of the microwave solid-state devices is very small (of the order $0.008 - 0.08$ mm^3), the technique of signal input to these devices and extracting output power from them uses microstrip lines on the surface on which they can be easily mounted. Figure 4.3 (b) shows a typical cross-section of a microstrip line. Due to absence of a top

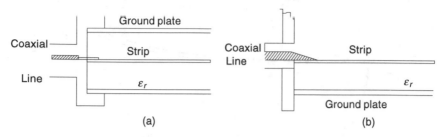

Fig. 4.8 *Strip line launcher connector (a) Strip line connector (b) Microstrip line connector*

ground plate and the dielectric substrate above the strip, the electric field lines remain partially in the air and partially in the lower dielectric substrate. This makes the mode of propagation not pure TEM but what is called quasi-TEM. Due to open structure and any presence of discontinuity, the microstrip line radiates electromagnetic energy. The radiation loss is proportional to the square of the frequency. The use of thin and high dielectric materials reduces the radiation loss of the open structure where the fields are mostly confined inside the dielectric.

4.3.2.1 Effective dielectric constant

Since the propagation field lines in a microstrip lie partially in air and partially inside the homogenous dielectric substrate, the propagation delay time for a quasi-TEM mode is related to an effective dielectric constant ε_{eff} given by

$$\varepsilon_{\text{eff}} = \frac{\varepsilon_r + 1}{2} + \frac{\varepsilon_r - 1}{2}\left[\left(1 + \frac{12h}{W}\right)^{-1/2} + 0.04\left(1 - \frac{W}{h}\right)^2\right]; \ W/h \le 1 \ (4.30)$$

$$\varepsilon_{\text{eff}} = \frac{\varepsilon_r + 1}{2} + \frac{\varepsilon_r - 1}{2}\left(1 + \frac{12h}{W}\right)^{-1/2}; \ W/h \gg 1 \tag{4.31}$$

where ε_r is the relative dielectric constant of the substrate material.

4.3.2.2 Characteristic impedance and guide wavelength

The characteristic impedance of microstrip lines can be expressed by

$$Z_0 = \frac{60}{\sqrt{\varepsilon_{\text{eff}}}} \ln\left[\frac{8h}{W} + \frac{W}{4h}\right] \text{ohm}, \ W/h \le 1 \tag{4.32}$$

$$Z_0 = \frac{376.7}{\sqrt{\varepsilon_{\text{eff}}}\left[\frac{W}{h} + 1.4 + 0.667 \ln\left(\frac{W}{h} + 1.444\right)\right]} \text{ohm; for } W/h > 1 \ (4.33)$$

$$Z_0 = \frac{376.7}{\sqrt{\varepsilon_{\text{eff}}}} \frac{h}{W} \text{ohm; for } W/h \gg 1 \tag{4.34}$$

The guide wavelength for the propagation of quasi-TEM mode is given by

$$\lambda_g = \lambda_0 / \sqrt{\varepsilon_{\text{eff}}} \tag{4.35}$$

The characteristic impedance for a microstrip line vs W/h with ε_r as a parameter is plotted in Fig. 4.9. It is seen that the value of Z_0 decreases with increase of W/h and also with increase of ε_r.

4.3.2.3 Losses in microstrip lines

For the non-magnetic substrate material, two types of losses exist in microstrip lines which provide attenuation of signal.

(1) dielectric loss in the substrate and
(2) ohmic loss in the strip conductor and the ground plane due to finite conductivity. The total attenuation constant α can be expressed as $\alpha = \alpha_d + \alpha_c$

where α_d, α_c are the dielectric and ohmic attenuation constants, given by

$$\alpha_d = \frac{\sigma_d}{2} \sqrt{\frac{\mu}{\varepsilon}}$$

$$= 27.3 \, \frac{(\varepsilon_{\text{eff}} - 1)}{(\varepsilon_r - 1)} \, \frac{\varepsilon_r}{\varepsilon_{\text{eff}}} \cdot \frac{\tan \delta}{\lambda_g} \text{ dB/m} \tag{4.36}$$

Here σ_d is the conductivity of dielectric substrate and $\tan \delta = \dfrac{\sigma_d}{\omega \varepsilon}$ is the dielectric loss tangent.

For a low-loss dielectric substrate, the major attenuation factor at microwave frequencies arises due to finite conductivity of the strip conductor. This contributes to the ohmic losses due to the current on the strip. The current distribution in the transverse plane is fairly uniform with minimum value at the central axis and shoots up to a maximum at the edges of the strip .

For simplicity assuming uniform current distribution in the region $-W/2 < x < W/2$, the attenuation constant due to ohmic loss of a wide line ($W/h > 1$) is

$$\alpha_c \approx \frac{8.686}{Z_0 \, W} \sqrt{\frac{\pi f \mu}{\sigma}} \text{ dB/m} \tag{4.37}$$

From Wheeler's work, the values of α_c for a wide range of W/h are plotted in Fig. 4.10.

Microstrip lines also have radiation loss due to thin open structure and any current discontinuities in the strip conductor.

The commonly used substrate materials for strip and microstrip lines are polytetrafluoroethylene (PTFE)/teflon, RT Duroid 5880, alumina, and sapphire, etc. whose electrical characteristics are given in Table 4.1. Alumina is most widely used for frequencies up to 20 GHz. At higher frequencies, sapphire is used.

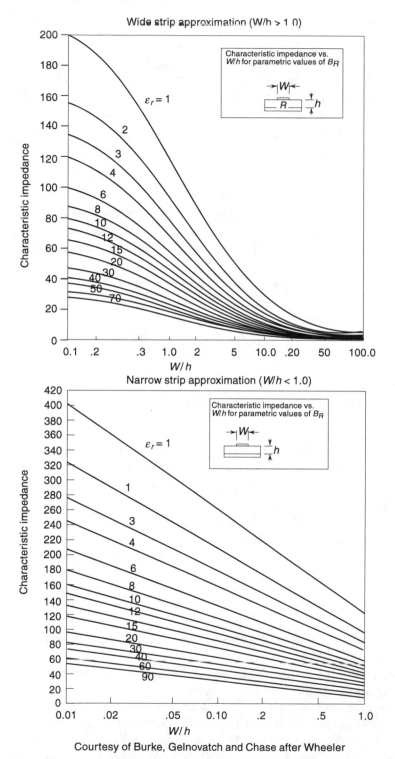

Courtesy of Burke, Gelnovatch and Chase after Wheeler

Fig 4.9 *Variation of Z_0 with W/h*

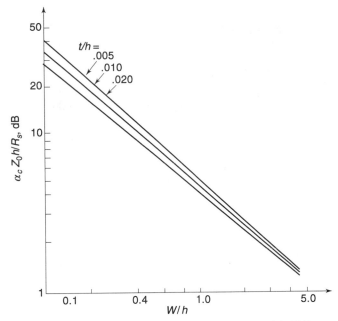

Fig. 4.10   *Variation of attenuation constant with W/h*

Table 4.1   *Properties of substrate materials*

Material	ε	$\tan \delta$
PTFE	2.2	0.0002–0.0005
RT/Duroid 5880	2.26	0.001
Alumina	9.6–10	0.0005–0.002
Sapphire	9.4	0.0002

4.3.2.4   Excitation of microstrip line

The radiation loss of microstrip lines are eliminated by enclosing the microstrip within a metallic box having first resonance frequency much above the signal frequency. Microstrip lines are excited by a coaxial line through a connector launcher as shown in Fig. 4.8 (b). Microstrip lines are suitable for the design of passive circuits and series mounting of active components across a gap in the strip.

Example 4.3   A microstrip line is composed of zero thickness copper conductors on a substrate having $\varepsilon_r = 8.4$, $\tan \delta = 0.0005$ and thickness 2.4 mm. If the line width is 1 mm, and operated at 10 GHz, calculate (a) the characteristic impedance, (b) the attenuation due to conductor loss and dielectric loss.

Solution

Given,

$$f = 10 \text{ GHz}, \ \lambda = c/f = 30/10 = 3 \text{ cm}$$

$$W = 1 \text{ mm}, h = 2.4 \text{ mm}, \varepsilon_r = 8.4$$
$$\tan \delta = 0.0005, t = 0$$

(a) $W/h = 1/2.4 = 0.417 < 1$.

From Eqs 4.30 and 4.32

$$\varepsilon_{\text{eff}} = \frac{\varepsilon_r + 1}{2} + \frac{\varepsilon_r - 1}{2} \left[(1 + 12h/W)^{-1/2} + 0.04 (1 - W/h)^2 \right]$$

$$= \frac{8.4 + 1}{2} + \frac{8.4 - 1}{2} \left[(1 + 12/0.417)^{-1/2} + 0.04 (1 - 0.417)^2 \right]$$

$$= 5.43$$

$$Z_0 = 60/\left(\sqrt{\varepsilon_{\text{eff}}} \right) \ln [8h/W + W/4h]$$

$$= 60/\left(\sqrt{5.43} \right) \ln [8/0.417 + 0.25/0.417]$$

$$= 76.2 \text{ ohm}$$

(b) $\quad \alpha_c = 8.686 \, R_s/Z_0 W$

$$R_s = \sqrt{(\pi f \mu / \sigma)}$$

$$= \frac{\sqrt{(3.14 + 10 \times 10^{10} + 4\pi \times 10^{-7})}}{\sqrt{[5.8 \times 10^7]}}$$

$$= 2.6 \times 10^{-2} \text{ ohm/sq.m}$$

$$\alpha_c = \frac{8.686 \times 2.6 \times 10^{-2}}{76.2 \times 1 \times 10^{-3}}$$

$$= 2.98 \text{ dB/m}$$

$$\alpha_d = 27.3 \left(\frac{\varepsilon_{\text{iff}} - 1}{\varepsilon_r - 1} \right) \frac{\varepsilon_r}{\varepsilon_{\text{eff}}} \frac{\tan \delta}{\lambda_g}$$

$$\lambda_g = \frac{\lambda}{\sqrt{\varepsilon_{\text{eff}}}} = \frac{3}{\sqrt{5.43}} = 1.287 \text{ cm}$$

Then

$$\alpha_d = 27.3 \left(\frac{5.43 - 1}{8.4 - 1} \right) \frac{8.4}{5.43} \frac{0.0005}{1.287 \times 10^{-2}}$$

$$= 0.98 \text{ dB/m}$$

4.3.3 Slot Lines

Some planar configurations used as transmission lines in MICs include slot lines as shown in Fig. 4.3 (c), where the conductors are in one plane on a dielectric substrate. Hence unlike strip lines and microstrip lines, active and passive components can easily be shunt mounted across the lines from the top. The propagating fields are concentrated in the dielectric regions at the gap between the two adjacent conductors. Since the magnetic field has both longitudinal and transverse components, the propagating mode is *TE*. The magnetic field is found circularly polarised at periodical locations which could be utilised to design ferrite isolators. The characteristic impedance Z_0 of slot lines increases with the increase of width of the slot. The Z_0 also varies rapidly with frequency as compared to microstrip lines.

4.3.4 Coplanar Lines

A coplanar line is a parallel three-conductor line consisting of a thin centre strip (W) and two thin ground strips (G) on a dielectric substrate as shown in Fig. 4.3 (d). Coplanar lines are very convenient for both series and shunt mounting of components across the gap in the centre strip and across the slots between the conductors, respectively. Like slot lines the propagating modes in coplanar lines are TE modes and ferrite components can be designed due to existence of circular polarisation of the magnetic field at given locations.

4.3.5 Advantages and Disadvantages of Planar Transmission Lines

The basic advantages of planar transmission lines are: (i) small size and weight, (ii) can be flash mounted on a metallic body, (iii) increased reliability, (iv) low cost, (v) easy access for component mounting, (vi) the Z_0 can be controlled by defining the dimensions in a single plane, (vii) passive circuit design is possible easily by changing the dimensions of the line in one plane only.

There are certain disadvantages in these lines. One of which is the low power handling capability due to small size. There are also radiation losses from the open structures like microstrip, slot and coplanar lines. This loss is reduced by confining the field lines more in the dielectric substrate of a high dielectric constant, alumina ($\varepsilon_r \cong 9 - 10$), sapphire ($\varepsilon_r \cong 9.3 - 11.7$) and beryllia ($\varepsilon_r \cong 6.6$). Strip lines also use polystyrene ($\varepsilon_r \cong 2.54$) as substrate. The circuit design using these configurations should be very accurate since matching screws and short circuit plungers cannot be used in planar transmission line circuits. The design of some active devices are also limited because of low Q (of the order 100) obtainable with ordinary microstrip configurations.

4.4 RECTANGULAR WAVEGUIDES

Since the losses in two conductor lines increase with frequency, microwaves are transmitted through hollow metallic tube (wave guide) of uniform cross-section with minimum loss. A rectangular (cross-section) waveguide is most commonly used and is shown in Fig. 4.11.

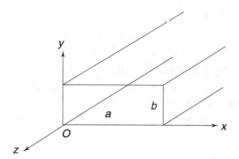

Fig. 4.11 *Rectangular waveguide geometry*

The fields inside the guide can be obtained from the solutions to the wave equation which satisfies the boundary conditions for a rectangular waveguide of infinite length. Two basic sets of solutions exist, each defining a set of modes. Thus there are two modes of propagation possible inside a hollow metal waveguide, *TE* waves in which the **E**-field is wholly transverse and *TM* waves in which **H**-field is wholly transverse. No TEM waves are possible in this case since it requires an axial conductor for axial current flow or an axial displacement current (axial **E**-field) to support a transverse magnetic field.

For an infinite guide and propagation along the z-direction ($e^{-jk_z z}$), the wave equations are

$$\Delta^2 \mathbf{E} + k^2\, \mathbf{E} = 0 \tag{4.38}$$

$$\Delta^2 \mathbf{H} + k^2 \mathbf{H} = 0 \tag{4.39}$$

which can be written as

$$\text{TM} : \Delta_t^2\, E_z + k_c^2\, E_z = 0 \tag{4.40}$$

$$\text{TE} : \Delta_t^2\, H_z + k_c^2 H_z = 0 \tag{4.41}$$

where $\quad \nabla_t = \dfrac{\partial^2}{\partial x^2} + \dfrac{\partial^2}{\partial y^2} \quad$ and $\quad k_c = \sqrt{k^2 - k_z^2} = \sqrt{k_x^2 + k_y^2} = \dfrac{2\pi}{\lambda_c} \quad$ (4.42)

and is called the cut-off wave number. The propagation constant in the guide is, therefore,

$$k_z = \pm j\, \sqrt{\omega^2\, \mu\,\varepsilon - k_c^2} \tag{4.43}$$

There are three special cases apparent

1. For no propagation (evanescent mode) to take place, $k_z = 0$ or $k_c = \omega_c \sqrt{\mu\varepsilon}$. Here w_c is the cut-off frequency in radian/sec.

2. For propagation of real power, $k_z > 0$, or $\omega \sqrt{\mu\varepsilon} > k_c$

3. If $\omega\sqrt{\mu\varepsilon} < k_c$, k_z is imaginary ($j\alpha$), so that the wave will be attenuated in accordance with $e^{-\alpha z}$.

4.4.1 TE Waves Solution

For TE waves propagating in +z direction, $E_z = 0$, and solution may be derived from the H_z component of Eq. 4.41

$$H_z = \cos\left(\frac{m\pi x}{a}\right)\cos\left(\frac{n\pi y}{b}\right) \tag{4.44}$$

satisfying the boundary conditions on the walls. Harmonic solution in the x and y directions show that the waves produce a standing wave between the longitudinal walls, giving the values $k_x = \dfrac{m\pi}{a}$ and $k_y = \dfrac{n\pi}{b}$. $e^{-jk_z z}$ factor is omitted but this is present in all field components. Here $m = 0, 1, 2,....$ denotes the number of half waves in the x direction and $n = 0, 1, 2, ...$ those in the y direction, when $m \neq n = 0$. To each set of integers m and n a solution exists, and these modes are designated as the TE_{mn} modes. The other field components are obtained from the Maxwell's equations as follows.

$$H_y = \frac{jk_z}{k_c^2}\frac{n\pi}{b}\cos\frac{m\pi x}{a}\sin\frac{n\pi y}{b} \tag{4.45}$$

$$H_x = \frac{jk_z}{k_c^2}\frac{m\pi}{a}\sin\frac{m\pi x}{a}\cos\frac{n\pi y}{b} \tag{4.46}$$

$$E_z = 0 \tag{4.47}$$

$$E_y = -\frac{\omega\mu}{k_z}H_x \tag{4.48}$$

$$E_x = \frac{\omega\mu}{k_z}H_y \tag{4.49}$$

The cut-off frequency and cut-off wavelength as defined in Eq. 4.42 for the TE_{mn} modes are,

$$f_c = \frac{1}{2\sqrt{\mu\varepsilon}}\sqrt{\left(\frac{m}{a}\right)^2 + \left(\frac{n}{b}\right)^2} \tag{4.50}$$

$$\lambda_c = \frac{2}{\sqrt{(m/a)^2 - (n/b)^2}} \tag{4.51}$$

The propagation constant is expressed by

$$k_z = \beta = \omega\sqrt{\mu\varepsilon}\sqrt{1 - (\omega_c/\omega)^2} \tag{4.52}$$

The phase velocity of the waves

$$u_p = \frac{\omega}{\beta} = \frac{u}{\sqrt{1 - (f_c/f)^2}} \tag{4.53}$$

where $u = \dfrac{1}{\sqrt{\mu\varepsilon}}$, the phase velocity in an unbounded dielectric (μ, ε).

The velocity of energy propagation, called the *group velocity* is given by

$$v_g = u^2/u_p \tag{4.54}$$

The propagating wavelength inside the guide is called the *guide wavelength* and is given by

$$\lambda_g = \frac{1}{\sqrt{1-(\lambda/\lambda_c)^2}} \tag{4.55}$$

where $\lambda = \mu/f$, the wavelength in an unbounded dielectric.
The characteristic wave impedance in the guide can be derived as, for TE_{mn} mode

$$Z_w = \frac{E_x}{H_y} = -\frac{E_y}{H_x} = \frac{\omega\mu}{\beta} = \frac{\eta k}{\beta} = \frac{\eta}{\sqrt{1-(f_c/f)^2}} = 377\frac{\lambda_g}{\lambda_0} \text{ ohm} \tag{4.56}$$

where $\eta = \sqrt{\mu/\varepsilon}$ is the intrinsic impedance in an unbounded dielectric.
The field configurations of different TE modes in a rectangular waveguide are shown in Fig. 4.12.

4.4.2 TM Waves Solution

The TM waves propagating in $+z$ direction are characterised by $H_z = 0$. Field solutions to Eq. 4.40 gives

$$E_z = \sin\left(\frac{m\pi x}{a}\right)\sin\left(\frac{n\pi y}{b}\right) \tag{4.57}$$

$$E_y = \frac{jn\pi k_z}{bk_c^2}\sin\left(\frac{m\pi x}{a}\right)\cos\left(\frac{n\pi y}{b}\right) \tag{4.58}$$

$$E_x = \frac{jm\pi k_z}{ak_c^2}\cos\left(\frac{m\pi x}{a}\right)\sin\left(\frac{n\pi y}{b}\right) \tag{4.59}$$

$$H_z = 0 \tag{4.60}$$

$$H_y = \frac{\omega\varepsilon}{k_z}E_x \tag{4.61}$$

$$H_x = \frac{-\omega\varepsilon}{k_z}E_y \tag{4.62}$$

where $m = 1, 2, 3, \ldots$ and $n = 1, 2, 3 \ldots$ Here m or $n \neq 0$, since either $m = 0$ or $n = 0$ leads to zero-field intensities. For TM_{mn} modes, the equations for the cut-off frequency f_c, propagation constant β, the phase and group velocities u_p and u_g, and the guide wavelength λ_g are the same as those for TE_{mn} modes, given in Eqs 4.50–4.55.

The characteristic wave impedance is given by

$$Z_w = \frac{E_x}{H_y} = \frac{-E_y}{H_x} = \frac{\beta}{\omega\varepsilon} = \frac{\eta\beta}{k} = \eta\sqrt{1-\left(\frac{f_c}{f}\right)^2} = 377\,\lambda_0/\lambda_g \text{ ohm} \tag{4.63}$$

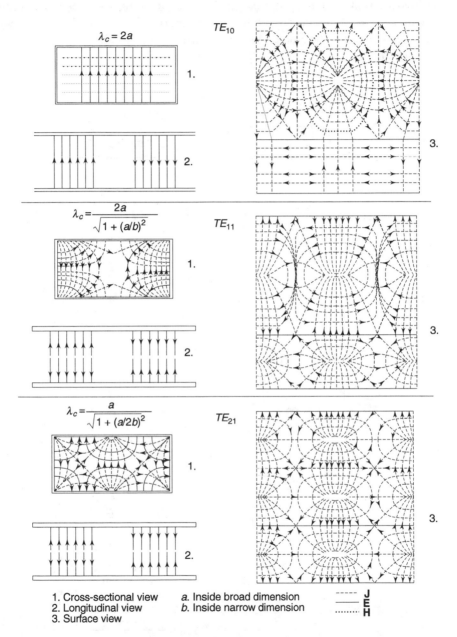

Fig. 4.12   *Field configuration of TE modes*

From Eqs 4.56 and 4.63 it is seen that the characteristic wave impedances of TE_{mn} and TM_{mn} modes are different and they exhibit a relation

$$Z_{w(TE)}. \; Z_{w(TM)}. = \eta^2 \qquad (4.64)$$

When the mode does not propagate, Z_w becomes imaginary and the mode is called *evanescent mode* characterised by no net energy flow through the guide–but energy is stored in the guide. The field configurations of different *TM* modes are shown in Fig. 4.13.

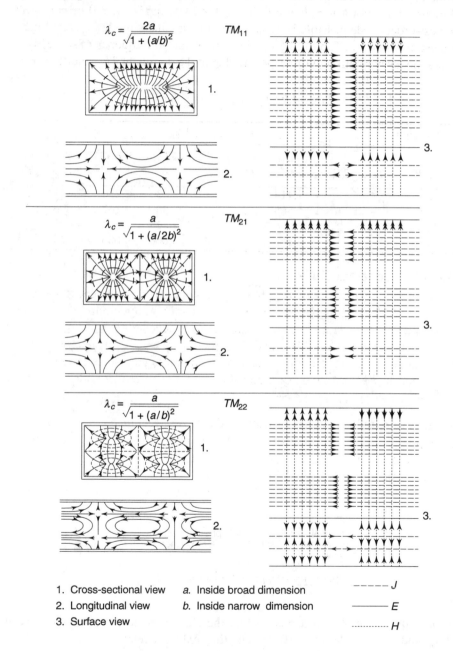

1. Cross-sectional view   a. Inside broad dimension $------ J$
2. Longitudinal view   b. Inside narrow dimension $—————E$
3. Surface view $\cdots\cdots\cdots H$

Fig. 4.13 *Field configuration of TM modes*

4.4.3 Dominant Mode

The cut-off frequency expressed by Eq. 4.50 shows that the physical size of the waveguide will determine the propagation of the modes of specific orders determined by m and n. The minimum cut-off frequency is obtained for a guide having dimension $a > b$, for $m = 1$, $n = 0$. Since for TM_{mn} modes, $n \neq 0$ or $m \neq 0$, the lowest-order-mode possible is TE_{10}, called the dominant mode in a rectangular waveguide for $a > b$. Table 4.2 gives the ratio of cut-off frequency of some higher order modes normalised with respect to that of the dominant TE_{10} mode.

Table 4.2 $(f_c)_{mn} / (f_c)_{TE_{10}}$ for $a > b$

Dimension											$(f_c)_{mn} / (f_c)_{TE_{10}}$
a/b	TE_{10}	TE_{01}	TE_{11} TM_{11}	TE_{20}	TE_{02}	TE_{21} TM_{21}	TE_{12} TM_{12}	TE_{22} TM_{22}	TE_{30}
1	1	1	1.414	2	2	2.236	2.236	2.828	3
1.5	1	1.5	1.803	2	3	2.500	3.162	3.606	3
2	1	2	2.236	2	4	2.828	4.123	4.472	3
3	1	3	3.162	2	6	3.606	6.083	6.325	3

4.4.4 Degenerate Modes

Some of the higher order modes, having the same cut-off frequency, are called *degenerate modes*. It is seen that in a rectangular waveguide possible TE_{mn} and TM_{mn} modes (both $m \neq 0$ and $n \neq 0$) are always degenerate. In a square guide $(a = b)$, all the TE_{pq}, TE_{qp}, TM_{pq}, and TM_{qp} modes are together degenerate. Waveguide dimensions are, therefore, selected such that higher-order-modes are not supported in the operating band and thus only desire mode/modes propagate through the guide.

Example 4.4 A rectangular waveguide of cross-section 5 cm × 2 cm is used to propagete TM_{11} mode at 10 GHz. Determine the cut-off wavelength and the characteristic impedance.

Solution For TM_{11} mode, $m = 1$, $n = 1$
Given $b = 2$ cm, $a = 5$ cm, $f = 10$ GHz, $\lambda = 30/10 = 3$ cm

$$\lambda_c = \frac{2}{\sqrt{[(m/a)^2 + (n/b)^2]}} = \frac{2ab}{\sqrt{a^2 + b^2}}$$

$$= \frac{2 \times 5 \times 2}{\sqrt{5^2 + 2^2}} = 3.714 \text{ cm}$$

$$Z_0 = 120\pi \sqrt{\left[1 - (\lambda / \lambda_c)^2\right]}$$

$$= 120\pi \sqrt{\left[1 - (3/3.714)^2\right]}$$

$$= 222.24 \text{ ohm}$$

4.4.5 Power Flow in Rectangular Waveguide

Assuming that the waveguide is infinitely long or it is terminated without any reflection from the receiving end, the power transmitted through a guide from the sending end is given by

$$P_{mn} = \frac{1}{2} R_e \int_0^a \int_0^b \mathbf{E} \times \mathbf{H}^* \cdot \mathbf{dS} \tag{4.65}$$

For a lossless dielectric, the time average power flow can be expressed by

$$P_{mn} = \frac{Z_w}{2} \int_0^a \int_0^b \left(|H_x|^2 + |H_y|^2 \right) \mathrm{d}x \, \mathrm{d}y \tag{4.66}$$

Appropriate expressions for field components and the characteristic wave impedance are substituted from Eqs 4.45, 4.46, 4.56, 4.61, 4.62 and 4.63 into Eq. 4.66 to evaluate the power flow.

Because of the orthogonality of the functions (eigen functions) that describe the transverse variation of the fields appear in the integrand of power expression, each possible mode of propagation in a lossless guide carries energy independently of all other modes that may be present. Even for small losses, power orthogonality applies with negligible error except for degenerate modes where coupling between the modes occur due to having the same propagation constant.

4.4.6 Attenuation in Rectangular Waveguide

Attenuation of signal transmission in metallic waveguide occurs due to following two types of power losses:

1. Ohmic losses in the highly conductive guide walls due to finite

   conductivity or non-zero surface resistance $R_s = \sqrt{\dfrac{\omega\mu}{2\sigma}} \neq 0$.

2. Losses in the dielectric of filled waveguide.

Ohmic losses: Since the field magnitudes decrease according to $-\alpha_c^Z$, the power decreases proportionately with $|\mathbf{E} \times \mathbf{H}|$ i.e. according to $P_0 e^{-2\alpha_c Z}$, where P_0 is the power at the sending end $z = 0$ and a_c is the attenuation contant due to conductor loss. The rate of decrease of power propagated for a given mode TE_{mn} and TM_{mn} is given by

$$-\frac{\mathrm{d}P_{mn}}{\mathrm{d}z} = 2\alpha_c P_{mn} = P_L \tag{4.67}$$

where P_L is the power loss per unit length of the guide and P_{mn} is the power transmitted through the guide. Therefore, the attenuation constant α_c is

$$\alpha_c = \frac{P_L}{2P_{mn}} \tag{4.68}$$

The power loss per unit length of the guide is

$$P_L = \frac{R_s}{2} \int_{\text{guide walls}} \mathbf{J}_s . \mathbf{J}_s^* \, dl \qquad (4.69)$$

Assuming that the currents on the lossy walls are same as the currents on the lossless walls, $\mathbf{J}_s = \hat{n} \times \mathbf{H}$, where $\hat{n}$ is the unit outward normal vector at the inner guide walls.

From Eqs 4.66, 4.68 and 4.69,

$$\alpha_c = \frac{R_s}{2Z_w} \frac{\oint_{\text{walls}} |\mathbf{J}_s|^2 \, dl}{\int_0^a \int_0^b \left(|H_x|^2 + |H_y|^2 \right) dx \, dy} \qquad (4.70)$$

Dieleetric loss: If the guide is completely filled with a low loss dielectric ($\sigma \ll \omega\varepsilon$), the attenuation constant in the guide due to dielectric loss becomes

$$\alpha_d = \frac{\alpha_0}{\sqrt{1 - (f_c/f)^2}} \quad \text{for } TE \text{ modes} \qquad (4.71)$$

$$= \alpha_0 \sqrt{1 - (f_c/f)^2} \quad \text{for } TM \text{ modes} \qquad (4.72)$$

When $f \gg f_c$, the dielectric attenuation approaches that α_0 for the unbounded dielectric given by Eq. 3.57.

For the dominant TE_{10} mode, the non-zero field components are

$$H_z = \cos\left(\frac{\pi x}{a}\right) \qquad (4.73)$$

$$H_x = \frac{jk_z}{k_c} \sin\left(\frac{\pi x}{a}\right) \qquad (4.74)$$

$$E_y = -\frac{\omega\mu}{k_z} H_x$$

$$= -j\frac{\omega\mu}{k_c} \sin\left(\frac{\pi x}{a}\right); \qquad (4.75)$$

where $k_z = \beta$ and $k_c = \dfrac{\pi}{a}$.

From Eq. 4.66, the rate of energy flow along the guide is

$$P_{10} = \frac{ab}{4} \left(\frac{k_z}{k_c}\right)^2 \cdot Z_w \qquad (4.76)$$

The surface current densities on the walls are

$$\mathbf{J}_s = \hat{x} \times \hat{z} H_z = -\hat{y} H_z \quad \text{at } x = 0 \text{ wall (} yz \text{ plane)} \qquad (4.77)$$

$$\mathbf{J}_s = \hat{x} \times \hat{z} H_z = +\hat{y} H_z \quad \text{at } x = a \text{ wall (} yz \text{ plane)} \qquad (4.78)$$

$$\mathbf{J}_s = \hat{y} \times \mathbf{H} = \hat{z}\, \frac{jk_z}{k_c^2}\, \frac{\pi}{a} \cdot \sin\left(\frac{\pi x}{a}\right) + \hat{x}\cos\left(\frac{\pi x}{a}\right) \quad \text{at } y = 0 \text{ wall } (xz \text{ plane})\quad (4.79)$$

$$\mathbf{J}_s = -\hat{y} \times \mathbf{H} = \hat{z}\, \frac{jk_z}{k_c^2}\, \frac{\pi}{a} \cdot \sin\left(\frac{\pi x}{a}\right) - \hat{x}\cos\left(\frac{\pi x}{a}\right) \quad \text{at } y = b \text{ wall } (xz \text{ plane})\quad (4.80)$$

The power loss in the walls per unit length of the guide is obtained from Eqs. 4.69, 4.73, 4.74, 4.77 – 4.80 as

$$P_L = R_s\left[b + \frac{a}{2}\left(\frac{\beta}{k_c}\right)^2 + \frac{a}{2}\right]\qquad (4.81)$$

The attenuation constant for the TE_{10} mode is, therefore, given by

$$\alpha_c = \frac{R_s}{ab\beta\omega\mu}\left(2bk_c^2 + a\omega^2\mu\varepsilon\right) \text{ neper/m}$$

$$= 8.686\left(\frac{R_s}{ab\beta k\mu}\right)\left(2bk_c^2 + ak^2\right) \text{ dB/m}$$

$$= 8.686\,\frac{R_s}{b\sqrt{\dfrac{\mu}{\varepsilon}\left[1 - \left(\dfrac{f_c}{f}\right)^2\right]}}\left[1 + \frac{2b}{a}\left(\frac{f_c}{f}\right)^2\right] \text{dB/m}\quad (4.82)$$

where $k = \omega\sqrt{\mu\varepsilon}$ and $\eta = \sqrt{\dfrac{\mu}{\varepsilon}}$. Figure. 4.14 shows that the attenuation for dominant mode due to finite conductivity of the waveguide walls is very high at $f = f_c$ and decreases with frequency to a lowest value and increases again with frequency beyond that.

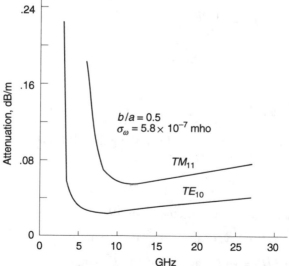

Fig. 4.14 *Attenuation of dominant mode in a rectangular waveguide*

Example 4.5 A rectangular air-filled copper waveguide with dimension 0.9 inch × 0.4 inch cross section and 12″ length is operated at 9.2 GHz with a dominant mode. Find (a) cut-off frequency, (b) guide wavelength, (c) phase velocity, (d) characteristic impedance and (e) the loss.
Solution

Given

a = 0.9 inch = 2.28 cm

b = 0.4 inch = 1.01 cm

f = 9.2 GHz

λ = 30/9.2 = 3.26 cm

l = 12 inch = 30.48 cm

Dominant mode is TE_{10}, $m = 1$, $n = 0$

(a) $f_c = c / \lambda_0 = c/2a = 30/(2 \times 2.28)$ GHz = 6.579 GHz

(b) $\lambda_g = \dfrac{\lambda}{\sqrt{\left[1-(f_c/f)^2\right]}}$

$= \dfrac{3.26}{\sqrt{\left[1-(6.579/9.2)^2\right]}}$

$= 4.664$ cm

(c) $u_p = \dfrac{1}{\sqrt{\mu_0 \varepsilon_0 \left[1-(f_c/f)^2\right]}}$

$= \dfrac{3 \times 10^8}{\sqrt{\left[1-(6.579/9.2)^2\right]}}$

$= 4.29 \times 10^8$ m/sec

(d) $Z_0 = \dfrac{377}{\sqrt{\left[1-(f_c/f)^2\right]}} = 539.3$ ohm

(e) $\alpha_c = \dfrac{R_s \left[1 + 2b/a \; (f_c/f)^2\right]}{b\sqrt{\mu_0/\varepsilon_0} \; \sqrt{1-(f_c/f)^2}}$ Np/m

where $R_s = \sqrt{\pi f \mu/\sigma}$

$= \sqrt{\dfrac{\left[3.14 \times 9.2 \times 10^9 \times 4\pi \times 10^{-7}\right]}{5.8 \times 10^7}}$

$= 2.5 \times 10^{-2}$ ohm

Therefore,

$$\alpha_c = \frac{2.5 \times 10^{-2}}{1.01 \times 377 \times \sqrt{\left[1 - (6.579/9.2)^2\right]}} \, [1 + 2 \times 1.01/2.286 \, (6.579/9.2)^2]$$

$$= 1.364 \times 10^{-4} \text{ Np/cm}$$

Therefore, total attenuation $= 1.364 \times 10^{-4} \times 30.48$

$$= 4.1566 \times 10^{-3} \text{ NP}$$

$$= 8.686 \times 4.1566 \times 10^{-3} \text{ dB}$$

$$= 0.036 \text{ dB}$$

4.4.7 Methods of Excitation of Modes

The waveguide modes are usually excited from a signal source through a coaxial cable. The outer conductor of the cable makes a 360° connection with the body of the waveguide and the centre conductor is projected inside the guide either as a small probe or forming a small loop as shown in Fig. 4.15. All TE_{mo} modes with m odd have a maximum field at the centre of the cross-section in transverse direction. These modes can be excited by a single probe at the centre as shown in Fig. 4.15 (a). All TE_{mo} modes with m even has a zero field at the centre and field polarities reverse alternately in the transverse direction as shown in Fig. 4.12. These modes can be excited by two probes as shown in Fig. 4.15 (d) having out of phase currents. One end of the guide is short circuited at a distance $\lambda/2$ from the probe so that the waves in the forward direction are enhanced by the waves which travel from the probe to the short and reflected towards the forward direction. Similarly, depending on the field configuration of the higher order modes, excitation methods change. Excitation of other higher order modes TM_{11} and TM_{21} are also shown in Figs 4.15(b) and (c).

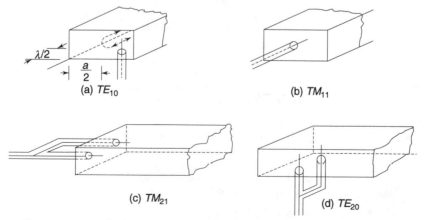

(a) TE_{10}

(b) TM_{11}

(c) TM_{21}

(d) TE_{20}

Fig. 4.15 *Methods of excitation of modes in a rectangular waveguide*

4.5 CIRCULAR WAVEGUIDES

A circular waveguide is a cylindrical hollow metallic pipe with uniform circular cross-section of a finite radius a as shown in Fig. 4.16. The general properties of the modes in the circular waveguides are similar to those for the rectangular guide. A unique property of the TM_{on} modes in circular waveguides is rapid decrease in attenuation with increasing frequency which makes their application in long low-loss communication links.

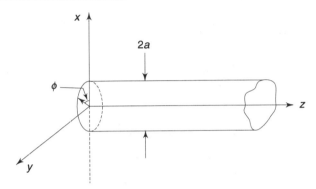

Fig. 4.16 *Uniform circular waveguide*

For an infinite length of a lossless uniform circular waveguide, the field solutions are obtained from the following homogeneous wave equation in the cylindrical coordinates since the geometry has a cylindrical symmetry.

$$\frac{\partial^2 \psi}{\partial \rho^2} + \frac{1}{\rho}\frac{\partial \psi}{\partial \rho} + \frac{1}{\rho^2}\frac{\partial^2 \psi}{\partial \varphi^2} + k_c^2 \psi = 0 \tag{4.83}$$

Here ψ stands for E_z for TM modes and H_z for TE modes. Using the method of separation of variables, the solution is obtained as

$$\psi = J_n (k_c \rho) \cos n\varphi e^{-jk_z z} \tag{4.84}$$

where $J_n (k_c \rho)$ is the nth order Bessel function of the first kind representing a standing wave in the radial direction for $\rho \le a$ with a finite field at $\rho = 0$, $\cos n\varphi$ represents φ symmetric harmonic solution.

The propagation constant in the guide is given by

$$k_z = \pm\sqrt{\omega^2 \mu\varepsilon - k_c^2} \tag{4.85}$$

4.5.1 TE Waves Solution

TE waves propagating in +z direction is characterised by $E_z = 0$. The field components of this mode are obtained from $\psi = H_z$ which is subjected to the boundary conditions $\mathbf{E} \times \hat{n} = 0$ and $\hat{n} \cdot \mathbf{H} = 0$. Thus using the Maxwell's equations, the field components can be written as

$$H_z = J_n(k_c \rho) \cos n\varphi \, e^{-jk_z z} \tag{4.86}$$

$$H_\rho = -j\frac{k_z}{k_c} J_n'(k_c \rho) \cos n\varphi \, e^{-jk_z z} \tag{4.87}$$

$$H_\varphi = +\frac{jnk_z}{\rho k_c^2} J_n(k_c \rho) \sin n\varphi \, e^{-jk_z z} \tag{4.88}$$

$$E_z = 0 \tag{4.89}$$

$$E_\rho = \frac{k}{k_z}\sqrt{\frac{\mu}{\varepsilon}} \, H\varphi \tag{4.90}$$

$$E_\varphi = -\frac{k}{k_z}\sqrt{\frac{\mu}{\varepsilon}} \, H_\rho \tag{4.91}$$

Here $k_c = \dfrac{x'_{np}}{a}$ represents the eigen values, where x'_{np} are the roots of $\dfrac{dJ_n(k_c \rho)}{d\rho} = 0$ at $\rho = a$ as given in Table 4.3.

Table 4.3 *Values of x'_{np} for TE modes*

n	x'_{n_1}	x'_{n_2}	x'_{n_3}
0	3.832	7.016	10.174
1	1.842	5.331	8.536
2	3.054	6.706	9.970

Subscripts $n = 0, 1, 2, 3, \ldots$ represents the number of full cycles of field variation in one angular revolution through $360°$ and $p = 1, 2, 3, 4 \ldots$ represents the number of zeroes of $J_n'(k_c \rho)$ along the radius excluding the zeros on the axis.

From Eq. 4.85, the cut-off frequency for *TE* modes in a circular waveguide is given by

$$f_c = \frac{x'_{np}}{2\pi a \sqrt{\mu\varepsilon}} \tag{4.92}$$

The phase velocity

$$v_p = \frac{1/\sqrt{\mu\varepsilon}}{\sqrt{1 - (f_c/f)^2}} \tag{4.93}$$

The guide wavelength

$$\lambda_g = \frac{\lambda}{\sqrt{1 - (f_c/f)^2}} \tag{4.94}$$

The wave impedance

$$Z_w = \frac{\omega\mu}{\beta} = \frac{\sqrt{\frac{\mu}{\varepsilon}}}{\sqrt{1-(f_c/f)^2}} \qquad (4.95)$$

where λ = wavelength in an unbounded dielectric.

4.5.2 TM Waves Solution

For a *TM* wave $H_z = 0$. The field solutions are obtained from *Helmholtz equation* with $\psi = E_z$ satisfying the boundary conditions. Thus

$$E_z = J_n (k_c \rho) \cos n \, \varphi \, e^{-jk_z z} \qquad (4.96)$$

$$E_\rho = \frac{-jk_z}{k_c} J^1_n (k_c \rho) \cos n \, \varphi \, e^{-jk_z z} \qquad (4.97)$$

$$E_\varphi = \frac{jnk_z}{\rho k_c^2} J_n (k_c \rho) \sin n\varphi \, e^{-jk_z z} \qquad (4.98)$$

$$H_z = 0$$

$$H_\rho = -\frac{k}{k_z} \sqrt{\frac{\varepsilon}{\mu}} E_\varphi \qquad (4.99)$$

$$H_\varphi = \frac{k}{k_z} \sqrt{\frac{\varepsilon}{\mu}} E_\rho \qquad (4.100)$$

Here $k_e = \dfrac{x_{np}}{a}$, where x_{np} are the roots of $J_n (k_c \rho) = 0$ at $\rho = a$ as given in Table 4.4, for $n = 0, 1, 2, 3 \dots$ and $p = 1, 2, 3, \dots$

Table 4.4 *Values of x_{np} for TM modes*

n	x_{n_1}	x_{n_2}	x_{n_3}
0	2.405	5.520	8.654
1	3.832	7.016	10.174
2	5.135	8.417	11.620

The cut-off frequency, phase velocity, guide wavelength and the wave impedance of *TM* modes are given by

$$f_c = \frac{x_{np}}{2\pi a\sqrt{\mu\varepsilon}} \qquad (4.101)$$

$$v_p = \frac{1/\sqrt{\mu\varepsilon}}{\sqrt{1-(f_c/f)^2}} \qquad (4.102)$$

$$\lambda_g = \frac{\lambda}{\sqrt{1-(f_c/f)^2}} \qquad (4.103)$$

$$Z_w = \frac{\beta}{\omega\varepsilon} = \eta\sqrt{1-(f_c/f)^2} \qquad (4.104)$$

Figure 4.17 shows the field configuration of different *TE* and *TM* modes.

4.5.3 Dominant Mode

From Tables 4.3 and 4.4 it is seen that the lowest order-cut off frequency is obtained from the root $x'_{11} = 1.841$ which corresponds to TE_{11} mode. Therefore, the dominant mode in a circular waveguide is TE_{11} mode.

4.5.4 Degenerate Modes

It is seen from Tables 4.3 and 4.4, $x'_{op} = x_{1p}$ and hence all the TE_{op} and TM_{1p} modes are degenerate in a uniform circular waveguide.

TM modes in circular waveguide

TM_{01}

1.

$\lambda_c = 2.613\ a$

2.

3.

l, s

TM_{11}

1.

l, s

$\lambda_c = 1.640\ a$

2.

3.

TM_{21}

l, s l, s 1.

$\lambda_c = 1.224\ a$

2.

3.

1. Cross-sectional view   a. inside radius   ----- J
2. Longitudinal view through plane l-l   ——— E
3. Surface view from s-s   ------- H

Fig. 4.17 *(Contd)*

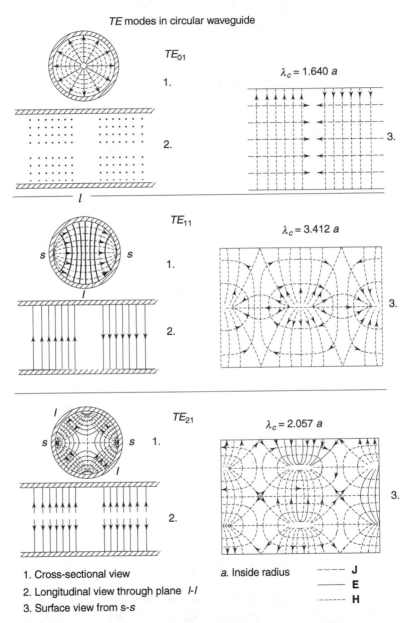

TE modes in circular waveguide

TE_{01}

1.

2.

l

$\lambda_c = 1.640\ a$

3.

TE_{11}

1.

2.

$\lambda_c = 3.412\ a$

3.

TE_{21}

1.

2.

$\lambda_c = 2.057\ a$

3.

1. Cross-sectional view
2. Longitudinal view through plane l-l
3. Surface view from s-s

a. Inside radius

----- J
——— E
------ H

Fig. 4.17 *Field configurations of different TE and TM modes*

4.5.5 Attenuation in a Circular Waveguide

The attenuation in a circular waveguide for *TE* and *TM* modes can be found in the same manner as in rectangular waveguides. For an air-filled guide the attenuation is due to finite conductivity of the waveguide walls and can be expressed by

$$\alpha = \frac{\text{power loss per unit length}}{2 \times \text{average power transmitted}}$$

Average power transmitted is given by

$$P_{np} = \frac{1}{2Z_w} \int_0^{2\pi} \int_0^a \left(|E_\varphi|^2 + |E_\rho|^2 \right) \rho d\rho \, d\varphi$$

$$= \frac{Z_w}{2} \int_0^{2\pi} \int_0^a \left(|H_\phi|^2 + |H_\rho|^2 \right) \rho d\rho \, d\varphi \qquad (4.105)$$

For TE_{np} modes

$$P_{np} = \frac{\sqrt{1 - (f_c/f)^2}}{2\eta} \int_0^{2\pi} \int_0^a \left(|E_\rho|^2 + |E_\varphi|^2 \right) \rho d\rho \, d\varphi \qquad (4.106)$$

For TM_{np} modes

$$P_{np} = \frac{1}{2\eta \sqrt{1 - (f_c/f)^2}} \int_0^{2\pi} \int_0^a \left(|E_\rho|^2 + |E_\varphi|^2 \right) \rho d\rho \, d\varphi \qquad (4.107)$$

Power loss P_L per unit length of the guide is given by

$$P_L = \frac{R_s}{2} \oint_{\text{guide walls}} \vec{J}_s \cdot \vec{J}_s^* \, dl \qquad (4.108)$$

The final expressions for the attenuation constant are obtained as

$$\alpha_{TE} = \frac{R_s}{a Z_0} \left(1 - f_c^2/f^2 \right)^{-1/2} \left(f_c^2 / f^2 + \frac{n^2}{x_{np}'^2 - n^2} \right) \qquad (4.109)$$

$$\alpha_{TM} = \frac{R_s}{a Z_0} \left(1 - f_c^2/f^2 \right)^{-1/2} \qquad (4.110)$$

For TE_{op} modes attenuation falls off as $f^{-3/2}$ according to

$$\alpha = \frac{R_s}{a Z_0} \frac{f_c^2}{f(f^2 - f_c^2)^{-1/2}} \qquad (4.111)$$

as shown in Fig. 4.18. Such rapid decrease in attenuation with increasing frequency makes it possible to use TE_{01} mode for very long low-loss waveguide communication links. But there are practical difficulties while operating the guide at a frequency well above the dominant mode TE_{11}. At these frequencies any small discontinuities in the guide converts the power in the TE_{01} mode to other modes with different propagation phase constants. At a sufficient distance, away from the discontinuities; these additonal modes again converted back into a TE_{01} mode leading to signal distortion.

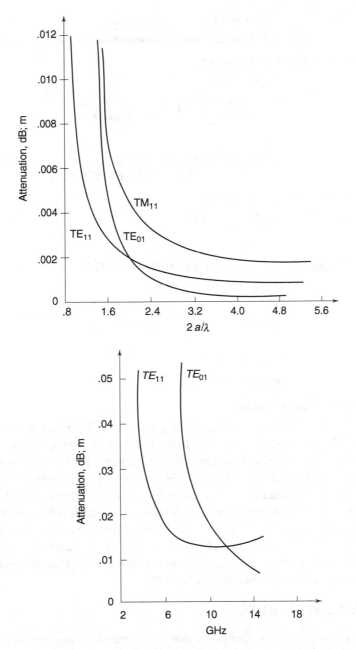

Fig. 4.18 *Attenuation characteristics in a circular waveguide*

Example 4.6 An air-filled circular waveguide having an inner radius of 1 cm is excited in dominant mode at 10 GHz. Find (a) the cut-off frequency of dominant mode, (b) guide wavelength, (c) wave impedance. Find the bandwidth for operation in dominant mode only.

Solution

Given $a = 1$ cm, $f = 10$ GHz. The dominant mode is TE_{11}, $m = 1$, $n = 1$. $\lambda = c/f$ $= 30/10 = 3$ cm.

(a) $$f_c = cx'_{11}/2\pi a$$

$$= \frac{3 \times 10^{10} \times 1.841}{2 \times 3.14 \times 1} = 8.795 \text{ GHz}$$

(b) $$\lambda_g = \frac{\lambda}{\sqrt{[1-(f_c/f)^2]}} = \frac{3}{\sqrt{[1-(8.795/10)^2]}} = 6.303 \text{ cm}$$

(c) $$Z_w = 377 \times \lambda_g/\lambda$$

$$= 377 \times 6.303 / 3$$

$$= 792 \text{ ohm}$$

$$\text{Bandwidth} = f_c \text{ of next higher order mode } TM_{01} - f_c \text{ of } TE_{11}$$

$$= c \times x_{01}/2\pi a - 8.795$$

$$= \frac{3 \times 10^{10} \times 2.405}{2 \times 3.14 \times 1} - (8.795 \times 10^9)$$

$$= 11.49 - 8.795$$

$$= 2.695 \text{ GHz}$$

4.5.6 Excitation of Modes

Since the electric field is generated with a polarisation parallel to the electric current and that magnetic field perpendicular to a loop of electric current, various *TE* and *TM* modes can be generated in a circular waveguide by coaxial line probes or loops as shown in Fig. 4.19. In Fig. 4.19(a) coaxial line probe excites the TE_{10} dominant mode in a rectangular waveguide which is converted to TE_{11} dominant mode in the circular waveguide through the transition length in between them. In Fig. 4.19(b) (a) longitudinal coaxial line probe directly excites the symmetric TM_{01} mode. TE_{01} mode is excited by means of two diametrically oppositely placed longitudinal narrow slots parallel to the broad wall of the connected rectangular waveguides as shown in Fig. 4.19(c).

4.6 RIDGE WAVEGUIDES

A ridge waveguide is formed with a rectangular ridge projecting inward from one or both of the wide walls in a rectangular waveguide as shown in Fig. 4.20.

The ridge has the effect of increasing the capacitance between the wide walls by reducing the dimension at the maximum E-field region parallel to E. This capacitance lowers the cut-off frequency of the dominant mode TE_{10} and increases the operating frequency range. Of course, ridge lowers the effective impedance of the guide. The ridge does not however, disturb the next higher order mode TE_{20}.

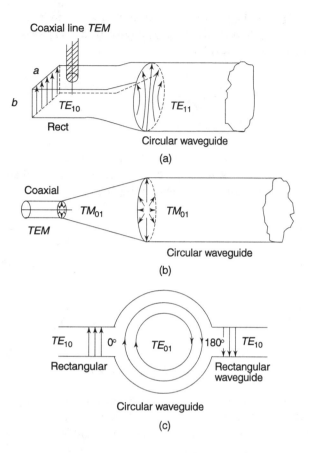

Fig. 4.19 *Methods of excitation of TE and TM modes in circular waveguides.*
(a) TE_{11} mode excitation (b) TM_{01} mode excitation
(c) TE_{01} mode excitation

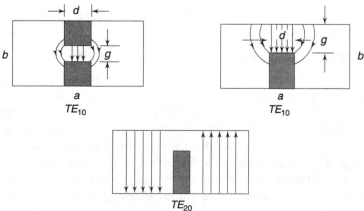

Fig. 4.20 *Ridge waveguides E-field*

The disadvantages of ridge guide over rectangular waveguides are (1) the reduction in the power it can handle, and (2) higher attenuation. An approximate expression of cut-off frequency for the ridge guide is

$$f_c = \frac{1}{2\pi} \sqrt{\frac{4g}{\mu \varepsilon bd(a-d)}} \tag{4.112}$$

4.7 SURFACE WAVEGUIDES

A surface waveguide is an open boundary structure, such as a dielectric rod, dielectric-coated conducting wire, dielectric sheet on a metal plane, a corrugated conducting plane or cylinder as shown in Fig. 4.21.

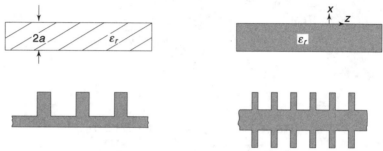

Fig. 4.21 *Surface waveguides*

These structures can guide an electromagnetic wave in longitudinal direction with a propagation function $e^{-j\beta z}$, but the field strength decay from the surface exponentially in transverse direction. Such waveguides are characterised by

1. Hybrid mode HE_{mn} of propagation ($TE_{mn} + TM_{mn}$), except axi-symmetric pure TM_{on} or TE_{on} modes.
2. Zero frequency cut-off for non axi-symmetric dominant hybrid mode HE_{11}.
3. A finite number of discrete modes at a given frequency, together with a continuous eigen value spectrum.

4.7.1 Dielectric Rod Waveguides

A uniform cylindrical low-loss high dielectric constant dielectric rod is commonly used for dielectric surface waveguide. Wave propagation takes place by total internal reflection of an obliquely incident wave at the interface between the dielectric and the air medium outside. The field intensity drops off exponentially with an increase in distance from the outer surface of the rod. Thus the propagation energy is confined inside and at the surface of the rod with decrease in radiation loss as the ratio diameter/wavelength and the dielectric constant increase. The axi-symmetric modes are pure TM or TE modes having a nonzero cut-off frequency whereas all modes with angular dependence are a combination of a TM and a TE mode, called the hybrid HE or EH modes, depending on whether the TE

or *TM* mode predominates, respectively. Hybrid modes consist of all components of **E** and **H** fields. HE_{11} mode is the dominant mode having zero cut-off frequency as shown in Fig. 4.22. The propagation constant $\beta > k_0$ inside the dielectric so that the phase velocity is less than that of light in free space.

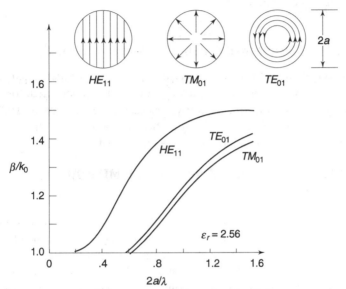

Fig. 4.22 *Propagation characteristics*

The cut-off wavelength for first higher order mode in a dielectric rod of radius *a* is given by

$$\lambda_c \cong 2.6\,a\sqrt{\varepsilon_r - 1} \qquad (4.113)$$

Thus for higher frequency (mm waves) dielectric waveguides must be very thin. At optical frequencies flexible dielectric waveguide or optical fibre is manufactured in the form of a thin filament of dielectric which is coated successively by a number of other dielectrics of lower dielectric constants to transmit optical signals (light).

4.7.1.1 *Excitation of a dielectric waveguide*

The easiest way to excite dominant mode in a dielectric rod is to use a circular waveguide with flared flanges as shown in Fig. 4.23. The rod is tapered inside the guide for impedance matching.

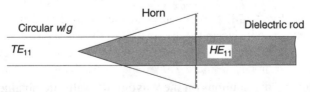

Fig. 4.23 *Excitation of hybrid modes in a dielectric guide*

For launching of waves from a circular waveguide to the dielectric rod, there is a discontinuity at the junction between the waveguide and the rod. This causes reflection of waves. A flared horn makes this launching smooth with gradual transition from the waveguide to the dielectric rod with minimum reflection. The launching efficiency is given by

$$\eta_L = \frac{\text{Surface Wave Power}}{\text{Input Power} - \text{Reflected Power}} = \frac{|S_{12}|^2}{1 - |S_{11}|^2} \qquad (4.114)$$

η_L can be determined by measuring S-parameters using *Deschamps method*. This utilises measured values of the complex input reflection coefficient for eight different lengths of the uniform dielectric rod changing in steps of $1/8^{th}$ guided wavelength with the free end of the rod short circuited by a shorting plate of dimensions much larger than the free space wavelength.

4.8 POWER HANDLING CAPABILITY OF MICROWAVE TRANSMISSION LINES

The maximum power that can be transmitted through a microwave transmission line is limited by the rf electric break down in the region where the electric field intensity is at maximum E_{bd} in the guide or line. At atmospheric pressures, electric breakdown occurs in the form of a spark discharge with a loud sound. At low pressures in the upper atmospheres, a glow discharge occurs. Large reflections of the propagating signal occur at the discharge location. The reflected wave produces instability of the signal source at the sending end and can even damage source. At the breakdown point the conductor of the line or guide becomes oxidized or even burnt out. The level of breakdown voltage or field depends on the nature of the medium (gas), gas pressure, frequency and degree of initial gas ionization.

4.8.1 Rectangular Waveguide

By normalising the dominant field components of equation (4.73)-(4.75), the electric and magnetic fields of dominant TE_{10} mode in a rectangular waveguide section are given by the following components.

$$E_y = E_0 \sin (px/a) \qquad (4.115)$$

$$Hx = -\frac{E_0}{Z_w} \sin \left(\frac{px}{a} \right) \qquad (4.116)$$

$$H_z = \frac{jk_c}{W\mu} \cos \left(\frac{px}{a} \right) \qquad (4.117)$$

$$E_x = H_y = E_z = 0 \qquad (4.118)$$

Wall currents

The electric current distributions on the waveguide walls are obtained from the relation $\mathbf{J} = \mathbf{n} \times \mathbf{H}$ and are shown in Fig. 4.24 for the dominant TE_{10} mode.

Electric breakdown

The time average power flow through the rectangular waveguide for dominant TE_{10} mode is given by

$$P_{10} = 1/2 \, \text{Re} \int \mathbf{E} \times \mathbf{H}^* . \, d\mathbf{S}$$

$$= \frac{E_0^2 \, ab}{4Z_w} \left(\frac{\beta}{k_c} \right)^2 = \frac{E_0^2 \, ab}{4} \frac{v \left[1 - (f_c/f)^2 \right]}{\eta} \tag{4.119}$$

where $\eta = v \, (\mu_o/e_o)$, the intrinsic impedance of free space and Z_w is the characteristic wave impedance for dominant mode. It is known experimentally that in dry air at normal atmospheric pressure the microwave electric breakdown takes place when the electric field level at the middle of the broad wall is of the order of 30 kV/cm. Substituting this value of field in equation 4.119, the corresponding breakdown power is obtained for the dominant TE_{10} mode as

$$(P_{bd})_{10} \approx 597 \, ab[1 - (\lambda_0/2a)^2]^{1/2} \text{ kW} \tag{4.120}$$

where a, b and λ_0 are in cm.

For a rectangular air-filled waveguide (2.3 cm $\times$ 1.0 cm) the breakdown power in dominant mode at 9.375 GHz is approximately 987 kW.

4.8.2. Circular Waveguide

For the dominant TE_{11} mode and the minimum attenuation TE_{01} mode in circular air-filled waveguide, the breakdown powers for a maximum field of 30 kV/cm, are respectively given by

$$(P_{bd})_{11} \approx 1790 \, a^2 \, v[1 - (f_{c11}/f)^2]^{1/2} \text{ kW} \tag{4.121}$$

$$(P_{bd})_{01} \approx 1805 \, a^2 \, v[1 - (f_{c01}/f)^2]^{1/2} \text{ kW} \tag{4.122}$$

where a is the radius of the circular waveguide in cm, and f_{c11} and f_{c01} are the cut-off frequencies of TE_{11} and TE_{01} modes, respectively.

4.8.3 Coaxial Line

In coaxial lines the maximum field strength (breakdown) is determined by the cable dielectric break down,

$$E_{bd} = \frac{V_{peak}}{2a} / \ln \frac{b}{a} \tag{4.123}$$

Where V_{peak} is the peak voltage at breakdown. For harmonic time variation in a matched line, the breakdown power is given by

$$P_{bd} = \left[\frac{V_{peak}}{\sqrt{2}} \right]^2 / z_0$$

$$= 4a^2 E_{bd}^2 \ln (b/a) \tag{4.124}$$

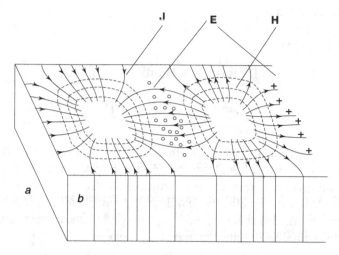

Fig. 4.24 *Rectangular waveguide wall currents*

For $E_{bd} = 30$ kV/cm, the breakdown power for air-filled coaxial line for TEM mode is given by

$$P_{bd} = 3600 \, a^2 \ln (b/a); \text{ kW} \qquad (4.125)$$

where all dimensions are expressed in cms.

Since the cross sectional dimensions of coaxial lines are restricted to satisfy the conditions $(b + a) < \lambda/\pi$ to exclude higher order modes, and for 50 ohm line, $b/a \approx 2.3$, for the single dominant mode propagation $a = 0.3\lambda/\pi$, $b = 0.7\lambda/\pi$ and the breakdown power becomes

$$(P_{bd})_{TEM} = 398 \text{ kW} < (P_{bd})TE_{10} \text{ at } 9.375 \text{ GHz.} \qquad (4.126)$$

Therefore, the breakdown power of coaxial lines is always lower than that of dominant waveguides operating at same frequency due to reduced separation distance between the conductors compared to that between waveguide walls. Due to similar reason microstrip lines have further lower breakdown voltage. The smaller the dimensions of a transmission line, the lower the breakdown power.

The breakdown is reduced at points of discontinuities in the line where intensities of the electric field is higher. In many applications transmission line ends are fitted with connector of various dimensions. Due to inherent constructional discontinuities in the connector, the breakdown power in line is limited not by the guide or line but by the connectors.

The breakdown field is highly dependent on the gas pressure inside the guide as shown schematically in Fig. 4.25 as a function of frequency and pressure. When the high power microwave equipment is designed to operate at high altitude (low pressure) the waveguide is usually pressurised to increase the breakdown voltage.

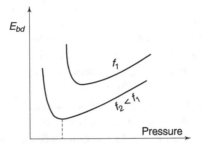

Fig. 4.25 E_{bd} *vs gas pressure*

Example 4.7 A rectangular waveguide of dimensions 1 cm × 2.3 cm is excited in dominant mode at 9.375 GHz. Calculate the breakdown power.

Solution Dominant mode is TE_{10}

$$a = 2.3 \text{ cm}$$
$$b = 1.0 \text{ cm}$$
$$f = 9.375 \text{ GHz}$$
$$\lambda_0 = 3.2 \text{ cm}$$

Assuming breakdown field = 30 KV/cm.

$$P_{bd} = 597 \, ab \sqrt{1 - (\lambda/2a)^2} \text{ kW}$$

$$= 597 \times 2.3 \times 1.0 \times \sqrt{1 - \left(\frac{3.2}{2 \times 2.3}\right)^2} = 986.4 \text{ kW}$$

Example 4.8 An air-filled coaxial line is operating at $\lambda = 3.2$ cm in *TEM* mode. Assuming the ratio of outer and inner radii of the conductor is $b/a = 3$ and $a = \dfrac{\lambda}{4\pi}$. Calculate the breakdown power.

Solution $\quad P_{bd} = 3600 \, a^2 \ln \dfrac{b}{a} \text{ kW} = 256.5 \text{ kW}$

EXERCISES

4.1 An air-filled rectangular copper waveguide of cross section 0.9 inch × 0.4 inch is operated at 10 GHz. Calculate the attenuation constant in dB/m. Find the frequency and attenuation at which the attenuation is minimum for TE_{10} mode.

4.2 A circular waveguide is filled with a lossless dielectric of $\varepsilon_r = 9$. If the cut-off frequency is 6 GHz, calculate the diameter and find the upper frequency limit over which only the dominant mode will propagate.

4.3 An air-filled X-band waveguide (0.9 inch × 0.4 inch) is fed from a signal source at 6 GHz. Calculate the wave impedance. Explain about the characteristic of single propagation through the guide.

4.4 Design a rectangular waveguide with filling by a dielectric of $\varepsilon_r = 4$, so that the cut-off frequency for the dominant mode is 14 GHz and the cut-off frequency for the TM_{11} mode is 30 GHz.

4.5 A rectangular waveguide (0.9 inch × 0.4 inch) made of brass ($\sigma = 1.1 \times 10^7$ mho/m) is excited at 9 GHz in dominant mode. If the guide is filled with Teflon ($\varepsilon_r = 2.2$, tan $\delta = 0.0005$), calculate α_c and α_d.

4.6 Calculate the cut-off frequencies for TE_{11}, TM_{01} and TE_{01} modes in a circular waveguide filled with a dielectric of $\varepsilon_r = 2.5$.

4.7 A microstrip line with $W = 2$ mm, $h = 0.4$ mm has a quartz substrate of $\varepsilon_r = 3.8$ and tan $\delta = 0.0001$. Find the (a) effective permittivity, (b) Z_0, (c) α_c and α_d at 9 GHz.

4.8 RG-141/U coaxial cable has attenuation 8.8 dB/100 ft at 1 GHz and power rating 200 watts. Calculate the attenuation coefficient, the dielectric constant and the power loss for a length of 5 m.

4.9 Design a 50 ohm strip line with PTFE substrate material of $\varepsilon_r = 9.6$ and $h = 1/8"$. Calculate the cut-off wavelengths for higher order TM_{11} and TE_{10} modes.

REFERENCES

4.1 Collin, RE, *Foundation for Microwave Engineering*, McGraw-Hill, Inc., International Editions, 1992.

4.2 Collin, RE, *Field Theory of Guided Waves*, McGraw-Hill Book Co., 1960.

4.3 Lebedev, I, *Microwave Engineering*, MIR Publishers, Moscow, 1973.

4.4 Harrington, PF, *Time-harmonic Electromagnetic Field*, McGraw-Hill Book Co., 1961.

4.5 Edwards, T, *Foundations for Microstrip Circuit Design*, Sec. Edition, John Wiley and Sons, 1992.

4.6 Ramo, JR, Whinnery, and TV Duzer, *Firleds and Waves in Communication Electronics*, Sec. Edition, John Wiley and Sons, 1984.

4.7 Liboff, RL and GC Dalman, *Transmission lines, Waveguides, and Smith Charts*, MacMillan Publishing Co., New York, 1985.

4.8 Bharathi Bhat and SK Koul, *Stripline-like Transmission Lines for Microwave Integrated Circuits*, Wiley Eastern Ltd., 1989.

4.9 Marcuvitz, N, *Waveguide Handbook,* Peter Peregrinus Ltd. (IEE), 1986.

4.10 Gupta, KC, G Ramesh and IJ Bahl, *Microstrip lines and slotlines*, Artech House, Inc., Dedham, Massachusetts, 1979.

4.11 Sander, KF, and GA Reed, *Transmission and Propagation of Electromagnetic Waves*, Cambridge University Press, Cambridge, 1978.

Chapter 5

IMPEDANCE TRANSFORMATIONS FOR MATCHING

5.1 INTRODUCTION

When a transmission line is not terminated by its characteristic impedance, maximum power is not transferred to the load and power will be wasted due to reflection from the load. This reduces the efficiency of transmission. A microwave circuit involves many different types of transmission lines and devices, each characterised by its own impedance. Impedance transformation between two reference planes along a microwave transmission line is made for the matching of an arbitrary load impedance at one plane to the line or the matching of two lines having different characteristic impedances for the purpose of maximum power transfer. There are numerous methods for matching a mismatched load or line.

5.2 GENERAL CONDITION FOR IMPEDANCE MATCHING

Consider a lossless line with real characteristic impedance Z_0 as shown in Fig. 2.2 in Chapter 2, where sending end and load end current and voltage parameters are given by

$$V(0) = V_s, V(l) = V_r \tag{5.1}$$

$$I(0) = I_s, I(l) = I_r \tag{5.2}$$

For any terminating impedance, $Z_L = R_L + jX_L$

$$V_s = V_r \cos \beta l + jI_r Z_0 \sin \beta l \tag{5.3}$$

$$I_s = I_r \cos \beta l + jV_r / Z_0 \sin \beta l \tag{5.4}$$

or,

$$|V_s| = |V_r| \left[\left(\cos \beta l + \frac{Z_0 X_L}{Z_L^2} \sin \beta l \right)^2 + \left(\frac{Z_0 R_L}{Z_L^2} \sin \beta l \right)^2 \right]^{1/2} \tag{5.5}$$

$$|I_s| = |I_r| \left[\left(\cos \beta l + \frac{X_L}{Z_0} \sin \beta l \right)^2 + \left(\frac{R_L}{Z_0} \sin \beta l \right)^2 \right]^{1/2} \qquad (5.6)$$

From the above equations it is apparent that the line is matched to produce uniform magnitudes of voltage and current throughout the line when

$$X_L = 0,$$
$$Z_L = R_L = Z_0 \qquad (5.7)$$

Therefore, the reactive component of the mismatched impedance should be cancelled by a suitable element and the real part of the mismatched impedance should be transformed to Z_0.

There are basically two types of matching to be considered: one is narrowband matching and other one is broadband matching.

5.3 NARROWBAND MATCHING

Impedance matching over a narrow band of frequencies can be accomplished by inserting reactive elements between the two mismatched impedances. The method for narrowband impedance matching uses shunt reactive elements, such as tuning screw, reactive irises and series and shunt stub matching.

5.3.1 Waveguide Windows/Irises/Diaphragms

When a flat thin metal obstacle partially fills the waveguide transversely as shown in Fig. 5.1, it offers inductive or capacitive reactance across the rectangu-

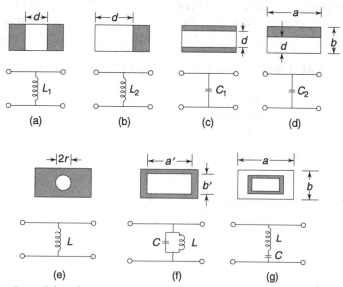

Fig. 5.1 *Irises in rectangular waveguide (a) Symmetrical inductive (b) Asymmetrical inductive (c) Symmetrical capacitive (d) Asymmetrical capacitive (e) Inductive (f) Parallel resonant iris (g) Series resonant iris*

lar guide , depending on the extension of the obstacle across the narrow or the broad dimensions, respectively, when excited in dominant mode. Such obstacles are called *irises* or *windows* or *diaphragms*.

When a TE_{10} dominant wave is incident on such a junction, higher order evanescent TE and TM waves are excited near the junction to satisfy the boundary condition of zero total tangential electric field on the irises. These higher order modes are of non-propagating or evanescent type, i.e. they die down within a distance of quarter wavelength from the junction and store reactive energy at the junction. The irises of Figs 5.1(a) and (b) affect mainly the magnetic field of the dominant TE_{10} mode and store predominantly magnetic energy resulting in inductive characteristics. The irises of Figs. 5.1 (c) and (d) mainly affect the E-field of TE_{10} mode, concentrating it in the gap and storing predominantly electric energy from the evanescent modes. Thus these irises attain capacitive characteristics. The normalised susceptances of commonly used thin irises are given below:

Inductive

Symmetrical

$$B/Y_0 = - (\lambda_g/a) \cot^2 (\pi d/2a)/[1 + 1/6 (\pi d/\lambda_g)^2]; d/a \ll 1 \qquad (5.8)$$

$$= - (\lambda_g/a) \cot^2 2\pi(a - d)/a]/[1 + 8/3 \pi^2 (a - d)^2/ \lambda_g^2];$$

$$(a - d)/a \ll 1 \qquad (5.9)$$

Asymmetrical

$$B/Y_0 = - [\lambda_g\pi^2 (a - d)^2/2a^3] [1 + \pi^2/\lambda_g^2 (a - d)^2 \ln(\pi/2) (a - d)/a];$$

$$(a - d) /a \ll 1 \qquad (5.10)$$

Capacitive

Symmetrical

$$B/Y_0 = (4b/\lambda_g) [\ln 2b/\pi d + b^2/2 \lambda_g^2]; d/b \ll 1 \qquad (5.11)$$

$$= \pi^2 (b - d)^2/ 2b \lambda_g ; (b - d)/b \ll 1 \qquad (5.12)$$

Asymmetrical

$$B/Y_0 = (8b/\lambda_g) [\ln 2b/\pi d + 2b^2/\lambda_g^2]; d/b \ll 1 \qquad (5.13)$$

$$= \pi^2 (b - d)^2/b \lambda_g ; (b - d)/b \ll 1 \qquad (5.14)$$

Figure 5.2 shows the variations of normalised susceptance with iris dimension. Irises are employed as impedance matching elements where the reactive impedance is used to cancel the opposite reactance in the mismatched line.

Inductive irises are also made with a small centred circular hole of radius r_0 as shown in Fig. 5.1(e). The normalised inductive susceptance of a small circular hole is

$$\frac{B}{Y_0} = \frac{3ab \lambda_g}{16 \pi r_0^3}; \quad r_0 \ll b \qquad (5.15)$$

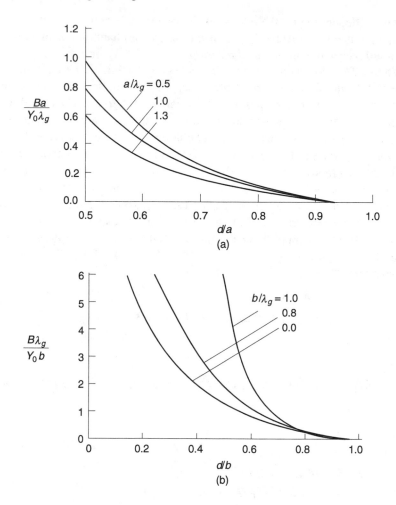

Fig. 5.2 *Normalised susceptances of irises*
(a) Symmetrical inductive (b) Capacitive

When the iris is extended both across the narrow and broad walls, it behaves as a parallel L-C circuit in shunt across the guide as shown in Fig. 5.1 (f). By choosing the dimensions of a' and b', resonant condition (zero susceptance) can be obtained at selective frequencies in the waveguide band. Quality factor of resonator decreases with increase in aperture size and Q of the order of 10 can be achieved for a small aperture size. The series resonant iris is a metal ring supported by a lossless dielectric as shown in Fig. 5.1 (g).

5.3.2 Tuning Screw/Sliding Screw Tuner

A tuning screw is a metal rod or probe inserted into a rectangular waveguide through the broad wall. This probe provides a reactance across the guide which can be varied from capacitive to inductive by varying the depth of penetration as shown in Fig. 5.3. The theoretical and experimental analysis shows that a thin screw of diameter $\ll \lambda_g/4$ possesses susceptance of following nature: (1) Capacitive when $h < \lambda_g/4$ (2) Infinite when $h = \lambda_g/4$ (3) Inductive when $h > \lambda_g/4$

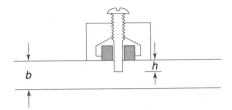

Fig. 5.3 *Tuning screw*

Tuning screw is used as a tuning device for impedance matching on account of this reactive nature of the screw.To avoid power leakage through the screw gap, half wavelength choke is used at the screw insertion junction, as shown in Fig. 5.3.

The above tuning screw can be slided along the axis of the waveguide through a narrow longitudinal slot, centred in the broad wall. This helps varying both the penetration and the position of the tuning screw along a longitudinal distance of half guide wavelength for better matching with ease. Such devices are called *slide screw tuners*.

A greater range of impedance matching can be achieved with three fixed screws 3 $\lambda_g/8$ apart (Fig. 5.4).

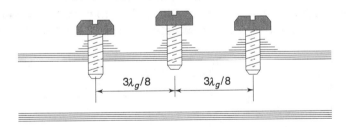

Fig. 5.4 *Three screws tuner*

5.3.3 Inductive and Capacitive Posts

A thin cylindrical post extending completely across the narrow width of the waveguide at the centre of the broad wall, provides an inductive susceptance across the guide operated in the dominant TE_{10} mode. When the post is placed across the waveguide at right angles to **E** of the TE_{10} mode, it offers capacitive susceptance across the guide. The value of the susceptance increases with the diameter of the post. Thus these posts can be used as matching elements. Figure. 5.5 shows the post and its susceptance characteristics.

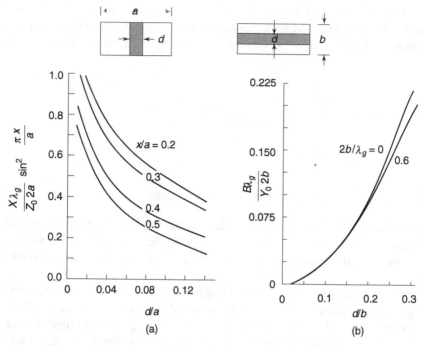

Fig. 5.5 *Post and its susceptance characteristics*

5.3.4 Stub Matching in Lossless Lines

The impedance of an open circuited or a short circuited lossless line of length l is a function of $\cot \beta l$ or $\tan \beta l$ respectively, and the line can have a wide range of inductive and capacitive reactances, depending on the length l, as shown in Fig. 5.6.

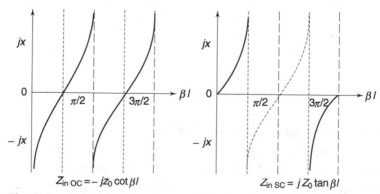

Fig. 5.6 *Variation of reactance of short and open circuited line with line length*

Thus any value of reactance ranging from $-\infty$ to ∞ can be obtained by proper choice of the length of O.C. or S.C. lines. Short circuited sections of such lines are called *stubs*, and are preferable to open circuited sections, because a good short is easily achieved in all kinds of transmission lines.

5.3.4.1 Single stub matching in a lossless line shunt stub

When a complex load $Z_L = R_L \pm jX_L$ is required to be matched using a single stub, where the load impedance and the stub impedance appear in parallel, as shown in Fig. 5.7, the matching concepts are better explained in terms of admittances $y_L = g_L \pm jb_L$ normalised by Y_0. The generator end is assumed to be matched to the characteristic impedance. The following section describes the design procedure for single stub matching.

If we move away from the complex load towards the generator along the line, there are voltage minima at which the reflection coefficient is a negative real quantity and the input admittance is pure conductance of value $Y_{in} = (1 - \Gamma) / (1 + \Gamma) = g = S$. Similarly, there are voltage maxima at which the admittance is pure conductance of value $Y_{in} = (1 + \Gamma) / (1 - \Gamma) = 1/S$. Therefore, in between these two points, there must be a point at a certain distance from the first voltage minimum where the real part of the normalised admittance is unity, so that $y_{in} = 1 \mp jb$. The reactive component of this admittance can be cancelled by an equal and opposite susceptance $\mp jb$ of a parallel single stub of length l, located at that point, and having characteristic impedance equal to that of the line i.e. Z_0.

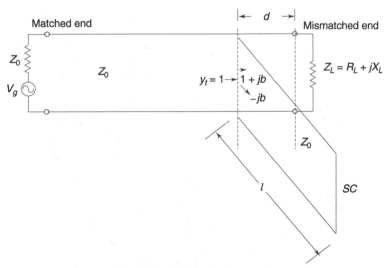

Fig. 5.7   *Single stub matching*

The total normalised admittance of the line and the stub looking towards the load side is then

$$y_t = 1 \pm jb \pm j \cot \beta l = 1; b = a + \beta l \tag{5.16}$$

for matching. Here

$$d = (\lambda/4\pi) \cos^{-1}[(S - 1)/(S + 1)];$$

$$l = (\lambda/2\pi) \tan^{-1}[\sqrt{S}/(S-1)] \qquad (5.17)$$

The following steps are involved in the design of single shunt stub using the Smith chart of Fig. 5.8.

1. The load admittance point is located at a point A corresponding to VSWR S. Join the centre O with A.
2. Location of the stub is found by rotating OA towards generator with O as centre, to intersect the unit resistance circle at B, where the real part of the normalised admittance is 1.0. The distance d/λ corresponding to the arc AB gives the stub position.
3. The normalised susceptance Y_t at B is read and an equal and opposite susceptance of the stub is located at C on the outer circle ($g = 0$).
4. The stub length l/λ is found by rotating in a clockwise direction around the unit circle to C, starting from the right-hand intersection between the unit circle and the real axis which corresponds to the short circuit position on the stub.

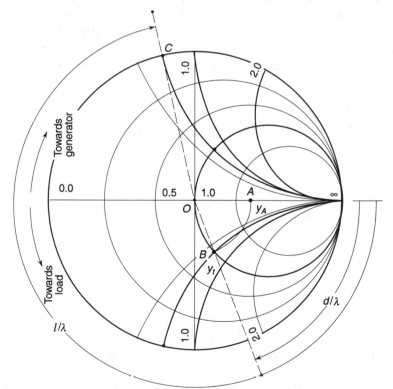

Fig. 5.8 *Single stub design using the Smith chart*

Series stub

At voltage minimum, input impedance of the line is $Z_{in} = 1/S$, and at the maximum, $Z_{in} = S$. Therefore, in between these two points, at a distance d from the load towards the generator, $z_{in} = 1 \pm jx$. A short circuited series stub of length l can cancel the reactive part to provide matching condition, $z_{in} = 1$. Here

$$d = (\lambda/4\pi) \cos^{-1} [(1 - S)/(1+S)];$$

$$l = (\lambda/2\pi) \tan^{-1} [(1 - S)/\sqrt{S}]$$

$$\pm jx = \mp j \tan \beta l \qquad (5.18)$$

The series stub is convenient for waveguide and strip line circuits. Shunt stub is easy to construct for coaxial lines (also for two wire lines).

5.3.4.2 *Double stub matching*

It is stated in the previous section that a single stub must be located at a fixed point near the load, where real part of the line admittance is Y_0. In microwave transmission lines, such as coaxial lines, or waveguides, it is not practical to find such an exact position to place the stub. Moreover, this position changes for every load. Under such circumstances, matching in coaxial cables can be better achieved by the use of double stubs, placed at fixed positions and having variable lengths as shown in Fig. 5.9. The two stubs are spaced at a distance of $3 \lambda/8$ apart where the first stub is arbitrarily placed at a distance of $d \geq 0$ from the load end. Double stub matching is conveniently performed by use of the Smith chart from the knowledge of the load Z_L and its VSWR.

The illustration of double stub matching is shown in Fig. 5.10 where it is desired that the input admittance at C should be Y_0 corresponding to a point C on the unit circle ($g = 1$) of the Smith chart having its centre at O'. It is necessary to be on the unit conductance circle at the second stub position C before the addition of the second stub. A movement toward the load from C by $3\lambda/8$ makes all points of the unit conductance circle to rotate by the same amount, so that in effect, the entire unit conductance circle rotates. Stub 1 adds susceptance to the load such that the resulting admittance lies on the rotated circle and consequently, a $3 \lambda/8$ movement towards the generator makes the admittance at C to lie on $g = 1$ circle. The second stub then cancels the susceptance and provides matching. The following steps are involved.

1. The load admittance y_L is located at A diametrically opposite to load impedance Z_L/Z_0
2. From the load admittance point A, a rotation towards generator on the load VSWR circle by an arbitrary amount, say, $d = 0.067 \lambda$ gives the position B of the first stub where the normalised line admittance is found to be

$$y_B = g_B + jb_B \qquad (5.19)$$

3. The unit conductance circle ($g = 1$) is rotated towards the load by $3 \lambda/8$, where the new centre of the circle becomes O'' on the line passing $+ j'$ and $-j$.
4. The point of intersection B' between g_B circle and the rotated unit conductance circle with centre O'' gives the total normalised admittance y_{BS} of the combination of the line at the right of point B and the short circuited first stub of length l_1:

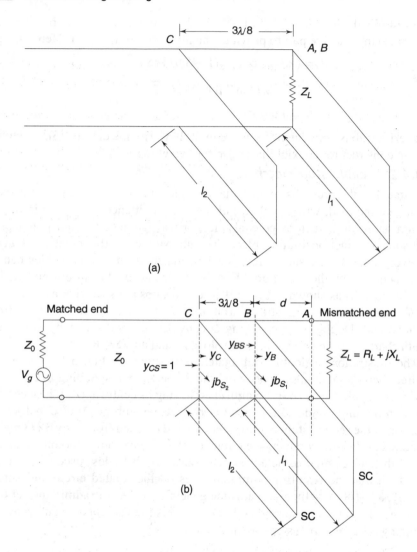

Fig. 5.9 *Double stub matching (a) First stub at load end (b) First stub at a distance from load*

$$y_{BS} = y_B + jb_{S1} = g_B + jb_B + jb_{S1} \tag{5.20}$$

Here b_{S1} is the susceptance of the first stub which can be determined by knowing y_{BS} and y_B from the Smith chart.

5. The point B' is then rotated towards generator by 3 $\lambda/8$ to the point C on the $g = 1$ circle with centre at O'. The total normalised admittance at C for the original line and first stub is

$$y_C = 1 + j\,b_C \tag{5.21}$$

6. For matching, the total admittance at C is

$$y_{CS} = 1.0 \qquad (5.22)$$

This is achieved by adjusting the length l_2 of the second stub to provide an equal and opposite susceptance of $b_{S2} = -b_C$ corresponding to C' on the Smith chart.

7. The stub lengths are found by rotating towards the generator around the unit reflection coefficient circle, starting from the right hand intersection of the unit circle and the real axis, which corresponds to the origin of infinite susceptance of the short termination of short-circuited stubs, up to the point of corresponding stub susceptance.

Limitation

The double stub system cannot match all impedances. If the normalised load conductance exceeds $1/\sin^2 \beta s$, double stub can not be used, since the matching

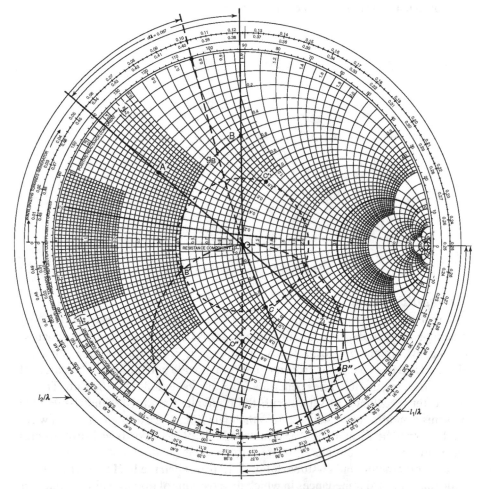

Fig. 5.10 *Double stub design using the Smith chart*

condition will not be satisfied for these values of g_I, where s is the separation distance between the stubs. Also alternatively, if $g_L > 2$ there will be no point of intersection between the rotated circle and $g = 1$ circle, and no solution will be possible.

5.3.4.3 Realisation of stubs in microwave circuit

Coaxial line

In two conductor coaxial lines, the stub is realised by a tee junction at the stub position as shown in Fig. 5.11, where a moveable short is used to adjust the length of the stub. The stub position on the line can be adjusted by a sliding arrangement, where a longitudinal slot on the outer conductor of the line allows movement of the centre conductor of the stub. However, this arrangement is a complicated process and therefore, the coaxial line matching requires two fixed double stubs of variable lengths. The double shunt stubs are easy to construct for coaxial line. Series stub cannot be realised easily in coaxial lines, because the circuit cannot be closed using a short in this configuration.

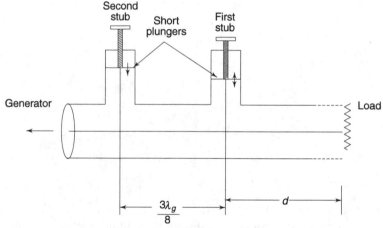

Fig. 5.11 *Coaxial line stubs*

Waveguide stub

In a rectangular waveguide, the stub may be connected in the **E** or **H** plane by placing a waveguide short circuit plunger, across the wide dimensions or across the narrow dimensions respectively, as shown in Fig. 5.12. For an **E** stub, the electric field lines penetrate from the main waveguide into the stub, whereas, in **H** stub, the magnetic field lines penetrate from the waveguide into the stub. Thus **E** stub offers reactance in series with the equivalent main line and the **H** stub presents a parallel susceptance across the main line. Since a single stub in a wave guide cannot be located very accurately at any place, two fixed stubs are preferred in waveguide circuits.

The combined **E** and **H**-stubs of Fig. 5.12 (c), called the **E-H** tuner, matches a wide range of load impedances in which the positions of the short circuit plungers in the E-and H-arms are varied.

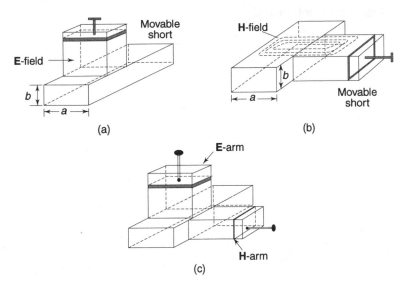

Fig. 5.12 *Waveguide stubs (a)* **E**-*stub (b)* **H**-*stub (c)* **E-H** *tuner*

Example 5.1 A load impedance of $73 - j80$ ohm is required to be matched to a 50 ohm coaxial line having a lossless dielectric of dielectric constant 4, using a short circuited shunt stub at 500 MHz. Determine the position and length of the stub.

Solution The guide wavelength in *TEM* mode coaxial line is

$$\lambda = \frac{30}{0.5\sqrt{4}} = 30 \text{ cm}$$

Normalised load impedance $z_L = (73 - j80)/50 = 1.46 - j1.6$. The position and the length of the stub are found from the Smith chart of Fig. 5.13 as follows.

1. z_L is located on the Smith chart.
2. A circle is drawn through z_L, with centre on unity of real axis.
3. The normalised load admittance y_L is located at A diametrically opposite to z_L.
4. A distance d/λ is moved from A towards generator to reach a point B on the circle where $y = 1 + jb$. The values $d/\lambda = 0.12$ and $jb = j\,1.38$ are read from the Smith chart.
5. To cancel the susceptance $j\,1.38$ by the stub at B of length l, the stub should provide a susceptance of $-j\,1.38$ corresponding to C. To find the stub length l a distance of $l/\lambda = 0.1$ is moved towards generator from the short circuit susceptance point $y = \infty$ to reach the stub susceptance point C, where $y = -j\,1.38$.

    Therefore, the required distance and length are

    $d = 0.12\ \lambda = 3.6$ cm.

    $l = 0.1\ \lambda = 3.0$ cm.

The coaxial line configuration of the stub is shown in Fig. 5.13.

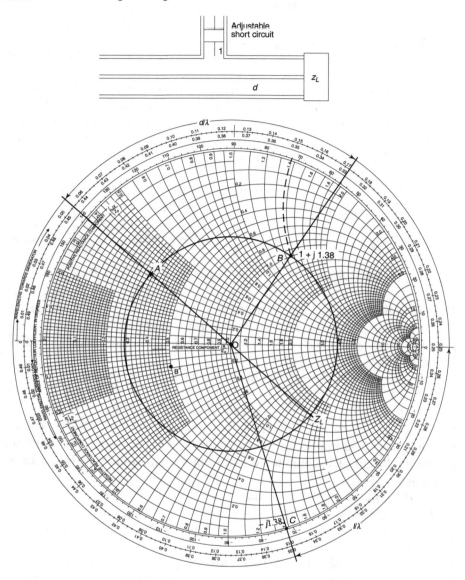

Fig. 5.13 *Single coaxial stub design*

Example 5.2 A 100 ohm line with air as dielectric is terminated by a load
impedance of $75 + j40$ ohm and is excited at 1 GHz by a matched generator.
Find the position of a single matching stub of 100 ohm impedance on the line,
and determine the length of the stub. Explain how such a stub is realised in a
rectangular waveguide circuit, excited in the dominant mode.

Solution

The normalised load impedance $z_L = \dfrac{75 + j40}{100} = 0.75 + j\,0.4$

The wavelength $\lambda = c/f = 30$ cm.

The stub design is made using the Smith chart Fig. 5.14 as follows.

1. z_L is located on the Smith chart and the corresponding VSWR circle is drawn through z_L.
2. The normalised load admittance y_L is located at a diametrically opposite point A on the VSWR circle to obtain the value $y_L = 1.04 - j\,0.55$.
3. A distance d/λ is moved towards the generator from y_L to reach a normalised admittance point of intersection B between the VSWR circle and the unit conductance circle where the normalised admittance is $y_B = 1 + j\,0.55$. This gives $d/\lambda = 0.3$.
4. For matching, the normalised stub susceptance is $-j\,0.55$ which is located at C on the outer circle.
5. To find the length of the short circuited stub a distance l/λ is moved towards the generator from the short circuit infinite admittance point to reach C so that $l/\lambda = 0.17$.

    Therefore,

    $$d = 0.3 \times 30 = 9 \text{ cm}$$

    $$l = 0.17 \times 30 = 5.1 \text{ cm}$$

A shunt stub can be realised in rectangular waveguide by joining it to the waveguide's narrow dimension so that magnetic loops can penetrate from the main waveguide into the stub as shown in Fig. 5.14. It is assumed that the tee junction formed between the waveguide and the stub do not create any field discontinuities d and l should be determined in terms of guide wavelength.

Example 5.3   A lossless 50 ohm air line has $V_{max} = 2.5$V and $V_{min} = 1$V when terminated with an unknown load. The distance between the successive voltage minima is 5 cm and the first voltage minimum from the load end is 1.25 cm. Design a short circuited single stub for impedance matching.

Solution

Distance between two successive minima is

$$\lambda/2 = 5 \text{ cm}, \quad \text{or} \quad \lambda = 10 \text{ cm}$$

$$\text{VSWR} = V_{max} / V_{min} = 2.5/1 = 2.5$$

Normalised distance of first minimum from load end is

$$d_{min}/\lambda = 1.25/10 = 0.125$$

1. Using Smith chart of Fig. 5.15, a VSWR circle $S = 2.5$ is drawn.
2. Locate first minimum point A on the real axis where normalised impedance is $1/S = 1/2.5 = 0.04$.
3. From A, a distance d_{min}/λ is moved towards load to reach the load point B where the normalised load impedance $z_L = 0.7 - j\,0.75$. An opposite point C is marked for load admittance $y_L = 0.7 + j\,0.73$.
4. From C, a distance d/λ is moved towards generator to reach the point D where S circle and $g = 1$ circle intersects. At D, the normalised line admit-

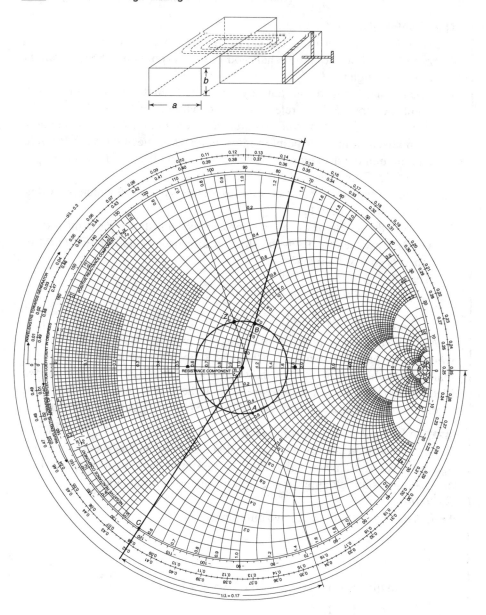

Fig. 5.14 *Single waveguide stub design*

tance $= 1 + jb = 1 + j\,0.95$. The distance of the stub from the load is $d/\lambda = 0.035$.

5. The stub must provide a susceptance of $-j0.95$ for matching ($y_{in} = 1$). From the short circuit admittance ($g = \infty$) point E, a distance l/λ is moved towards generator to reach a point F on the outer circle where the normalised susceptance is $-j\,0.95$. $l/\lambda = 0.13$, the required stub length. Therefore,

$d = 0.035 \times 10 = 0.35$ cm, and $l = 0.13 \times 10 = 1.3$ cm

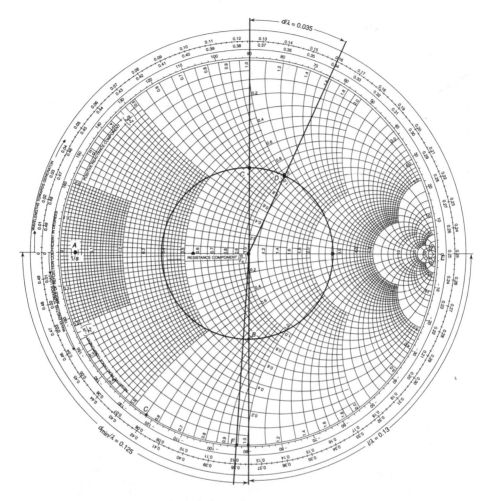

Fig. 5.15 *Single stub design from load VSWR and voltage minimum position*

Example 5.4   A 75 ohm air-filled coaxial line is terminated with a complex load of $109.5 - j120$ ohm. Design a double stub matching system using short circuited coaxial lines of characteristic impedance 75 ohm.

Solution

The normalised load impedance

$$z_L = (109.5 - j120)/75 = 1.46 - j1.6;$$

The load reflection coefficient

$$|\Gamma| = \left| \frac{Z_L - Z_0}{Z_L + Z_0} \right| = \left| \frac{109.5 - j120 - 75}{109.5 - j120 + 75} \right|$$

$$= \left| \frac{34.5 - j120}{184.5 - j120} \right| = 0.567$$

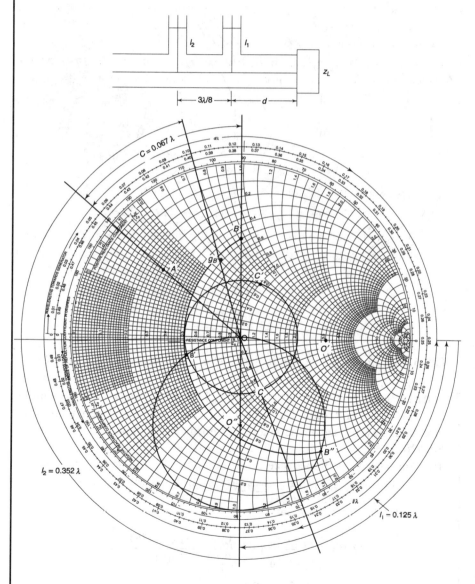

$$\text{VSWR} = \frac{1+|\Gamma|}{1-|\Gamma|}$$

$$= \frac{1+0.567}{1-0.567} = 3.6$$

The double stub system is designed using the Smith chart of Fig. 5.16 and the following steps.

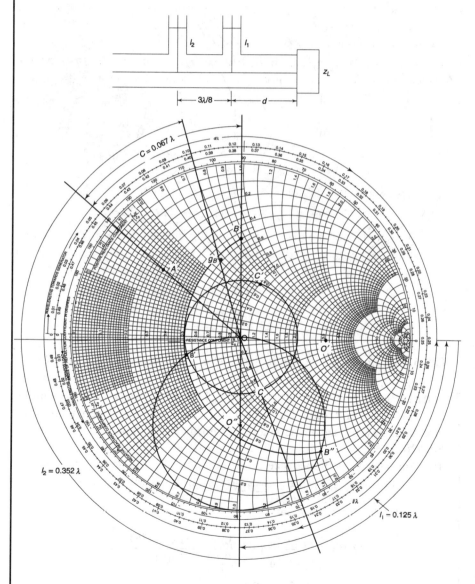

Fig. 5.16 *Coaxial line double stub design*

1. Load impedance z_L is located on the Smith chart. A circle is drawn through z_L with center at $y = 1$ and normalised load admittance y_L is located on the circle at A diametrically opposite z_L, where $y_L = 0.31 + j\,0.35$.

2. At an arbitrary distance $d/\lambda = 0.067$ from y_L towards the generator along the circle of radius $S = OA$ gives the position B of the first stub where the normalised admittance is

$$y_B = g_B + jb_B = 0.5 + j\,0.85$$

3. The unit conductance circle ($g = 1$) with centre O' is rotated towards load by $3\lambda/8$ to reach the new centre O'' of the rotated circle on the line passing through the points $+j$ and $-j$.

4. The point of intersection B' between g_B circle and the rotated unit conductance circle of centre O'' gives the total admittance

$$y_{BS} = g_B + jb_B + jb_{S1}$$
$$= 0.5 + j\,0.85 + jb_{S1}$$
$$= 0.5 - j\,0.14$$

or, $\quad j\,0.85 + jb_{S1} = -j\,0.14$

or the susceptance of the first stub is

$$jb_{S1} = -j\,0.14 - j\,0.85 = -j\,0.99$$

5. The length l_1 of the first stub is found by rotating towards the generator along the periphery of outer circle from the right-hand intersection of the unit circle and the real axis, which corresponds to the short circuit to obtain admittance of the stub, to its input susceptance value $-j\,0.99$ on the outer circle

$$l_1/\lambda = 0.125$$

6. The point B' is rotated towards generator by $3\,\lambda/8$ to C on the original $g = 1$ circle (centre O') to get the total normalised admittance at C due to the line and the first stub:

$$y_C = 1 - jb_C = 1 - j\,0.75$$

7. Matching by second stub is obtained by its susceptance $jb_{S2} = -jb_C = +j\,0.75$ corresponding to the opposite point C' on the outer circle.

8. The length of the second stub l_2 is found by rotation along the outer circle towards generator from the right hand intersection of the unit circle and the real axis to the point $+j\,0.75$:

$$l_2/\lambda = 0.352$$

Hence the required stub lengths are

$$l_1/\lambda = 0.125 \quad l_2/\lambda = 0.352$$

for distance of the first stub from the load $d/\lambda = 0.067$.

Example 5.5 Design a double stub tuner with a stub separation of $3\lambda/8$ and first stub position at the load $z_L = 80 + j\,60$ ohm, to match the load Z_L to a 50 ohm line. The generator end is matched.

Solution The normalised load impedance $z_L = (80 + j\,60)/50 = 1.6 + j1.2$. The normalised load admittance y_L is at the diametrically opposite point A on the load VSWR circle, where $y_L = 0.4 - j\,0.3$. The first stub should be located at the load end and the stub separation $= 3\lambda/8$.

The following steps are performed on the Smith chart.

1. The point of intersection B' or B'' between $g = 0.4$ circle and the rotated unit conductance circle of centre O'' gives the total susceptance

$$y_{BS} = 0.4 - j\,0.2 \quad \text{or} \quad 0.4 - j\,1.8 = y_L + jb_{S1}$$

respectively. Therefore, the first stub must add a susceptance of

$$jb_{S1} = -j\,0.2 + j\,0.3 = +j\,0.1 \quad \text{or} \quad -j\,1.8 + j\,0.3 = -j\,1.5$$

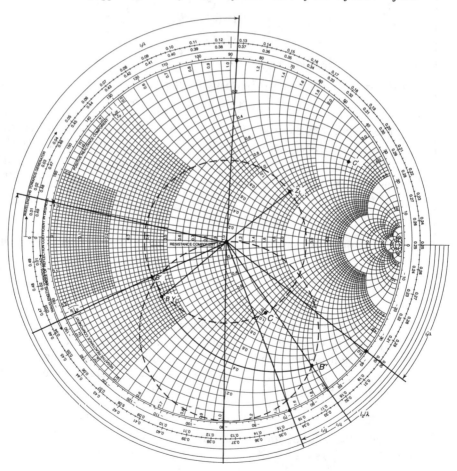

Fig. 5.17 *Double stub design*

to the load admittance to result y_{BS} value as above at the first stub point.

2. The length of the first stub is found by rotating towards the generator along the outer circle starting from the $y = \infty$ point to the jb_{S1}. For the two solutions of jb_{S1}, the lengths are obtained as

$$l_1/\lambda = 0.268$$
$$l_2/\lambda = 0.094$$

3. The points B' and B'' are then rotated towards generator by $3\lambda/8$ to reach the point C or C' on the $g = 1$ circle where total admittance

$$y_C = 1 + jb = 1 - j\,1.05 \quad \text{or} \quad 1.0 + j\,3.0$$

4. Matching will be achieved when total admittance at C or C' is $y_{cs} = 1.0$. Therefore, the second stub must offer a susceptance of $+j\,1.05$ or $-j\,3.0$.

5. The length of the second stub is obtained in similar way as in (2):

$$l_2/\lambda = 0.378$$
or, $$l_2'/\lambda = 0.052$$

5.3.5 Strip Line and Microstrip Line Matching

In microwave integrated circuits, such as strip lines and microstrip lines, realisation of stubs is difficult because short or open circuit configurations of the lines can not be achieved. Lumped-parameter elements such as inductors and capacitors are built which are conveniently used for impedance matching up to several GHz as shown in Fig. 5.18.

For a spiral inductor the total length of the spiral must be a small fraction of the wavelength to function as a lumped element. The impedance of the element must be large for an inductor, because $Z = \sqrt{(L/C)}$. Therefore, a narrow width is required for a lumped inductor design.

A lumped shunt capacitor of the order of limited value up to about 1 pF can be achieved by using a small open stub, as shown in Fig. 5.18(c). For higher values an interdigital configuration provides a series capacitance. Values up to several pF can be obtained depending on the number of fingers and their length. A high capacitance also can be obtained by using a small gap between two lines or inserting a chip capacitor as shown in Fig. 5.18(e).

5.4 BROADBAND MATCHING

5.4.1 Quarterwave Impedance Transformers

The equations (5.3) and (5.4) show that a lossless line can be used as voltage and current transformers with the secondary values V_r or I_r at the load end and the primary values V_s or I_s at the sending end, by the proper selection of l and Z_0. If a line of electrical length $\beta l = \pi/2$ having characteristic impedance Z_2 is terminated by an impedance Z_3, the input impedance of the line is given by

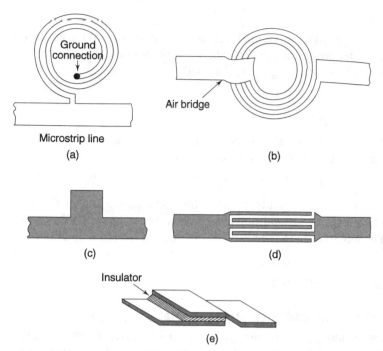

Fig. 5.18 *Lumped matching elements in microstrip lines (a) A spiral shunt inductor (b) A spiral series inductor (c) A short open-circuited stub (d) An interdigital capacitor (e) A gap capacitor*

$$Z_{in} = Z_2 \frac{Z_3 + j Z_2 \tan \beta l}{Z_2 + j Z_3 \tan \beta l}, \beta l = \pi/2$$

$$= Z_2^2 / Z_3 \tag{5.23}$$

This Z_{in} is matched to the generator of impedance $Z_1 = Z_{in}$, so that

$$Z_2 = [Z_1 Z_3]^{1/2} \tag{5.24}$$

Therefore, a generator of impedance Z_1 can be matched to a load $Z_3 \neq Z_1$ by inserting a length $l = \lambda_g/4$ of the transmission line between them, whose characteristic impedance is $[Z_1 Z_3]^{1/2}$. This matching section of the line is called the *quarterwave matching transformer* as shown in Fig. 5.19 (a).

When Z_1 and Z_3 are widely different, better matching can be achieved by using multisection quarterwave transformers each having characteristic impedance equal to the square root of the product of those on both sides as shown in Fig. 5.19(b).

5.4.2 Frequency Response of a Quarterwave Transformer

A single section quarterwave transformer provides a perfect matching at the single designed frequency f_0 (guide wavelength λ_{g0}) and at all the odd harmonics for which the length becomes equal to an odd multiple of $\lambda_{g0}/4$. The bandwidth of such a section can be determined by considering the electrical length $\beta l = \theta$ at any

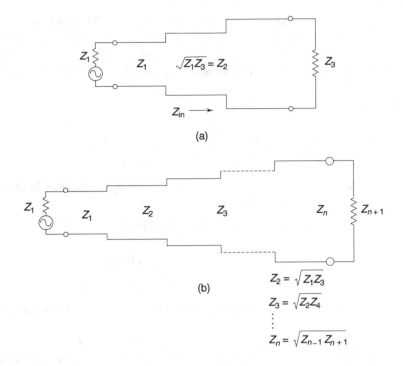

Fig. 5.19 *Quarterwave impedance transformer (a) One section (b) Multisections*

other frequency $f \neq f_0$ and finding the input reflection coefficient Γ. Let a quarter-wave transformer match a load $Z_3 = Z_L$ to a line of impedance Z_1. Then

$$\Gamma = \frac{Z_{in} - Z_1}{Z_{in} + Z_1} \tag{5.25}$$

where,

$$Z_{in} = Z_2 \frac{Z_L + jZ_2 \tan \theta}{Z_2 + jZ_L \tan \theta} \tag{5.26}$$

Substituting this Z_{in} and $Z_2 = \sqrt{(Z_1 \, Z_L)}$, Eq. 5.25 becomes

$$|\Gamma| = \frac{1}{\left[1 + \left(\dfrac{2\sqrt{Z_1 \, Z_L}}{Z_L - Z_1} \sec \theta\right)^2\right]^{1/2}}$$

$$\approx |(Z_L - Z_1) \cos \theta| / 2\sqrt{(Z_1 \, Z_L)} \tag{5.27}$$

when

$$\theta = 0, |\Gamma| = \frac{Z_L - Z_1}{2\sqrt{(Z_L \, Z_1)}} \quad \text{since } Z_{in} = Z_L \tag{5.28}$$

and

$$\theta = \pi/2, |\Gamma| = 0 \text{, since } Z_{in} = Z_2^2/Z_L \tag{5.29}$$

Thus $|\Gamma|$ varies periodically with θ as shown in Fig. 5.20, which in turn shows the variation of the magnitude of reflection coefficient with frequency. The bandwidth of the transformer is defined for the range of $|\Gamma|$ having the maximum allowable limit $|\Gamma_m|$ around the midband frequency f_0, where $|\Gamma| = 0$ and $\theta = \pi/2$ or $f = f_0$.

Therefore, for $|\Gamma| = |\Gamma_m|$,

$$\theta_m = \cos^{-1}\left|\frac{2|\Gamma_m|\sqrt{(Z_1 Z_L)}}{(Z_L - Z_1)\sqrt{(1-|\Gamma_m|)^2}}\right| \tag{5.30}$$

For *TEM* line, $\theta = \beta l = \pi f/2f_0$, so that at θ_m, $f_m = 2f_0\,\theta_m/\pi$. The bandwidth of the single section transformer is, therefore,

$$\Delta f = 2\,(f_0 - f_m) = 2f_0\,(1 - 2\theta_m/\pi) \tag{5.31}$$

or,

$$\Delta\frac{f}{f_0} = 2 - \frac{4}{\pi}\cos^{-1}\left|\frac{2|\Gamma_m|\sqrt{(Z_1 Z_L)}}{(Z_L - Z_1)\sqrt{(1-|\Gamma_m|)^2}}\right| \tag{5.32}$$

At the band edge, the electrical length, frequency and wavelength parameters are defined by

$\theta = \theta_1 = \pi f_1/2f_0$, $\lambda_g = \lambda_{g1}$ and $\theta = \theta_2 = \pi - \theta_1 = \pi f_2/2f_0$. $\lambda_g = \lambda_{g2}$, for $m = 1$ and 2 at both sides of f_0. The length of the transformer and the fractional bandwidth are then given by

$$l = \lambda_{g0}/4 = \lambda_{g1}\,\lambda_{g2}/2\,(\lambda_{g1} + \lambda_{g2}) \tag{5.33}$$

$$\frac{\Delta f}{f_0} = 2\frac{\lambda_{g1} - \lambda_{g2}}{\lambda_{g1} + \lambda_{g2}} = 2\frac{f_2 - f_1}{f_2 + f_1} \tag{5.34}$$

For a non-dispersive transmission line $\lambda_g = \lambda_0$, the free space wavelength.

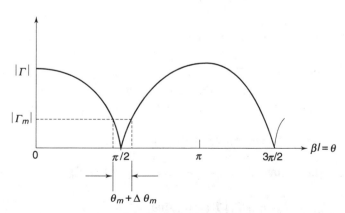

Fig. 5.20 *Response of a single section quarterwave transformer*

The above analysis shows that the bandwidth of a single section of a quarterwave transformer is small. This analysis assumes that the characteristic impedances Z_1 and Z_L are frequency independent. This is a good approximation for *TEM* wave lines. But for waveguides, these impedances vary with frequency and the bandwidth characteristic of quarterwave transformer becomes complicated, and is not discussed here. Moreover, the equivalent shunt susceptances which appear at the junctions of the transformer ends, due to sudden change of dimensions of the lines with different impedances, are neglected in this theory.

5.4.3 Realisation of Waveguide Impedance Transformer

Waveguide impedance transformers can be easily designed with steps in either the magnetic plane or electric plane for which the mathematical expressions for impedance ratio is simpler. Figure. 5.21 shows a typical waveguide transformer consisting of two waveguides of narrow wall dimensions b_1 and b_3 joined by a quarterwave long, third waveguide of narrow wall dimension

$$b_2 = \sqrt{(b_1/b_3)} \tag{5.35}$$

For *E*-plane transformer of Fig. 5.21(b), the impedance condition is given by

$$a_2/\lambda_g = \sqrt{(a_1\, a_3)/(\lambda_{g1}\, \lambda_{g3})} \tag{5.36}$$

where λ_{g2}, λ_{g1}, and λ_{g3} are the guide wavelengths of the transformer and the two waveguide sections, respectively.

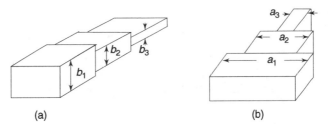

(a) (b)

Fig. 5.21 *Quarterwave waveguide transformers (a) H-plane (b) E-plane*

5.4.4 Effect of Multiple Reflections at Junctions

The overall reflection coefficient at the input can be evaluated by considering all the multiple reflections at the junction of the transformer. It can be shown that for small reflections, the total reflection coefficient can be obtained only from the first order reflections.

Figure 5.22 shows two discontinuities of a single quarterwave transformer at which the reflection and transmission coefficients are Γ and T, respectively, which are indicated with appropriate suffices in Γ and T. These are expressed by

$$\Gamma_1 = \frac{Z_2 - Z_1}{Z_2 + Z_1}, \quad \Gamma_2 = -\Gamma_1, \quad \Gamma_3 = \frac{Z_3 - Z_2}{Z_3 + Z_2} \tag{5.37}$$

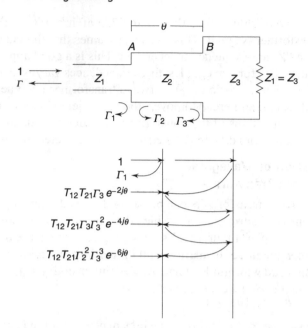

Fig. 5.22 *Multiple reflection in a quarterwave transformer*

$$T_{21} = 1 + \Gamma_1, \quad T_{12} = 1 + \Gamma_2 = 1 - \Gamma_1 \tag{5.38}$$

Let us assume that a unit amplitude wave is incident at the input. Inspection of the figure shows that the net reflected wave amplitude due to infinite number of multiple reflection is

$$\Gamma = \Gamma_1 + T_{21} T_{12} \Gamma_3 e^{-j2\theta} + T_{12} T_{21} \Gamma_3^2 \Gamma_2 e^{-j4\theta} + \dots$$

$$= \Gamma_1 + T_{12} T_{21} \Gamma_3 e^{-j2\theta} \sum_{n=0}^{\infty} \Gamma_2^n \Gamma_3^n e^{-j2n\theta};$$

$$= \Gamma_1 + \frac{T_{12} T_{21} \Gamma_3 e^{-j2\theta}}{1 - \Gamma_2 \Gamma_3 e^{-j2\theta}}; \text{ since } \sum_{n=0}^{\infty} x^n = \frac{1}{1-x} \tag{5.39}$$

Eliminating T_{12} and T_{21} by $1 - \Gamma_1$ and $1 + \Gamma_1$, respectively,

$$\Gamma = \frac{\Gamma_1 + \Gamma_3 e^{-j2\theta}}{1 + \Gamma_1 \cdot \Gamma_3 e^{-j2\theta}} \tag{5.40}$$

Assuming small reflections, $|\Gamma_1|$ and $|\Gamma_3| \ll 1$, so that

$$\Gamma \approx \Gamma_1 + \Gamma_3 e^{-j2\theta} \tag{5.41}$$

Thus for small reflections, the total reflection coefficient can be obtained only from the first order reflection with a maximum error of 4% for $|\Gamma_1| = |\Gamma_3| \le 0.2$.

Example 5.6 What are the required length and impedance of a $\lambda_g/4$ transformer that will match a 100 ohm load to a 50 ohm air-filled line at 10 GHz ? Consider both rectangular waveguide and coaxial line cases. What is the frequency band of operation for coaxial line over which the reflection coefficient remains less than 0.1 ?

Solution   Given,

$Z_0 = 50$ ohm, $Z_L = 100$ ohm, $f_0 = 10$ GHz, and $\Gamma_m \leq 0.1$, $\lambda_0 = c/f_0 = \dfrac{30}{10} = 3$ cm.

(a) For x-band rectangular w/g in the TE_{10} mode

$$\lambda_g = \lambda_0 / \sqrt{\left[1 - (\lambda_0/2a)^2\right]}$$

$$= 3 / \sqrt{\left[1 - (3/2 \times 2.286)^2\right]}$$

$$\approx 3 / \sqrt{[1 - 0.43]}$$

$$\approx 4 \text{ cm}$$

Therefore, the length of the matching section is $l = \lambda g/4 = 1$ cm.

(b) For air-filled coaxial line, $\lambda_g = \lambda_0 = c / f_0 = 30/10 = 3$ cm. Therefore, the length of the matching section is

$$l = \lambda_g /4 = 3/4 = 0.75 \text{ cm}.$$

The characteristic impedance of the matching section

$$Z_2 = \sqrt{(Z_1 \cdot Z_L)} = \sqrt{(50 \times 100)} = 50\sqrt{2} = 70.71 \text{ ohm}$$

The fractional bandwidth for *TEM* line is

$$\Delta f / f_0 \ = 2 - 4/\pi \cos^{-1} \left| \frac{2\,\Gamma_m \,\sqrt{Z_1\, Z_L}}{(Z_L - Z_1)\,\sqrt{\left(1 - \Gamma_m^2\right)}} \right|$$

$$= 2 - \frac{4}{180} \cos^{-1} \left| \frac{2 \times 0.1 \,\sqrt{(100 \times 50)}}{(100 - 50) \,\sqrt{\left[1 - (0.1)^2\right]}} \right|$$

$$= 2 - 4/180 \cos^{-1} \left| \frac{0.2 \times 50\sqrt{2}}{50 \times \sqrt{0.99}} \right|$$

$$= 2 - 4/180 \cos^{-1} 0.282/0.99$$

$$= 2 - 4/180 \times 73.5$$

$$= 0.367$$

Therefore, $\Delta f = 0.367 \times 10$ GHz $= 3.67$ GHz.

Example 5.7   An empty rectangular waveguide is matched to a dielectric ($K = 2.56$) filled waveguide in TE_{10} mode at 10 GHz by means a quarterwave transformer. Find the length and dielectric constant of the matching section. The broader dimension of the waveguide is $a = 2.5$ cm.

Solution   Let l and K_1 be the length and dielectric constant for the matching section guide as shown in Fig. 5.23. Given

$$a = 2.5 \text{ cm}$$
$$f = 10 \text{ GHz}$$
$$K = 2.56$$
$$K_0 = 1, \text{ empty guide dielectric constant}$$
$$\lambda_0 = 30/10 = 3 \text{ cm}$$

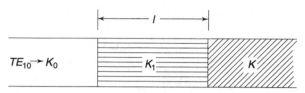

Fig. 5.23   *Quarterwave waveguide matching section*

Let Z, Z_1, Z_0 are the wave impedances of the sections for the dominant TE_{10} mode. We will assume that due to introduction of the matching section, no higher order modes are generated.

Wave impedances of the dielectric filled waveguide and air-filled waveguide excited in TE_{10} mode are

$$Z = \frac{377}{\sqrt{\left[2.56 - (\lambda_0/2a)^2\right]}}$$

$$= \frac{377}{\sqrt{\left[2.56 - (3/2 \times 2.5)^2\right]}}$$

$$= 254.17 \text{ ohm}$$

$$Z_0 = \frac{377}{\sqrt{\left[1 - (\lambda_0/2a)^2\right]}}$$

$$= \frac{377}{\sqrt{\left[1 - (3/2 \times 2.5)^2\right]}}$$

$$= 471.25 \text{ ohm}$$

Therefore, impedance of the matching section at the designed frequency 10 GHz is

$$Z_1 = \sqrt{[Z_0 \ Z]} = \sqrt{[254.17 \times 471.25]}$$
$$= 346.09 \text{ ohm}$$

Dielectric constant K_1 of the matching section guide is obtained from

$$Z_1 = \frac{377}{\sqrt{\left[K_1 - (3/2 \times 2.5)^2\right]}} \text{ ohm} = 346.09 \text{ ohm}.$$

or,
$$\sqrt{\left[K_1 - (3/5)^2\right]} = 377/346.09$$

or,
$$K_1 = 1.5466$$

Length of matching section at the designed frequency 10 GHz is 1/4 of guide wavelength in this section i.e.,

$$l = \lambda_{g1}/4$$

$$= \frac{1}{4} \frac{\lambda_0}{\sqrt{\left[K_1 - (\lambda_0/2a)^2\right]}}$$

$$= \frac{3}{4} \frac{1}{\sqrt{\left[1.5466 - (3/2 \times 2.5)^2\right]}}$$

$$= 0.6885 \text{ cms.}$$

5.4.5 Multisection Quarterwave Transformer

In order to increase the bandwidth of quarterwave impedance matching transformer, multisections are used with gradual change of step sizes. The input reflection coefficient can be obtained from partial reflection coefficients at various junctions/steps of the transformer as shown in Fig. 5.24.

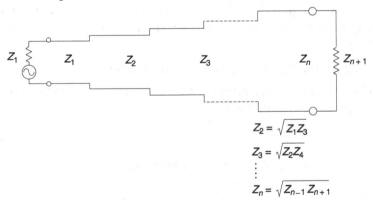

$$Z_2 = \sqrt{Z_1 Z_3}$$
$$Z_3 = \sqrt{Z_2 Z_4}$$
$$\vdots$$
$$Z_n = \sqrt{Z_{n-1} Z_{n+1}}$$

Fig. 5.24 *Multisection quarterwave transformer*

The problem here is to match the two impedances Z_1 and Z_{n+1} by $n-1$ sections of the quarterwave transformers. There are now n steps which generate n reflection coefficients $\Gamma_1, \Gamma_2, \dots, \Gamma_n$. Because of the discontinuities at the junctions, each component wave will undergo multiple reflections back and forth, between any two successive steps. Therefore, for $n-1$ sections, the total reflection coefficient Γ_T at any frequency f, can be written as

$$\Gamma_T = \Gamma_1 + \Gamma_2 e^{-j2\theta} + \Gamma_3 e^{-j4\theta} + \dots + \Gamma_n e^{-j2(n-1)\theta} \qquad (5.42)$$

where $\Gamma_i = (Z_{i+1} - Z_i)/(Z_{i+1} + Z_i)$, the reflection coefficient at the junction between $i+1$ and ith sections for a small reflection.

Assuming that the transformer is symmetrical with respect to a transverse reference plane at the centre,

$$\Gamma_1 = \Gamma_n, \Gamma_2 = \Gamma_{n-1}, \Gamma_3 = \Gamma_{n-2}, \text{etc.} \tag{5.43}$$

Then,

$$\Gamma_T = 2 \left[\Gamma_1 \cos (n-1)\theta + \Gamma_2 \cos (n-3)\theta + \Gamma_3 \cos (n-5)\theta + ...\right] \tag{5.44}$$

$$= 2 \sum_{k=1}^{m} \Gamma_k \cos\left(n - \overline{2k-1}\right) \theta; \, m = n/2 \text{ for even } n \tag{5.45}$$

$$= 2 \sum_{k=1}^{m} \Gamma_k \cos\left(n - \overline{2k-1}\right) \theta + \Gamma_{(n+1)/2}; \, m = (n-1)/2 \text{ for odd } n$$
$$\tag{5.46}$$

At $\quad f = f_0, \, \theta = \pi/2,$

so that $\Gamma_T = 0$, n is even $\tag{5.47}$

$$= \Gamma_{(n+1)/2} - 2 \Gamma_{(n-1)/2} + 2\Gamma_{(n-3)/2} + ... \pm 2\Gamma_1, \, n \text{ is odd} \tag{5.48}$$

By a proper choice of the reflection coefficient Γ_k, $k = 1, 2,..., n$, a variety of passband characteristics of a multisection quarterwave transformer can be obtained. The impedances Z_k's for various sections are calculated from Γ_k's. The most common passband characteristics are obtained from the *binomial* (maximally flat) and *Chebyschev* (equal ripple) distribution functions which are described below.

5.4.6 Maximally Flat Quarterwave Transformer

Maximally flat passband characteristic is obtained if Γ_k's are made proportional to the binomial coefficients, that is,

$$\Gamma_k = \Gamma_1 C_{k-1}^{n-1} \tag{5.49}$$

The reflection coefficient Γ_T can now be expressed as

$$\Gamma = 2\Gamma_1 \sum_{k=1}^{m} C_{k-1}^{n-1} \cos (n - \overline{2k-1}) \theta; \, m = \frac{n}{2} \text{ for even } n \tag{5.50}$$

$$= 2\Gamma_1 \sum_{k=1}^{m} C_{k-1}^{n-1} \cos (n - \overline{2k-1}) \theta + C_{(n-1)/2}^{n-1}; \, m = \frac{n-1}{2} \text{ for odd } n$$
$$\tag{5.51}$$

Expanding $\cos p\theta$ in terms of $\cos \theta$, where $p = n - \overline{2k-1}$, we get

$$\cos p\theta = 2 \cos (p-1)\theta \cos \theta - \cos (p-2)\theta \tag{5.52}$$

Substituting the expansion of $\cos p\theta$ into the expression of reflection coefficient, only one term containing $\cos^{(n-1)}\theta$ with coefficient 2^{n-2} for expansion of $\cos(n-1)\theta$ remains, to give

$$\Gamma = 2\Gamma_1 2^{n-2} \cos^{n-1}\theta = \Gamma_1 2^{n-1} \cos^{n-1}\theta \qquad (5.53)$$

Since $|\Gamma|$ and its first $(n-2)$ derivatives with respect to frequencies (or θ) vanish at the centre frequency $f = f_0$, where $\theta = \pi/2$, the binomial distribution provides a maximally flat response as shown in Fig. 5.25.

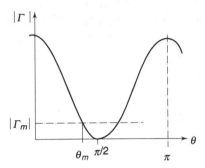

Fig. 5.25 *Maximally flat response*

In the design of the transformer, a maximum value of $|\Gamma| = |\Gamma_m|$, may be prescribed over a specified band of frequencies from f_1 to f_2 with $f_0 = (f_1 + f_2)/2$ and given number of sections n. The band edge reflection coefficient Γ_1 is then computed from the equation

$$|\Gamma_1| \le \frac{|\Gamma_m|}{2^{n-1}\cos^{n-1}\theta_1}; \; \theta = \theta_1 \text{ or } \theta_2 \text{ at } |\Gamma| = |\Gamma_m| \qquad (5.54)$$

$$|\Gamma_m| = \Gamma_1 2^{n-1} \cos^{n-1}\theta_1 = \Gamma_1 2^{n-1} \cos^{n-1}\theta_2 \qquad (5.55)$$

For symmetry of the response curve, $\cos\theta_1 = |\cos\theta_2|$, or, $\theta_2 = \pi - \theta_1$. If r is the ratio of the wavelengths at the band edge frequencies f_1 and f_2, some of the design formulae are given by

$$r = \frac{\lambda_{g1}}{\lambda_{g2}} = \frac{\theta_2}{\theta_1} = \frac{\pi - \theta_1}{\theta_1} = \frac{\theta_2}{\pi - \theta_2} \qquad (5.56)$$

$$\theta_1 = \frac{\pi}{1+r} = \frac{2\pi l}{\lambda_{g1}} = \frac{\pi\lambda_{g0}}{2\lambda_{g1}} \qquad (5.57)$$

$$\theta_2 = \frac{\pi r}{1+r} = \frac{2\pi l}{\lambda_{g2}} = \frac{\pi\lambda_{g0}}{2\lambda_{g2}} \qquad (5.58)$$

$$l = \frac{\lambda_{g0}}{4} = \frac{\theta_1\lambda_{g1}}{2\pi} = \frac{\theta_2\lambda_{g2}}{2\pi} \qquad (5.59)$$

$$\lambda_{g0} = \frac{2\lambda_{g1}\lambda_{g2}}{\lambda_{g1} + \lambda_{g2}} \qquad (5.60)$$

If n is not specified, it is selected by trial and error method to find $|\Gamma| \le |\Gamma_m|$ over the passband.

The characteristic impedances of the sections are determined from the relation

$$\Gamma_k = \frac{Z_{k+1} - Z_k}{Z_{k+1} + Z_k} = C_{k-1}^{n-1} \tag{5.61}$$

where Z_{i+1} is the characteristic impedance of ith the section and Γ_i is the reflection coefficient at the input of ith section as shown in Fig. 5.23. Further, if the reflections are small ($\Gamma_k \leq 0.35$ for an error of less then 4%), we can write

$$\ln\left(\frac{Z_{k+1}}{Z_k}\right) = 2\left[\frac{Z_{k+1} - Z_k}{Z_{k+1} + Z_k} + \frac{1}{3}\left(\frac{Z_{k+1} - Z_k}{Z_{k+1} + Z_k}\right)^3 + \dots\right]$$

$$= 2\left[\Gamma_k + 1/3\,\Gamma_k^3 + \dots\right]$$

$$\approx 2\,\Gamma_k \tag{5.62}$$

Therefore,

$$\Gamma_k = \frac{1}{2}\ln\frac{Z_{k+1}}{Z_k} = C_{k-1}^{n-1}\frac{1}{2}\ln\frac{Z_2}{Z_1} \tag{5.63}$$

or,

$$\ln(Z_{k+1}/Z_k) = C_{k-1}^{n-1}\ln\frac{Z_2}{Z_1} \tag{5.64}$$

Thus the impedances can be computed from the corresponding reflection coefficients

$$\Gamma_1 = 1/2 \ln Z_2 / Z_1, Z_2 = Z_1 \text{ [anti ln } (2\Gamma_1)]$$
$$\Gamma_2 = 1/2 \ln Z_3/Z_2, Z_3 = Z_2 \text{ [anti ln } (2\Gamma_2)]$$
$$\vdots$$
$$\vdots$$
$$\Gamma_k = 1/2 \ln Z_{k+1}/Z_k, Z_n = Z_{n-1} \text{ [anti ln } (2\Gamma_{n-1})] \tag{5.65}$$

By adding,

$$\Gamma_1 + \Gamma_2 + \dots + \Gamma_n = \frac{1}{2}\ln\frac{Z_{n+1}}{Z_1}$$

or,

$$2(\Gamma_1 + \Gamma_2 + \dots + \Gamma_{n/2}) = \frac{1}{2}\ln\frac{Z_{n+1}}{Z_1} \text{ for even } n \tag{5.66}$$

and

$$2(\Gamma_1 + \Gamma_2 + \dots) + \Gamma_{(n+1)/2} = \frac{1}{2}\ln\frac{Z_{n+1}}{Z_1} \text{ for odd } n \tag{5.67}$$

Example 5.8 Design a two-section binomial transformer to match the load given in Problem 5.6. What bandwidth is obtained for $\Gamma_m = 0.1$?

Solution Given number of sections $N = 2$, number of steps $n = N + 1 = 3$. $Z_1 = 50$ ohm, $Z_3 = 100$ ohm, $\Gamma_m = 0.1$, $f_0 = 10$ GHz, $\lambda_0 = 3$ cm.

Therefore,

$$\Gamma_1 + \Gamma_2 + \Gamma_3 = 1/2 \ln 100/50 = 1/2 \ln 2 \approx 0.6931/2$$
$$= 0.3465$$

For symmetrical binomial transformer

$$\Gamma_1 = \Gamma_3;$$
$$\Gamma_2 = N\,\Gamma_1 = 2\,\Gamma_1$$

Therefore, $\Gamma_1 + 2\Gamma_1 + \Gamma_1 = 0.3465$

or, $\Gamma_1 = 0.3465/4 = 0.0866$

$$\Gamma_2 = 2\,\Gamma_1 = 2 \times 0.0866 = 0.1732$$

Therefore, $Z_2 = Z_1 e^{2\Gamma_1} = 50\, e^{0.1732} = 59.46$ ohm

$$Z_3 = Z_2 e^{2\Gamma_2} = 59.46\, e^{2 \times 0.1732} = 84.1 \text{ ohm}$$

Length of the sections is $\lambda_g/4 = \lambda_0/4 = 3/4 = 0.75$ cm

$$|\Gamma_m| = \Gamma_1\, 2^{n-1} \cos^{n-1}\theta_1 = \Gamma_1\, 2^{n-1} \cos^{n-1}\theta_2$$
$$0.1 = \Gamma_1\, 2^2 \cos^2\theta_1 = 0.0866 \times 4 \times \cos^2\theta_1$$

or, $\cos^2\theta_1 = 0.1 / (0.0866 \times 4) = 0.2887$

or, $\cos\theta_1 = 0.5373$

$$\theta_1 = 57.5°$$
$$\theta_2 = \pi - \theta_1 = 180 - 57.5 = 122.5°$$

Therefore, $\theta = \theta_2 - \theta_1 = 122.5 - 57.5 = 65°;$

$$r = \theta_2/\theta_1 = 122.5/57.5 = 2.13$$

For *TEM* line the bandwidth
$$\Delta f/f_0 = 2/\pi\,(\theta_2 - \theta_1) = (2/180°) \times 65° = 130/180 = 0.7222$$
or,
$\Delta f = 0.722 \times 10$ GHz $= 7.22$ GHz which is double the bandwidth of Problem 5.6.

Example 5.9   Design a five section quarterwave, impedance transformer, to match the two lines of impedances 100 ohm and 1000 ohm. Assume binomial distribution passband response of the reflection coefficients for the wavelength ratio of 2, corresponding to the band edges. Find the maximum VSWR at the input of the transformer.

Solution   For five sections, number of steps $n = 6$ (even), wavelength ratio, $\lambda_{g1}/\lambda_{g2} = r = 2$, original input and output line impedances are,

$$Z_1 = 100 \text{ ohm}$$
$$Z_{n+1} = 1000 \text{ ohm}$$

Assuming small reflections at each step

$$2(\,\Gamma_1 + \Gamma_2 + \Gamma_3\,) = \frac{1}{2} \ln \frac{1000}{100} = 1.15$$

For binomial distribution

$$\Gamma_k = C_{k-1}^{n-1} \, \Gamma_1 = C_{k-1}^{5} \, \Gamma_1$$

$$\Gamma_1 = C_0^{5} \, \Gamma_1 = \Gamma_1 = \Gamma_6$$

$$\Gamma_2 = C_1^{5} \, \Gamma_1 = 5 \, \Gamma_1 = \Gamma_5$$

$$\Gamma_3 = C_2^{5} \, \Gamma_1 = 10 \, \Gamma_1 = \Gamma_4$$

Therefore, $\quad 2 \, (\Gamma_1 + 5 \, \Gamma_1 + 10 \, \Gamma_1) = 1.15$

or, $\qquad \Gamma_1 = \dfrac{1.15}{32} = 0.036 = \Gamma_6$

$$\Gamma_2 = 5 \times 0.036 = 0.180 = \Gamma_5$$
$$\Gamma_3 = 10 \times 0.036 = 0.360 = \Gamma_4$$

The characteristic impedances of the sections are

$$Z_2 = 100 \, [\text{ anti ln } (2 \times 0.036)] = 107.5 \text{ ohm}$$
$$Z_3 = 107.5 \, [\text{ anti ln } (2 \times 0.180)] = 154 \text{ ohm}$$
$$Z_4 = 154 \, [\text{ anti ln } (2 \times 0.360)] = 316 \text{ ohm}$$
$$Z_5 = 316 \, [\text{ anti ln } (2 \times 0.360)] = 648 \text{ ohm}$$
$$Z_6 = 648 \, [\text{ anti ln } (2 \times 0.180)] = 930 \text{ ohm}$$

At the band edge

$$\theta_1 = \frac{\pi}{1+r} = \frac{\pi}{1+2} = \frac{\pi}{3}$$

Therefore $\cos \theta_1 = 0.5 = x$

$$|\Gamma_m| = \Gamma_1 2^{n-1} \cos^{n-1} \theta_1 = 0.036 \times 2^5 \times \cos^5 \theta_1$$

or, $\qquad \cos^5 \theta_1 = (0.5)^5 = 0.03125$

Therefore, $\quad |\Gamma_m| = 0.072 \times 16 \times 0.03125$

$$= 0.036$$

Maximum VSWR $= \dfrac{1+|\Gamma_m|}{1-|\Gamma_m|} = \dfrac{1.036}{0.964} = 1.075$

If a condition that maximum allowable $|\Gamma_m| \le 0.03$ or less is imposed, more number of matching sections are needed and can be found by trial and error method by increasing $n > 6$.

Example 5.10 Design a three-section binomial transformer to match a 100 ohm load to a 50 ohm line. The maximum VSWR that can be tolerated is 1.1. What bandwidth can be obtained ?

Solution Given $Z_1 = 50$ ohm, $Z_L = Z_5 = 100$ ohm, $N = 3$, $n = N + 1 = 4$
$\text{VSWR}_{max} = [1 + |\Gamma_m|] / [1 - |\Gamma_m|] = 1.1$

or, $\qquad |\Gamma_m| = (S - 1) / (S + 1)$

$\qquad\qquad\qquad = (1.1 - 1) / (1.1 + 1)$

$\qquad\qquad\qquad = 0.1/2.1 = 0.0476$

Sum of reflection coefficients at the steps:

$\Gamma_1 + \Gamma_2 + \Gamma_3 + \Gamma_4 = 1/2 \ln Z_5/Z_1 = 1/2 \ln 100/50$

$\qquad\qquad\qquad = 0.3465$

For symmetrical binomial transformer : $\Gamma_1 = \Gamma_4$, $\Gamma_2 = \Gamma_3$

$$\Gamma_2 = N \Gamma_1 = 3 \Gamma_1 = \Gamma_3$$

Therefore, $\qquad \Gamma_1 + 3 \Gamma_1 + 3\Gamma_1 + \Gamma_1 = 0.3465$

or, $\qquad 8 \Gamma_1 = 0.3465$

or, $\qquad \Gamma_1 = 0.0434$

$\qquad\qquad \Gamma_2 = 3 \Gamma_1 = 0.1302$

Therefore, $\quad Z_2 = Z_1 e^{2\Gamma_1} = 50 \times e^{0.0868} = 50 \times 1.09 = 54.5$ ohm

$\qquad\qquad Z_3 = Z_2 e^{2\Gamma_2} = 54.5 \times e^{0.26} = 54.5 \times 1.297 = 70.5$ ohm

$\qquad\qquad Z_4 = 70.5 \, e^{2\Gamma_3} = 91.2$ ohm

$\qquad\qquad l = \lambda_g/4$ at centre frequency f_0

Bandwidth

$$|\Gamma_m| = \Gamma_1 \, 2^{n-1} \cos^{n-1}\theta_1$$

or, $\qquad 0.0476 = 0.0434 \times 2^3 \times \cos^3 \theta_1$

or, $\qquad \cos^3 \theta_1 = 0.0476/(0.0434 \times 8) = 0.1378$

or, $\qquad \cos \theta_1 = (0.1378)^{1/3}$

$\qquad\qquad\qquad = 0.5165$

Therefore, $\quad \theta_1 = 58.9°$

$\qquad\qquad \theta_2 = \pi - \theta_1 = 121.1°$

Therefore, $\Delta f / f_0 = 2/\pi (\theta_2 - \theta_1)$ for TEM waves

$$= \frac{2}{180} (121.1 - 58.9) = 0.69$$

Example 5.11 Design a two-section rectangular waveguide binomial transformer to match an empty rectangular waveguide with a dielectric filled guide in Problem 5.7 excited in dominant mode.

Solution Number of steps $n = 2 + 1 = 3$

$\qquad\qquad N = n - 1 = 2$

$$a = 2.5 \text{ cms}$$
$$f = 10 \text{ GHz}$$
$$K_4 = 2.56$$
$$K_1 = 1$$

$Z_4 = 254.17$ ohm, $Z_1 = 471.25$ ohm (from Example 5.7)

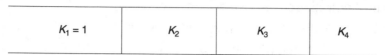

$K_1 = 1$	K_2	K_3	K_4

Fig. 5.26 *Multisection waveguide transformer design*

Assuming 1st order reflection at the steps, $\Gamma_1 + \Gamma_2 + \Gamma_3 = 1/2 \ln (Z_4 / Z_1) = 1/2 \ln (254.17/471.25) = -0.3087$

Assuming a symmetrical transformer

$$\Gamma_1 = \Gamma_3$$
$$\Gamma_2 = N\Gamma_1 = 2\Gamma_1$$

Therefore, $\Gamma_1 + 2\Gamma_1 + \Gamma_1 = 4\Gamma_1 = 1/2 \ln Z_4/Z_1$

$$= -0.3087$$

or, $\Gamma_1 = -0.3087/4$

$$= -0.07717 = \Gamma_3$$
$$\Gamma_2 = \Gamma_1 \times 2 = -0.1543$$

Now, $Z_2 = Z_1 [\text{anti ln } (2\Gamma_1)] = 403.725$ ohm

$Z_3 = Z_2 [\text{anti ln } (2\Gamma_2)] = 295.79$ ohm

Therefore, $403.725 = \dfrac{377}{\sqrt{\left[K_2 - (3/2 \times 2.5)^2\right]}}$ ohm

$$295.79 = \dfrac{377}{\sqrt{\left[K_3 - (3/2 \times 2.5)^2\right]}} \text{ ohm}$$

or, $K_2 = 1.232$

$K_3 = 1.984$

The lengths of the sections are

$$l_2 = \lambda_{gK2} /4$$

$$= \frac{1}{4} \frac{\lambda_0}{\sqrt{\left[K_2 - (\lambda_0/2a)^2\right]}}$$

$$= \frac{3}{4 \sqrt{\left[1.232 - (3/5)^2\right]}} = 0.804 \text{ cm}$$

$$l_3 = \lambda_{gK3}/4$$

$$= \frac{1}{4} \frac{\lambda_0}{\sqrt{\left[K_3 - (\lambda_0/2a)^2\right]}}$$

$$= \frac{3}{4\sqrt{\left[1.984 - (3/5)^2\right]}} = 0.588 \text{ cm}$$

5.4.7 Equal Ripple Transformer

Equal ripple characteristics of a multisection transformer is obtained by choosing the reflection coefficient Γ proportional to a *Chebyshev polynomial* $T_{n-1}(x)$ of order $n-1$ for $n-1$ number of sections

$$\begin{aligned}
T_{n-1}(x) &= \cos\left[(n-1)\cos^{-1}(x)\right]; -1 \le x \le 1 \\
&= \cosh\left[(n-1)\cosh^{-1}(x)\right]; x < -1, x > 1
\end{aligned} \tag{5.68}$$

where x is the parameter for change of variable according to

$$x = \frac{\cos\theta}{\cos\theta_1} = \frac{\cos\theta}{|\cos\theta_2|} \tag{5.69}$$

and is a function of frequency. The polynomials have the following properties as shown in Fig. 5.27 and are as follows.

1. All polynomials pass through the point $(1,1)$
2. $-1 \le T(x) \le 1$ for $-1 \le x \le 1$
3. All roots occurs within $-1 \le x \le 1$, and all maxima and minima have values of 1 and −1, respectively.
4. $T_k(x) = 2x\, T_{k-1}(x) - T_{k-2}(x)$

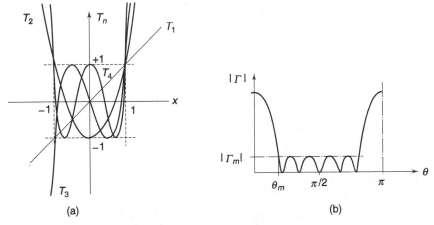

Fig. 5.27 *Chebyshev response*
(a) Chebyshev polynomial
(b) Reflection coefficient characteristics

An equal ripple characteristic results in a larger bandwidth compared to a binomial transformer for the same number of sections. The total reflection coefficient Γ behaves according to a Chebyshev response. This makes Γ have equal ripple characteristic over the range θ_1 to $\pi - \theta_1 = \theta_2$ with a specified maximum ripple height $|\Gamma_m|$. Thus with reference to a transverse plane at the centre,

$$\Gamma = |\Gamma_m| T_{n-1}(x) = 2 \sum_{k=1}^{m} \Gamma_k \cos\left(n - 2\overline{k-1}\right)\theta; \; m = n/2 \text{ for even } n \qquad (5.70)$$

$$= 2 \sum_{k=1}^{m} \Gamma_k \cos\left(n - 2\overline{k-1}\right)\theta + \Gamma_{(n+1/2)}; \; m = (n-1)/2 \text{ for odd } n \qquad (5.71)$$

This ensures that $|\Gamma_m T_{n-1}(x)| \le |\Gamma_m|$, since $|T_{n-1}(x)| \le 1$. The Γ_k's are obtained by expanding the right hand side of Eqs 5.70 and 5.71 in terms of powers of $\cos\theta$ and substituting $x \cos\theta_1$ for $\cos\theta$ and equating the coefficient of like powers of x with those on left hand side. Chebyshev transformer gives the largest bandwidth for a specified $|\Gamma_m|$ and number of sections and for a given bandwidth gives the lowest ripple $|\Gamma_m|$.

Example 5.12 Design a two-section quarter wave impedance transformer to match a line with characteristic impedance 100 ohm to a 200 ohm line. The maximum tolerable reflection coefficient is 0.05.
Solution

Here $\qquad n$ = no of sections $+1 = 2 + 1 = 3$; (odd)

$$T_{n-1}(x) = T_2(x) = 2x^2 - 1$$

Therefore, $\qquad \Gamma = 2[\Gamma_1 \cos(3-1)\theta] + \Gamma_2 = 2\Gamma_1 \cos 2\theta$

$$= |\Gamma_m| T_2(x) = 0.05\, T_2(x)$$

or, $\qquad 2\Gamma_1 \cos 2\theta + \Gamma_2 = 0.05(2x^2 - 1)$

or, $\qquad 2\Gamma_1(2\cos^2\theta - 1) + \Gamma_2 = 0.1x^2 - 0.05$

Since, $\qquad x = \dfrac{\cos\theta}{\cos\theta_1}, \cos\theta = x \cos\theta_1$

$$2\Gamma_1(2\cos^2\theta_1 x^2 - 1) + \Gamma_2 = 0.1\, x^2 - 0.05$$

Equating the coefficients of equal power

$$4\,\Gamma_1 \cos^2\theta_1 = 0.1, \; \Gamma_2 - 2\Gamma_1 = -0.05$$

Therefore, $\qquad \Gamma_1 = \dfrac{0.1}{4\cos^2\theta_1}$

$$\Gamma_2 = 2\Gamma_1 - 0.05 = (0.1/2\cos^2\theta_1) - (0.05)$$

For small reflections,

$$2\Gamma_1 + \Gamma_2 = \frac{1}{2} \ln \frac{200}{100} = \frac{1}{2} \ln 2 = 0.3466$$

or, $\quad 2 * \dfrac{0.1}{4 \cos^2 \theta_1} + \dfrac{0.1}{2 \cos^2 \theta_1} - 0.05 = 0.3466$

or, $\quad \dfrac{0.1}{\cos^2 \theta_1} = 0.3466 + 0.05 = 0.3966$

or, $\quad \cos^2 \theta_1 = \dfrac{0.1}{0.3966} = 0.2521$

or, $\quad \cos \theta_1 = 0.5021$

Therefore, $\quad \theta_1 = 59.86°$

$$\Gamma_1 = \frac{0.1}{4*0.2521} = 0.09916 = \Gamma_3$$

$$\Gamma_2 = \frac{0.1}{2*0.2521} - 0.05 = 0.1483$$

$$Z_2 = 100 \,[\, \text{anti ln} \,(2 * 0.09916)] = 121.94 \text{ ohm}$$

$$Z_3 = 121.94 \,[\, \text{anti ln} \,(2 * 0.1483)] = 164.04 \text{ ohm}$$

Summary

Therefore, the sequence of impedances are $Z_1 = Z_0 = 100$ ohm

$$Z_2 = 121.9 \text{ ohm}$$
$$Z_3 = 164 \text{ ohm}$$
$$Z_4 = Z_L = 200 \text{ ohm}$$

Maximum reflection coefficient at band edges is $|\Gamma_m| = 0.05$, VSWR $= 1.105$. Wavelength ratio $r = (\pi - \theta_1)/\theta_1 = 2$.

Example 5.13 Design a two-section Chebychev transformer to match a 100 ohm load to a 50 ohm line. The maximum VSWR allowed is 1.2. Find the corresponding bandwidth.

Solution

Number of section $N = n - 1 = 2$, Number of steps $n = 3$, $Z_L = Z_4 = 100$ ohm, $Z_1 = 50$ ohm

$$|\Gamma_m| = \frac{1.2 - 1}{1.2 + 1} = \frac{0.2}{2.2} = 0.0909$$

Total reflection coefficient for symmetrical transformer (n odd)

$$\Gamma = |\Gamma_m| \, T_{n-1}(x) = 0.0909 \, T_2(x) = 2 \,[\, \Gamma_1 \cos 2\theta] + \Gamma_2$$

or, $\quad 0.0909 \,[\, 2x^2 - 1\,] = 2 \,[\, \Gamma_1 \,(2 \cos^2 \theta - 1)\,] + \Gamma_2$

$$= 2 \, \Gamma_1 \,(2x^2 \cos^2 \theta_1 - 1) + \Gamma_2$$

Equating the coefficient of x^2, x, etc.

$$2 \times 0.091 = 4\, \Gamma_1 \cos^2 \theta_1; \text{ or, } \Gamma_1 = 2 \times 0.091 / 4 \cos^2 \theta_1$$

$$-0.091 = -2\Gamma_1 + \Gamma_2; \text{ or, } \Gamma_2 = 2\,\Gamma_1 - 0.091$$

Now, $(Z_4 - Z_1)/(Z_4 + Z_1) = \Gamma_m\, T_{n-1} (\sec\, \theta_1)$

or, $(100 - 50)/(100 + 50) = 0.091\, T_2 (\sec\, \theta_1)$

or, $50/150 = 0.091\, (2\, \sec^2 \theta_1 - 1)$

or, $2\, \sec^2 \theta_1 - 1 = 1/3 \times 0.091$

$$= 1 / 0.273$$

$$= 3.68$$

$$2\, \sec^2 \theta_1 = 3.68 + 1$$

$$= 4.68$$

$$\sec^2 \theta_1 = 4.68/2 = 2.34$$

$$\cos\, \theta_1 = \sqrt{0.428} = 0.652;\ \theta_1 = \cos^{-1} (0.652) = 49.3°$$

Therefore, $\Gamma_1 = (2 \times 0.091) / (4 \times 0.428) = 0.106$

$$\Gamma_2 = 2\Gamma_1 - 0.091 = 0.212 - 0.091 = 0.121$$

Now, $\Gamma_1 = (Z_2 - Z_1)/ (Z_2 + Z_1),\ \Gamma_2 = (Z_3 - Z_2)/(Z_3 + Z_2)$

Therefore, $Z_2 = (1 + \Gamma_1)/(1 - \Gamma_1) \cdot Z_1$

$$= \frac{1 + 0.106}{1 - 0.106} \times 50$$

$$= \frac{1.106}{0.894} \times 50 \text{ ohm} = 62 \text{ ohm}$$

$$Z_3 = (1 + \Gamma_2)/(1 - \Gamma_2) \cdot Z_2$$

$$= \frac{1 + 0.121}{1 - 0.121} \times 62 = 79 \text{ ohm}$$

The fractional bandwidth $\Delta f/f_0 = 4/\pi\, (\pi/2 - \theta_1)$

$$= 4/\pi\, (\pi/2 - 49.3°)$$

$$= \frac{4}{180} \times \left[\frac{180}{2} - \frac{4 \times 49.3°}{180°} \right]$$

$$= 2 - 1.095 = 0.905$$

5.5 TAPERED TRANSMISSION LINES

The error introduced in neglecting the equivalent shunt reactive elements and higher order reflections at the discrete steps of multisection quarterwave impedance transformers can be minimised by using a continuously tapered

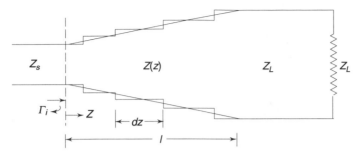

Fig. 5.28 *Tapered transmission line matching section*

transmission line between the two mismatch impedances Z_s and Z_L as shown in Fig. 5.28. Because of continuous tapering of the transverse dimensions of the line in the longitudinal direction z, the impedance of the line is a function of z. Assuming that the taper is gradual, it can be approximated by a large number of elementary sections dz with incremental impedance change dZ and a differential reflection coefficient

$$d\Gamma = \frac{(Z + dZ) - Z}{(Z + dZ) + Z} = \frac{dZ}{2Z + dZ} = \frac{dZ}{2Z} \approx \frac{1}{2} d \, (\ln Z) \qquad (5.72)$$

Since $dZ \ll Z$

The input reflection coefficient for a tapered line of length l can be given by

$$\Gamma_i = 1/2 \int_0^l d \, (\ln Z) \, e^{-j2\beta z} \qquad (5.73)$$

Γ_i can be determined if $Z(z)$ is known, or for a given response characteristic of Γ_i as a function of frequency, $Z(z)$ can be determined.

5.5.1 Exponential Taper Line Section

For the exponential taper, Z varies exponentially and hence $\ln Z$ varies linearly from Z_s to Z_L with z. Thus

$$\ln Z = (z/l) \ln Z_L$$

or,

$$Z = e^{(z/l)} \ln Z_L \qquad (5.74)$$

Therefore, assuming that $\beta = 2\pi/\lambda$, is independent of z,

$$\Gamma_i = (1/2) \int_0^l d(\ln Z) \, e^{-j2\beta z} \qquad (5.75)$$

$$= \frac{\ln Z_L}{2l} \int_0^l e^{-j2\beta z} \cdot dz$$

$$= \frac{\ln Z_L}{2l} \left[(e^{-j2\beta z}/-j2\beta) \right]_0^l$$

$$= \frac{\ln Z_L}{2\beta l} e^{-j\beta l} \cdot \frac{e^{j\beta l} - e^{-j\beta l}}{2j}$$

$$= \frac{\ln Z_L}{2} e^{-j\beta l} \cdot \sin \beta l / \beta l$$

or, $$|\Gamma_i| = \frac{\ln Z_L}{2} \left| \frac{\sin \beta l}{\beta l} \right| \qquad (5.76)$$

Since $\beta \left(= \frac{2\pi l}{c} f \right)$ is proportional to frequency, the variation of $|\Gamma_i|$ with frequency is sin x/x pattern. It is seen that at specific frequencies for which $\beta l = n\pi$ ($n = 1, 2, 3, \ldots$), $\Gamma_i = 0$ and impedance matching becomes perfect for lossless lines. At any frequency when the length of the taper section $l > \lambda/2$, Γ_i in very small and decreases with increase in length l. This shows that a minimum length of $\lambda/2$ is preferred for good matching at lowest frequency of interest (Fig. 5.29).

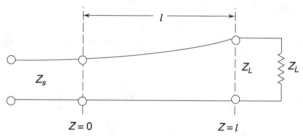

Fig. 5.29 *Exponential taper line*

5.5.2 Waveguide Taper Matching

When two waveguide sections of different shapes are required to be joined together, such as a rectangular waveguide at TE_{10} mode and a circular waveguide at TE_{11} mode, a gradual taper as shown in Fig. 5.30 is required over a distance of at least $2\lambda_g$ for impedance matching between them. This also reduces abrupt discontinuity at the junction. A taper is also used to connect two similar waveguide sections of different dimensions.

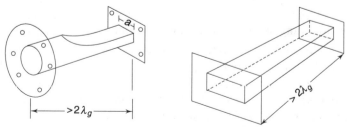

Fig. 5.30 *Waveguide tapers*

The characteristic impedance of a rectangular ($a \times b$) waveguide is defined by

$$Z_0 = \eta_0 \, \pi b \lambda_g / (2a\lambda_0) \qquad (5.77)$$

The impedance match due to two different dimensions can be achieved by using a taper to adjust the values of a and b until the characteristic impedances of the two sections are equal.

Matching a waveguide to free space impedance can also be achieved by flaring the guide in the form of a horn until the factor $\pi b \lambda_g / (2a \lambda_0)$ becomes unity. However, the final dimension will be reasonably large.

EXERCISES

5.1   A 50 ohm lossless line is terminated with a load of $30 - j40$ ohm. A short circuited 50 ohm stub is connected parallel to this line at a distance d from the load. Find d and the length of the stub for matching of the load.

5.2   A lossless line is terminated with a load of VSWR = 3 and the first voltage minimum and maximum are located at distance 50 cm and 250 cm from the load, respectively. Design a single short circuited stub for making VSWR = 1.

5.3   Design a single stub tuner to match a load $Z_L = 110 - j\,70$ ohm to a lossless 50 ohm transmission line.

5.4   The 50 ohm lossless transmission line is terminated with a loading impedance of 75 ohm. Design a single stub matching system using a Smith chart.

5.5   A complex load $Z_L = 100 + j100$ ohm is connected to a 50 ohm lossless line. Design a double stub 50 ohm system for matching. The first stub is placed 0.4λ away from load and the spacing between the stubs is $3\lambda/8$.

5.6   Explain what terminations cannot be matched with a lossless line by the double stub system. Why?

5.7   A lossless 100 ohm transmission line is terminated with an admittance $0.003 - j0.003$ mho. Design a double stub tuner with spacing $\lambda/8$ when the first stub is at the position of the load.

5.8   A 50 ohm lossless transmission line is required to be matched with the load impedance $75 - j60$ ohm by a double stub tuner with separation distance $3\lambda/8$ and the distance of the first stub from the load is $0.01\,\lambda$. Determine the lengths of the matching stubs.

5.9   A load is required to be matched to a 100 ohm line by means of fixed double stub tuner separated by $\lambda/4$. At the position of the first stub the admittance is $0.008 - j0.014$ mho. Find the length of the stubs when the line is excited at 100 MHz. Explain if there is any limitations on the use of double stub $\lambda/4$ apart.

5.10 A lossless 100 ohm coaxial line is terminated by a load of admittance $0.016 - j0.008$ mho. The line is matched by means of two fixed stubs separated by $3\lambda/8$, with first stub placed at the plane of the load. Calculate the length of each stub by using the Smith chart.

5.11 Design a Chebyshev quarterwave transformer for overall impedance ratio 2.5 to have a VSWR < 1.03 over a 20 percent bandwidth.

5.12 A three-section Chebyshev quarterwave transformer has a bandwidth 80 percent and overall impedance ratio 200. Find the VSWR at each step and the maximum passband VSWR.

5.13 Design a six-section Chebychev quarterwave transformer of 40 percent bandwidth for an impedance ratio of 10.

5.14 Design a six-step binomial quarterwave transformer between lines of impedances 200 ohms and 1000 ohms.

5.15 Design a four-section maximally flat transformer for load ratio 3 and a normalised frequency band 2.

5.16 A quarterwave section is used to match a 100 ohm resistive load to a 50 ohm line at a given frequency. (a) Calculate the characteristic impedance of the matching section, (b) The VSWR on the main line with matching transformer when the frequency is increased by 20%.

REFERENCES

1. Collin, RE, "Theory and design of wide band multisection quarterwave transformers", *Proc. IRE,* Vol. 43, pp. 179-185, February, 1955.

2. Cohn, SB, "Optimum design of stepped transmission line transformers", *IRE trans.,* Vol. MTT-3, pp. 16-21, April, 1955.

3. Riblet, HJ, "General synthesis of quarterwave impedance transformers", *IRE trans.,* Vol. MTT-5, pp. 36-43, January,1957.

4. Young, L, "Tables for cascaded homogeneous quarterwave transformers", *IRE trans.,* Vol. MTT-7, April 1959.

5. Collin, RE, "The optimum tapered transmission line matching section", *Proc. IRE,* Vol. 44, pp. 539-548, April, 1956.

6. Johnson, RC, "Design of linear double tapers in rectangular waveguides", *IRE trans.,* Vol. MTT-7, pp. 374-378, July 1959.

7. Collin, RE, "General synthesis of quarterwave impedance transformers," *IRE trans.* PGMIT-7, pp. 233-237 April 1959, and PGMIT-8, pp. 243-244 March 1960.

8. Leo Young, "Inhomogeneous quarterwave," *IRE trans.* PGMIT-8, pp. 478-482, September 1960.

9. Leo Young, "Stepped impedance transformers and filter prototypes," *IRE trans.* PGMIT-10, pp. 339-359 Sept.1962.

10. Leo Young, " Practical design of a wide band quarterwave transformer in waveguide," *The microwave journal,* 6, No. 10, pp. 76-79 (October 1963).

11. Leo Young, "Practical design of a wide band quarterwave transformer in Waveguide," The Microwave Journal, 6, No. 10, pp 76-79, (October 1963).

12. Matthaei, GL, Young and Jones, EMT, *Microwave filters, impedance matching networks and coupling structures,* McGraw-Hill Book Company, 1984.

13. Levy, R, "Directional couplers, advances in microwave", Vol. 1, Academic Press, 1966, p.186.

14. Howe, H, *"Stripline circuit design",* Artech House, 1974.

Chapter 6

MICROWAVE NETWORK THEORY AND PASSIVE DEVICES

6.1 INTRODUCTION

A microwave network is formed when several microwave devices and components such as sources, attenuators, resonators, filters, amplifiers, etc., are coupled together by transmission lines or waveguides for the desired transmission of a microwave signal. The point of interconnection of two or more devices is called a *junction.*

For a low frequency network, a port is a pair of terminals whereas for a microwave network, a port is a reference plane transverse to the length of the microwave transmission line or waveguide. At low frequencies the physical length of the network is much smaller than the wavelength of the signal transmitted. Therefore, the measurable input and output variables are voltage and current which can be related in terms of the impedance Z-parameters, or admittance *Y*-parameters, or hybrid *h*-parameters, or *ABCD* parameters. For a two port network as shown schematically in Fig. 6.1, these relationships are given by

$$\begin{bmatrix} V_1 \\ V_2 \end{bmatrix} = \begin{bmatrix} Z_{11} & Z_{12} \\ Z_{21} & Z_{22} \end{bmatrix} \begin{bmatrix} I_1 \\ I_2 \end{bmatrix} \tag{6.1}$$

$$\begin{bmatrix} I_1 \\ I_2 \end{bmatrix} = \begin{bmatrix} Y_{11} & Y_{12} \\ Y_{21} & Y_{22} \end{bmatrix} \begin{bmatrix} V_1 \\ V_2 \end{bmatrix} \tag{6.2}$$

$$\begin{bmatrix} V_1 \\ I_2 \end{bmatrix} = \begin{bmatrix} h_{11} & h_{12} \\ h_{21} & h_{22} \end{bmatrix} \begin{bmatrix} I_1 \\ V_2 \end{bmatrix} \tag{6.3}$$

$$\begin{bmatrix} V_1 \\ I_1 \end{bmatrix} = \begin{bmatrix} A & -B \\ C & -D \end{bmatrix} \begin{bmatrix} V_2 \\ I_2 \end{bmatrix} \tag{6.4}$$

where Z_{ij}, Y_{ij}, and A, B, C and D are suitable constants that characterise the junction. A, B, C and D parameters are convenient to represent each junction when a number of circuits are connected together in cascade. Here the resultant matrix, which describes the complete cascade connection, can be obtained by multiplying the matrices describing each junction:

$$\begin{bmatrix} A & B \\ C & D \end{bmatrix} = \begin{bmatrix} A_1 & B_1 \\ C_1 & D_1 \end{bmatrix} \begin{bmatrix} A_2 & B_2 \\ C_2 & D_2 \end{bmatrix} \cdots \begin{bmatrix} A_n & B_n \\ C_n & D_n \end{bmatrix} \tag{6.4a}$$

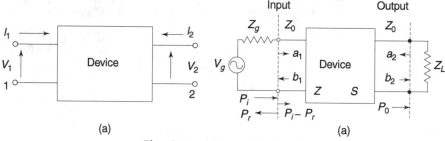

Fig. 6.1 *A two-port network*

These parameters can be measured under short or open circuit condition for use in the analysis of the circuit.

At microwave frequencies the physical length of the component or line is comparable to or much larger than the wavelength. Thus the voltage and current are not well-defined at a given point for a microwave circuit, such as a waveguide system. Furthermore, measurement of Z, Y, h and $ABCD$ parameters is difficult at microwave frequencies due to following reasons.

1. Non-availability of terminal voltage and current measuring equipment.
2. Short circuit and especially open circuit are not easily achieved for a wide range of frequencies.
3. Presence of active devices makes the circuit unstable for short or open circuit.

Therefore, microwave circuits are analysed using scattering or S-parameters which linearly relate the reflected waves' amplitude with those of incident waves. However, many of the circuit analysis techniques and circuit properties that are valid at low frequencies are also valid for microwave circuits. Thus, for circuit analysis S-parameters can be related to the Z or Y or $ABCD$ parameters. The properties of the parameters are described in the following sections.

6.2 SYMMETRICAL Z AND Y MATRICES FOR RECIPROCAL NETWORK

In a reciprocal network, the impedance and the admittance matrices are symmetrical and the junction media are characterised by scalar electrical parameters μ and ε. For a multiport (N ports) network, let the incident wave amplitudes V_n^+ be so chosen that the total voltage $V_n = V_n^+ + V_n^- = 0$ at all ports $n = 1, 2, \ldots, N$, except the ith port where the fields are $\mathbf{E}_i$, $\mathbf{H}_i$. Similarly, let $V_n = 0$ at all ports except jth one where the fields are $\mathbf{E}_j$, $\mathbf{H}_j$. Then from the Lorentz reciprocity theorem

$$\int_S (\mathbf{E}_i \times \mathbf{H}_j - \mathbf{E}_j \times \mathbf{H}_i) \cdot dS = 0 \tag{6.5}$$

where S is the closed surface area of the conducting walls enclosing the junction and N ports in the absence of any source. Since the integral over the perfectly conducting walls vanishes, the only non-zero integrals are those taken over the reference planes of the corresponding ports, so that

$$\sum_{n=1}^{N} \int_{t_n} (\mathbf{E}_i \times \mathbf{H}_j - \mathbf{E}_j \times \mathbf{H}_i) \cdot dS = 0 \tag{6.6}$$

Since all V_n except V_i and V_j are zero, $\mathbf{E}_{ti} = \mathbf{n} \times \mathbf{E}_i$ and $\mathbf{E}_{tj} = \mathbf{n} \times \mathbf{E}_j$ are zero on all reference planes at the corresponding ports except t_i and t_j, respectively. Therefore, Eq. 6.6 reduces to

$$\int_{t_i} (\mathbf{E}_i \times \mathbf{H}_j) \cdot dS = \int_{t_j} (\mathbf{E}_j \times \mathbf{H}_i) \cdot dS \tag{6.7}$$

or,
$$P_{ij} = P_{ji} \tag{6.8}$$

where P_{ij} represents the power at reference plane i due to an input voltage at plane j.

From the admittance matrix representation $[I] = [Y][V]$ and power relation $P = VI$, Eq. 6.8 reduces to

$$V_i V_j Y_{ij} = V_j V_i Y_{ji}$$

or,
$$Y_{ij} = Y_{ji} \tag{6.9}$$

and
$$Z_{ij} = Z_{ji} \tag{6.10}$$

This proves that the impedance and admittance matrices are symmetrical for a reciprocal junction.

6.3 SCATTERING OR S-MATRIX REPRESENTATION OF MULTIPORT NETWORK

As discussed in Sec. 6.1 the incident and reflected amplitudes of microwaves at any port are used to characterise a microwave circuit. The amplitudes are normalised in such a way that the square of any of these variables gives the average power in that wave in the following manner

Input power at the nth port, $\qquad P_{in} = 1/2 \, |a_n|^2 \qquad (6.11)$

Reflected power at the nth port, $\qquad P_{rn} = 1/2 \, |b_n|^2 \qquad (6.12)$

where a_n and b_n represent the normalised incident wave amplitude and normalised reflected wave amplitude at the nth port.

In a two-port network we can express the normalised waves by.

$$a_1 = \frac{V_{i1}}{\sqrt{Z_0}} = \frac{V_1 - V_{r1}}{\sqrt{Z_0}}, \ u_2 = \frac{V_{i2}}{\sqrt{Z_0}} = \frac{V_2 - V_{r2}}{\sqrt{Z_0}} \qquad (6.13)$$

$$b_1 = \frac{V_{r1}}{\sqrt{Z_0}} = \frac{V_1 - V_{i1}}{\sqrt{Z_0}}, \ b_2 = \frac{V_{r2}}{\sqrt{Z_0}} = \frac{V_2 - V_{i2}}{\sqrt{Z_0}} \qquad (6.14)$$

where a's represents normalised incident wave and b's represent normalised reflected wave at the corresponding ports. Here the total voltage wave is the sum of incident and reflected voltage waves V_i and V_r, respectively

$$V_1 = V_{i1} + V_{r1} \qquad (6.15)$$
$$V_2 = V_{i2} + V_{r2} \qquad (6.16)$$

The numeric suffices represent the port number. The total or net power flow into any port is given by

$$P = P_i - P_r = 1/2 \ (|a|^2 - |b|^2) \qquad (6.17)$$

Therefore, in this normalisation process, the characteristic impedance is normalised to unity. For a two-port network (Fig. 6.1) the relation between incident and reflected waves are expressed in terms of scattering paramters S_{ij}'s

$$b_1 = S_{11} \ a_1 + S_{12} \ a_2 \qquad (6.18)$$
$$b_2 = S_{21} \ a_1 + S_{22} \ a_2 \qquad (6.19)$$

The normalisation process leads to a symmetrical scattering matrix for reciprocal structures. The physical significance of S-parameters can be described as follows:

$S_{11} = b_1/a_1 \ | \ a_2 = 0 \ = $ reflection coefficient Γ_1 at port 1 when port 2 in terminated with a matched load ($a_2 = 0$)

$S_{22} = b_2/a_2 \ | \ a_1 = 0 \ = $ reflection coefficient Γ_2 at port 2 when port 1 in terminated with a matched load ($a_1 = 0$)

$S_{12} = b_1/a_2 \ | \ a_1 = 0 \ = $ attenuation of wave travelling from port 2 to port 1

$S_{21} = b_2/a_1 \ | \ a_2 = 0 \ = $ attenuation of wave travelling from port 1 to port 2.

In general, since the incident and reflected waves have both amplitude and phase, the S-parameters are complex numbers.

For a multiport (N) networks or components, the S-parameters equations are expressed by

$$\begin{bmatrix} b_1 \\ b_2 \\ \vdots \\ \vdots \\ b_N \end{bmatrix} = \begin{bmatrix} S_{11} & S_{12} & \cdots & S_{1N} \\ S_{21} & S_{22} & \cdots & S_{2N} \\ \vdots & & & \vdots \\ \vdots & & & \vdots \\ S_{N1} & S_{N2} & \cdots & S_{NN} \end{bmatrix} \begin{bmatrix} a_1 \\ a_2 \\ \vdots \\ \vdots \\ a_N \end{bmatrix} \qquad (6.20)$$

In microwave devices or circuits it is important to express several losses in terms of S-parameters when the ports are matched terminated. In a two port network if power fed at port 1 is P_i, power reflected at the same port is P_r and the output power at port 2 is P_o, then following losses are defined in terms of S-parameters

$$\text{Insertion loss (dB)} = 10 \log \frac{P_i}{P_o} = 10 \log \frac{|a_1|^2}{|b_2|^2}$$

$$= 20 \log \frac{1}{|S_{21}|}$$

$$= 20 \log \frac{1}{|S_{12}|} \tag{6.21}$$

$$\text{Transmission loss or attenuation (dB)} = 10 \log \frac{P_i - P_r}{P_o}$$

$$= 10 \log \frac{1 - |S_{11}|^2}{|S_{12}|^2} \tag{6.22}$$

$$\text{Reflection loss (dB)} = 10 \log \frac{P_i}{P_i - P_r}$$

$$= 10 \log \frac{1}{1 - |S_{11}|^2} \tag{6.23}$$

$$\text{Return loss (dB)} = 10 \log P_i / P_r$$

$$= 20 \log \frac{1}{|\Gamma|}$$

$$= 20 \log \frac{1}{|S_{11}|} \tag{6.24}$$

6.3.1 Properties of S-Parameters

In general the scattering parameters are complex quantities having the following properties for different characteristics of the microwave network.

(a) Zero diagonal elements for perfect matched network

For an ideal N-port network with matched termination, $S_{ii} = 0$, since there is no reflection from any port. Therefore, under perfect matched conditions the diagonal elements of $[S]$ are zero.

(b) Symmetry of $[S]$ for a reciprocal network

A reciprocal device has the same transmission characteristics in either direction of a pair of ports and is characterised by a symmetric scattering matrix,

$$S_{ij} = S_{ji} \ (i \neq j) \tag{6.25}$$

which results

$$[S]_t = [S] \tag{6.26}$$

This condition can be proved in the following manner. For a reciprocal network with the assumed normalisation, the impedance matrix equation is

$$[V] = [Z][I] = [Z]([a] - [b]) = [a] + [b]$$

or, $$([Z] + [U])[b] = ([Z] - [U])[a]$$

or, $$[b] = ([Z] + [U])^{-1}([Z] - [U])[a] \tag{6.27}$$

where $[U]$ is the unit matrix. The S-matrix equation for the network is

$$[b] = [S][a] \tag{6.28}$$

Comparing Eqs 6.27 and 6.28, we have

$$[S] = ([Z] + [U])^{-1}([Z] - [U]) \tag{6.29}$$

Let $$[R] = [Z] - [U], [Q] = [Z] + [U] \tag{6.30}$$

For reciprocal network, the Z-matrix is symmetric. Hence

$$[R][Q] = [Q][R]$$

or, $$[Q]^{-1}[R][Q][Q]^{-1} = [Q]^{-1}[Q][R][Q]^{-1}$$

or, $$[Q]^{-1}[R] = S = [R][Q]^{-1} \tag{6.31}$$

Now the transpose of $[S]$ is

$$[S]_t = ([Z] - [U])_t ([Z] + [U])_t^{-1} \tag{6.32}$$

Since the Z-matrix is symmetrical

$$([Z] - [U])_t = [Z] - [U] \tag{6.33}$$

$$([Z] + [U])_t = [Z] + [U] \tag{6.34}$$

Therefore,

$$[S]_t = ([Z] - [U])([Z] + [U])^{-1}$$
$$= [R][Q]^{-1} = [S] \tag{6.35}$$

Thus it is proved that $[S]_t = [S]$, for a symmetrical junction.

(c) Unitary property for a lossless junction

For any lossless network the sum of the products of each term of any one row or of any column of the S-matrix multiplied by its complex conjugate is unity.

For a lossless n-port device, the total power leaving N-ports must be equal to the total power input to these ports, so that

$$\sum_{n=1}^{N} |b_n|^2 = \sum_{n=1}^{N} |a_n|^2$$

or,
$$\sum_{n=1}^{N} \left| \sum_{i=1}^{n} S_{ni}\, a_i \right|^2 = \sum_{n=1}^{N} |a_n|^2 \tag{6.36}$$

If only ith port is excited and all other ports are matched terminated, all $a_n = 0$, except a_i, so that,

$$\sum_{n=1}^{N} |S_{ni}\, a_i|^2 = \sum_{n=1}^{N} |a_i|^2 \tag{6.37}$$

$$\sum_{n=1}^{N} |S_{ni}|^2 = 1 = \sum_{n=1}^{N} S_{ni}\, S_{ni}^{*} \tag{6.38}$$

Therefore, for a lossless junction

$$\sum_{n=1}^{N} S_{ni} \cdot S_{ni}^{*} = 1 \tag{6.39}$$

If all $a_n = 0$, except a_i and a_k,

$$\sum_{n=1}^{N} S_{nk} \cdot S_{ni}^{*} = 0 \; ; \; i \neq k \tag{6.40}$$

In matrix notation, these relations can be expressed as

$$[S^{*}]\,[S]_t = [U]$$

or,
$$[S^{*}] = [S]_t^{-1} \tag{6.41}$$

Here $[U]$ is the identity matrix or unit matrix. A matrix $[S]$ for lossless network which satisfies the above three conditions 6.39 – 6.41 is called a *unitary matrix*.

(d) Phase shift property

Complex S-parameters of a network are defined with respect to the positions of the port or reference planes. For a two-port network with unprimed reference planes 1 and 2 as shown in Fig. 6.2, the S-parameters have definite complex values

$$[S] = \begin{bmatrix} S_{11} & S_{12} \\ S_{21} & S_{22} \end{bmatrix} \tag{6.42}$$

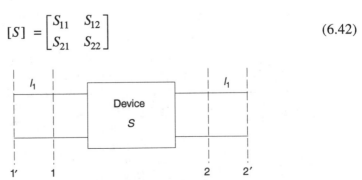

Fig. 6.2 *Phase shift property of S*

If the reference planes 1 and 2 are shifted outward to 1' and 2' by electrical phase shifts $\phi_1 = \beta_1 \, l_1$ and $\phi_2 = \beta_2 \, l_2$, respectively, then the new wave variables are $a_1 e^{j\phi_1}$, $b_1 e^{-j\phi_1}$, $a_2 e^{j\phi_2}$, $b_2 e^{-j\phi_2}$. The new S-matrix S' is then given by

$$[S'] = \begin{bmatrix} e^{-j\phi_1} & 0 \\ 0 & e^{-j\phi_2} \end{bmatrix} [S] \begin{bmatrix} e^{-j\phi_1} & 0 \\ 0 & e^{-j\phi_2} \end{bmatrix} \tag{6.43}$$

This property is valid for any number of ports and is called the *phase shift property* applicable to a shift of reference planes.

6.3.2 S-parameters of a Two-port Network with Mismatched Load

A two-port network or junction is formed when there is a discontinuity between the input and output ports of a transmission line. Many configurations of such junctions practically exist some of which are shown in Fig. 6.3.

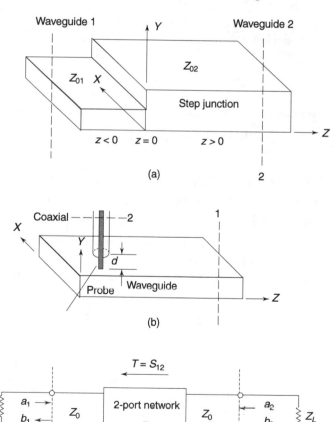

Fig. 6.3 *Two-port junctions (a) waveguide step junction (b) coaxial to waveguide transition (c) a two-port network*

During propagation of microwaves through the junction from one port to other, evanescent modes are excited at each discontinuity which contain reactive energy. Evanescent Modes decay very fast away from the junction and become negligible after a distance of the order of one wavelength. The terminal reference planes 1 and 2 are chosen beyond this distance so that the equivalent voltage and currents at these positions are proportional to the total transverse electric and magnetic fields, respectively, for the propagating mode only. These circuits are analysed using S-matrix formulation.

Consider a two-port network terminated by normalised load and generator impedances $\dfrac{Z_L}{Z_o}$ and $\dfrac{Z_g}{Z_o} = 1$. Then the load reflection coefficient

$$\Gamma_2 = \frac{a_2}{b_2} = \frac{Z_L/Z_o - 1}{Z_L/Z_o + 1} \tag{6.44}$$

Now,
$$b_1 = S_{11}\,a_1 + S_{12}\,a_2 = S_{11}\,a_1 + S_{12}\,b_2\,\Gamma_2 \tag{6.45}$$
$$b_2 = S_{21}\,a_1 + S_{22}\,a_2 = S_{21}\,a_1 + S_{22}\,b_2\,\Gamma_2 \tag{6.46}$$

Solving for the input reflection coefficient

$$\Gamma_1 = b_1/a_1 = S_{11} + \frac{S_{12}\,S_{21}\,\Gamma_2}{1 - S_{22}\,\Gamma_2} \tag{6.47}$$

Therefore, for a mismatch load, input reflection coefficient $\Gamma_1 \neq S_{11}$. For a reciprocal network, $S_{12} = S_{21}$ so that

$$\Gamma_1 = S_{11} + \frac{S_{12}^2\,\Gamma_2}{1 - S_{22}\,\Gamma_2} \tag{6.48}$$

Further, if the junction is lossless, from Eqs 6.39 and 6.40

$$S_{11}\,S_{11}^* + S_{12}\,S_{12}^* = 1 \tag{6.49}$$
$$S_{22}\,S_{22}^* + S_{12}\,S_{12}^* = 1 \tag{6.50}$$
$$S_{11}\,S_{12}^* + S_{12}\,S_{22}^* = 0 \tag{6.51}$$

Therefore, for a lossless, reciprocal two-port network, terminated by a mismatch load, Eqs 6.49 and 6.50 give

$$|S_{11}| = |S_{22}| \tag{6.52}$$

$$|S_{12}| = \sqrt{\left(1 - |S_{11}|^2\right)} \tag{6.53}$$

and the input reflection coefficient

$$\Gamma_1 = S_{11} + \frac{S_{12}^2\,\Gamma_2}{1 - S_{22}\,\Gamma_2} \tag{6.54}$$

The last equation is the working equation for the computation of the S-parameters. By measuring Γ_1 for known values of Γ_2 (0, –1 and 1) a set of simultaneous equations are obtained which will give the S-parameters of a reciprocal junction.

6.3.3 Comparison between [S], [Z] and [Y] Matrices

From the analysis of S, Z and Y parameters it is known that $[S]$, $[Z]$ and $[Y]$ represent unique intrinsic properties of the device, supplying all the circuit characteristics of the device at a given frequency. Therefore, terminal conditions may be changed for convenience without affecting the matrices themselves. The $[S]$ can be expressed in terms of $[Z]$ and $[Y]$ as given below.

$$[S] = ([Z]/Z_o - [U]) \, ([Z]/Z_o + [U])^{-1}$$
$$= ([U] - [Y]/Y_o) \, ([U] + [Y]/Y_o)^{-1} \qquad (6.55)$$

Thus the following properties are common for $[S]$, $[Z]$ and $[Y]$

(1) Number of elements are equal.
(2) For reciprocal devices both $[Z]$ and $[S]$ satisfy reciprocity properties, $Z_{ij} = Z_{ji}$, $S_{ij} = S_{ji}$.
(3) If $[Z]$ is symmetrical, $[S]$ is also symmetrical.
(4) The following are the advantages of $[S]$ over $[Z]$ or $[Y]$.
  (a) In microwave techniques the source remains ideally constant in power, regardless of circuit changes. Besides frequency measurements, the only other possible measurement parameters are VSWR, power and phase. These are essentially measurements of b/a, $|a|^2$ and $|b|^2$. Such a direct correspondence is not possible with $[Z]$ or $[Y]$ representations.
  (b) The unitary property of $[S]$ helps a quick check of the power balance of lossless structures. No such immediate check is possible with $[Z]$ or $[Y]$.
  (c) $[S]$ is defined for a given set of reference planes only. If the reference planes are changed, the S-coefficients vary only in phase. This is not the case in $[Z]$ or $[Y]$, because voltage and current are functions of complex impedance and therefore both magnitude and phase change in $[Z]$ and $[Y]$.

Example 6.1 Two transmission lines of characteristic impedance Z_1 and Z_2 are joined at plane pp'. Express S-parameters in terms of impedances.

Solution The incident and scattered wave amplitude are related by $[b] = [S] [a]$.

(i) Assuming that the output line to be matched ($a_2 = 0$), the input impedance Z_{in} at the junction $= Z_2 = $ load of line Z_1.

Therefore,

$$S_{11} = \frac{Z_2 - Z_1}{Z_2 + Z_1} = \text{reflection coefficient on the input side.}$$

(ii) Similarly, for symmetry, assuming input side is matched ($a_1 = 0$).

$$S_{22} = \frac{Z_1 - Z_2}{Z_1 + Z_2} = -S_{11}$$

(iii) In general

$$b_1 = S_{11} a_1 + S_{12} a_2$$
$$b_2 = S_{21} a_1 + S_{22} a_2$$

With output line matched ($a_2 = 0$) for lossless line, Z_{in} at the junction $= Z_2$, a pure shunt element.

Therefore,

$$b_2 = a_1 + b_1 = a_1 + S_{11} a_1 = a_1 (1 + S_{11})$$

or, $$b_2/a_1 = 1 + S_{11}$$

or, $$S_{21} = 1 + S_{11}$$

$$= 1 + \frac{Z_2 - Z_1}{Z_2 + Z_1}$$

$$= \frac{2Z_2}{Z_2 + Z_1}$$

(iv) With input line matched ($a_1 = 0$),

$$b_1 = a_2 + b_2 = a_2 + S_{22} a_2 = a_2 (1 + S_{22})$$

or, $$b_1/a_2 = 1 + S_{22}$$

or, $$S_{12} = 1 + S_{22}$$

$$= 1 + \frac{Z_1 - Z_2}{Z_1 + Z_2}$$

$$= \frac{2Z_1}{Z_1 + Z_2}$$

Therefore,

$$[S] = \begin{bmatrix} \dfrac{Z_2 - Z_1}{Z_2 + Z_1} & \dfrac{2Z_1}{Z_1 + Z_2} \\ \dfrac{2Z_2}{Z_1 + Z_2} & \dfrac{Z_1 - Z_2}{Z_1 + Z_2} \end{bmatrix}$$

6.3.4 Relations of Z, Y and ABCD Parameters with S-Parameters

Rearranging Eq. 6.1 and comparing with Eq. 6.4 gives

$$A = Z_{11}/Z_{21}, B = - (Z_{11}Z_{22} - Z_{12}Z_{21})/Z_{21}$$
$$C = 1/Z_{21}, D = -Z_{22}/Z_{21}$$

Similarly from Eqs 6.42 and 6.4

$$Y_{11} = D/B, Y_{22} = - A/B, Y_{12} = 1/B, Y_{21} = C + AD/2$$

When the characteristic impedances of input and output lines at the junction are same

$$S_{11} = \frac{A - B - C + D}{A - B + C - D} \qquad (6.56)$$

$$S_{22} = \frac{-A - B - C - D}{A - B + C - D} \qquad (6.57)$$

$$S_{12} = \frac{-2(AD + BC)}{A - B + C - D} \qquad (6.58)$$

$$S_{21} = \frac{2}{A - B + C - D} \qquad (6.59)$$

6.4 MICROWAVE PASSIVE DEVICES

Microwave passive devices and components are designed using sections of coaxial line, waveguides, strip lines and microstrip lines for use in both, laboratory and in microwave communication and radar systems. These components can be considered as one-port or multiport networks characterised by the basic parameters, like the VSWR, reflection coefficient, and various losses defined by Eqs 6.20–6.23, under output matched conditions. In this section, the basic operating principles for a number of most commonly used devices such as line sections, connectors, terminators, attenuators, phase shifters, directional couplers, power dividers, T-junctions, hybrides, etc. are described.

6.4.1 Coaxial Cables

A length of coaxial cable is used for interconnecting several microwave components. The theory of coaxial lines was described in Chapter 4. In this section some practical aspects of these lines are described. The outer conductor of the coaxial line is used to guide the signal through *TEM* mode and shields the external or internal signal leakage through it. The standard characteristic impedance of these cables are 50 ohms and 75 ohms. There are three basic types of coaxial cables with increasing order of shielding, i.e., flexible, semi-rigid and rigid. Flexible coaxial cables use low loss solid or foam polyethylene dielectrics. The outer single braid or double braid of the flexible cable is constructed for electromagnetic shielding by using knitted metal wire mesh as shown in Fig. 6.4. Rigid cables have air dielectric and conductors are supported by small dielectric spacers such that they do not produce significant discontinuities to the signal flow. Semi-rigid cables have solid dielectric and use thin copper outer conductor so that it can be bent for convenient routing. Coaxial cables are used in the frequency range from dc to microwaves. Since the attenuation in these cables increases with frequency, the upper frequency of operation is limited.

The shielding effectiveness of outer conductor is expressed in terms of transfer impedance Z_T of the cable, which is defined as

$$Z_T = \frac{\text{Longitudinal voltage } V_i \text{ induced per unit length on one side of the shield (outside)}}{\text{The leakage current } I_s \text{ flowing on the other side (inside) of the shield}}$$

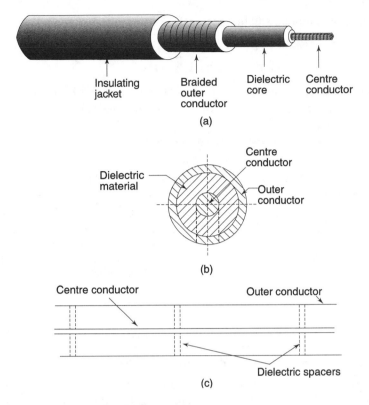

Fig. 6.4 *Coaxial cable (a) Flexible (b) Semi-rigid (c) Rigid*

$$= \frac{V_i}{I_s} \tag{6.60}$$

If both ends of a short cable are terminated into matched loads, as shown in Fig. 6.5, the load voltage is

$$V_o = V_i/2 = I_s Z_T/2 \tag{6.61}$$

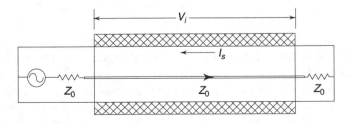

Fig. 6.5 *Transfer impedance of a coaxial cable*

Therefore,

$$Z_T = 2 V_o/I_s \tag{6.62}$$

For a thin-walled tubular shield, the transfer impedance is given by

$$Z_T = \frac{1}{2\pi b \sigma t} \frac{(1+j)\,t/\delta}{\sinh\,[(1+j)\,t/\delta]}\,;\,t \ll b \ll \delta \qquad (6.63)$$

where b is the inner radius of the shield, t is the shield wall thickness, σ is the conductivity of the shield, δ is the skin depth in the shield. For a better shielding Z_T should be smaller.

Some standard coaxial cables are given in Table 6.1 with their *radio guide* (RG) and *universal* (U) numbers and specifications.

Table 6.1 Some standard coaxial cables

Number RG-	Z_o Ohm	Inner and outer conductor
9/U	50	Silver-plated copper
11/U	75	Tinned copper
58A/U	50	Tinned copper
59/U	75	Copper
141A/U	50	Silver-plated copper
179B/U	75	Silver-plated copper
214/U	50	Silver-plated copper

6.4.2 Coaxial Connectors and Adapters

Coaxial cables are terminated or connected to other cables and components by means of shielded standard connectors. The outer shield makes a 360 degree extremely low impedance joint to maintain shielding integrity. These connectors are of various types depending on the frequency range and the cable diameter. Commonly used microwave connectors are type N (male/female), BNC (male/female), TNC (male/female), APC (sexless), etc. Adapters, having different connectors at the two ends, are also made for interconnection between two different ports in a microwave system. The basic schematic diagrams of these connectors and adapters are shown in Fig. 6.6. The type N (Navy) connector is 50 and 75 ohms connector which was designed for military system applications during World War II. This is suitable for flexible or rigid cables in the frequency range of 1–18 GHz. The BNC (Bayonet Navy Connector) is suitable for 0.25 inch 50 ohm or 75 ohm flexible cables used up to 1 GHz. The TNC (Threaded Navy Connector) is like BNC, except that, the outer conductor has thread to make firm contact in the mating surface to minimise radiation leakage at higher frequencies. These connectors are used up to 12 GHz.

The SMA (Sub-Miniature A) connectors are used for thin flexible or semi-rigid cables. The higher frequency is limited to 24 GHz because of generation of higher order modes beyond this limit. All the above connectors can be of male or female configurations except the APC-7(Amphenol Precision Connector-7 mm) which provides coupling without male or female configurations. The APC-7 is a very accurate 50 ohm, low VSWR connector which can operate up to 18 GHz.

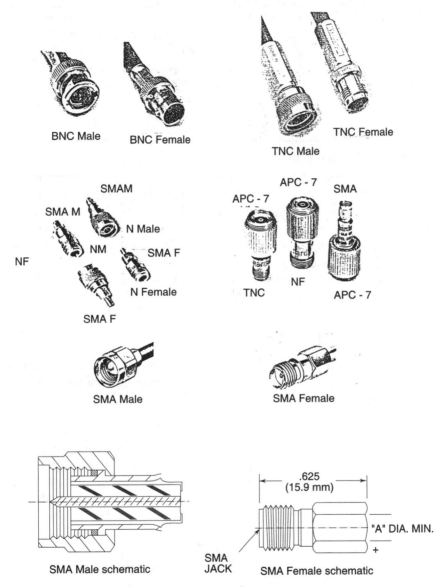

Fig. 6.6   *Coaxial connectors and adapters*

Another APC-3.5 connector is a high precision 50 ohm, low VSWR connector which can be either the male or female and can operate up to 34 GHz. It can mate with the oppositely sexed SM connector. Table 6.2 shows the type, dielectric in mating space and impedance of some of the above standard connectors.

Table 6.2 *Mating connectors*

Type	Sex	Dielectric in mating space	Impedance (ohm)
N	M/F	Air	50 / 75
BNC	M/F	Solid	50 / 75
TNC	M/F	Solid	50 / 75
SMA	M/F	Solid	50
APC-7	Sexless	Air	50
APC-3.5	Sexed	Air	50

6.4.3 Waveguide Sections

Waveguide sections are used for low-loss transmission, change of direction and polarisation of the signal. Sections may be straight with rectangular or circular cross-sections, rectangular twisted section in either **E** or **H** planes, rectangular bends, circular flexible section, rectangular or circular tapered section, and transition from circular to rectangular cross-sections. Some of them are as shown in Fig. 6.7. The standard waveguides are made of brass, bronze or aluminium. For higher frequency bands, the surface is finished with silver coatings to minimise ohmic loss. The S-matrix for ideal sections are

$$[S] = \begin{bmatrix} 0 & 1 \\ 1 & 0 \end{bmatrix} \tag{6.64}$$

The general theory of waveguides is given in Chapter 4. This section describes some practical aspects of rectangular and circular waveguides for the construction of waveguide components.

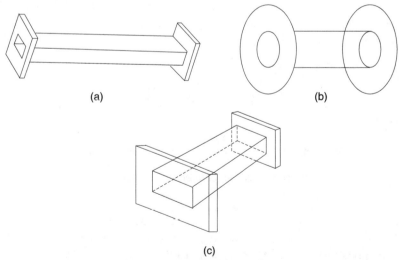

(a)　　　　　　　(b)

(c)

Fig. 6.7 *Waveguide sections.*
(a) *Rectangular*
(b) *Circular*
(c) *Rectangular taper*

6.4.3.1 Rectangular waveguide section

It has been shown that attenuation due to finite conductivity of waveguide walls decreases with increase of frequency at $f > f_c$ to a minimum and increases again for a given b/a ratio. The attenuation also increases when b/a reduces. The design requirements of rectangular waveguide for the dominant mode are

1. Energy transmission through a single mode.
2. The breakdown power should be higher than the transmitted power.
3. Low attenuation constant.

For a waveguide of dimension $b/a < 0.5$, the successive higher order modes are TE_{10}, TE_{20}, TE_{01}, etc. Therefore, the dimensions of the cross-section which ensure only the dominant mode propagation are determined as follows.

$$\lambda_{max} < (\lambda_c) \; TE_{10} = 2a$$

$$\lambda_{min} > (\lambda_c) \; TE_{20} = a$$

$$\lambda_{min} > (\lambda_c) \; TE_{01} = 2b \tag{6.65}$$

Hence, the dimensions for dominant mode rectangular waveguides are to be selected according to

$$0 < b < \lambda/2, \; \lambda/2 < a < \lambda \tag{6.66}$$

To provide adequate power handling capability, usually the dimensions are chosen as $b = 0.3\lambda$ to 0.4λ, $a = 0.7\lambda$ to 0.8λ.

6.4.3.2 Circular waveguide sections

Circular waveguide sections are used in some waveguide components where circularly polarised waves are involved. Both TE and TM modes can be excited depending on the requirements. The dominant mode in a circular waveguide is the TE_{11} mode. Of the multitude of possible modes in circular waveguides, the modes of highest practical interest are TE_{11}, TE_{01} and TM_{01}. The field configurations of these modes are shown in Fig. 6.8.

The electric currents on the waveguide walls for the TM_{01} mode are purely longitudinal, whereas for the TE_{01} mode it flows along closed circular paths and does not have longitudinal components.

The attenuation in the circular waveguide for TE_{01} mode is very low and decreases with frequency as $f^{-3/2}$ according to Eq. 4.111.

Therefore, although TE_{01} mode is not dominant, it is used for long telecommunication lines and also for resonant cavities displaying very high Q wavemeters. Figure 6.9 shows the distribution of the cut-off wavelengths of a circular waveguide for few modes.

Selection of the guide radius is made for propagation of a single dominant mode such that

$$a \le \frac{1.841}{2\pi f_{c(Hz)} \sqrt{\mu \varepsilon}} \tag{6.67}$$

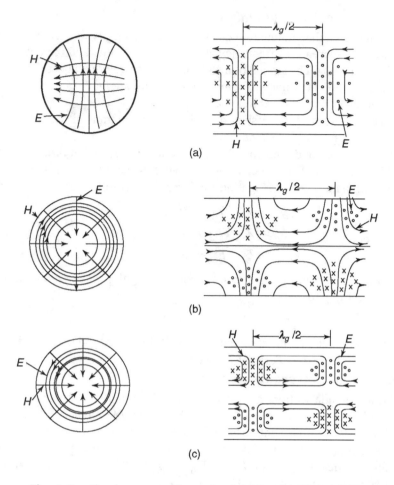

Fig. 6.8 *Circular waveguide modes (a) TE_{11} (b) TM_{01} (c) TE_{01}*

For other higher modes the dimensions are selected based on the corresponding root of Bessel functions as described in Chapter 4 and by an appropriate excitation method.

6.4.4 Waveguide Flanges

Waveguide sections are fitted with flanges as shown in Fig. 6.10, which are connected using nut-bolt arrangement with other section or waveguide devices. In order to avoid leakage of the high power signal between two flange joints, choke flanges are sometimes used which consist of a circular groove of depth d in the flange such that the depth d plus the distance to the waveguide joint on wide wall is half the wavelength. This arrangement makes an ideally short circuited half wavelength line providing zero resistance contact between the inner walls of the two waveguide sections at joint. Inadequate contact at the joint results in power loss, due to leakage, Joule heating, partial reflection of waves and even burn out of the contact for high power applications.

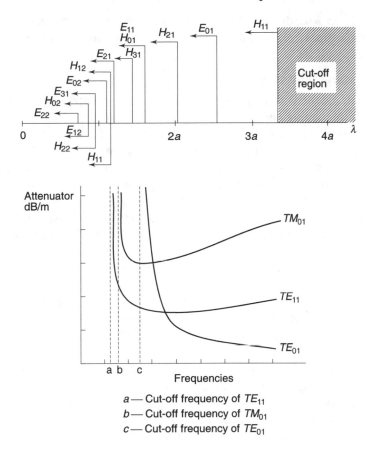

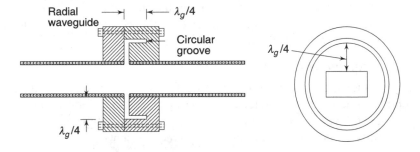

a — Cut-off frequency of TE_{11}
b — Cut-off frequency of TM_{01}
c — Cut-off frequency of TE_{01}

Fig. 6.9 *Distribution of cut-off wavelength in a circular waveguide
and attenuation characteristics*

Fig. 6.10   *Choke flange*

At the joint, the first waveguide is equipped with a choke flange and the second
one with a plane flange such that the choke provides a low characteristic impedance line. The choke flange is frequency selective. Frequency sensitivity is mini-
miscd by making the low characteristic impedance line half wavelength. At the

edge of a 10–15% bandwidth, VSWR of the order of 1.02 to 1.05 can be achieved. Choke connections are used in microwave oscillator and amplifier tubes and coaxial rotary joints.

6.4.5 Rotary Joints

Rotary joints are used to change the direction of microwave propagation between two guides by rotating one with respect to other. Both coaxial line and circular waveguide versions are used. In radar applications a fixed waveguide is connected to a horn antenna by a rotary joint for feeding a paraboloid reflector which rotates for tracking. The most important design considerations of rotary joints are low reflection, negligible power leakage through, and mechanical strength at the joint. Figure 6.11 shows coaxial line and waveguide rotary joints.

The sliding joint is designed by incorporating a half wavelength short-circuited lines or choke in such a manner that the actual point of sliding contact is at a current minimum with zero impedance.

In the waveguide rotary joint, input and output rectangular waveguides operated in the dominant TE_{10} mode are connected through a section of circular waveguide operated in ϕ-symmetric TM_{01} mode. The diameter of the circular waveguide is selected in such a way that modes higher than TM_{01} cannot propagate. The lowest dominant mode TE_{11} is suppressed by a ring filter. Moreover, the diameter of the circular waveguide and the end length h are chosen such that h is an odd multiple of the quarter guide wavelength for the TE_{11} mode, but an even multiple of the quarter guide wavelength for the TM_{01} mode. Then the series junction between the rectangular and circular guides produces infinite input impedance of the section h at the junction for the TE_{11} mode. This establishes weak TE_{11} mode in the circular guide. The section h effectively produces zero input impedance at the junction for the TM_{01} mode to be strongly excited in the circular guide. The distance between the input and output rectangular guides is selected such that resonance in the TE_{11} mode is avoided.

The junctions between the rectangular and the circular guides are matched by using a reactive tuner, such as inductive irises in the rectangular waveguide side. Such rotary joints produce a good VSWR < 1.1.

6.4.6 Strip and Microstrip Line Sections

Strip and microstrip line sections are used for interconnection and impedance matching in microwave integrated circuits. Impedance and attenuation characteristics of such lines are described in Chapter 4. In this section several forms of discontinuity are described for circuit design requirements.

Open ended section

The following three phenomena are associated with a physical open end
(a)   Fringing fields extended beyond the physical end of the strip.
(b)   Radiation from the open end
(c)   Launching of surface waves from the open end of the strip.

The phenomenon (a) can be accounted for by replacing the fringing fields by means of an equivalent shunt capacitance C_f. This capacitance is also equivalent

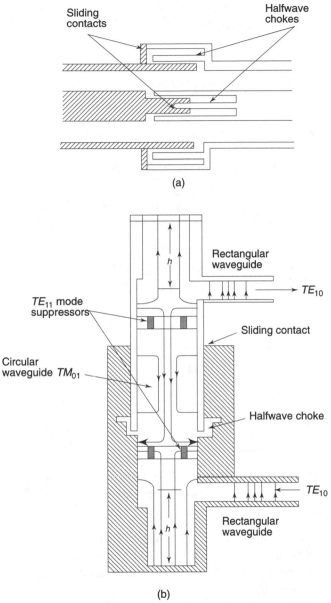

Fig. 6.11 *Rotary joints (a) Coaxial (b) Waveguide*

to an extra length $\Delta l \ll \lambda_g$ of the same line with electrically open circuit end as shown in Fig. 6.12.

For a given structure the end effect length is given in the book, *Foundations for Microstrip Circuit Design* (Reference 6.3) as

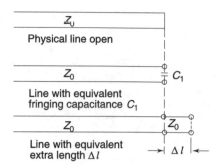

Fig. 6.12 *Equivalent end effect length*

$$\Delta l = 0.412\, h \left(\frac{\varepsilon_{\text{eff}} + 0.3}{\varepsilon_{\text{eff}} - 0.258} \right) \left(\frac{W/h + 0.262}{W/h + 0.813} \right) \qquad (6.68)$$

Short circuit

In a microstrip line a short circuit can be achieved by a short-circuiting post between the strip and the ground plane as shown in Fig. 6.13.

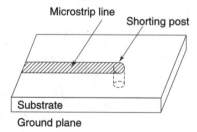

Fig. 6.13 *Short circuit in a microstrip line*

At frequencies below 3 GHz, this shorting post could be a short metallic wire bonded to the microstrip and the ground plane.

Right-angled bend

In order to divert the signal path, a right-angled bend in the microstrip line often appears. The additional charge accumulation at the corners and interruption of current flow provide equivalent capacitance and inductance, respectively to form a T-network as shown in Fig. 6.14.

The bend capacitance and inductance are expressed by

$$C_b/W = \frac{(14\,\varepsilon_r + 12.5)\, W/h - (1.83\,\varepsilon_r - 2.25)}{\sqrt{W/h}} \quad \text{pF/m ; for } W/h < 1$$

$$= (9.5\,\varepsilon_r - 1.25)W/h + 5.2\,\varepsilon_r + 7.0 \text{ pF/m ; for } W/h > 1 \quad (6.69)$$

$$L_b/h = 100 \left[4\sqrt{W/h} - 4.21 \right] \text{ nH/m} \qquad (6.70)$$

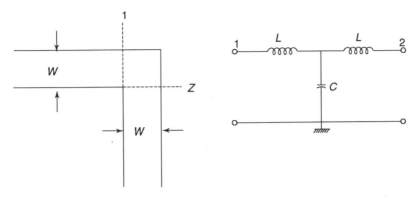

Fig. 6.14 *Right-angled bend and its electrical equivalent circuit*

In order to reduce the bend effect, the corner is chamfered or mitred as shown in Fig. 6.15, where $b = 0.57$ W for $\varepsilon_r = 9.8$.

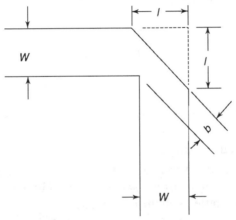

Fig. 6.15 *Chamfered bend*

6.4.7 Matched Terminations

Matched terminations are used in coaxial lines, strip lines and waveguides to absorb the incident power without appreciable reflection and radiation. A tapered lossy dielectric is placed at the end of a shorted line as shown in Fig. 6.16 to form a matched termination.

The length of the tapered section is kept about one to two guide wavelengths at the lowest frequency of operation for effective absorption of power. To increase the power dissipation, aquadag-coated sand is used as lossy material. High power (>1W) terminations use outer cooling fins for heat dissipation. Practical VSWR of these loads is in the range of $1.02 - 1.05$ over a frequency bandwidth of the order of 20–30 % of the centre frequency.

A matched load is a single port device having its ideal parameters:

$$Z_{in} = Z_o = 50 \text{ ohm or } 75 \text{ ohm}$$

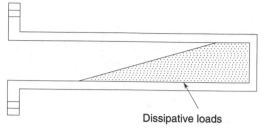

Fig. 6.16 *Matched termination*

$$\Gamma = 0 = S_{11}$$
$$\text{VSWR} = 1.0$$

The load is located at the centre of the Smith chart as the reference point.

Example 6.2   A waveguide termination having VSWR of 1.1 is used to dissipate 100 watts of power. Find the reflected power.

Solution

$$|\Gamma| = \frac{S-1}{S+1} = \frac{1.1-1}{1.1+1} = \frac{0.1}{2.1}$$
$$= 0.04762$$
$$P_{\text{ref}} = |\Gamma|^2 \, P_i = (0.04762)^2 \times 100 = 0.2268 \text{ W}$$

6.4.8 Short Circuit Plunger

A short circuit plunger is a variable short circuit to provide an adjustable reactive load $Z_{in} = jZ_0 \tan \beta l$, ranging from $-j\infty$ to $j\infty$, depending on the physical length l of the line and hence reflects all the incident power. These terminations are used for measurements of impedance or scattering parameters of a microwave device or circuit element. The short is achieved by a sliding block of copper as shown in Fig. 6.17(a).

The main requirements for the design of a short circuit plunger are
1. low contact resistance at the guide wall
2. constant contact resistance for movement along the line.

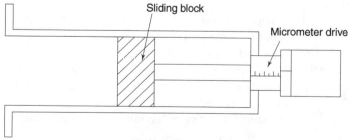

Fig. 6.17(a)   *(Cont.)*

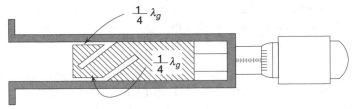

Fig. 6.17(b)   *Short circuit plungers*

These requirements can be met better by employing quarter wavelength long spring contact assemblies to place the tips of spring contact at a node of longitudinal electric current as shown in Fig. 6.17(b).

Since there is difficulty in achieving an adequate contact throughout the entire perimeter of the plunger, choke plungers are often used as shown in Fig. 6.18.

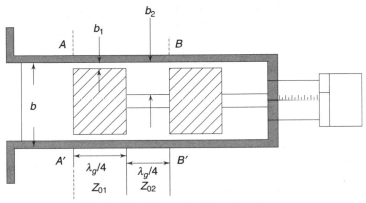

Fig. 6.18   *Short circuit choke plunger*

A choke plunger uses the impedance transformation properties of a quarterwave transformer. In the waveguide choke plunger, the front choke is a quarterwave section. The second section is also a quarter guide wavelength long. The widths of the plungers are uniform and slightly less than the interior guide width of broad wall. The height of the plungers are not uniform, the front one is $b - 2b_1$ and second one is $b - 2b_2$, where b_1 is made as small as possible and b_2 is made as large as possible. The back section makes a sliding fit in the guide with almost zero gap. In this case the contact resistance is in a current antinode and the input impedance is zero at the plane of reflection. If Z_{01} and Z_{02} are the characteristic impedances of the quarter wavelength coaxial sections $b - 2b_1$ and $b - 2b_2$, the impedance seen at the input plane AA' is

$$Z_{in} = (Z_{01}/Z_{02})^2 \, Z_i' = (b_1/b_2)^2 \, Z_i' \qquad (6.71)$$

where Z_i' is the input impedance at the second section plane BB'. To make $Z_{in} \to 0$, $Z_{01} \ll Z_{02}$ and accordingly the gaps b_1 and b_2 are to be selected where $b_2 \gg b_1$. If by good mechanical design, b_2 is made equal to $5b_1$, $Z_{in} = Z_i'/25$. Therefore, the shorting effect is improved 25 times as compared to a single non-choke type sliding short. For a coaxial line, shorting plungers are used to provide impedance transforming properties which do not have physical contact with the line conductors.

The disadvantage of all non-contact plungers and choke plungers is the bandwidth limitation of 20 to 30 % of the mid-frequency. In case of a circular waveguide in the TE_{01} mode, a short circuit plunger does not require any choke connections since there is no longitudinal current. A simple sliding metal disc is adequate. The locus of reflection coefficient of this plunger is the outermost circle of the Smith chart.

6.4.9 Rectangular to Circular Waveguide Transition

This transition section is used to convert the dominant TE_{10} mode in the rectangular waveguide to TE_{11} dominant mode in circular waveguide and vice-versa as shown in Fig. 6.19. The minimum length of the transition should be quarter wavelength to avoid abrupt dimensional changes and generation of higher order modes.

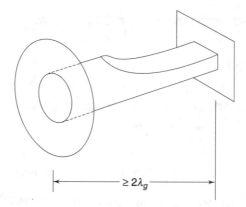

Fig. 6.19 *Rectangular to circular waveguide transition*

6.4.10 Waveguide Corners, Bends and Twists

The waveguide E and H plane corners and bends are used to alter the direction of the guide to any convenient angle, as shown in Fig. 6.20.

Such arrangements produce discontinuities and as a result reflection of propagating waves takes place at any point of the discontinuities. In order to cancel the reflected waves from both ends of the waveguide corner, the mean length L between the junction of corners is kept an odd multiple of the quarter wavelength. For waveguide bend, the discontinuity effects are reduced by keeping the minimum radius of curvature $R = 1.5b$ for an E-bend and $R = 1.5a$ for a H-bend.

A waveguide twist is used to change the polarisation of propagating wave by 90 degrees. The length of the twist is again kept equal to an odd multiple of the quarter guide wavelength.

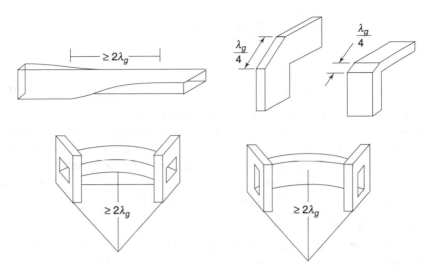

Fig. 6.20 *Waveguide corners, bends and twists*

6.4.11 Coaxial Line to Waveguide Adapters

In a rectangular waveguide, the electromagnetic fields are fed by means of a small coaxial line probe driven by a generator as shown in Fig. 6.21. For the dominant TE_{10} mode of excitation, the probe is inserted from the broad wall perpendicular or parallel to the maximum **E**-field. The depth d of the probe is kept small so that a constant current on the probe radiates electromagnetic field inside the guide.

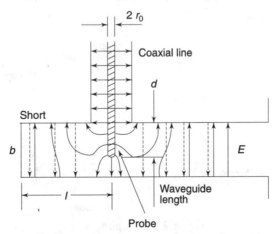

Fig. 6.21 *Coaxial to waveguide adapter*

In order to have radiation in one propagating direction only, a short circuit is placed at a distance l approximately equal to a quarter wavelength in the backward direction. By designing the values of l and d, the input impedance $Z_{in} = R_{in} + jX_{in}$ of the probe can be made pure resistive which is equal to the characteristic impedance Z_o of the coaxial line feeding the signal to the guide:

$$Z_{in} = R = \frac{2\eta_o}{ab\ \beta_{10}\ k_o}\ \sin^2 \beta_{10}\ l\ \tan^2 (k_o\ d/2) \tag{6.72}$$

where,

$$\eta_o = \sqrt{\mu_o/\varepsilon_o}\ ;\ \beta_{10} = k_o\ \sqrt{1 - (\lambda_o/2a)^2}\ ;\ k_o = 2\ \pi/\lambda_o \tag{6.73}$$

Since the electric field in the coaxial line and in the waveguide for TE_{10} mode in the vicinity of the probe are orthogonal to each other, higher-order-modes (evanescent) are also excited near the probe which are highly attenuated within one wavelength distance. By making the probe diameter very small ($< 0.15\ a$) the reactance X_{in} of the probe resulting from higher order mode is made negligible small (or ideally zero) with proper choice of l and d.

6.4.12 Coupling Loops

Electromagnetic fields can also be excited in waveguides by a coupling loop placed midway between the top and bottom walls of the guide with its plane transverse to the waveguide as shown in Fig. 6.22. Thus maximum **H**-field for TE_{10} mode is coupled to the loop. For propagation in one propagating direction only, a short circuit plate is placed at a distance l on the back side. The input impedance of the probe is made pure resistive, equal to the characteristic impedance Z_o of the feeder coaxial line:

$$Z_{in} = Z_o = R = \frac{k_o\ \eta_o}{2ab\ \beta_{10}}\ (\pi/a)^2\ (\pi d^2)^2\ \sin^2 \beta_{10} l \tag{6.74}$$

Z_{in} is made equal to Z_o by adjusting the diameter of the thin loop and l such that $d < \lambda_o/10$ and $\lambda_g/2 < l < \lambda_g/4$.

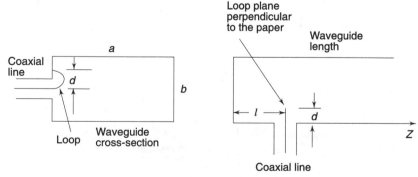

Fig. 6.22 *Coupling loops*

6.4.13 Coupling Aperture

The electromagnetic signal can be coupled into a waveguide from another guide by cutting a small aperture of radius $r_o \ll \lambda_o$ in the common wall. If the aperture plane is perpendicular to the electric field, the normal component of the electric field lines and the tangential component magnetic field lines fringe through the aperture as shown in Fig. 6.23. Under this condition, the small aperture becomes equivalent to a combination of an electric dipole perpendicular to the aperture with its dipole moment being proportional to the normal component of the electric field in the guide, and a magnetic dipole in the plane of the aperture with its dipole moment proportional to the tangential component of the magnetic field. The constants of proportionality depend on the aperture size and shape. For **E** and **H**-field couplings, the constants of proportionality are called the *electric* and *magnetic polarisabilities* α_e and α_m, respectively, given by

$$\alpha_e = -2/3 \, r_o^3, \; \alpha_m = 4/3 \, r_o^3 \tag{6.75}$$

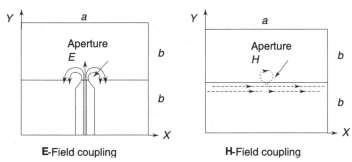

E-Field coupling   H-Field coupling

Fig. 6.23   *Aperture coupling*

When a small circular aperture of radius $r_o \ll \lambda_o$, exists in a transverse wall in a rectangular waveguide operating in the TE_{10} mode, it offers an inductive susceptance B given by

$$B = \frac{-3 \, ab}{8 \, r_o^3 \, \beta_{10} \, Z_w} \tag{6.76}$$

The transmission and reflection coefficients of the aperture are given by

$$T = \frac{j \, 16 \, r_o^3 \, k_o \, \eta_o}{3 \, ab \, Z_w} \tag{6.77}$$

$$\Gamma = \frac{j \, 16 \, r_o^3 \, \beta_{10}}{3 \, ab} - 1 \tag{6.78}$$

A circular aperture of radius r_o in the common broad wall between two rectangular waveguide couples the energy from the one guide to the other. For the dominant TE_{10} mode of the input wave propagation, the aperture will be excited by a y-directed electric dipole and x- and z-directed magnetic dipoles. The normal (y) electric dipole and the axial (z) component of the magnetic dipole radiate in the upper guide symmetrically in both directions $(\pm z)$. The transverse (x) component of the magnetic dipole radiates asymmetrically. The amplitude of the field in the upper guide can be controlled by adjusting the angle θ between the two waveguides (Fig. 6.39a) or by adjusting the aperture position d in transverse direction. When d is chosen to satisfy $\sin(\pi d/a) = \lambda_o/\sqrt{6}\,a$, there will be zero power at port 4 in the upper guide and input power at port 1 will be coupled into ports 2 and 3 only. The 4-port device is then called a *directional coupler* or *Bethe hole coupler* which is described in Section 6.4.18.1 A (p. 191).

6.4.14 Attenuators

Attenuators are passive devices used to control power levels in a microwave system by partially absorbing the transmitted signal wave. Both fixed and variable attenuators are designed using resistive films (aquadag).

A coaxial fixed attenuator uses a film with losses on the centre conductor to absorb some of the power as shown in Fig. 6.24a. The fixed waveguide type consists of a thin dielectric strip coated with resistive film and placed at the centre of the waveguide parallel to the maximum E-field. Induced current on the resistive film due to the incident wave results in power dissipation, leading to attenuation of microwave energy. The dielectric strip is tapered at both ends up to a length of more than half wavelength to reduce reflections. The resistive vane is supported by two dielectric rods separated by an odd multiple of quarter wavelength and perpendicular to the electric field (Fig. 6.24b).

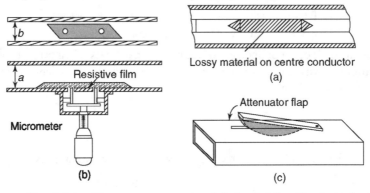

Fig. 6.24 *Microwave attenuator (a) Coaxial line fixed attenuator (b) & (c) Waveguide attenuators*

A variable type attenuator can be constructed by moving the resistive vane by means of micrometer screw from one side of the narrow wall to the centre where the E-field is maximum (Fig. 6.24b) or by changing the depth of insertion of a resistive vane at an E-field maximum through a longitudinal slot at the middle of the broad wall as shown in Fig. 6.24c. A maximum of 90 dB attenuation is possible with VSWR of 1.05. The resistance card can be shaped to give a linear variation of attenuation with the depth of insertion.

A precision type variable attenuator makes use of a circular waveguide section (C) containing a very thin tapered resistive card (R_2), to both sides of which are connected axisymmetric sections of circular to rectangular waveguide tapered transitions (RC_1 and RC_2) as shown in Fig. 6.25. The centre circular section with the resistive card can be precisely rotated by 360° with respect to the two fixed sections of circular to rectangular waveguide transitions. The induced current on the resistive card R_2 due to the incident signal is dissipated as heat producing attenuation of the transmitted signal. The incident TE_{10} dominant wave in the rectangular waveguide is converted into a dominant TE_{11} mode in the circular waveguide. A very thin tapered resistive card is placed perpendicular to the E-field at the circular end of each transition section so that it has a negligible effect on the field perpendicular to it but absorbs any component parallel to it. Therefore, a pure TE_{11} mode is excited in the middle section.

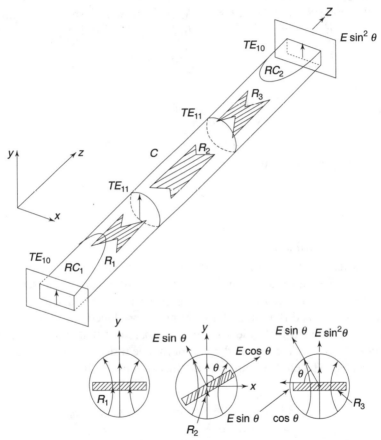

Fig. 6.25 *Precision type variable attenuator*

R_1, R_2, R_3 – Tapered resistive cards
RC_1 & RC_2 – Rectangular-to-circular waveguide transitions
C – Circular Waveguide Section

If the resistive card in the centre section is kept at an angle θ relative to the E-field direction of the TE_{11} mode, the component $E \cos \theta$ parallel to the card get absorbed while the component $E \sin \theta$ is transmitted without attenuation. This later component finally appears as electric field component $E \sin^2 \theta$ in rectangular output guide. Therefore, the attenuation of the incident wave is

$$\alpha = \frac{E}{E \sin^2 \theta} = \frac{1}{\sin^2 \theta} = \frac{1}{|S_{21}|}$$

or,

$$\alpha \, (\text{dB}) = -40 \log (\sin \theta) = -20 \log |S_{21}| \tag{6.79}$$

Therefore, the precision rotary attenuator produces attenuation which depends only on the angle of rotation θ of the resistive card with respect to the incident wave polarisation. Attenuators are normally matched reciprocal devices, so that

$$|S_{21}| = |S_{12}| \tag{6.80}$$

and

$$|S_{11}| \quad \text{or} \quad |S_{22}| = \frac{\text{VSWR} - 1}{\text{VSWR} + 1} \ll 0.1 \tag{6.81}$$

where the VSWR is measured at the port concerned. The S-matrix of an ideal precision rotary attenuator is

$$[S] = \begin{bmatrix} 0 & \sin^2 \theta \\ \sin^2 \theta & 0 \end{bmatrix} \tag{6.82}$$

6.4.15 Phase Shifters

A phase shifter is a two-port passive device that produces a variable change in phase of the wave transmitted through it. A phase shifter can be realised by placing a lossless dielectric slab within a waveguide parallel to and at the position of maximum **E**-field. A differential phase change is produced due to the change of wave velocity through the dielectric slab compared to that through an empty waveguide. Two ports are matched by reducing the reflections of the wave from the dielectric slab tapered at both ends as shown in Fig. 6.26.

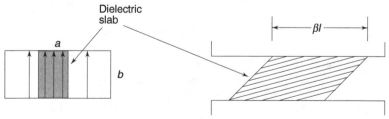

Fig. 6.26 *Phase shifters*

The propagation constant through a length l of a dielectric slab and of an empty guide are, respectively,

$$\beta_1 = \frac{2\pi}{\lambda_{g1}} = \frac{2\pi\sqrt{\left[1-\left(\lambda_o/(2a\sqrt{\varepsilon_r})\right)^2\right]}}{\lambda_o/\sqrt{\varepsilon_r}} \tag{6.83}$$

$$\beta_2 = \frac{2\pi}{\lambda_{g2}} = \frac{2\pi\sqrt{\left[1-(\lambda_o/2a)^2\right]}}{\lambda_o/\sqrt{\varepsilon_r}} \tag{6.84}$$

Thus the differential phase shift produced by the phase shifter is $\Delta\phi = (\beta_1 \sim \beta_2)\, l$.

By adjusting the length l, different phase shifts can be produced. The S-matrix of an ideal phase shifter can be expressed by

$$[S] = \begin{bmatrix} 0 & e^{-j\Delta\phi} \\ e^{-j\Delta\phi} & 0 \end{bmatrix} \tag{6.85}$$

6.4.15.1 *Precision phase shifter*

A precision phase shifter can be designed as a rotary type as shown in Fig. 6.27. This uses a section of circular waveguide containing a lossless dielectric plate of length $2l$ called halfwave (180°) section. This section can be rotated over 360° precisely between two sections of circular to rectangular waveguide transitions each containing lossless dielectric plates of length l called quarterwave (90°) sections oriented at an angle of 45° with respect to the broad wall of the rectangular waveguide ports at the input and output. The incident TE_{10} wave in the rectangular guide becomes a TE_{11} wave in the circular guide. The halfwave section produces a phase shift equal to twice its rotation angle θ with respect to the quarterwave section. The dielectric plates are tapered through a length of quarter wavelength at both ends for reducing reflection due to discontinuity.

The principle of operation of the rotary phase shifter can be explained as follows. The TE_{11} mode incident field E_i in the input quarterwave section can be decomposed into two transverse components, one E_1, polarised parallel and other, E_2 perpendicular to quarterwave plate. After propagation through the quarter wave plate these components are

$$E_1 = E_i \cos 45°\ e^{-j\beta_1 l} = E_o\ e^{-j\beta_1 l} \tag{6.86}$$

$$E_2 = E_i \sin 45°\ e^{-j\beta_2 l} = E_o\ e^{-j\beta_2 l} \tag{6.87}$$

where, $E_o = E_i/\sqrt{2}$. The length l is adjusted such that these two components will have equal magnitude but a differential phase change of $(\beta_1 - \beta_2)\,l = 90°$. Therefore, after propagation through the quarterwave plate these field components become

$$E_1 = E_o\ e^{-j\beta_1 l} \tag{6.88}$$

$$E_2 = jE_o\ e^{-j\beta_1 l} = jE_1 = E_1\ e^{j\pi/2} \tag{6.89}$$

Thus the quarterwave sections convert a linearly polarised TE_{11} wave to a circularly polarised wave and vice-versa.

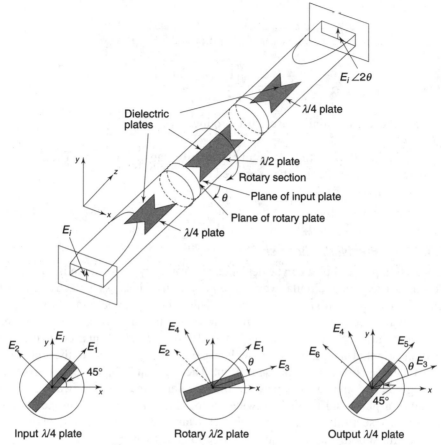

Fig. 6.27 *Precision rotary phase shifter*

After emergence from the halfwave section, the field components parallel and perpendicular to the halfwave plate can be represented as

$$E_3 = (E_1 \cos\theta - E_2 \sin\theta) \, e^{-j2\beta_1 l} = E_o \, e^{-j\theta} \, e^{-j3\beta_1 l} \qquad (6.90)$$

$$E_4 = (E_2 \cos\theta + E_1 \sin\theta) \, e^{-j2\beta_2 l} = E_o \, e^{-j\theta} \, e^{-j3\beta_1 l} \, e^{-j\pi/2} \qquad (6.91)$$

since,

$$2 \, (\beta_1 - \beta_2)l = \pi \quad \text{or} \quad -2 \, \beta_2 \, l = \pi - 2 \, \beta_1 l \qquad (6.92)$$

After emergence from the halfwave section the field components E_3 and E_4 may again be decomposed into two TE_{11} modes, polarised parallel and perpendicular to the output quarterwave plate. At the output end of this quarterwave plate the field components parallel and perpendicular to the quarterwave plate can be written as

$$E_5 = (E_3 \cos\theta + E_4 \sin\theta) \, e^{-j\beta_1 l} = E_o \, e^{-j2\theta} \, e^{-j4\beta_1 l}; \qquad (6.93)$$

$$E_6 = (E_4 \cos\theta - E_3 \sin\theta) \, e^{-j\beta_2 l} = E_o \, e^{-j2\theta} \, e^{-j4\beta_1 l}; \qquad (6.94)$$

Therefore, the parallel component E_5 and perpendicular component E_6 at the output end of the quarterwave plate are equal in magnitude and in phase to produce a resultant field which is a linearly polarised TE_{11} wave

$$
\begin{aligned}
E_{\text{out}} &= \sqrt{2}\ E_o\ e^{-i2\theta}\ e^{-j4\beta_1 l} \\
&= E_i\ e^{-j2\theta}\ e^{-j4\beta_1 l}
\end{aligned}
\tag{6.95}
$$

having the same direction of polarisation as the incident field E_i with a phase change of $2\theta + 4\beta_1 l$. Since θ can be varied and $4\beta_1 l$ is fixed at a given frequency and structure, a phase shift of 2θ can be obtained by rotating the halfwave plate precisely through an angle of θ with respect to the quarterwave plates.

6.4.16 Waveguide Tees

Waveguide tees are three-port components. They are used to connect a branch or section of the waveguide in series or parallel with the main waveguide transmission line for providing means of splitting, and also of combining power in a waveguide system. The two basic types, E-plane (series) T and H-plane (shunt) T, are constructed as shown in Fig. 6.28. These are named according to the axis of the side arm which is parallel to the **E**-field or the **H**-field in the collinear arms, respectively.

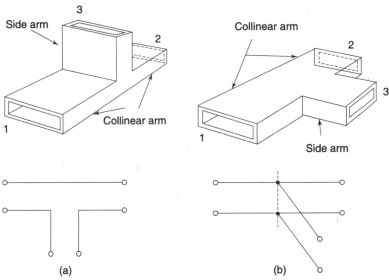

Fig. 6.28 *Waveguide tees (a) E-Tee or Series-T (b) H-Tee or Shunt-T*

Because of the junction, waveguide tees are poorly matched devices. Adjustable matching reactance can be introduced by means of a tuning screw at the centre. Because of symmetry and absence of non-linear elements in the junction, the S-matrix is symmetric: $S_{ij} = S_{ji}$; $i = 1, 2$; $j = 1, 2$. The general S-matrix for a tee junction is

$$[S] = \begin{bmatrix} S_{11} & S_{12} & S_{13} \\ S_{12} & S_{11} & S_{23} \\ S_{13} & S_{23} & S_{33} \end{bmatrix} \tag{6.96}$$

6.4.16.1 E-Plane tee

From considerations of symmetry and the phase relationship of the waves in each of the arms, it can be seen that a wave incident at port 3 will result in waves at ports 1 and 2, which are equal in magnitude but opposite in phase, i.e., $S_{31} = S_{13} = -S_{23} = -S_{32}$, $S_{12} = S_{21}$. If two input waves are fed into ports 1 and 2 of the collinear arm, the output wave at port 3 will be opposite in phase and subtractive. Sometimes this third port is called the *difference arm*. By analogy with the voltage relationship in the series circuit, E-plane junction is also called a *series junction*. All diagonal elements of the S-matrix of a E-plane T junction cannot be zero simultaneously since the tee junction cannot be matched to all the three arms simultaneously. Considering the as matched port 3, the S-matrix of a E-plane T can be derived as follows.

Denoting the incident and outgoing signal variables at the ith port by a_i and b_i, respectively, for an input power at port 3, the net input power to port 3 is $|a_3|^2 - |b_3|^2 = |a_3|^2 (1 - |S_{33}|^2)$, and the output power is $|b_1|^2 + |b_2|^2 = 2 |a_3|^2 |S_{13}|^2$, since $|S_{31}| = |S_{32}|$ by symmetry. Since the junction is assumed lossless, the input power must be equal to the output power i.e.,

$$(1 - |S_{33}|^2) = 2 |S_{13}|^2$$

By a suitable matching element we can make $S_{33} = 0$, so that $|S_{13}| = 1/\sqrt{2}$. From the symmetry characteristics described above,

$$S_{13} = S_{31} = 1/\sqrt{2}, \; S_{23} = S_{32} = -1/\sqrt{2}$$

After matching the port 3, if one attempts to match either port 1 or 2 by similar method, the matching elements, such as irises or tuning screws will interact with each other and matching at port 3 would be disturbed. Based on power consideration it can also be shown that $S_{11} = S_{22} = 1/2$ and $S_{12} = S_{21} = 1/2$ for $S_{33} = 0$. Therefore, with matching at port 3, the S-matrix of a E-plane T can be expressed by real values with proper choice of reference plane:

$$[S] = \begin{bmatrix} 1/2 & 1/2 & 1/\sqrt{2} \\ 1/2 & 1/2 & -1/\sqrt{2} \\ 1/\sqrt{2} & -1/\sqrt{2} & 0 \end{bmatrix} \tag{6.97}$$

6.4.16.2 H-Plane tee

In a H-plane tee if two input waves are fed into port 1 and port 2 of the collinear arm, the output wave at port 3 will be in phase and additive. Because of this, the third port is called the *sum arm*. Reversely, an input wave at port 3 will be equally divided into ports 1 and 2 in phase. Because the magnetic field loops get divided into two arms 1 and 2 in a manner similar to currents between branches in parallel circuit, H-plane junction is also called a shunt junction. For a symmetrical and lossless junction, in absence of non-linear elements at the H-plane junction, the S-parameters are obtained in similar manner as in the case of E-plane junction:

$$[S] = \begin{bmatrix} 1/2 & -1/2 & 1/\sqrt{2} \\ -1/2 & 1/2 & 1/\sqrt{2} \\ 1/\sqrt{2} & 1/\sqrt{2} & 0 \end{bmatrix} \qquad (6.98)$$

when matched at side port 3.

Because of mismatch at any two ports, the VSWR at the mismatch port of either E or H-plane tee junction is very high

$$\text{VSWR} = \frac{1 + 1/2}{1 - 1/2} = 3.0$$

Example 6.3 A 20 mW signal is fed into one of collinear port 1 of a lossless H-plane T-junction. Calculate the power delivered through each port when other ports are terminated in matched load.

Solution

Since ports 2 and 3 are matched terminated, $a_2 = a_3 = 0$, $S_{11} = 1/2$. The total effective power input to port 1 is

$$P_1 = |a_1|^2 (1 - |S_{11}|^2)$$
$$= 20 (1 - 0.5^2) = 15 \text{ mW}$$

The power transmitted to port 3 is

$$P_3 = |a_1|^2 |S_{31}|^2$$
$$= 20 \times (1/\sqrt{2})^2 = 10 \text{ mW}$$

The power transmitted to port 2 is

$$P_2 = |a_1|^2 |S_{21}|^2$$
$$= 20 \times (1/2)^2 = 5 \text{ mW.}$$

Therefore, $P_1 = P_3 + P_2$

Example 6.4 In a H-plane T-junction, compute power delivered to the loads 40 ohm and 60 ohm connected to arms 1 and 2 when 10 mW power is delivered to matched port 3.

Solution

With port 3 matched, the scattering matrix for H-plane T is

$$[S] = 1/2 \begin{bmatrix} 1 & 1 & \sqrt{2} \\ 1 & 1 & \sqrt{2} \\ \sqrt{2} & \sqrt{2} & 0 \end{bmatrix}$$

Therefore, input power at port 3 is equally divided in arms 1 and 2. Since input at port 3 = 10 mW = 0.01 W, power towards ports 1 and 2 = 0.005 W = 1/2 $|b_1|^2$ = 1/2 $|b_2|^2$. Considering 1st order reflection, reflected power from ports 1 and 2 are

$$1/2\ |\Gamma_1 b_1|^2 \text{ and } 1/2\ |\Gamma_2 b_2|^2$$

Therefore, power delivered to load Z_1 = 40 ohm and Z_2 = 60 ohm are

$$P_1 = 1/2\ |b_1|^2 - 1/2\ |\Gamma_1 b_1|^2 = 1/2\ |b_1|^2\ (1 - |\Gamma_1|^2)$$

and $$P_2 = 1/2\ |b_2|^2 - 1/2\ |\Gamma_2 b_2|^2 = 1/2\ |b_2|^2\ (1 - |\Gamma_2|^2)$$

Now taking the characteristic impedance of the line = 50 ohm

$$|\Gamma_1| = |40 - 50|\ /\ |40 + 50| = 1/9\ ;\ |\Gamma_1|^2 = 0.01234$$

$$|\Gamma_2| = |60 - 50|\ /\ |60 + 50| = 1/11;\ |\Gamma_2|^2 = 8.2694 \times 10^{-3}$$

Therefore,

$$P_1 = 0.005\ (1 - 0.01234) = 4.938 \times 10^{-3} = 4.9383\ \text{mW}$$

$$P_2 = 0.005\ (1 - 8.2694 \times 10^{-3}) = 4.9586 \times 10^{-3}\ \text{W}$$

$$= 4.9586\ \text{mW}$$

6.4.16.3 Hybrid or magic-T

A combination of the *E*-plane and *H*-plane tees forms a hybrid tee, called a magic–*T*, having 4 ports as shown in Fig. 6.29.

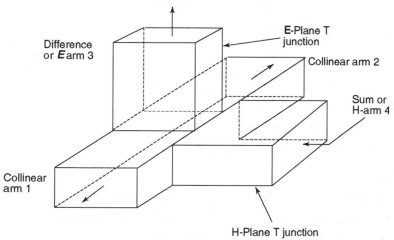

Fig. 6.29 *Magic-T*

The magic-*T* has the following characteristics when all the ports are terminated with matched load.

1. If two inphase waves of equal magnitude are fed into ports, 1 and 2, the output at port 3 is subtractive and hence zero and total output will

appear additively at port 4. Hence port 3 is called the difference or E-arm and 4 the sum or H-arm.

2. A wave incident at port 3 (E-arm) divides equally between ports 1 and 2 but opposite in phase with no coupling to port 4 (H-arm). Thus

$$S_{13} = S_{31} = 1/\sqrt{2} = S_{24} = S_{42} \text{ and } S_{34} = 0 \qquad (6.99)$$

3. A wave incident at port 4 (H-arm) divides equally between ports 1 and 2 in phase with no coupling to port 3 (E-arm). Thus

$$S_{14} = S_{41} = 1/\sqrt{2} = S_{24} = S_{42} \text{ and } S_{34} = 0 \qquad (6.100)$$

4. A wave fed into one collinear port 1 or 2, will not appear in the other collinear ports 2 or 1. Hence two collinear ports 1 and 2 are isolated from each other, making

$$S_{12} = S_{21} = 0 \qquad (6.101)$$

A magic-T can be matched by putting tuning screws suitably in the E and H arms without destroying the symmetry of the junction. Therefore, for an ideal lossless magic-T matched at ports 3 and 4, $S_{33} = S_{44} = 0$. The procedure of derivation of the S-matrix considers the symmetry property at the junction for which $S_{14} = S_{41} = S_{24} = S_{42}$, $S_{31} = S_{13} = -S_{23} = -S_{32}$, $S_{34} = S_{43} = 0$, $S_{12} = S_{21} = 0$. Therefore, the S-matrix for a magic-T, matched at ports 3 and 4 is given by

$$[S] = \begin{bmatrix} S_{11} & S_{12} & S_{13} & S_{14} \\ S_{12} & S_{22} & -S_{13} & S_{14} \\ S_{13} & -S_{13} & 0 & 0 \\ S_{14} & S_{14} & 0 & 0 \end{bmatrix} \qquad (6.102)$$

From the unitary property applied to rows 1 and 2, we get

$$|S_{11}|^2 + |S_{12}|^2 + |S_{13}|^2 + |S_{14}|^2 = 1 \qquad (6.103)$$

$$|S_{12}|^2 + |S_{22}|^2 + |S_{13}|^2 + |S_{14}|^2 = 1 \qquad (6.104)$$

Subtracting these two equations:

$$|S_{11}|^2 - |S_{22}|^2 = 0$$

or,
$$|S_{11}| = |S_{22}| \qquad (6.105)$$

From the unitary property applied to rows 3 and 4

$$2|S_{13}|^2 = 1, \text{ or } |S_{13}| = 1/\sqrt{2} \qquad (6.106)$$

$$2|S_{14}|^2 = 1, \text{ or } |S_{14}| = 1/\sqrt{2} \qquad (6.107)$$

Substituting these values in Eq. 6.103.

$$|S_{11}|^2 + |S_{12}|^2 + 1/2 + 1/2 = 1$$

or,
$$|S_{11}|^2 + |S_{12}|^2 = 0 \qquad (6.108)$$

which is valid if $S_{11} = S_{12} = 0$ (6.109)

From Eqs 6.105 and 6.109, $S_{22} = 0$ (6.110)

Therefore,

$$[S] = \begin{bmatrix} 0 & 0 & S_{13} & S_{13} \\ 0 & 0 & -S_{13} & S_{13} \\ S_{13} & -S_{13} & 0 & 0 \\ S_{13} & S_{13} & 0 & 0 \end{bmatrix} \tag{6.111}$$

where $|S_{13}| = 1/\sqrt{2} = |S_{14}|$

By proper choice of reference planes in arms 3 and 4, it is possible to make both S_{13} and S_{14} real, resulting in the final form of S-matrix of magic-T

$$[S] = 1/\sqrt{2} \begin{bmatrix} 0 & 0 & 1 & 1 \\ 0 & 0 & -1 & 1 \\ 1 & -1 & 0 & 0 \\ 1 & 1 & 0 & 0 \end{bmatrix} \tag{6.112}$$

Example 6.5 A magic-T is terminated at collinear ports 1 and 2 and difference port 4 by impedances of reflection coefficients $\Gamma_1 = 0.5$, $\Gamma_2 = 0.6$ and $\Gamma_4 = 0.8$, respectively. If 1W power is fed at sum port 3, calculate the power reflected at port 3 and power transmitted to other three ports.

Solution

S-matrix for a matched magic-T with collinear ports 1 and 2 and sum and difference ports 3 and 4, respectively, is given by

$$[S] = 1/\sqrt{2} \begin{bmatrix} 0 & 0 & 1 & 1 \\ 0 & 0 & 1 & -1 \\ 1 & 1 & 0 & 0 \\ 1 & -1 & 0 & 0 \end{bmatrix}$$

If a_1, a_2, a_3 and a_4 be the normalised input voltages and b_1, b_2, b_3 and b_4 are the corresponding output voltage at ports 1, 2, 3 and 4 respectively, then

$$a_1 = \Gamma_1 b_1, a_2 = \Gamma_2 b_2, a_3 = \text{input applied voltage, and}$$
$$a_4 = \Gamma_4 b_4.$$

Now, $P_i = |a_3|^2 = 1\text{W}$,   or, $a_3 = 1$ V

Therefore,

$$\begin{bmatrix} b_1 \\ b_2 \\ b_3 \\ b_4 \end{bmatrix} = 1/\sqrt{2} \begin{bmatrix} 0 & 0 & 1 & 1 \\ 0 & 0 & 1 & -1 \\ 1 & 1 & 0 & 0 \\ 1 & -1 & 0 & 0 \end{bmatrix} \begin{bmatrix} .5b_1 \\ .6b_2 \\ 1.0 \\ .8b_4 \end{bmatrix}$$

or,
$$b_1 + 0 + 0 - 0.8\, b_4/\sqrt{2} = 1/\sqrt{2}$$

$$0 + b_2 + 0 + 0.8\, b_4/\sqrt{2} = 1/\sqrt{2}$$

$$-0.5 b_1/\sqrt{2} - 0.6\, b_2/\sqrt{2} + b_3 + 0 = 0$$

$$-0.5 b_1/\sqrt{2} + 0.6\, b_2/\sqrt{2} + 0 + b_4 = 0$$

The unknown quantities b's may be solved by Cramer's rule

$$b_1 = \frac{\begin{bmatrix} 1 & 0 & 0 & -0.8 \\ 1 & \sqrt{2} & 0 & 0.8 \\ 0 & -0.6 & \sqrt{2} & 0 \\ 0 & 0.6 & 0 & \sqrt{2} \end{bmatrix}}{\begin{bmatrix} \sqrt{2} & 0 & 0 & -0.8 \\ 0 & \sqrt{2} & 0 & 0.8 \\ -0.5 & -0.6 & \sqrt{2} & 0 \\ -0.5 & 0.6 & 0 & \sqrt{2} \end{bmatrix}} = \sqrt{2}\ \frac{1 - 0.6 \times 0.8}{2 - 0.8\,(0.5 + 0.6)}$$

$$= \sqrt{2}\ \frac{1 - 0.48}{2 - 0.88} = 0.6566\ \text{V}$$

Similarly,

$$b_2 = \sqrt{2}\ \frac{1 - 0.5 \times 0.8}{2 - 0.8\,(0.5 + 0.6)} = 0.7576\ \text{V}$$

$$b_3 = \frac{0.5 + 0.6 - 2 \times 0.5 \times 0.6 \times 0.8}{2 - 0.8\,(0.5 + 0.6)} = 0.5536\ \text{V}$$

$$b_4 = \sqrt{2}\ \frac{0.5 - 0.6}{2 - 0.8\,(0.5 + 0.6)} = -0.0893\ \text{V}$$

Therefore,
Power transmitted at port 1 = $|b_1|^2 = 0.4309$ W
Power transmitted at port 2 = $|b_2|^2 = 0.5738$ W
Power transmitted at port 4 = $|b_4|^2 = 0.00797$ W
Power reflected at port 3 = $|b_3|^2 = 0.3065$ W

Note: Power absorbed at port $i = 1/2\,(|b_i|^2 - |a_i|^2)$, $i = 1, 2, 4$. Total power absorbed by the system $= 1/2\,(|a_3|^2 - |b_3|^2)$.

6.4.16.4 Application of magic-T

The magic-T has a number of applications in various microwave circuits, such as the *E-H* tuner for impedance matching, balanced mixer in a microwave superheterodyne receiver to balance out the local oscillator noise at the IF amplifier input, power combiner and duplexer.

In an *E-H* tuner (Fig. 6.30.) both the *E* and *H* arms are terminated by movable shorts which act as *E*-plane and *H*-plane stubs. The position of the shorts can be adjusted so that a wide range of load impedance may be matched to reduce the VSWR of a waveguide system connected through the collinear arms.

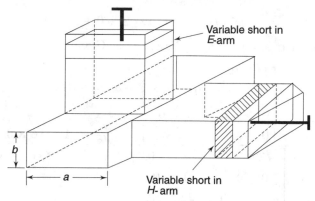

Fig. 6.30   *E-H tuners*

In a balanced microwave mixer configuration, an incoming signal is fed to the *E*-arm and a local oscillator signal is fed to the *H*-arm as shown in Fig. 6.31. These two signals when enter the collinear arms, the crystal diodes placed in these arms produce the IF signal or difference signal in the following manner.

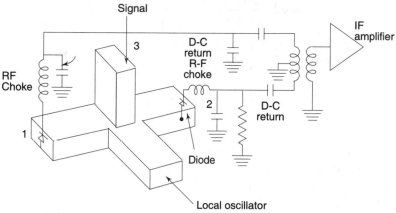

Fig. 6.31   *Microwave mixer*

The local oscillator signal from the *H*-arm will arrive at the diodes in phase, whereas the incoming signal from *E*-arm will arrive at the diodes out-of-phase. These signals are mixed in the nonlinear diodes and produce IF signals in the collinear arms which are out-of-phase by 180°. Since local oscillator noise will be in phase at the diodes, this gets cancelled at the balanced IF input whereas, the IF signals are added up in phase for amplification in IF amplifier. Moreover, IO and RF signals are uncoupled due to magic-T properties of *E* and *H* arms. For equal power inputs at isolated ports 3 and 4, $P_3 = P_4 = P = |a|^2$, where $a_3 = a_4 = a$. If port 1 is matched, $a_1 = 0$. Therefore,

$$b_2 = (0 + 0 - a + a)\sqrt{2} = 0.$$

and hence $a_2 = 0$. Consequently $b_3 = b_4 = 0$. Thus

$$b_1 = S_{13}\, a + S_{14}\, a = \sqrt{2}\, a$$

Therefore, output power at port 1 is $P_1 = |b_1|^2 = 2\, |a|^2 = 2P = $ sum of two equal input powers.

In a duplexer circuit of a radar system a common antenna is connected to port 1 while the transmitter and receiver are connected to isolated E and H-arms and a dummy load is connected at port 2. 50 % of the power transmitted is coupled to the antenna and 50 % of the received power get into the receiver, remaining 50 % power is absorbed in dummy load.

6.4.17   Isolators

An isolator is a two-port non-reciprocal device which produces a minimum attenuation to wave propagation in one direction and very high attenuation in the opposite direction. Thus when inserted between a signal source and load almost all the signal power can be transmitted to the load and any reflected power from the load is not fed back to the generator output port. This eliminates variations of source power output and frequency pulling due to changing loads.

An isolator can be constructed in a rectangular waveguide $(a \times b)$ operating in dominant mode as shown in Fig. 6.32. The non-reciprocal characteristics are obtained by establishing a steady magnetic field $\mathbf{H}_o$ in y direction and placing a ferrite slab at any of the longitudinal planes $x = x_1$ near and parallel to the narrow waveguide wall, where the magnetic field exhibits circular polarisation. This occurs at $x_1 = a/4$ or, $3a/4$.

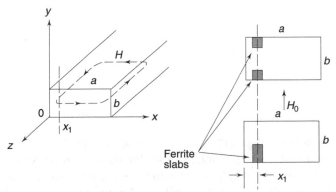

Fig. 6.32   *Waveguide isolator*

For the propagation of waves in $+z$ direction, direction of rotation of $\mathbf{H}$ in the planes at $x_1 = a/4$ and $3a/4$ are opposite to each other. The non-reciprocal characteristic is achieved by placing a ferrite slab at any one of these two planes. The required steady state magnetic field $\mathbf{H}_o$ in the y-direction is established by placing permanent magnetic poles between the two broad walls.

It is known that the attenuation in ferrite for negative/clockwise circular polarisation is very small whereas for positive/counter clockwise circular polarisation is very large at and near the resonance frequency $f \approx f_o$. Therefore, the ferrite slab is placed in such a way that while transmission it encounters negative circular polarisation and a positive circular polarisation in the reverse direction. The steady magnetic field is set to be equal to the resonant value. The isolation of the order of $20 - 30$ dB in the backward direction and a transmission loss of 0.5 dB in the forward direction can be achieved with a VSWR of the order of 1.1.

Since the reverse power is absorbed in the ferrite and dissipated as heat, the maximum power handling capability of an isolator is limited. To increase the capacity of heat dissipation, two ferrite slabs of smaller heights are used instead of one with a larger height.

The requirement of very high steady magnetic field (10,000 oersteds at $\lambda = 1$ cm) is the main drawback of waveguide resonance isolator at higher frequencies.

6.4.17.1 *Faraday rotation isolator*

A Faraday rotation isolator is a circular waveguide section axially loaded with a ferrite rod of smaller diameter as shown in Fig. 6.33.

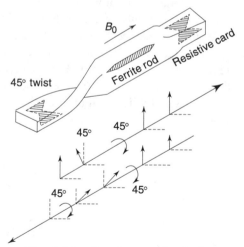

Fig. 6.33 *Faraday rotation isolator*

The ferrite rod is subjected to a steady axial magnetic field $\mathbf{H}_o$ of strength much smaller than the resonant intensity so that the dissipative loss in the ferrite is neglected. The dominant TE_{11} mode in the circular section can be decomposed into two oppositely rotating circularly polarised waves of equal magnitude. These waves encounter different permeabilities μ'_+ and μ'_- for the clockwise and anti-clockwise directions of field rotation and exhibit changes in the phase velocities. This will result in a change in the plane of polarisation of the main mode TE_{11} which will experience gradual rotation θ during propagation to other end. The rotation angle θ is proportional to the length of the ferrite rod. In this case for the reverse wave the direction of rotation remains the same confirming the non-reciprocal characteristics of the ferrite.

The isolator input is a 45° twist where a tapered resistive card is mounted parallel to the broad wall of the rectangular waveguide part. The dominant TE_{10} mode does not get attenuated while transmission and is rotated 45° at the twist output and enters the circular waveguide through rectangular to circular waveguide transition as the TE_{11} mode. The length of the ferrite rod is selected so as to obtain Faraday rotation $\theta = 45°$ at the output and regain its original polarisation. The plane of polarisation of the reflected wave from the load is again rotated by the same angle $\theta = 45°$ and at the emergence through the 45° twist becomes aligned with the surface of the absorbing plate and gets absorbed. Thus non-reciprocal isolation action takes place. Typical insertion loss and isolation are approximately 1dB and 20-30dB, respectively, for these isolators. Isolators are also available in the coaxial and microstrip forms.

For an ideal lossless, matched isolator

$$|S_{21}| = 1, |S_{12}| = |S_{11}| = |S_{22}| = 0 \tag{6.113}$$

i.e.,

$$[S] = \begin{bmatrix} 0 & 0 \\ 1 & 0 \end{bmatrix} \tag{6.114}$$

Example 6.6   A matched isolator has insertion loss of 0.5 dB and isolation 25 dB. Find the scattering coefficients.

Solution

$$\text{Insertion loss} = 0.5 \text{ dB} = -20 \log |S_{21}|$$

or,

$$S_{21} = 10^{-.5/20} = 10^{-0.025}$$

The isolation is 25 dB $= -20 \log |S_{12}|$

or,

$$S_{12} = 10^{-25/20} = 10^{-1.2}$$

Since there is no reflection, $S_{11} = S_{22} = 0$. Therefore, the S-matrix for the isolator is

$$[S] = \begin{bmatrix} 0 & 10^{-1.2} \\ 10^{-0.025} & 0 \end{bmatrix}$$

6.4.17.2 Circulators

A circulator is a multiport junction in which the wave can travel from one port to next immediate port in one direction only as shown in Fig. 6.34. Commonly used circulators are three-port or four-port devices although more number of ports is possible.

(a) *Four-port circulator*   A four port circulator can be constructed from two magic-T's and a non-reciprocal 180° phase shifter or a combination of two 3-dB side hole directional couplers and a rectangular waveguide with two non-reciprocal phase shifters as shown in Figs 6.35(a) and (b).

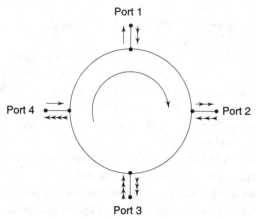

Fig. 6.34 *Schematic diagram of a four-port circulator*

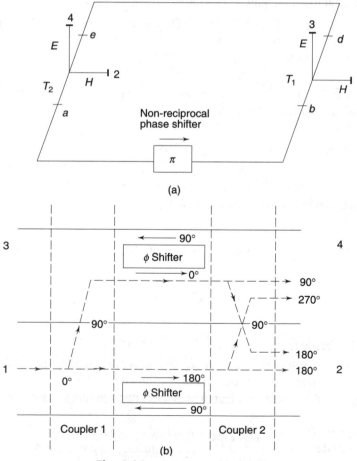

Fig. 6.35 *Four-port circulators*

In Fig. 6.35(a), an input signal at port 1 is split into two in phase and equal amplitude waves in the collinear arms b and d of the magic-tee, T_1 and added up to emerge from port 2 in the magic tee, T_2. On the other hand a signal at port 2 will be split into two equal amplitude and equiphase waves in the collinear arms of the magic-tee, T_2 and appears at point b and d out of phase due to presence of the non-reciprocal 180° phase shifter. These out of phase waves add up and appear from port 3 in the magic-tee, T_1. In a similar manner, an input signal at port 3 will emerge from 4, an input at port 4 will appear at port 1. Thus the circulator property is exhibited.

In Fig. 6.35(b), each of the two 3-dB couplers introduces a 90° phase shift. An input signal at port 1 is split into two components by the coupler 1 and the coupled signals are again split into two components by the coupler 2 with a 90° phase shift in each. Each of the two phase shifters produces additional phase shift so that the signal components at port 2 are in phase, and at port 4 they are out of phase. Since port 3 is the decoupled port for the directional coupler, the input signal at port 1 appears in port 2. Similarly, signals flow from port 2 to port 3, from port 3 to port 4 and from port 4 to port 1.

A perfectly matched, lossless, and non-reciprocal four-port circulator has S-matrix

$$[S] = \begin{bmatrix} 0 & 0 & 0 & 1 \\ 1 & 0 & 0 & 0 \\ 0 & 1 & 0 & 0 \\ 0 & 0 & 1 & 0 \end{bmatrix} \tag{6.115}$$

(b) Three-port circulator   A three-port circulator is formed by a 120° H-plane waveguide or strip line symmetrical Y-junction with a central ferrite post. A steady magnetic field $\mathbf{H}_o$ is applied along the axis of the post as shown in Fig. 6.36.

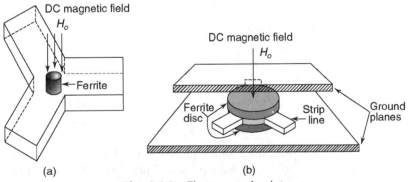

Fig. 6.36   *Three-port circulator*
(a) Waveguide type
(b) Stripline type

For a perfectly matched, lossless, non-reciprocal three-port circulator, the S-matrix is

$$[S] = \begin{bmatrix} 0 & 0 & S_{13} \\ S_{21} & 0 & 0 \\ 0 & S_{32} & 0 \end{bmatrix} \qquad (6.116)$$

If the terminal planes are properly chosen to make the phase angles of S_{13}, S_{21} and S_{32} zero,

$$S_{13} = S_{21} = S_{32} = 1$$

so that

$$[S] = \begin{bmatrix} 0 & 0 & 1 \\ 1 & 0 & 0 \\ 0 & 1 & 0 \end{bmatrix} \qquad (6.117)$$

The matching of the junction can be achieved by placing suitable tuning elements in each arm.

Since in practice losses are always present, the performance is limited by finite isolation and non-zero insertion loss. Typical characteristics can be represented by

Insertion loss < 1 dB
Isolation ≈ 30–40 dB
VSWR < 1.5

Example 6.7 Prove that it is impossible to construct a perfectly matched, lossless, reciprocal three-port junction.

Solution A perfectly matched three-port junction has a scattering matrix

$$[S] = \begin{bmatrix} 0 & S_{12} & S_{13} \\ S_{12} & 0 & S_{23} \\ S_{13} & S_{23} & 0 \end{bmatrix}$$

For a lossless junction S-matrix is unitary

$$S_{12} S_{12}{}^* + S_{13} S_{13}{}^* = 1$$

$$S_{12} S_{12}{}^* + S_{23} S_{23}{}^* = 1$$

$$S_{13} S_{13}{}^* + S_{23} S_{23}{}^* = 1$$

$$S_{13} S_{23}{}^* = S_{12} S_{23}{}^* = S_{12} S_{13}{}^* - 0$$

If S_{12} is not equal to zero, the fourth equation from above gives $S_{13} = 0 = S_{23}$. But this does not satisfy the third equation. Therefore, a reciprocal lossless three-port junction cannot be perfectly matched.

Example 6.8 A three-port circulator has an insertion loss of 1 dB, isolation 30 dB and VSWR = 1.5. Find the S-matrix.
Solution The S-matrix of three-port circulator is

$$[S] = \begin{bmatrix} S_{11} & S_{12} & S_{13} \\ S_{21} & S_{22} & S_{23} \\ S_{31} & S_{32} & S_{33} \end{bmatrix}$$

Insertion Loss = 1 dB = $-20 \log |S_{21}|$

or, $\qquad |S_{21}| = 10^{-1/20} = 0.89$

For same insertion loss between ports 1 and 2, 2 and 3, and 3 and 1, $|S_{21}| = |S_{32}| = |S_{13}| = 0.89$.
The isolation between the ports is 30 dB = $-20 \log |S_{31}|$

or, $\qquad |S_{31}| = 10^{-30/20} = 10^{-1.5} = 0.032$

$$= |S_{23}| = |S_{12}|$$

Since VSWR $S = 1.5$, reflection coefficient

$$|\Gamma| = \frac{S-1}{S+1} = \frac{1.5-1}{1.5+1} = 0.2 = |S_{11}|$$

$$= |S_{22}| = |S_{33}|$$

By placing reference planes suitably to make the phase of S-parameters zero,

$$[S] = \begin{bmatrix} 0.200 & 0.032 & 0.890 \\ 0.890 & 0.200 & 0.032 \\ 0.032 & 0.890 & 0.200 \end{bmatrix}$$

6.4.17.3 YIG filters and oscillators

Yttrium iron garnet (YIG) is a complex solid having ferromagnetic properties and is represented by $Me_3 Fe_5 O_{12}$ or $5 Fe_2 O_3 \, 3Me_2O_3$ or $Me_3Fe_2(FeO_4)_3$, where Me is a rare-earth-metal. It has much smaller losses than ordinary ferrites. A small sphere (40 mil diameter) of YIG usually has a very small loss and can be used as a resonator in microwave filters and oscillators with resonant frequency at $f_o = 2.8 \, \mathbf{H}_o$, where f_o is in MH_z and $\mathbf{H}_o$ is steady external field in oersteds. Thus the resonant frequency can be controlled electronically by changing $\mathbf{H}_o$. The low frequency limit (about 3.5 GHz) of the YIG resonator is controlled by the minimum $\mathbf{H}_o$ applied for saturation of magnetisation. This frequency can be brought down by doping pure YIG with gadolinium (Gd).

A small YIG sphere is placed at the centre of two mutually perpendicular loops in yz and xz planes (Fig. 6.37). A dc magnetic field $\mathbf{H}_o$ along z-axis is created perpendicular to the two loops. Input RF current is fed to the input loop and the output loop is connected to the load. In the steady $\mathbf{H}_o$ field, the YIG has a net magnetisation in the $\mathbf{H}_0$ direction. The resulting magnetic field component H_x makes the magnetisation vector to precess about the z-axis resulting in a y-component of magnetisation which in turn induces a voltage at the output loop. The off-frequency components of RF input couples a small power to the output loop. Thus the system acts like a tuned band-pass filter at frequency f_0.

In a waveguide, the YIG sphere is placed in the coupling hole of two rectangular waveguides having dominant mode magnetic fields perpendicular to each other for filtering action.

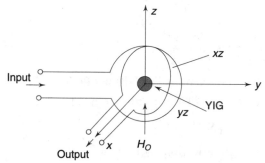

Fig. 6.37 *YIG filters and oscillators*

6.4.18 Directional Couplers

A directional coupler is a four-port passive device commonly used for coupling a known fraction of the microwave power to a port (coupled port) in the auxiliary line while flowing from input port to output port in the main line. The remaining port is ideally isolated port and matched terminated. There are three basic types of directional couplers. One is a multiple aperture waveguide type, second one is a coupled coaxial, or strip or microstrip line, and the third one is branch line couplers as shown schematically in Fig. 6.38.

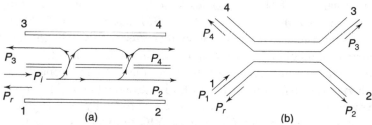

Fig. 6.38 *Directional couplers (a) Waveguide (b) Stripline*

The theory of the waveguide coupler was first established by 'Bethe', using a single hole in the common broad wall of two rectangular waveguides, commonly known as *Bethe-hole* directional coupler. Practical waveguide directional couplers are multi hole couplers in which the desired coupling response vs frequency can be achieved by proper selection of the number of holes and size of the holes.

Waveguide couplers are forward coupler since the coupled power in the ancillary guide flows in the same direction as the input power in the main guide, whereas the coaxial, strip and microstrip couplers are backward couplers because the coupled energy travels in the opposite direction of the input energy flow.

The performance of a directional coupler is measured in terms of four basic parameters, i.e., coupling (C), transmission loss (T), directivity (D), and the return loss (R) when all the ports are matched. These are defined as

$$C(\text{dB}) = 10 \log P_1/P_4 \tag{6.118}$$

$$T(\text{dB}) = 10 \log P_1/P_2$$

$$D(\text{dB}) = 10 \log P_4/P_3 \tag{6.119}$$

$$R(\text{dB}) = 10 \log P_1/P_r \tag{6.120}$$

where P's are the powers at the ports shown in Fig. 6.38.

6.4.18.1 Waveguide directional coupler

A waveguide directional coupler commonly consists of two waveguides coupled together through one (Bethe-hole) or a number of small openings (multi hole) in a common broad wall as shown in Fig. 6.39.

A. Bethe-hole coupler   The Bethe-hole coupler is a single hole broad wall aperture coupled waveguide coupler, where the hole is located at the centre of common broad wall of two waveguides inclined at an angle θ, or at an offset position d of two parallel waveguides as shown in Fig. 6.39(a). If the aperture is small compared to the wavelength, it can be considered as an electric dipole normal to the aperture with dipole moment proportional to the normal component of the electric field in the main guide at the aperture, plus a magnetic dipole in the plane of the aperture with dipole moment proportional to the tangential component of the exciting magnetic field at the aperture. The coupling to the auxiliary guide is due to radiation from these dipoles. The electric dipole radiates equally in both the directions longitudinally. But the magnetic dipole radiates asymmetrically in longitudinal directions. By varying the angle θ between the two waveguides, or by adjusting the distance d, powers in port 3 and port 4 can be controlled. Ideally power at port 4 can be zero (isolated) whereas that at port 3 is maximum to achieve directional coupling.

Centre hole   For the centre coupling hole, since practically $P_4 \neq 0$, the directivity is finite. Coupling and directivity of this Bethe-hole coupler are given by

$$C = -20 \log \frac{4}{3} \frac{\beta r_o^3}{ab} \left(\cos\theta + \lambda_g^2/2 \, \lambda_o^2\right) \text{dB} \tag{6.121}$$

$$D = 20 \log \frac{2\beta^2 \cos\theta + k_o^2}{2\beta^2 \cos\theta - k_o^2} \text{dB} \tag{6.122}$$

where r_o = radius of hole

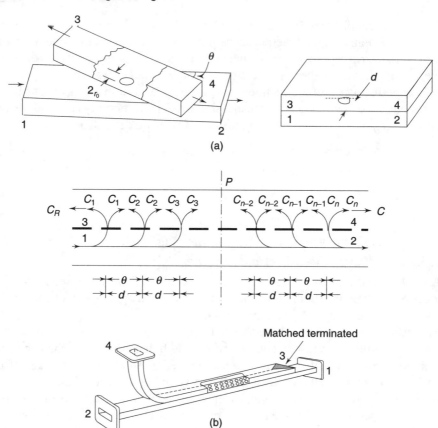

Fig. 6.39 *Waveguide directional couplers (a) Bethe-hole coupler (b) Multihole coupler*

$a \times b$ = waveguide cross-sectional dimension

$$\beta = 2\pi d \, \lambda_g$$

$$\lambda_g = \frac{\lambda_o}{\sqrt{[1 - (\lambda_o/2a)^2]}}$$

$$k_o = 2\pi/\lambda_o$$

Here the guide wall thickness is assumed negligibly small. The optimum angle θ (for $P_4 = 0$) is obtained from the relation $\cos \theta = \lambda_g^2 / 2 \lambda_o^2$.

Off-set hole For the off-set hole, the optimum forward coupling ($P_4 = 0$) is obtained for the off-set value d given by

$$\sin (\pi d/a) = \frac{\lambda_o}{\sqrt{6} \, a} \tag{6.123}$$

The maximum coupling and directivity are given by

$$C = 20 \log [(1 + X^2)/X] \qquad (6.124)$$
$$D = 20 \log 1/X \qquad (6.125)$$

where $X = (16\pi r_o^3/3ab\, \lambda_g) \sin^2(\pi d/a)$.

For non-zero finite wall thickness ($t \approx 0.51$ mm), the coupling decreases and will be 1 to 2 dB less compared to the zero thickness value.

Example 6.9 Design a centre hole Bethe-hole directional coupler with air-filled rectangular waveguide of dimensions 0.9 in × 0.4 in at 9.8 GHz for 20 dB coupling and 40 dB directivity, respectively.

Solution

$$a = 0.9 \text{ in} = 2.286 \text{ cm}, b = 0.4 \text{ in} = 1.016 \text{ cm}, f_o = 9.8 \text{ GH}_z$$

$$\lambda_0 = c/f = \frac{30}{9.8} = 3.06 \text{ cm}, k_o = \frac{2\pi}{\lambda_o} = \frac{2 \times 3.1415}{3.06}$$

$$= 2.0533 \text{ rad/cm}$$

$$\lambda_g/\lambda_o = \frac{1}{\sqrt{\left[1 - (\lambda_o/2a)^2\right]}}$$

$$= \frac{1}{\sqrt{\left[1 - (3.06/2 \times 2.286)^2\right]}}$$

$$= 1.3458$$

$$\lambda_g = 1.3458 \times 3.06 = 4.118 \text{ cm},$$

$$\beta = 2\pi/\lambda_g = 1.525 \text{ rad/cm}$$

$$\cos\theta = 1/2\,(\lambda_g/\lambda_o)^2 = 1/2 \times (1.3458)^2 = 0.9055$$

$$\theta = \cos^{-1} 0.9055 = 25.09°$$

Therefore, $$C = 20\text{dB} = -20 \log\left[\frac{4}{3} \times \frac{1.525}{2.286 \times 1.016} \times 2 \times 0.9055\right] r_o^3$$

$$= -20 \log 1.5855 \, r_o^3$$

or, $$\log 1.5855 \, r_o^3 = -1$$

or, $$1.5855 r_o^3 = 10^{-1} = 0.1$$

or, $$r_o^3 = 0.1/1.5855 = 0.0631$$

Therefore,

$$r_o = (0.0631)^{1/3} = 0.398 \text{ cm}$$

$$d = a/2 = 1.143 \text{ cm}$$

$$D = 20\log \frac{2\beta^2 \cos\theta + (2\pi/\lambda_o)^2}{2\beta^2 \cos\theta - (2\pi/\lambda_o)^2}$$

At the band edge, f_1

$$40 = 20\log \frac{2\beta_1^2 \times 0.9055 + (2\pi/\lambda_o)^2}{2\beta_1^2 \times 0.9055 - (2\pi/\lambda_o)^2}$$

or, $\quad \dfrac{2\beta_1^2 \times 0.9055 + (2\pi/\lambda_o)^2}{2\beta_1^2 \times 0.9055 - (2\pi/\lambda_o)^2} = 10^2 = 100$

or, $\quad 1.811 \times \left(\dfrac{2\pi}{\lambda_{g1}}\right)^2 + \left(\dfrac{2\pi}{\lambda_{o1}}\right)^2 = 100\left[1.811\left(\dfrac{2\pi}{\lambda_{g1}}\right)^2 - \left(\dfrac{2\pi}{\lambda_{o1}}\right)^2\right]$

or, $\quad \dfrac{1.811}{\lambda_{g1}^2} + \dfrac{1}{\lambda_{o1}^2} = \dfrac{181.1}{\lambda_{g1}^2} - \dfrac{100}{\lambda_{o1}^2}$

or, $\quad \dfrac{1}{\lambda_{g1}^2}[1.811 - 181.1] = -\dfrac{1}{\lambda_{o1}^2}(100 + 1)$

or, $\quad \dfrac{179.89}{\lambda_{g1}^2} = \dfrac{101}{\lambda_{o1}^2}$

or, $\quad \dfrac{\lambda_{g1}^2}{\lambda_{o1}^2} = \dfrac{179.89}{101} = 1.775$

$$\frac{1}{\sqrt{\left[1 - \left(\lambda_{o1}/2 \times 2.286\right)^2\right]}} = 1.775$$

or, $\quad 1 - \left(\dfrac{\lambda_{o1}}{4.572}\right)^2 = \left(\dfrac{1}{1.775}\right)^2$

or, $\quad \lambda_{o1}^2 = \left[1 - \left(\dfrac{1}{1.775}\right)^2\right](4.575)^2 = 9.1276$

or, $\quad \lambda_{o1} = 3.02$ cm

Therefore, $\quad f_1 = \dfrac{c}{\lambda_{o1}} = \dfrac{30}{3.02} = 9.929$ GHz

$\qquad\qquad f_2 = 2f_o - f_1 = 2 \times 9.8 - 9.929 = 9.671$ GHZ

Therefore, bandwidth for $D > 40$ dB is $\Delta f = f_1 \sim f_2 = 258$ MHz

(B) Multi hole waveguide coupler The structure of multi hole coupler is symmetrical with respect to a transverse plane. For an input at port 1 of the main guide, a fraction of input power is coupled into the auxiliary guide in forward direction at port 4. The successive openings are spaced quarter wavelength apart at designed frequency so that the coupled waves, travel back towards port 3 will be out of phase and cancelled; whereas the forward coupled waves are in phase and thus reinforce each other. Thus port 3 is ideally isolated port and matched terminated for absorbing any coupled power flows in that port in practical situation. Such coupler is called a forward coupler. For an input to port 2 coupled power gets absorbed at port 3 and port 4 is isolated.

Since in practice due to non-ideal mechanical configurations there is some coupled power available in the isolated port, the matched directional coupler is designed from specifications of two basic parameters-coupling and finite directivity.

For the reciprocal, matched coupler, the scattering matrix of a directional coupler can be derived from the following general S-matrix

$$[S] = \begin{bmatrix} 0 & S_{12} & 0 & S_{14} \\ S_{12} & 0 & S_{23} & 0 \\ 0 & S_{23} & 0 & S_{34} \\ S_{14} & 0 & S_{34} & 0 \end{bmatrix} \tag{6.126}$$

Since for a lossless network the S-matrix is unitary

$$S_{12} S_{12}{}^* + S_{14} S_{14}{}^* = 1 \tag{6.127}$$

$$S_{12} S_{12}{}^* + S_{23} S_{23}{}^* = 1 \tag{6.128}$$

$$S_{23} S_{23}{}^* + S_{34} S_{34}{}^* = 1 \tag{6.129}$$

$$S_{14} S_{14}{}^* + S_{34} S_{34}{}^* = 1 \tag{6.130}$$

$$S_{12} S_{23}{}^* + S_{14} S_{34}{}^* = 0 \tag{6.131}$$

From the first two equations, we get

$$|S_{14}| = |S_{23}| \tag{6.132}$$

and from the second and third equations, we get

$$|S_{12}| = |S_{34}| \tag{6.133}$$

Further by choosing reference plane of port 1 with respect to that of port 2 and the reference plane of port 3 with respect to that of 4, we can make S-parameters real

$$S_{12} = S_{34} = \alpha, \text{ say} \tag{6.134}$$

where α is a positive real number. Then from the fifth Eq. 6.131

$$\alpha (S_{23}{}^* + S_{14}) = 0$$

Since, $\quad \alpha \neq 0, S_{23}{}^* + S_{14} = 0 \tag{6.135}$

Further, selecting the reference plane of port 4 with respect to port 1, we can make S_{14} real, so that

$$S_{23} = -S_{23}{}^* = S_{14} = \beta, \text{ say} \tag{6.136}$$

Therefore,
$$[S] = \begin{bmatrix} 0 & \alpha & 0 & \beta \\ \alpha & 0 & \beta & 0 \\ 0 & \beta & 0 & \alpha \\ \beta & 0 & \alpha & 0 \end{bmatrix} \qquad (6.137)$$

where $\alpha^2 + \beta^2 = 1$ for conservation of energy. Here α is called the transmission factor and β is the coupling factor.

From above, the coupling, directivity and transmission loss for a matched directional coupler can be expressed in terms of S-parameters as

$$C = 10 \log \frac{P_1}{P_4} = -20 \log |S_{41}| \qquad (6.138)$$

$$D = 10 \log \frac{P_4}{P_3} = 20 \log \frac{|S_{41}|}{|S_{31}|} = 20 \log \frac{|S_{41}|}{|S_{42}|} \qquad (6.139)$$

$$T = 10 \log P_1/P_2 = 20 \log |S_{21}| \qquad (6.140)$$

where $|S_{31}| = |S_{42}|$ for symmetry.

The amount of coupling (3 dB, 6 dB, 10 dB, 20 dB, 30 dB, etc.) and the directivity (30 – 40 dB) depend upon the sizes and locations of the holes in the common wall.

Since the phase cancellation in the reverse direction between successive holes can occur only at the designed frequency which satisfy $d = \lambda_{go}/4$, the bandwidth and the frequency response of the coupler are determined by designing the coupler following a suitable reverse voltage response function, such as, maximally flat or binomial response, Chebychev response, etc., as shown in Fig. 6.40.

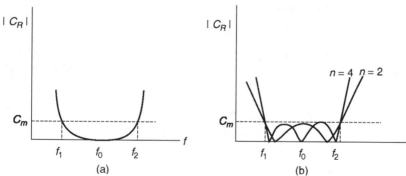

Fig. 6.40 *Coupling characteristics of directional couplers*
(a) Binomial (b) Chebychev

For simplicity we assume that the total power coupled is a small fraction of the incident power so that the incident wave amplitude can be considered to be same at each coupling aperture and only phase will change at different holes. The total reverse and forward voltages at any frequency f can be represented for a unit input voltage at port 1 by,

$$C_R' = \sum_{k=1}^{n} C_k \, e^{-j2(k-1)\theta} \tag{6.141}$$

$$C' = e^{-j(n-1)\theta} \sum_{k=1}^{n} C_k \tag{6.142}$$

respectively. Here $\theta = 2\pi d/\lambda_g$, with $\theta = \pi/2$ at $f = f_o$, the design frequency. The coupling and directivity are expressed by

$$\text{Coupling } C \text{ (dB):} \quad = -20 \log \left| \sum_{k=1}^{n} C_k \right| \tag{6.143}$$

$$\text{Directivity } D \text{ (dB):} = -20 \log \frac{\left| \sum_{k=1}^{n} C_k \, e^{-j2(k-1)\theta} \right|}{\left| \sum_{k=1}^{n} C_k \right|}$$

$$= -C - 20 \log \left| \sum_{k=1}^{n} C_k \, e^{-j2(k-1)\theta} \right| \tag{6.144}$$

By changing the reference plane to the plane of symmetry P (Fig. 6.39b) and choosing a symmetrical array ($C_1 = C_n$, $C_2 = C_{n-1}$, etc.), the reverse voltage is given by

$$C_R = C_R' \, e^{j(n-1)\theta}$$

$$= 2 \sum_{k=1}^{n/2} C_k \cos \left(n - \overline{2k-1} \right) \theta \, ; n \text{ even}$$

$$= 2 \sum_{k=1}^{(n-1)/2} C_k \cos \left(n - \overline{2k-1} \right) \theta + C_{(n+1)/2} \, ; n \text{ odd} \tag{6.145}$$

since $e^{jn\theta} + e^{-jn\theta} = 2 \cos n\theta$.
The forward voltage

$$C = e^{-j(n-1)\theta/2} \sum_{k=1}^{n} C_k$$

$$= e^{-j(n-1)\theta/2} \, 2 \sum_{k=1}^{n/2} C_k; n \text{ even}$$

$$= e^{-j(n-1)\theta/2} \left[2 \sum_{k=1}^{(n-1)/2} C_k + C_{(n+1)/2} \right]; n \text{ odd} \tag{6.146}$$

The multi hole directional couplers are designed for a desired coupling and directivity from the type of distribution by keeping the reverse voltage magnitude $|C_R|$ under a specified value C_m over the frequency band between the edges f_1 and f_2 (Fig 6.40).

Binomial coupler Binomial couplers are designed by choosing the coupling co-efficient C_k at hole k proportional to the corresponding binomial coefficient of $(a + b)^{n-1}$ for total n holes. For a unit input voltage, the total reverse coupled voltage $C_R = 0$ at $f = f_o$ and increases slowly at either side of f_o with a maximally flat realisable response, so that

$$C_k = a_k C_1 = {}^{n-1}C_{k-1} C_1 \qquad (6.147)$$

where

$$ {}^{n-1}C_{k-1} = \frac{n-1!}{(k-1)! \ (n-k)!} \qquad (6.148)$$

C_k can be adjusted by varying the aperture size while maintaining its shape and position fixed. For the reverse wave for $f_1 \le f \le f_2$.

$$|C_R| = \left| 2C_1 \sum_{k=1}^{n/2} {}^{n-1}C_{k-1} \cos(n - 2\overline{k-1})\theta \right| \le C_m; \ n \text{ even} \qquad (6.149)$$

$$|C_R| = \left| 2C_1 \sum_{k=1}^{(n-1)/2} {}^{n-1}C_{k-1} \cos(n - 2\overline{k-1})\theta + {}^{n-1}C_{(n-1)/2} \right| \le C_m; \ n \text{ odd} \quad (6.150)$$

Expanding $\cos m\theta$ in each term of the above equation, in terms of $\cos\theta$, only one term containing $\cos^{n-1}\theta$ with coefficient A_{n-1} does not cancel, so that

$$C_R = 2C_1 A_{n-1} \cos^{n-1}\theta \qquad (6.151)$$

where A_{n-1} is a constant determined from the expansion of $\cos(n-1)\theta$. Since all the derivatives up to the $(n-2)$ the terms of C_R at mid frequency $f = f_o$ are zero, the binomial distribution results in a maximally flat response without any ripple. Since the maximum value of C_R at the band edges is specified to be C_m,

$$C_m = 2C_1 A_{n-1} \cos^{n-1}\theta; \ \theta = \theta_1, \theta_2 \text{ for } f = f_1, f_2$$

or,

$$|C_1|_{\max} \le \frac{C_m}{2A_{n-1}\cos^{n-1}\theta}; \ \theta_1 \le \theta \le \theta_2 \qquad (6.152)$$

since $|C_R| = C_m$ at f_1 and f_2, $\theta_2 = \pi - \theta_1$. If the band edge frequency ratio is defined by,

$$r = \lambda_{g1}/\lambda_{g2} = \theta_2/\theta_1, \text{ and } d = \lambda_{go}/4,$$
$$\theta_1 = \pi/(1 + r) = 2\pi d/\lambda_{g1}$$
$$\theta_2 = \pi r/(1 + r) = 2\pi d/\lambda_{g2}$$

The coupling coefficients for a binomial coupler having specified number of holes are given as follows.

No. of holes n	Coefficients
2	$a_1 = {}^{2-1}C_o = 1 = a_2$
3	$a_1 = {}^{3-1}C_o = 1 = a_3, \ a_2 = {}^{3-1}C_1 = 2$
4	$a_1 = {}^{4-1}C_o = 1 = a_4, \ a_2 = {}^{4-1}C_1 = 3 = a_3$
5	$a_1 = {}^{5-1}C_o = 1 = a_5, \ a_2 = {}^{5-1}C_1 = 4 = a_4,$
	$a_3 = {}^{5-1}C_2 = 6$

The coefficients for other number of holes can be determined in a similar manner. C_1 can be computed from the desired coupling C for given number of holes.

A major practical disadvantage of a binomial coupler is the wide variation between the coupling coefficients of different holes of the coupler, especially for a large number of holes. This leads to fabrication difficulties while maintaining equal hole spacings of quarter wavelength.

Chebyshev coupler The Chebychev coupler response is an equal ripple charac-teristic where the reverse coupling C_R is made proportional to a Chebychev poly-nomial of order $n - 1$ (for number of holes $= n$)

$$T_p(x) = \cos[p\cos^{-1}(x)]; -1 \le x \le +1, p = n - 1$$
$$= \cosh[p\cosh^{-1}(x)]; x < -1, x > +1, p = n - 1 \quad (6.153)$$

where x is the parameter for change of variable and is a function of frequency:

$$x = \cos\theta/\cos\theta_1 = \cos\theta/|\cos\theta_2| \quad (6.154)$$

The polynomials have the following properties as shown in Chapter 5.
1. All polynomials pass through the point $(1,1)$
2. $-1 \le T_p(x) \le +1$, for $-1 \le x \le 1$,
3. All roots occur within $-1 \le x \le 1$, and all maxima and minima have values of $+1$ and -1, respectively.

Thus we have within the band

$$C_R = 2\sum_{k=1}^{n/2} C_k \cos\left(n - \overline{2k - 1}\right)\theta; n \text{ even}$$

$$= 2\sum_{k=1}^{(n-1)/2} C_k \cos\left(n - \overline{2k - 1}\right)\theta + C_{(n+1)/2}; n \text{ odd}$$

$$= C_m T_{n-1}(x) \quad (6.155)$$

The C_k's are obtained by expanding the left hand side in term of powers of cos θ and substituting $x\cos\theta_1$ for $\cos\theta$ and equating the coefficients of like powers of x with those on the right hand side.

Chebychev coupler gives the largest bandwidth for a specified C_m and n. Alternately, for a given bandwidth it gives the lowest ripple C_m.

Coupler design The design procedure of the multi hole directional couplers follows the steps given below:
1. Select the type of distribution (Binomial or Chebychev)
2. ΣC_k is calculated from the type of distribution and C_1 is computed from the coupling C for a chosen number of holes.
3. Test $|C_R| \le C_m$ for this n.
4. Select n to satisfy this condition over the band by repeating the procedure 2 to 3.
5. From the designed values of C_k's, the hole dimensions are determined from the empirical formulas given in Eq. 6.121.

Example 6.10 Design a five-hole 30 db directional coupler with Chebychev distribution for wavelength ratio of 2 at the band edges.

Solution

For $n = 5$ and $r = 2$, the reverse voltage

$$C_R = 2 [C_1 \cos (5-1)\theta + C_2 \cos (5-3)\theta] + C_3$$
$$= 2 [C_1 \cos 4\theta + C_2 \cos 2\theta] + C_3$$
$$= 2 [C_1(8 \cos^4 \theta - 8 \cos^2\theta + 1) + C_2 (2 \cos^2\theta - 1)] + C_3$$

At band edge $f = f_1$ and $\theta = \theta_1 = \dfrac{\pi}{1+r} = \pi/3$

A change of variable according to $\cos \theta = x \cos \theta_1$ gives $\cos \theta = x \cos(\pi/3) = x/2$. Therefore,

$$C_R = 2 [C_1 (x^4/2 - 2x^2 + 1) + C_2 (x^2/2 - 1)] + C_3$$
$$= C_1 x^4 + (C_2 - 4C_1) x^2 + (C_3 + 2C_1 - 2C_2)$$
$$= C_m T_{5-1}(x)$$
$$= C_m (8x^4 - 8x^2 + 1)$$

Equating the coefficients

$$C_1 = 8 C_m$$
$$C_2 = 4 C_1 - 8 C_m = 32 C_m - 8 C_m = 24 C_m$$
$$C_3 + 2 C_1 - 2 C_2 = C_m$$

or, $\quad\quad C_3 = C_m - 16 C_m + 48 C_m = 33 C_m$

Now coupling

$$C = 2(C_1 + C_2) + C_3 = 30 \text{ dB} = 10^{-30/20} = 0.0316.$$

or, $\quad\quad 2(8 + 24) C_m + 33 C_m = 0.0316.$

or, $\quad 97 C_m = 0.0316$ or, $C_m = 3.26 \times 10^{-4}$

Therefore, coupling values of the holes are

$$C_1 = 8 C_m = 26.08 \times 10^{-4}$$
$$C_2 = 24 C_m = 78.24 \times 10^{-4}$$
$$C_3 = 33 C_m = 107.58 \times 10^{-4}$$

Directivity $D = -C + 20 \log 1/C_m = 39.7$ db

In case the coupling coefficients make the hole size large, holes may overlap to make the centre-to-centre distance equal to the quarter wavelength. Otherwise, also for large holes, the coupling coefficients become more frequency sensitive. Under such a situation, the number of holes are increased to meet the desired specifications of coupling and directivity.

Example 6.11 Design a maximally flat 20 db directional coupler so that $D \geq 40$ db in the band $r = 2$.

Solution

Directivity $= 20 \log C/C_R = 20 \log C - 20 \log C_R$

or,

$$|\text{Directivity}| + |\text{Coupling}| = -20 \log C_R$$

Here in the band, $D + C \geq 60$ db $= -20 \log C_m$

Therefore, the maximum tolerance in reverse voltage $C_m = 10^{-3}$

Let $n = $ number of holes to be selected. If $C_1, C_2, C_3, \ldots$, are the coupled wave at corresponding holes

$$C_1 \, 2^{n-1} \cos^{n-1} \theta_1 \leq C_m; \, \theta_1 = \pi/(1+2) = \pi/3 \tag{A}$$

$$C = 20 \text{ db} = -20 \log C$$

Therefore,

$$C = 10^{-1} = 0.1 = C_1 + C_2 + C_3 + \ldots + C_n \tag{B}$$

For symmetrical coupler with binomial type,

$$C_1 = C_n$$
$$C_2 = (n-1)C_1 = C_{n-1}$$
$$C_3 = [(n-1)(n-2)/2] \, C_1 = C_{n-2}, \text{ etc.}$$

Selecting $n = 3, C_1 = C_3$

$$C_2 = (n-1)C_1 = 2C_1$$

Therefore, (B) reduces to

$$C = 0.1 = C_1 + 2C_1 + C_1 = 4C_1$$
$$C_1 = 0.1/4 = 0.025$$

Therefore, the left hand side of (A) is

$$\frac{0.1}{4} \times 2^2 \cos^2\left(\frac{\pi}{3}\right) = \frac{0.1}{4} > C_m$$

Therefore, $n > 3$

Selecting $n = 5, C_1 = C_5$

$$C_2 = (n-1)C_1 = 4C_1$$
$$C_3 = [(n-1)(n-2)/2]C_1 = 6C_1,$$

Therefore,

$$C = 0.1 = 2(C_1 + C_2) + C_3 = 2(C_1 + 4C_1) + 6C_1 = 16C_1$$

or, $$C_1 = 0.1/16$$

Therefore, the left hand side of (A) is

$$\frac{0.1}{16} \times 2^4 \cos^4\left(\frac{\pi}{3}\right) = 6.25 \times 10^{-3} > C_m$$

Therefore, $n > 5$

Selecting $n = 6, C_1 = C_6$

$$C_2 = (n-1)C_1 = 5C_1$$
$$C_3 = [(n-1)(n-2)/2]C_1 = 10C_1,$$

Therefore,

$$C = 0.1 = 2\,(C_1 + C_2 + C_3) = 2\,(C_1 + 5C_1 + 10C_1) = 32C_1$$

or, $\qquad C_1 = 0.1/32$

Therefore, the left hand side of (A) is

$$\frac{0.1}{32} \times 2^5 \cos^5\left(\frac{\pi}{3}\right) = 3.125 \times 10^{-3} > C_m$$

Therefore, $n > 6$

Selecting $n = 7$, similarly, the left hand side of (A) is

$$0.1 \cos^6\left(\frac{\pi}{3}\right) = 1.5625 \times 10^{-3} > C_m$$

Selecting $n = 8$, in a similar way, the left hand side of (A) is

$$0.1 \cos^7\left(\frac{\pi}{3}\right) = 7.8125 \times 10^{-4} < C_m$$

Therefore, since the condition (A) is satisfied for $n = 8$, the total number of holes are 8 with the coupling coefficients:

$$C_1 = 0.1/2^7, C_2 = 7C_1, C_3 = 21C_1, C_4 = 35\,C_1$$

Example 6.12 Design a three-hole Chebychev rectangular waveguide coupler of $C = 20$ db and $D = 40$ db at $f = 10$ GHz. The waveguide dimensions are $a = 2b = 2.4$ cm and the holes are cut at the centre of the broadwall. Find the frequency band.

Solution

$$- 20 \log C_m = \text{directivity} + \text{coupling} = 40 + 20 = 60 \text{ db}$$

Therefore,

$$C_m = 10^{(-60/20)} = 10^{-3}$$
$$C = C_1 + C_2 + C_3 = 2C_1 + C_2; C_1 = C_3 \text{ for symmetry.}$$

Reverse voltage

$$\begin{aligned}
C_R &= C_1\, 2 \cos 2\theta + C_2 \\
&= 2C_1\,(2 \cos^2\theta - 1) + C_2 \\
&= 2C_1\,(2x^2\cos^2\theta_1 - 1) + C_2 \\
&= C_m\, C_{n-1}\,(x);\ n = 3 \text{ holes} \\
&= C_m\, C_2(x) = C_m\,(2x^2 - 1);
\end{aligned}$$

Therefore, equating the coefficients,

$$4C_1\cos^2\theta_1 = 2C_m$$

$$C_2 - 2C_1 = -C_m = -10^{-3}$$

Directivity $D = 40 \text{ db} = 20 \log |C_2 (\sec\theta_1)|$

or, $\log |C_2 (\sec \theta_1)| = 2$

Therefore, $C_2 (\sec \theta_1) = 10^2 = 100$

or, $2 \sec^2\theta_1 - 1 = 100$

or, $\sec^2\theta_1 = 99/2$

or, $\cos \theta_1 = \sqrt{(2/99)} = 0.142134$

or, $\theta_1 = 81.83°$

Therefore, $4 C_1 \times 0.02 = 2 \times 10^{-3}$

or, $C_1 = \dfrac{2 \times 10^{-3}}{0.02 \times 4} = 0.025$

$$C_2 = -C_m + 2C_1 = -10^{-3} + 0.05 = 0.049$$

At 10 GHz, $\lambda_0 = 30/10 = 3$ cm

$$k_0 = \frac{2\pi}{\lambda_0} = \frac{2 \times 3.14}{3}$$

$$= 2.0943 \text{ rad/cm}$$

$$\lambda_{g0} = 4d = 3.84 \text{ cms}$$

The distance between the holes is

$$d = \lambda_{g0}/4 = \frac{1}{4} \frac{\lambda_0}{\sqrt{\left[1 - (\lambda_0/2a)^2\right]}}$$

$$= \frac{1}{4} \frac{3}{\sqrt{\left[1 - (3/(2 \times 2.4))^2\right]}}$$

$$= 0.96 \text{ cm}$$

The electrical distance at one of the band edges θ_1 and θ_2

$$\beta_1 d = \theta_1$$

or, $\dfrac{2\pi}{\lambda_{g1}} \times 0.96 = 81.83°$

or, $\lambda_{g1} = \dfrac{360° \times 0.96}{81.83°} = 4.226 \text{ cm.}$

Therefore, $4.226 = \dfrac{1}{\sqrt{\left[1 - (\lambda_1/2a)^2\right]}}$

or, $\qquad [1 - (\lambda_1/2a)^2] = (\lambda_1/4.226)^2$

or, $\qquad \lambda_1^2\left[\left(\dfrac{1}{2 \times 2.4}\right)^2 + \left(\dfrac{1}{4.226}\right)^2\right] = 1$

or, $\qquad \lambda_1^2 = 10.06$

$\qquad\qquad \lambda_1 = 3.172$ cm

Therefore, $\qquad f_1 = c/\lambda_1 = 30/3.172 = 9.458$ GHz

$\qquad\qquad f_0 = 10$ GHz

$\qquad\qquad \lambda_2 = \pi - \theta_1 = 98.17°$

Therefore,

$$\beta_2 d = \theta_2$$

or, $\qquad \lambda_{g2} = \dfrac{360° \times 0.96}{98.17°} = 3.521$ cm

Therefore,

$$\lambda_2^2\left[\left(\dfrac{1}{2 \times 2.4}\right)^2 + \left(\dfrac{1}{3.521}\right)^2\right] = 1$$

or, $\qquad\qquad \lambda_2 = 2.839$ cm

Therefore, $f_2 = 30/2.839 = 10.567$ GHz
Therefore, the bandwidth

$$\Delta f = f_2 - f_1 = 10.567 - 9.458 = 1.1088 \text{ GHz}$$

Aperture radius r_0 of the holes are obtained from C_1 and C_2 values from the formula (Eq. 6.121)

$$\beta_{10} = \sqrt{\left[k_0^2 - (\pi/a)^2\right]} = 1.6348 \text{ rad/cm}$$

$$C = \dfrac{4}{3}\dfrac{\beta_{10} r_0^3}{ab}\left(1 + \dfrac{1}{2}\dfrac{\lambda_{g0}^2}{\lambda_0^2}\right) = \dfrac{4}{3} \times \dfrac{1.6348 \, r_0^3}{2.4 \times 1.2}\left(1 + \dfrac{3.84^2}{2 \times 3^2}\right)$$

$$= 1.3769 \, r_0^3$$

or, $\qquad r_0 = (C/1.3769)^{1/3}$

Therefore, the radii of the holes are

$\qquad r_{01} = (C_1/1.3769)^{1/3} = (0.025/1.3769)^{1/3} = 0.263$ cm $= r_{03}$

$\qquad r_{02} = (C_2/1.3769)^{1/3} = (0.049/1.3769)^{1/3} = 0.329$ cm

Hole spacing $= \lambda_{g0}/4 = 0.96$ cm.

Example 6.13 A rectangular waveguide binomial coupler coupled by 5 circular holes in a common side/narrow wall to produce 30 db coupling at 10 GHz. The guide width $a = 2.5$ cm and the height $b = 1.2$ cm. The dominant input mode of unit amplitude radiates a field of amplitude $-j\frac{4}{3} r_0^3 (\pi/a)^2 \frac{1}{ab\beta}$ in both directions in the other guide. Find the required hole radii and the frequency ratio for which the directivity is greater than 50 db.

Solution

Given,

$$|C| = \frac{4}{3} r_0^3 \left(\frac{\pi}{a}\right)^2 \frac{1}{ab\beta}, \text{ for narrow wall waveguide coupler with number of}$$

hole $n = 5$

Binomial coupling $C = 30$ db at $f = f_0 = 10$ GHz.

Waveguide dimension $a = 2.5$ cm and $b = 1.2$ cm

(a) To find C

Assuming symmetrical binomial type coupler

$$C = C_1 + C_2 + C_3 + C_4 + C_5 = 2 (C_1 + C_2) + C_3$$

where $C_1 = C_5$

$$C_2 = C_4 = (n - 1)C_1 = 4C_1$$

$$C_3 = [(n - 1)(n - 2)/ 2] C_1 = 6C_1$$

Therefore $C = 2 (C_1 + C_2) + C_3 = 16 C_1$

Given, 30 db $= - 20 \log C$

$$C = 10^{-30/20} = 10^{-1.5} = 0.0316$$

Therefore, $C_1 = C/16 = 1.976 \times 10^{-3} = C_5$

$$C_2 = C_4 = 4C_1 = 7.905 \times 10^{-3}$$

$$C_3 = 6C_1 = 11.858 \times 10^{-3}$$

(b) To find hole spacing

$$\lambda_0 = c/f = 30/10 = 3 \text{ cm}$$

$$\lambda_{g0} = \frac{\lambda_0}{\sqrt{[1 - (\lambda_0/2a)^2]}}$$

$$= \frac{3}{\sqrt{[1 - 3/(2 \times 2.5)^2]}} = 3.75 \text{ cm}$$

Therefore, hole spacing

$$d = \lambda_{g0}/4 = 0.9375 \text{ cms}$$

(c) To find hole radius r_0

$$\beta = 2\pi / \lambda_{g0} = \frac{2 \times 3.14}{3.75} = 1.6755 \text{ rad / cm.}$$

Therefore, $\quad |C| = r_0^3 \times \frac{4}{3} \times \frac{\pi^2}{(2.5)^2} \times \frac{1}{2.5 \times 1.2 \times 1.6755}$

$$= r_0^3 \times 0.4188$$

or, $\quad r_0 = \left(\frac{|C|}{0.4188}\right)^{1/3}$

Thus radii of the holes are

$$r_{01} = \left(\frac{|C_1|}{0.4188}\right)^{1/3} = 0.1677 \text{ cm} = r_{05}$$

$$r_{02} = \left(\frac{|C_2|}{0.4188}\right)^{1/3} = 0.26625 \text{ cm} = r_{04}$$

$$r_{03} = \left(\frac{|C_3|}{0.4188}\right)^{1/3} = 0.3047 \text{ cm}$$

(d) Frequency range $r = \theta_2 / \theta_1$

$$D = 50 \text{ db} = 20 \log C/C_m; \; C_m = \text{total tolerable back wave amplitude.}$$

or, $\quad 20 \log C_m = - (\text{Directivity} + \text{Coupling}) = - 80 \text{ db}$

$$C_m = 10^{-4}$$

If θ_1 and θ_2 corresponds to band edges f_1 and f_2

$$C_1 \, 2^{n-1} \cos^{n-1} \theta_1 = C_m$$

or, $\quad \dfrac{10^{-1.5}}{16} \times 2^4 \times \cos^4 \theta_1 = C_m = 10^{-4};$

or, $\quad \cos^4 \theta_1 = 10^{-4}/10^{-1.5} = 3.16227 \times 10^{-3}$

$$\cos \theta_1 = 0.056$$

Therefore, $\quad \theta_1 = 86.77° = 180°/(1 + r)$

Therefore, $\quad r = \dfrac{180°}{\theta_1} - 1 = 1.074$

6.4.18.2 *Coupled transmission line coupler*

Coupled co-axial lines The electromagnetic field from one coaxial transmission line can be coupled to a second adjacent line when a narrow longitudinal slot is cut between the line on the common outer conductor joint as shown in Fig. 6.41. The electric field in the input or primary line induces an equal and opposite charge on the centre conductor of the two lines.

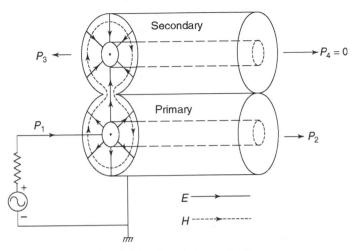

Fig. 6.41 *Coupled coaxial lines*

This results in an electric field in the coupled or secondary line directed oppositely as compared with that in the primary guide. Whereas, the magnetic field follows the same direction in both the lines. Hence the directions of power density flow $\mathbf{P} = \mathbf{E} \times \mathbf{H}$ in the two guides are opposite. Therefore, the coupled power flow is backward compared to forward power flow in the primary guide. For this reason this coupler is called the *backward coupler* where port 4 is ideally isolated. The coupling slot is made equal to the quarter wavelength in length for achieving maximum coupling as can be seen in the analysis section given below.

Microstrip and stripline coupler   Of the various types of strip line and microstrip line directional couplers, the edge-coupled parallel conductor configuration is most extensively used in practical circuits. Detailed analysis and design data on the characteristic impedances and effective dielectric constants of such structures are derived by S.B. Cohn [6.23] using the conformal mapping method and even and odd mode excitations. These couplers are shown schematically in Fig. 6.42. The lines are coupled over a length l with spacing between the adjacent edges. The width W of the lines are designed for desired impedance. The bends are metered by a length d, nearly equal to the line width W to reduce impedance mismatch and charge accumulation due to abrupt discontinuity. For 50 ohm stripline, $d = 1.131$ W, and for 50 ohm microstrip, $d = 1.194$ W. As in a coaxial system, these couplers are also backward couplers.

Analysis of the transmission line coupler   The main properties of the coupled line coupler can be analysed by decomposing the actual excitation into even and odd symmetry modes with reference to the plane of symmetry PP'. Response for actual excitation can be obtained by superimposing the responses of even and odd modes for a symmetrical structure. The even and odd mode configurations along with the actual circuit are shown in Fig. 6.43. In general, the characteristic impedances of these modes are not equal. Let us denote the characteristic imped-

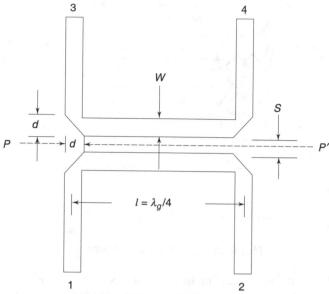

Fig. 6.42 *Strip line coupler*

ances of the even and odd modes to be Z_{oe} and Z_{oo}, respectively. The subscript 'e' will be used for even and 'o' for odd mode.

For odd mode of excitation, the instantaneous voltages at two centre conductors or strips are out of phase by 180° so that a majority of the electric field lines start from one conductor at positive potential to the other at negative potential. On the plane of symmetry PP', **E** is normal with zero tangential component, and **H** is tangential with zero normal component. Therefore, PP' can be replaced by an electric wall ($\mathbf{E}_t = 0$). For even mode of excitation, both the conductors are in phase of potential (positive, say) so that the **E**-field lines originate from both of them and terminate on the ground planes/plane. On the plane of symmetry PP', the electric field lines are completely tangential and magnetic field lines are completely normal. Therefore, PP' can be replaced by a magnetic wall ($\mathbf{H}_t = 0$). Because of different field distributions of even and odd modes, the characteristic impedances for these modes are different, denoted by Z_{oe} and Z_{oo}. These impedances are the major design parameters for any parallel coupled lines and are obtained in terms of coupling C and the single-line terminating characteristic impedance Z_o in the following manner.

The voltage outputs at the four ports of the coupler can be expressed by

$$V_1 = V_1 e + V_{10}, \; I_1 = I_{1e} + I_{10} \tag{6.156}$$

$$V_2 = V_{2e} + V_{20}, \; I_2 = I_{2e} + I_{20} \tag{6.157}$$

$$V_3 = V_{3e} + V_{30} = V_{1e} - V_{10}, \; I_3 = I_{1e} - I_{10} \tag{6.158}$$

$$V_4 = V_{4e} + V_{40} = V_{2e} - V_{20}, \; I_4 = I_{2e} - I_{20} \tag{6.159}$$

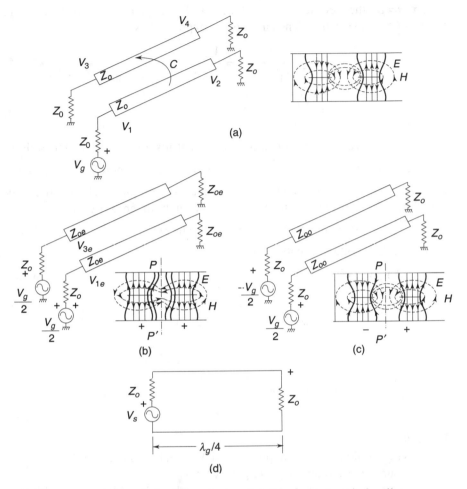

Fig. 6.43 *Transmission line couplers (a) Coupled transmission line (b) Even-mode symmetry (c) Odd-mode symmetry (d) Single equivalent circuit*

The input impedance of the coupler for TEM / quasi-TEM, even and odd modes can be obtained from

$$Z_{\text{in}} = \frac{V_{1e} + V_{10}}{I_{1e} + I_{10}} \tag{6.160}$$

$$Z_{1e} = \frac{Z_{oe}\left(Z_o + jZ_{oe}\tan\theta\right)}{Z_{oe} + jZ_o\tan\theta} \tag{6.161}$$

$$Z_{10} = \frac{Z_{oo}\left(Z_o + jZ_{oo}\tan\theta\right)}{Z_{oo} + jZ_o\tan\theta} \tag{6.162}$$

Where $\theta = \beta l$, the electrical length of the parallel coupled section of length l of a lossless ($\alpha = 0$) structure. The currents are expressed by

$$I_{1e} = \frac{1}{2}\frac{V_g}{Z_{1e} + Z_o} = \frac{V_{1e}}{Z_{1e}} \qquad I_{10} = \frac{1}{2}\frac{V_g}{Z_{10} + Z_o}$$

$$= \frac{V_{10}}{Z_{10}} \tag{6.163}$$

From Eqs 6.163–6.166, it can be shown that for a perfectly matched directional coupler, $Z_{in} = Z_o = \sqrt{Z_{oe}\, Z_{oo}}$ for all values of θ. Therefore, the problem reduces to find the even and odd mode terminal voltages in a section of transmission line of electrical length θ, terminated at both ends by an impedance $\sqrt{(Z_{oe}\, Z_{oo})}$, and excited by a voltage $1/2\, V_g$. The results when substituted in the actual terminal voltage equation we get

$$2\, V_1/V_g = 1 \tag{6.164}$$

$$2\, V_2/V_g = \frac{2}{2\cos\theta + j\sin\theta\left[\sqrt{(Z_{oe}/Z_{oo})} + \sqrt{(Z_{oo}/Z_{oe})}\right]} \tag{6.165}$$

$$2\, V_3/V_g = \frac{j\sin\theta\left[\sqrt{(Z_{oe}/Z_{oo})} - \sqrt{(Z_{oo}/Z_{oe})}\right]}{2\cos\theta + j\sin\theta\left[\sqrt{(Z_{oe}/Z_{oo})} + \sqrt{(Z_{oo}/Z_{oe})}\right]} \tag{6.166}$$

$$2\, V_4/V_g = 0 \tag{6.167}$$

These equations show that
   (1)   The coupled voltages at ports 2 and 3 are out of phase by 90°.
   (2)   Coupling is a function of frequency or θ.
   (3)   Coupled voltage at port 3 is maximum, when $\theta = $ odd   multiple of $\pi/2$.
        This corresponds to a minimum coupling length of $l = \lambda_{go}/4$ at the midband of frequency.
If C_o is the midband coupling coefficient ($\theta = \pi/2$)

$$C_o = \frac{V_3}{V_1} = \frac{Z_{oe} - Z_{oo}}{Z_{oe} + Z_{oo}} \tag{6.168}$$

Therefore, the impedances required are

$$Z_{oe} = Z_o^2/Z_{oo} = Z_o\sqrt{[(1 + C_o)/(1 - C_o)]} \tag{6.169}$$

and $\qquad\qquad Z_{oe} = Z_o^2/Z_{oe} = Z_o\sqrt{[(1 - C_o)/(1 + C_o)]} \tag{6.170}$

The coupling coefficient, transmission coefficient and directivity at any frequency can be written as

$$C = V_3 / V_1 = \frac{j\, C_o \sin\theta}{\sqrt{\left[\left(1 - C_o^2\right)\cos\theta + j\sin\theta\right]}} \tag{6.171}$$

$$T = V_2 / V_1 = \frac{\sqrt{\left(1 - C_o^2\right)}}{\sqrt{\left[\left(1 - C_o^2\right)\cos\theta + j\sin\theta\right]}} \tag{6.172}$$

$$D = V_4 / V_3 = 0 \tag{6.173}$$

Since T is 90° out of phase with both V_1 and V_3, parallel coupled line is called 90° hybrid or quadrature coupler. $D = 0$ ideally, but in practice, θ is not same for even and odd modes. Therefore, the relation $Z_o = \sqrt{\left(Z_{oe}\, Z_{oo}\right)}$ is only approximate. Moreover, due to imperfection in matching, directivity becomes non-zero. For loose coupling (more than 10 db) the present approximation is sufficiently good. The frequency response of this coupler is obtained by writing

$$\theta = 2\pi l / \lambda_g$$
$$= (2\pi l / u)\, f \tag{6.174}$$

Since $l = \lambda_{go}/4$, $\theta = \pi\, \lambda_{go}\, f / 2u$, when u is the velocity of propagation. Assuming u a constant, θ is proportional to f. The ideal frequency response of this coupler is shown in Fig. 6.44(a).

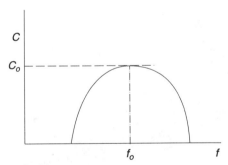

Fig. 6.44(a)   *Typical frequency response of a coupler*

The closed form expression for the even and odd mode characteristic imped-ances Z_{oe} and Z_{oo}, respectively, of edge-coupled homogeneous symmetric stripline coupler with negligible thickness strip conductors are given by

$$Z_{oe}\sqrt{\varepsilon_r} = \frac{30\pi\, K\!\left(k_e{}'\right)}{K(k_e)} \tag{6.175}$$

$$Z_{oo}\sqrt{\varepsilon_r} = \frac{30\pi\, K\!\left(k_o{}'\right)}{K(k_o)} \tag{6.176}$$

where
$$k_e = \tanh\!\left(\frac{\pi W}{2b}\right)\tanh\!\left(\frac{\pi}{2}\cdot\frac{W+s}{b}\right) \tag{6.177}$$

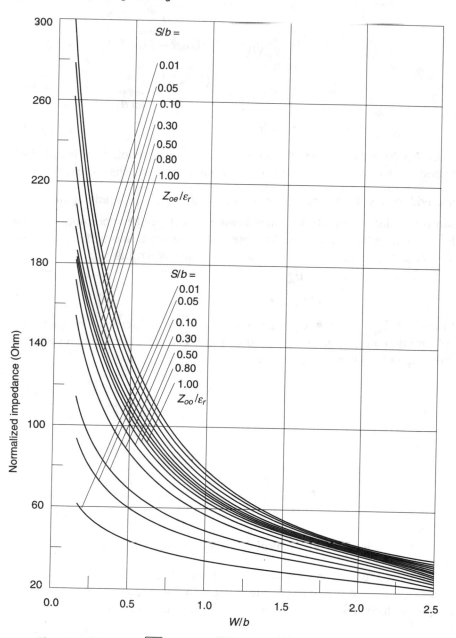

Fig. 6.44(b) $Z_{oe}\sqrt{\varepsilon_r}$ and $Z_{oo}\sqrt{\varepsilon_r}$ with W/b and s/b as parameters

$$k_o = \tanh\left(\frac{\pi W}{2b}\right) \coth\left(\frac{\pi}{2} \cdot \frac{W+s}{b}\right) \tag{6.178}$$

$$k_{e,\,o'} = \sqrt{\left(1 - k_{e,\,o^2}\right)} \tag{6.179}$$

Here $K(k)$ and $K(k')$ are the complete elliptic integrals of the first kind. The variation of $Z_{oe}\sqrt{\varepsilon_r}$ and $Z_{oo}\sqrt{\varepsilon_r}$ as a function of the dimensional parameters W/b and s/b is shown in Fig. 6.44(b).

The values of Z_{oe} and Z_{oo} can be obtained from the coupling factor. The design data of W/b and s/b are obtained from the graph in Fig. 6.44(b) for a given substrate.

Example 6.14 Design a 10 db stripline coupler at midband frequency 5 GHz with single feedline characteristic impedance $Z_o = 50$ ohm, substrate permittivity $\varepsilon_r = 9$, substrate thickness $h = 1$ mm.
Solution
1. Length of coupled section

$$\lambda_o = \frac{c}{f_o} = \frac{30}{5} = 6\ cms$$

$$l = \frac{\lambda_{go}}{4} = \frac{\lambda_o}{4\sqrt{\varepsilon_r}} = \frac{6}{4\sqrt{9}} = 0.5\ cm$$

2. Even and odd mode impedances

$$-10\ db = 20\log C_o$$

or, $C_o = 10^{-1/2} = 0.3162$, $Z_o = 50$ ohm

$$Z_{oe} = Z_o\frac{\sqrt{(1+C_o)}}{\sqrt{(1-C_o)}} = 69.5\ ohm$$

$$Z_{oo} = Z_o\frac{\sqrt{(1-C_o)}}{\sqrt{(1+C_o)}} = 36\ ohm$$

which satisfy $Z_o \approx \sqrt{Z_{oe}\,Z_{oo}} = 50.02$ ohms.

3. Line spacing s and width W
s and W are obtained from impedance vs line dimensions graphs from Fig. 6.44(b). Hence $s = 0.85$ mm, $W = 0.25$ mm.

Multisection symmetrical coupled line coupler
The bandwidth of coupled line coupler can be increased by cascading multiple sections of coupled lines as explained in Fig. 6.45. For perfect matching and increased directivity the even and odd mode impedances of various sections satisfy the following relation

$$\sqrt{(Z_{oe1}\,Z_{oo1})} = \sqrt{(Z_{oe2}\,Z_{oo2})} = \ldots = \sqrt{(Z_{oen}\,Z_{oon})} = Z_o \qquad (6.180)$$

The midband coupling coefficient of the kth section is given by

$$C_{ok} = \frac{Z_{oek}^2 - 1}{Z_{ook}^2 + 1} \qquad (6.181)$$

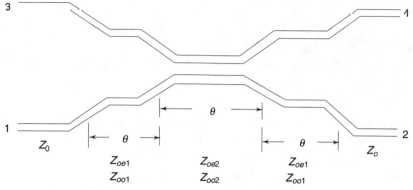

Fig. 6.45 *Multisection coupled line coupler*

The coupling coefficients can be determined from the binomial or Chebychev distributions in a similar manner as done in multisection quarter wave transformer or multihole directional couplers for a given coupling.

6.4.18.3 Branch-line directional couplers

Branch-line couplers are direct-coupled transmission line structures in which the main line is directly bridged to the secondary line by means of two shunt branches as shown in Fig. 6.46. These couplers can provide tight coupling and can handle high power.

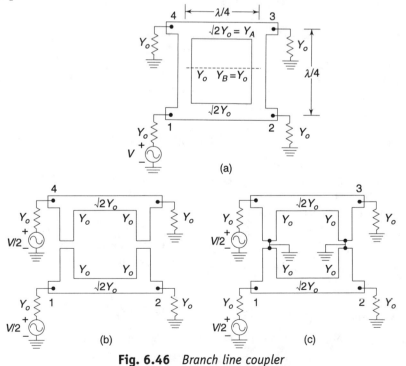

Fig. 6.46 *Branch line coupler*
(a) Coupler
(b) Even symmetry
(c) Odd symmetry

The branch lengths and spacings are quarter wavelength long at the midband frequency f_o. The characteristic admittances of the series and shunt branches are Y_A and Y_B, respectively, for a symmetrical coupler and all output and input lines have the same characteristic admittance Y_o.

The analysis of a single section branch coupler can be carried out from even and odd symmetry circuits of Figs 6.46(b) and (c). The port voltages can be found as follows

$$V_1 = V/2, \ V_2 = \frac{-jV}{2\sqrt{2}}, \ V_3 = \frac{-V}{2\sqrt{2}} \text{ and } V_4 = 0 \tag{6.182}$$

The coupling, transmission loss and the directivity are, therefore, given by

$$C = V_3/V_1 = -1/\sqrt{2} \to 3 \text{ dB} \tag{6.183}$$

$$T = V_2/V_1 = -j/\sqrt{2} \to 3 \text{ dB} \tag{6.184}$$

$$D = V_4/V_1 = 0 \tag{6.185}$$

Thus single section branch coupler is a 3 dB forward coupler.

6.4.18.4 Branching synthesis of couplers and hybrids

It can be shown that any device, whose admittance matrix is such that diagonal elements are zero and off diagonal elements are pure imaginary numbers may be synthesised by means of $\lambda/4$ or $3\lambda/4$ TEM elements of the appropriate characteristic admittance. Consider a two-port network consisting of a $\lambda/4$ line of characteristic admittance Y_{oA} inserted in a transmission line of characteristic admittance Y_o. The system is operating in TEM mode (coaxial, strip or two conductor line) so that the characteristic admittance is defined on a real voltage-current basis.

The admittance matrix of a section of line of length z is given by

$$[Y] = \frac{jY_o}{\sin \gamma z} \begin{bmatrix} -\cos \gamma z & 1 \\ 1 & -\cos \gamma z \end{bmatrix} \tag{6.186}$$

The admittance matrix of a $\lambda/4$ line of admittance Y_{oA} normalised on a Y_o basis is

$$[Y] = \frac{Y_{oA}}{Y_o} \begin{bmatrix} 0 & +j \\ +j & 0 \end{bmatrix} \tag{6.187}$$

For $l = 3\lambda/4$

$$[Y] = \frac{Y_{oA}}{Y_o} \begin{bmatrix} 0 & -j \\ -j & 0 \end{bmatrix} \tag{6.188}$$

From Eqs 6.187 and 6.188 it follows that any device, whose admittance matrix is such that diagonal elements are zero and off diagonal elements are pure imaginary numbers, can be synthesised by means of $\lambda/4$ or $3\lambda/4$ TEM elements of the appropriate characteristic admittance. A few examples for hybrid or directional coupler are given in the following paragraphs. The basic method of synthesis is to preselect the admittance matrix of a hybrid or a directional coupler and manufac ture such a matrix by branching quarter wavelength elements to obtain proto- types.

Synthesis of rat-race hybrid coupler from magic-T

The S-matrix for a matched magic-T with collinear ports 1 and 2, E and H ports 4 and 3, respectively, is

$$[S] = 1/\sqrt{2} \begin{bmatrix} 0 & 0 & 1 & 1 \\ 0 & 0 & 1 & -1 \\ 1 & 1 & 0 & 0 \\ 1 & -1 & 0 & 0 \end{bmatrix} \tag{6.189}$$

When all the reference planes are moved away from the junction by $\theta = \pi/4$ without altering the isolation property between 1, 2, 3 and 4,

$$S' = S_{12}e^{-j(\theta_1 + \theta_2)}$$
$$= S_{12}e^{-j\pi/2} = -j\, S_{12}, \text{ etc.} \tag{6.190}$$

The new S-matrix becomes

$$[S] = -j/\sqrt{2} \begin{bmatrix} 0 & 0 & 1 & 1 \\ 0 & 0 & 1 & -1 \\ 1 & 1 & 0 & 0 \\ 1 & -1 & 0 & 0 \end{bmatrix} \tag{6.191}$$

Here $[S]^* = -[S]$, $[S]_t = [S]$ (symmetry property). For a lossless network, the unitary property of $[S]$ gives:

$$[S]^* [S]_t = [U] \tag{6.192}$$

or, $$[S]^2 = -[U] \tag{6.193}$$

Now, $$[Y] = Y_o [U - S] [U + S]^{-1}; \tag{6.194}$$

By associative property of matrix operations

$$[Y] = Y_o [U - S] [U - S] [U - S]^{-1} [U + S]^{-1}$$
$$= [U - 2S + S^2] [U - S^2]^{-1} \tag{6.195}$$

So that $$[Y] = -Y_o [S]; \text{ since } [S]^2 = [U]$$

or, $$[Y] = +Y_o j/\sqrt{2} \begin{bmatrix} 0 & 0 & 1 & 1 \\ 0 & 0 & 1 & -1 \\ 1 & 1 & 0 & 0 \\ 1 & -1 & 0 & 0 \end{bmatrix} \tag{6.196}$$

Thus admittance matrix (6.196) may be synthesised by $\lambda/4$ lines for the positive coefficients and 3 $\lambda/4$ lines for the negative coefficients of admittance

$$Y_{oA}/Y_o = 1/\sqrt{2} \tag{6.197}$$

Here the self admittances y_{11}, y_{22}, y_{33} and $y_{44} = 0$. Since $y_{12} = y_{34} = 0$, and $y_{21} = y_{43} = 0$, there is no direct element between ports 1 and 2 and between ports 3 and 4. The following elements are required to be inserted:

(i) Between ports 1 and 3, a $\lambda/4$ element of normalised characteristic admittance $1/\sqrt{2}$.

(ii) Between 1 and 4, a $\lambda/4$ element of normalised characteristic admittance $1/\sqrt{2}$.

(iii) Between 2 and 3, a $\lambda/4$ element of normalised characteristic admittance $1/\sqrt{2}$.

(iv) Between 2 and 4, a $3\lambda/4$ element of normalised characteristic admittance $1/\sqrt{2}$.

A schematic circuit and its practical realisation in coax line form is shown in Fig. 6.47. This circuit is called a *rat-race hybrid*.

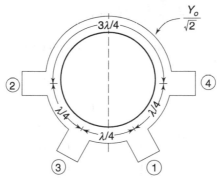

Fig. 6.47 *Rat-race*

Example 6.15 For the rat-race hybrid shown in Fig. 6.47, (1) calculate input admittance at port 1 when all others are matched, (2) when input is fed at port 1, find signal distributions in all other ports, (3) if input is fed to port 3 and all others are matched, calculate power distribution in all ports.

Solution

(1) When all ports are matched the admittance matrix equation is

$$\begin{bmatrix} I_1 \\ I_2 \\ I_3 \\ I_4 \end{bmatrix} = j/\sqrt{2} \begin{bmatrix} 0 & 0 & 1 & 1 \\ 0 & 0 & 1 & -1 \\ 1 & 1 & 0 & 0 \\ 1 & -1 & 0 & 0 \end{bmatrix} \begin{bmatrix} V_1 \\ V_2 \\ V_3 \\ V_4 \end{bmatrix}$$

when all ports except port 1 are matched terminated by normalised impedance $Z_o = 1$,

$$V_2 = -I_2 Z_o = -I_2$$
$$V_3 = -I_3$$
$$V_4 = -I_4$$

Therefore,

$$\begin{bmatrix} I_1 \\ I_2 \\ I_3 \\ I_4 \end{bmatrix} = j/\sqrt{2} \begin{bmatrix} 0 & 0 & 1 & 1 \\ 0 & 0 & 1 & -1 \\ 1 & 1 & 0 & 0 \\ 1 & -1 & 0 & 0 \end{bmatrix} \begin{bmatrix} V_1 \\ -I_2 \\ -I_3 \\ -I_4 \end{bmatrix}$$

or,

$$I_1 = -j/\sqrt{2} \ (I_3 + I_4)$$

$$I_2 = +j/\sqrt{2} \ (-I_3 + I_4)$$

$$I_3 = -j/\sqrt{2} \ (V_1 - I_2)$$

$$I_4 = +j/\sqrt{2} \ (V_1 + I_2)$$

Adding, $\quad I_3 + I_4 = j/\sqrt{2} \ (2V_1) = j\sqrt{2} \ V_1$

Substituting in the first equation above,

$$I_1 = -\frac{j}{\sqrt{2}} \cdot \frac{j}{\sqrt{2}} \cdot 2V_1 = V_1$$

Therefore, $I_1/V_1 = 1 =$ normalised input admittance at port 1. Thus if all ports are matched, port 1 also matched.

(2) When input is fed at port 1 the S-matrix equation is

$$\begin{bmatrix} b_1 \\ b_2 \\ b_3 \\ b_4 \end{bmatrix} = -j/\sqrt{2} \begin{bmatrix} 0 & 0 & 1 & 1 \\ 0 & 0 & 1 & -1 \\ 1 & 1 & 0 & 0 \\ 1 & -1 & 0 & 0 \end{bmatrix} \begin{bmatrix} a_1 \\ 0 \\ 0 \\ 0 \end{bmatrix}$$

Therefore, $\quad b_1 = b_2 = 0; \qquad P_1 = P_2 = 0$

$$b_3 = -ja_1/\sqrt{2}; \quad P_3 = |b_3|^2 = |a_1|^2/2$$

$$b_4 = -ja_1/\sqrt{2}; \quad P_4 = |b_4|^2 = |a_1|^2/2$$

Thus the input to port 1 is split equally and in phase at ports 3 and 4, with no power reflected back to port 1 and no output at port 2. If there is mismatch at port 4 it will not affect port 3 and vice-versa since $S_{34} = 0$.

(3) When input is fed to port 3 and all others are matched, the S-matrix equation is

$$\begin{bmatrix} b_1 \\ b_2 \\ b_3 \\ b_4 \end{bmatrix} = -j/\sqrt{2} \begin{bmatrix} 0 & 0 & 1 & 1 \\ 0 & 0 & 1 & -1 \\ 1 & 1 & 0 & 0 \\ 1 & -1 & 0 & 0 \end{bmatrix} \begin{bmatrix} 0 \\ 0 \\ a_3 \\ 0 \end{bmatrix}$$

$$b_1 = -ja_3/\sqrt{2}, \quad P_1 = |a_3|^2/2$$

$$b_2 = -ja_3/\sqrt{2}, \qquad P_2 = |a_3|^2/2$$
$$b_3 = b_4 = 0, \qquad P_3 = P_4 = 0$$

Thus input at port 3 is equally split with the same phase at ports 1 and 2, with no power reflected back to port 3 and no output at port 4.

A practical application of rat-race is shown in Fig. 6.48 as a hybrid transformer.

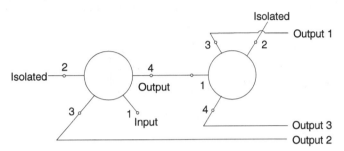

Fig. 6.48 *Hybrid transformer*

Synthesis of 3 dB, 90° hybrid from magic-T By choosing the shift of reference planes 1 and 3 as $\theta_1 = \theta_3 = \pi/2$; and those 2 and 4 as $\theta_2 = \theta_4 = \pi$, the S-matrix of the magic-T reduces to

$$[S] = -1/\sqrt{2}\begin{bmatrix} 0 & 0 & 1 & j \\ 0 & 0 & j & 1 \\ 1 & j & 0 & 0 \\ j & 1 & 0 & 0 \end{bmatrix} \tag{6.198}$$

such that isolation between 1 and 2, 3 and 4 are not violated. Here, new parameters are

$$S_{13} = \frac{b_1\, e^{-j\theta_1}}{a_3\, e^{j\theta_3}} = \frac{b_1}{a_3}\, e^{-j(\theta_1 + \theta_3)};$$

or,
$$S_{13} = e^{-j(\pi/2 + \pi/2)} = -1 \tag{6.199}$$
$$S_{14} = e^{-j(\pi/2 + \pi)} = -j \tag{6.200}$$
$$S_{23} = e^{-j(\pi/2 + \pi)} = -j \tag{6.201}$$
$$S_{24} = e^{-j(\pi + \pi)} = 1, \text{ etc.} \tag{6.202}$$

The corresponding normalised admittance matrix

$$[Y] = [U - S]\,[U + S]^{-1}$$

$$= j\begin{bmatrix} 0 & 1 & 0 & \sqrt{2} \\ 1 & 0 & \sqrt{2} & 0 \\ 0 & \sqrt{2} & 0 & 1 \\ \sqrt{2} & 0 & 1 & 0 \end{bmatrix} \tag{6.203}$$

Thus $Y_{11} - Y_{13} - Y_{22} - Y_{24} = Y_{31} = Y_{33} = Y_{42} = Y_{44} = 0$. The physical line elements to be inserted are as follows:

1. Between ports (3, 4) and (1, 2): a $\lambda/4$ element of characteristic admittance $= Y_o$.
2. Between (2, 3) and (1, 4): a $\lambda/4$ element of characteristic admittance $= \sqrt{2}\, Y_o$.

The implementation of this hybrid is shown in Fig. 6.49. If power is fed at port 1 with all the other ports matched, power distribution in all ports can be found as follows

$$
\begin{bmatrix} b_1 \\ b_2 \\ b_3 \\ b_4 \end{bmatrix} = -1/\sqrt{2}
\begin{bmatrix} 0 & 0 & 1 & j \\ 0 & 0 & j & 1 \\ 1 & j & 0 & 0 \\ j & 1 & 0 & 0 \end{bmatrix}
\begin{bmatrix} a_1 \\ 0 \\ 0 \\ 0 \end{bmatrix}
\tag{6.204}
$$

$$
b_1 = b_2 = 0,\ b_3 = (-1/\sqrt{2})\, a_1,\ b_4 = (-j/\sqrt{2})\, (a_1)
\tag{6.205}
$$

Therefore, b_3 and b_4 are 90° out of phase but the magnitude of power are same. This coupler is called 3 dB 90° hybrid coupler.

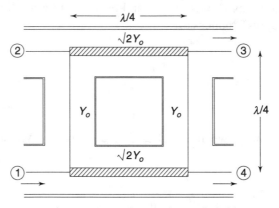

Fig. 6.49 *3 dB, 90 degree hybrid*

Example 6.16 Find the coupling to port 2 when signal is fed to port 1 with 3 and 4 connected to $\lambda/4$ short circuited line sections.

Solution

Power distribution can be found from the matrix equation $\{b\} = [S]\{a\}$.

or,
$$
\begin{bmatrix} b_1 \\ b_2 \\ b_3 \\ b_4 \end{bmatrix} = -1/\sqrt{2}
\begin{bmatrix} 0 & 0 & 1 & j \\ 0 & 0 & j & 1 \\ 1 & j & 0 & 0 \\ j & 1 & 0 & 0 \end{bmatrix}
\begin{bmatrix} a_1 \\ 0 \\ b_3 \\ b_4 \end{bmatrix}
\qquad \because \text{ Port 3 and 4 are opened, i.e. } \Gamma = 1
$$

$$b_1 = -\frac{1}{\sqrt{2}}(b_3 + j\,b_4)$$

$$b_2 = -\frac{1}{\sqrt{2}}(jb_3 + b_4)$$

$$b_3 = -\frac{1}{\sqrt{2}}a_1$$

$$b_4 = -\frac{1}{\sqrt{2}}j\,a_1$$

$$b_1 = -\frac{1}{\sqrt{2}}\left[-\frac{1}{\sqrt{2}}a_1 + j\left(-j/\sqrt{2}\right)a_1\right]$$

$$= -\frac{1}{\sqrt{2}}\left[-\frac{1}{\sqrt{2}}a_1 + \frac{a_1}{\sqrt{2}}\right] = 0$$

$$b_2 = -\frac{1}{\sqrt{2}}[jb_3 + b_4]$$

$$= -\frac{1}{\sqrt{2}}\left[-\frac{j}{\sqrt{2}}a_1 - \frac{j}{\sqrt{2}}a_1\right] = 0$$

$$= j\frac{1}{2}\,2a_1 = ja_1 = a_1 e^{j\pi/2}$$

Now in terms of current and voltage representation

$$V_2 = -I_2 Z_o = -I_2,\ Z_o = 1$$

$$\{I\} = \{Y\}\{V\}$$

or,

$$\begin{bmatrix} I_1 \\ I_2 \\ 0 \\ 0 \end{bmatrix} = j \begin{bmatrix} 0 & 1 & 0 & \sqrt{2} \\ 1 & 0 & \sqrt{2} & 0 \\ 0 & \sqrt{2} & 0 & 1 \\ \sqrt{2} & 0 & 1 & 0 \end{bmatrix} \begin{bmatrix} V_1 \\ -I_2 \\ V_3 \\ V_4 \end{bmatrix}$$

$$I_1 = j[-I_2 + \sqrt{2}\,V_4] = j[-jV_1 - j\sqrt{2}\,V_3 + \sqrt{2}\,V_4]$$

$$I_2 = j[V_1 + \sqrt{2}\,V_3]$$

$$0 = -V_2 I_2 + V_4$$

$$0 = +\sqrt{2}\,V_1 + V_3$$

Therefore, $V_3 = -\sqrt{2}\,V_1$

and

$$I_2 = j(V_1 - \sqrt{2}\,V_2 V_1) = j(V_1 - 2V_1) = -jV_1$$

$$V_4 = \sqrt{2}\,I_2 = -j\sqrt{2}\,V_1$$

Therefore, $I_1 = j \left[-jV_1 - j\sqrt{2}\,(-\sqrt{2}\,)V_1 + \sqrt{2}\,(-j\sqrt{2}\,)V_1 \right]$

Normalised input admittance $Y_1 = I_1/V_1 = j\left[-j + 2j - 2j\right] = 1$

Therefore, the port must be matched also. Actual stripline configuration of the coupler is shown in Fig. 6.50.

If V_1 and V_2 are the input voltages at ports 1 and 2, output voltage from ports 3 and 4 are $-j(V_1 + V_2)/\sqrt{2}$ and $-j(V_1 - V_2)/\sqrt{2}$ or power $(V_1 + V_2)^2/2Z_o$ and $(V_1 - V_2)^2/Z_o$. Thus sum and difference arms of magic-T are obtained.

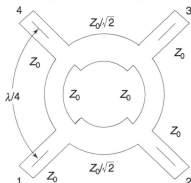

Fig. 6.50 *Strip branch line coupler*

6.4.19 Power Dividers and Combiners

A power divider is a device to split the input power into a number of smaller amounts of power at multiple ports (N) to feed N number of branching circuits with isolation between the output ports. For simplicity, a two-way equal power divider is shown in Fig. 6.51 which is a lossless three port junction. For equal power division, the device consists of two quarterwave sections with characteristic impedances Z_o connected in parallel with the input line, which also has a characteristic impedance Z_o. A resistor $R = 2Z_o$ is connected between ports 2 and 3 which are matched terminated. Since the input impedance at port 1 is now $Z_o/2$, a quarterwave matching transformer with characteristic impedance $Z_o/\sqrt{2}$ is used to transform the port 1 input impedance into the feeder line impedance Z_o. This also maintains zero current in the resistance R when port 2 and 3 are matched terminated.

With the help of even and odd mode analysis it can be shown that the power applied to port 1 divides equally between ports 2 and 3 with zero loss in the balancing resistor R, and the voltage at either output port lags that at the input port by 90°. Thus the device is a 3 dB, 90° power divider. It can be shown that the configuration also acts as a 3 dB power combiner when fed from the ports 2 and 3 with the output taken at port 1.

6.4.20 Turnstile Junction

A turnstile junction is a symmetrical six-port device as shown in Fig. 6.52. Ports 1, 2, 3 and 4 correspond to rectangular waveguide operating in dominant TE_{10}

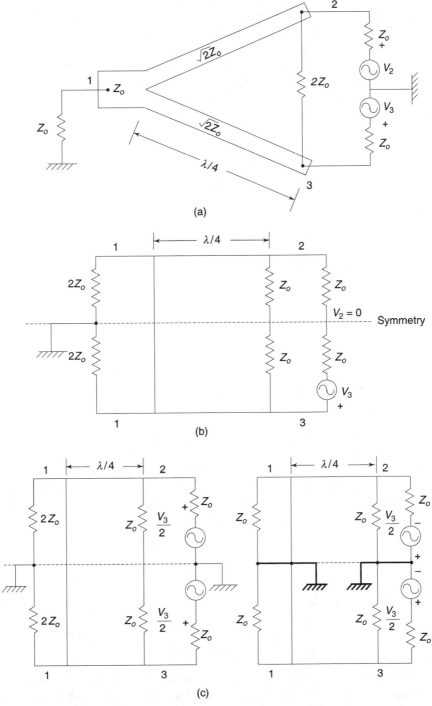

Fig. 6.51 *Two way power divider (a) Microstrip configuration (b) Equivalent circuit (c) Even and odd modes symmetries*

mode. At the centre a circular waveguide is placed with its axis perpendicular to the H-plane of the rectangular waveguide. Because of the rotational symmetry of the circular waveguide operating in TE_{11} mode, it can support two cross-polarised waves, each independent of the other, and thus exhibits two ports 5 and 6 correspond to each polarisation. Therefore, the circular waveguide can deliver power through any one of these waves for which **E** is parallel to the polarisation of the waves excited in the guide. If the polarisation is somewhere in between then both the ports 5 and 6 draw power.

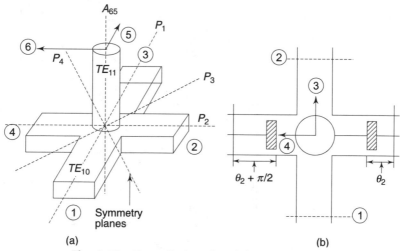

(a)   (b)

Fig. 6.52 *Turnstile junction (a) Isometric view*
(b) Top view of 4 port hybrid junction

There are four planes of symmetry P_1, P_2, P_3 and P_4 and one axis of symmetry A_{56}. Coupling between the circular and rectangular waveguides is achieved through a circular hole at the junction in the plane transverse to the circular guide. Plane P_1 interchanges the fields at ports 2 and 4, reverses the fields at 6, but leaves fields at 1, 3 and 5 unchanged. The reference planes are set to make the scattering coefficients real for a matched turnstile junction:

$$[S] = 1/2 \begin{bmatrix} 0 & 1 & 0 & 1 & \sqrt{2} & 0 \\ 1 & 0 & 1 & 0 & 0 & \sqrt{2} \\ 0 & 1 & 0 & 1 & -\sqrt{2} & 0 \\ 1 & 0 & 1 & 0 & 0 & -\sqrt{2} \\ \sqrt{2} & 0 & -\sqrt{2} & 0 & 0 & 0 \\ 0 & \sqrt{2} & 0 & -\sqrt{2} & 0 & 0 \end{bmatrix} \qquad (6.206)$$

Ports 5 and 6 can be matched by tuning screws in the corresponding arms independently, since $S_{56} = 0$. The ports 1, 2, 3 and 4 are matched by inserting a tuning post along the axis A_{56} within the junction from the bottom.

Application of turnstile junction

There are many applications of a turnstile junction as described below.

(a) Four port hybrid junction

By placing a short circuit at port 2 at distance θ_2 from the reference plane and another at port 4 at $\theta_4 = \theta_2 + \pi/2$, and with all other ports matched, the S-matrix equation for an input at port 1 is

$$
\begin{bmatrix} b_1 \\ b_2 \\ b_3 \\ b_4 \\ b_5 \\ b_6 \end{bmatrix} = [S] \begin{bmatrix} a_1 \\ -b_2\, e^{j2\theta_2} \\ 0 \\ b_4\, e^{j2\theta_2} \\ 0 \\ 0 \end{bmatrix}
\tag{6.207}
$$

This reduces the S matrix to

$$
S = \frac{1}{\sqrt{2}} \begin{bmatrix} 0 & 0 & 1 & -e^{j2\theta_2} \\ 0 & 0 & -1 & -e^{j2\theta_2} \\ 1 & -1 & 0 & 0 \\ -e^{j2\theta_2} & -e^{j2\theta_2} & 0 & 0 \end{bmatrix}
\tag{6.208}
$$

If $\theta_2 = n\pi$, $n = 0, 1, 2, \dots$; $e^{j\theta_2} = 1$, the $[S]$ becomes that of hybrid junction or magic-T with some phase change due to negative sign in the elements ($e^{j2n\pi} = 1$):

$$
[S] = 1/\sqrt{2} \begin{bmatrix} 0 & 0 & 1 & -1 \\ 0 & 0 & -1 & -1 \\ 1 & -1 & 0 & 0 \\ -1 & -1 & 0 & 0 \end{bmatrix}
\tag{6.209}
$$

Here ports 3 and 4 are contained in the same physical circular waveguide through cross polarisation replacing the ports 5 and 6, respectively, of turnstile junction as shown in Fig. 6.52(b). An input signal at port 1 produces equal and opposite signals at new ports 3 and 4. An input at port 2 produces equal and inphase signals at 3 and 4. Ports 1 and 2, 3 and 4 are isolated to each other. This satisfies the properties of a hybrid four-port coupler.

(b) Matched four-way power divider

When an equal signal is fed into ports 5 and 6 in the turnstile junction with all other ports match terminated, the output signal can be expressed by

$$
\begin{bmatrix} b_1 \\ b_2 \\ b_3 \\ b_4 \\ b_5 \\ b_6 \end{bmatrix} = [S] \begin{bmatrix} 0 \\ 0 \\ 0 \\ 0 \\ a \\ a \end{bmatrix} = 1/\sqrt{2} \begin{bmatrix} a \\ a \\ -a \\ -a \\ 0 \\ 0 \end{bmatrix}
\tag{6.210}
$$

This acts as a four-way (ports 1, 2, 3 and 4) equisignal distribution network. Here the phase of signals b_1 and b_2 is opposite to that of b_3 and b_4.

(c) Matched three-way power divider

When a short circuit is placed in circular waveguide ports at a distance θ from the reference plane and input power is fed to port 1, the output power distribution can be expressed by

$$\begin{bmatrix} b_1 \\ b_2 \\ b_3 \\ b_4 \\ b_5 \\ b_6 \end{bmatrix} = [S] \begin{bmatrix} a_1 \\ 0 \\ 0 \\ 0 \\ -b_5 e^{j2\theta} \\ -b_6 e^{j2\theta} \end{bmatrix} = 1/\sqrt{2} \begin{bmatrix} -\sqrt{2}\, b_5\, e^{j2\theta} \\ a_1 - \sqrt{2}\, b_6\, e^{j2\theta} \\ \sqrt{2}\, b_5\, e^{j2\theta} \\ a_1 + \sqrt{2}\, b_6\, e^{j2\theta} \\ \sqrt{2}\, a_1 \\ 0 \end{bmatrix} \qquad (6.211)$$

Substituting the elements of the S-matrix and solving the above equation yields

$$b_1 = -a_1 e^{j2\theta}/2, \ b_2 = a_1/2, \ b_3 = a_1 e^{j2\theta}/2, \ b_4 = a_1/2,$$

$$b_5 = a_1/\sqrt{2} \ \text{and} \ b_6 = 0$$

Since b_5 and b_6 are not used due to short circuit terminations, and if $\theta = n\pi$, b_2, b_3 and b_4 are equiphase and equi-amplitudes always. Since b_1 is the reflected component at the input port, this configuration acts as a three-way power divider. Since the reflection coefficient at port 1 is $e^{j2\theta}/2$,

$$\text{Input VSWR at port 1} = \frac{|a_1| + |b_1|}{|a_1| - |b_1|} = 3.0$$

Because of high VSWR an auxiliary matching structure is required at the input port. After matching, the power will be equally divided into the three rectangular waveguide ports 2, 3 and 4.

EXERCISES

6.1 A shunt susceptance $j0.5$ mho is connected across a lossless transmission line with characteristic impedance 50 ohm, terminated by matched impedances. Find the S-matrix of the junction.

6.2 A series reactance j 40 ohm is connected between two lossless transmission lines of characteristic impedances 50 ohm and 75 ohm. Find the S-matrix of the junction. Assume matched termination at both ends of these lines.

6.3 A reciprocal two port microwave device has a VSWR of 1.5 and an insertion loss of 2 dB. Find the magnitudes of S-parameters for the device.

6.4 A waveguide load has a VSWR of 1.1 and is used to absorb an average power of 5 W. Find the reflected power and the return loss.

6.5 Find the scattering matrix for an ideal short circuited section of waveguide with the reference ports chosen. Show that the coefficients are real numbers. How does the S-matrix change if the short circuit position of port 2 is variable?

6.6 A 10 dB attenuator having an input VSWR of 1.2 is terminated by matched load. Find the reflected power, the absorbed power and the transmitted power.

6.7 Plot the dB attenuation versus angle for an ideal rotary attenuator.

6.8 An ideal rotary phase shifter is terminated with a matched load. Find the scattering matrix for the device. Express the transmitted voltage at the output   in terms of the input for angles 0, 30, 45 and 90 degrees. Assume that the input voltage is real.

6.9 In a precision rotary phase shifter the output quarter wave plate, the transition and the rectangular waveguides are rotated by an angle of 30°. Show that phase of the transmitted wave will be changed by 30°.

6.10 The input power to the sum arm of an ideal matched magic-T is 1 W. Find the output powers from the other arms.

6.11 Find the magnitude of scattering coefficient for a directional coupler having a coupling coefficient of 3 dB and directivity of 25 dB and VSWR of 1.2 for the main guide.

6.12 The input power to a lossless matched directional coupler is 100 mW. If the coupling coefficient is 20 dB and the directivity is 30 dB, find the output powers at other ports.

6.13 An ideal three-port circulator is fed at port 1 with average power 100 W. If the power reflected by the antenna at the next port 2 is 100 μW, find the power outputs at all ports assuming they are match terminated.

6.14 A three-port circulator has an insertion loss of 1 dB, an isolation of 20 dB, and a VSWR of 1.2. Find the output power at ports 2 and 3 for an input power of 100 mW at port 1.

6.15 An E-plane tee is matched terminated at all the ports with an input power of 5 mW fed at port 2 (E arm). Determine the power flow through the junction. What changes in the power distribution will occur if the power is fed at the collinear arm 1.

6.16 A directional coupler of 10 dB coupling and 40 dB directivity produces a transmission loss of 1 dB. For an input power of 10 mW at the input port of the main arm, determine the power at the other ports.

6.17 A microstrip edge-coupled directional coupler is designed for a coupling factor of 10 dB and a characteristic impedance   of 50 ohm. Determine the even and odd mode characteristic impedances.

6.18 A Bethe-hole directional coupler with a centre circular aperture using the rectangular waveguide of size 0.9 in $\times$ 0.4 in is designed to operate at 10 GHz with 30 dB coupling. Find the aperture radius and the frequency band over which the directivity remains   greater than 20 dB.

6.19 Design a three-hole Chebychev directional coupler using centre apertures in the common broad wall between two rectangular waveguides of dimensions 0.9″ $\times$ 0.4″ to be operated at 9 GHz. Find the aperture radii, spacing and the bandwidth of the coupler for a coupling of 30 dB and directivity of 30 dB.

6.20 Design a stripline edge-coupled 10 dB directional coupler for ground plane spacing of 1 cm at 5 GHz. Find the strip width and spacing for obtaining input and output line characteristic impedances equal to 50 ohm.

6.21 A branch line 3 dB microstrip directional coupler at 5 GHz is designed with input and output line impedances of 50 ohm. Find the characteristic impedance of the through lines and the branch lines and their lengths in terms of the wavelength. The dielectric substrate is 20 mm thick and has a dielectric constant of 10. Find the width of the lines and the lengths of the through lines and branch lines.

6.22 A microstrip hybrid ring is constructed on 1 mm thick substrate with a dielectric constant of 2.5 at 3 GHz. Determine the widths of the transmission line and the ring as well as the radius of the ring for input and output line impedances of 50 ohm.

6.23 Show using *S*-matrix theory that a lossless non reciprocal two-port microwave device cannot be constructed.

REFERENCES

6.1 Lebedev, I, *Microwave Engineering*, MIR Publishers, Moscow,1973.

6.2 Bhat, Bharathi and Shiban K Koul, *Stripline-like Transmission Lines for Microwave Integrated Circuits*, Wiley Eastern limited, New Delhi, 1989.

6.3 Edwards, Terry, *Foundations for Microstrip Circuit Design*, John Wiley and Sons, New York,1981.

6.4 Seeger, John A, *Microwave Theory Components and Devices*, Prentice - Hall, Englewood Cliffs, New Jersey, 1988.

6.5 Altman, J L, *Microwave Circuits*, D. Van Nostrand Company, Inc. 1964.

6.6 Veley, V F, *Modern Microwave Technology*, Prentice-Hall, Inc., 1987.

6.7 Liao, S Y, *Microwave Devices and Circuit*, Prentice-Hall of India Private Limited, New Delhi, 1995.

6.8 Collin, R E, *Foundations for Microwave Engineering*, McGraw-Hill Inc., 1992.

6.9 Roddy, D, *Microwave Technology*, Prentice-Hall, Inc., 1986.

6.10 Montogomery, C G R H, Dicke, and E M, Purcell, *Principles of Microwave Circuits*, McGraw-Hill Book Company, New York, 1948.

6.11 Marcuvitz, N, *Waveguide Handbook*, McGraw-Hill Book Company, New York, 1951.

6.12 Bahl, I, and P Bhartia, *Microwave Circuit Design*, John Wiley and Sons, Inc., New York, 1988.

6.13 Sooboo, R F, *Theory and Application of Ferrites*, Prentice-Hall, Inc., Englewood Cliffs, N.J., 1960.

6.14 Clarricoats, P, *Microwave Ferrites*, John Wiley and Sons, Inc., New York, 1961.

6.15 Von Aulock, H Wilhelm, and E F Clifford, *Linear Ferrite Devices for Microwave Applications,* New York, Academic Press, Inc, 1968.

6.16 Bosma, H, On Stripline Y-Circulators at UHF, *IRE Trans.,* Vol. MTT-12, pp. 61-72, 1964.

6.17 Wu, Y S, and F J Rosenbaum, "Wide-Band Operation of Microstrip Circulators", *IEEE Trans.,* Vol. MTT-13, pp. 15–27, 1965.

6.19 "The Microwave Engineers Handbook and Buyers", *Guide, 1963 Ed.,* pp. T-107 to T-111, Horizon House-Microwave, Inc., Brookline, Mass.

6.20 Wilkinson, E, "An N-way Hybrid Power Divider " , *IEEE Trans.* MTT, pp 116-118, 1960.

6.21 Lance, A L, *Introduction to Microwave Theory and Measurements,* New York, McGraw-Hill, 1964.

6.22 Wheeler, G J, *Introduction to Microwaves,* Englewood Cliffs, N.J., Prentice-Hall, 1963.

6.23 Cohn, S B, and R Levy, " History of microwave passive components with particular attention to directional couplers", *IEEE Trans.* MTT, Vol-32 No.9, 1984.

6.24 Young, Leo, *Advances in Microwaves,* Vol.1, New York, Academic Press, Inc., 1966.

6.25 Gupta, K C, and Amarjit Singh, *Microwave Integrated Circuits,* John Wiley and Sons, New York, Inc., 1974.

6.26 Howe, JR, *Stripline Circuit Design*, Dedham, Mass. Artech House, Inc. 1974.

Chapter 7

Microwave Resonators

7.1 INTRODUCTION

Microwave resonators are tunable circuits used in microwave oscillators, amplifiers, wavemeters and filters. At the tuned frequency the circuit resonates where the average energies stored in the electric field (or in the capacitor), W_e and magnetic field (or in the inductor), W_m are equal and the circuit impedance becomes purely real. The total energy is therefore twice the electric or magnetic energy stored in the resonator. The parameters which describe the performance of a resonator are

1. Resonant frequency f_r at which the energy in the cavity attains maximum value $= 2W_e$ or $2W_m$.
2. Quality factor Q which is a measure of the frequency selectivity of a cavity, defined by

$$Q = \frac{2\,\pi \times \text{maximum energy stored}}{\text{energy dissipated per cycle}}$$

$$= f_r / 3 \text{ dB bandwidth}$$

3. Input impedance of the resonator specifying the matching with the input/output circuits.

The Q-factor relates a resonant circuit's capacity for electromagnetic energy storage with its energy dissipation through heat. In order to obtain high Q, the circuits are made highly reactive and can be simulated by using short circuited or open circuited sections of low loss coaxial transmission lines in the frequency range of 100 to 1000 MHz. Typical values of Q up to about 10,000 can be achieved from such transmission sections.

At higher frequencies transmission line resonators do not give very high Q due to skin effect and radiation loss in braided cables. Therefore, most of the microwave resonators are constructed with rectangular or circular metallic cavities, which have a natural resonant frequency and behave like a *LCR* circuit. The electric and magnetic energies stored in the volume of the cavity give rise to equivalent C and L, while the walls, having a finite conductivity, give rise to power loss which provides for the equivalent resistance R. The resonator bandwidth is inversely proportional to the Q factor.

The following sections describe the analysis of microwave resonators, their performances and applications.

7.2 COAXIAL RESONATORS

There are three basic configurations of coaxial resonant cavities: (1) quarterwave coaxial cavity; (2) halfwave coaxial cavity; and (3) coaxial cavity with a shortening capacitance as shown in Fig. 7.1 along with their equivalent circuits.

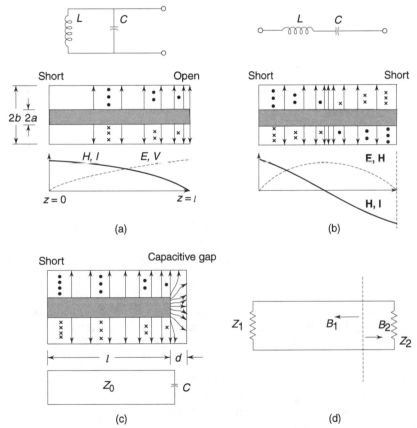

Fig. 7.1 *Coaxial cavities and equivalent circuits (a) Quarterwave cavity*
(b) Halfwave cavity (c) Capacitive end cavity (d)Resonance condition
(e) $\times \times \times$ H

The general method to determine the resonant length of these cavities is based on the fact that the total susceptance of the oscillatory system becomes zero at resonance, i.e., $B_1 + B_2 = 0$, where, B_1 and B_2 are the susceptances of the sections of the transmission lines looking towards left and right, respectively, from an arbitrary reference plane on the line. By taking the reference plane at right hand side end of these cavities, with reference to Fig. 7.1 (d), the resonant length l is obtained as follows.

7.2.1 Quarterwave Coaxial Cavity

From the resonant condition

$$j \, Z_0 \tan \beta l = \text{infinite}$$

or, $\qquad\qquad l = \lambda/4 \, (2n - 1)$ \hfill (7.1)

This cavity is essentially equivalent to a parallel resonant circuit.

7.2.2 Halfwave Coaxial Cavity

From the resonant condition

$$j Z_0 \tan \beta l = 0$$

or, $\qquad\qquad l = \lambda/2 \, (2n - 1)$ \hfill (7.2)

This is equivalent to series resonance.

7.2.3 Capacitive End Coaxial Cavity

From the resonant condition

$$j Z_0 \tan \beta l = - \, 1/j \omega_r C$$

or, $\qquad\qquad l = \lambda/2\pi \tan^{-1} (1/Z_0 \omega_r C)$ \hfill (7.3)

Here $n = 1, 2, 3, \cdots$; Z_0 is the characteristic impedance of the coaxial line and C is the gap capacitance between the centre conductor and the shorting termination of Fig. 7.1(c). The dimensions a and b of all the coaxial cavities in *TEM* modes are restricted by the generation of next higher order *TE* and *TM* mode, so that $\pi \, (b + a) < \lambda$.

Since the diameters of the coaxial cavities determine the power loss in the cavity, Q varies with b/a and attains a maximum value at $b/a = 3.6$. Due to microwave radiation from the open end of the quarterwave coaxial cavity, halfwave coaxial resonators are preferred over quarterwave sections and are used in microwave resonant wavemeters.

The coaxial cavity can be tuned either by changing the capacitive gap d, by means of a capacitive ridge at a fixed length l, or by changing length l by a variable shorting plunger at a fixed value of gap length d.

7.3 WAVEGUIDE CAVITY RESONATORS

Waveguide cavity resonators are formed by shorting two ends of a section of a waveguide as shown in Fig. 7.2. The field components inside the cavity can be computed from the wave equations which satisfy the boundary condition of the zero tangential E field at all conducting walls. Within the cavity, various TE_{mnp} and TM_{mnp} modes exist. Very high Q factors and accompanying narrow bandwidths can be obtained with these resonators. External circuits are coupled to the cavity through transmission line probes or loops or apertures. The cavity fields

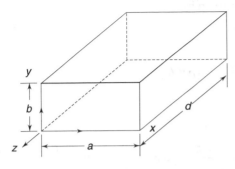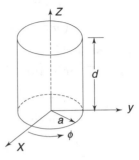

Fig. 7.2 *Waveguide cavity resonators (a) Rectangular cavity (b) Circular cavity*

can be very large compared to the input value because the constant stored energy oscillates between the electric and magnetic fields.

7.3.1 Rectangular Cavity

Because of the metallic surfaces at the boundaries, the field distributions possess standing wave patterns in all the three directions.

The following types of field components can exist in a rectangular cavity:

TE$_{mnp}$ **mode field:**

$$H_z = H_0 \cos (m\pi x/a) \cos(n\pi y/b) \sin (p\pi z/d); \tag{7.4}$$

$$H_y = \frac{1}{k_c^2} \frac{\partial^2 H_z}{\partial y \partial z}$$

$$= - H_0/k_c^2 \, (p\pi/d)(n\pi/b) \cos (m\pi x/a) \sin (n\pi y/b) \cos (p\pi z/d) \tag{7.5}$$

$$H_x = \frac{1}{k_c^2} \frac{\partial^2 H_z}{\partial x \partial z}$$

$$= - H_0/k_c^2 \, (p\pi/d)(m\pi/a) \sin (m\pi x/a) \cos (n\pi y/b) \cos (p\pi z/d) \tag{7.6}$$

$$E_z = 0 \tag{7.7}$$

$$E_y = j \frac{\omega \mu H_0}{k_c^2} \frac{\partial H_z}{\partial x}$$

$$= -j \frac{\omega \mu H_0}{k_c^2} \left(\frac{m\pi}{a} \right) \sin \left(\frac{m\pi x}{a} \right) \cos \left(\frac{n\pi y}{b} \right) \sin \left(\frac{p\pi z}{d} \right) \tag{7.8}$$

$$E_x = -j \frac{\omega \mu H_0}{k_c^2} \frac{\partial H_z}{\partial y}$$

$$= j \frac{\omega \mu H_0}{k_c^2} \left(\frac{n\pi}{b} \right) \cos \left(\frac{m\pi x}{a} \right) \sin \left(\frac{n\pi y}{b} \right) \sin \left(\frac{p\pi z}{d} \right) \tag{7.9}$$

where, $m = 0, 1, 2, 3, \cdots$, $n = 0, 1, 2, 3, \cdots$, and $p = 1, 2, 3, 4, \cdots$

Here k_c is the cut-off wave number given by

$$k_c^2 = (m\pi/a)^2 + (n\pi/b)^2 \tag{7.10}$$

TM$_{mnp}$ mode field:

$$E_z = E_0 \sin(m\pi x/a) \sin(n\pi y/b) \cos(p\pi z/d) \tag{7.11}$$

$$E_y = E_0/k_c^2 \frac{\partial^2 H_z}{\partial y \partial z}$$

$$= \frac{-E_0}{k_c^2} \left(\frac{n\pi}{b} \right)\left(\frac{p\pi}{d} \right) \sin\left(\frac{m\pi x}{a} \right) \cos\left(\frac{n\pi y}{b} \right) \sin\left(\frac{p\pi z}{d} \right) \tag{7.12}$$

$$E_x = E_0/k_c^2 \frac{\partial^2 E_z}{\partial x \partial z}$$

$$= \frac{-E_0}{k_c^2} \left(\frac{m\pi}{a} \right)\left(\frac{p\pi}{d} \right) \cos\left(\frac{m\pi x}{a} \right) \sin\left(\frac{n\pi y}{b} \right) \sin\left(\frac{p\pi z}{d} \right) \tag{7.13}$$

$$H_z = 0 \tag{7.14}$$

$$H_y = \frac{-j\omega\varepsilon E_0}{k_c^2} \frac{\partial E_z}{\partial x}$$

$$= \frac{-j\omega\varepsilon E_0}{k_c^2} \left(\frac{m\pi}{a} \right) \sin\left(\frac{m\pi x}{a} \right) \cos\left(\frac{n\pi y}{b} \right) \cos\left(\frac{p\pi z}{d} \right) \tag{7.15}$$

$$H_x = \frac{j\omega\varepsilon E_0}{k_c^2} \frac{\partial E_z}{\partial y}$$

$$= \frac{j\omega\varepsilon E_0}{k_c^2} \left(\frac{n\pi}{b} \right) \sin\left(\frac{m\pi x}{a} \right) \cos\left(\frac{n\pi y}{b} \right) \cos\left(\frac{p\pi z}{d} \right) \tag{7.16}$$

where, $m = 1, 2, 3, \cdots$, $n = 1, 2, 3, \cdots$ and $p = 0, 1, 2, \cdots$

Here, k_c is given by the same Eq. 7.10.

For either TE_{mnp} or TM_{mnp} mode, the resonant frequency is given by

$$f_r = \frac{1}{2\sqrt{\mu\varepsilon}} \sqrt{(m/a)^2 + (n/b)^2 + (p/d)^2 \, H_z} \tag{7.17}$$

The dominant mode of such a resonator depends on the dimensions of the cavity. For $b < a < d$, the dominant mode is the TE_{101} mode. The field configurations of the dominant mode are shown in Fig. 7.3.

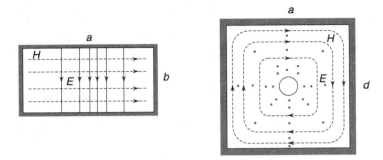

Fig. 7.3 *Rectangular cavity dominant mode field configuration (TE_{101})*

Example 7.1 Find the resonant frequencies of first five lowest modes of an air-filled rectangular cavity of dimensions 5 cm × 4 cm × 2.5 cm. List them in ascending order.

Solution

Given $a = 5$ cm, $b = 4$ cm, $d = 2.5$ cm, resonant frequencies of TE_{mnp} and TM_{mnp} modes can be calculated from equation

$$f_r = \frac{1}{2\sqrt{(\mu\varepsilon)}}\sqrt{(m/a)^2 + (n/b)^2 + (p/d)^2}\ H_z$$

and given as follows:

Modes	TE_{110}	TE_{101}	TE_{011}	TE_{210}	TE_{111}	TE_{120}
	TM_{110}	TM_{101}	TM_{011}	TM_{210}	TM_{111}	TM_{120}
Frequency (GHz)	4.80	6.71	7.08		7.69	8.08

7.3.2 Circular Cavity

Electromagnetic field analysis shows that due to ϕ-symmetric structure of circular cylindrical cavity, field solutions possess harmonic solution in ϕ and standing waves in the radial and z-directions. The field components inside the cavity are described in terms of TE_{nmp} and TM_{nmp} modes as follows:

TE_{nmp} *mode field*

$$H_z = H_0 J_n (x'_{nm}\rho/a) \cos n\phi \sin (p\pi z/d) \tag{7.18}$$

$$H_\phi = -H_0\left(\frac{p\pi}{d}\right)\left(\frac{n}{\rho}\right)\left(\frac{a}{x'_{nm}}\right)^2 J_n(x'_{nm}\rho/a)\sin n\phi\cos(p\pi z/d) \tag{7.19}$$

$$H_\rho = H_0\frac{p\pi}{d}\left(\frac{a}{x'_{nm}}\right) J'_n(x'_{nm}\rho/a)\cos n\phi\cos(p\pi z/d) \tag{7.20}$$

$$E_z = 0 \tag{7.21}$$

$$E_\phi = jH_0\,\omega\mu\,(a/x'_{nm})\,J'_n(x'_{nm}\rho/a)\cos n\phi\sin(p\pi z/d) \tag{7.22}$$

$$E_\rho = jH_0\,\omega\mu\,(n/\rho)\,(a/x'_{nm})^2 J_n(x'_{nm}\rho/a)\sin n\phi\sin(p\pi z/d) \tag{7.23}$$

where, $n = 0, 1, 2, \cdots$ is the number of periodicity in ϕ, $m = 1, 2, 3, \cdots$ is the number of zeroes of the field in ρ direction and $p = 1, 2, 3, \cdots$ is the number of halfwaves in z direction. Here $k_\rho = x'_{nm}/a$ and x'_{nm} is the mth root of equation $J'_n(k_\rho a) = 0$, such that $x'_{11} = 1.841$, $x'_{21} = 3.054$, $x'_{01} = 3.832$, $x'_{12} = 5.331$, etc.

TM$_{nmp}$ mode field

$$E_z = E_0 J_n (x_{nm} \rho/a) \cos n\phi \cos (p\pi z/d) \tag{7.24}$$

$$E_\phi = E_0 \frac{p\pi}{d} (n/\rho)(a/x_{nm})^2 J_n (x_{nm} \rho/a) \sin n\phi \sin (p\pi z/d) \tag{7.25}$$

$$E_\rho = - E_0 \frac{p\pi}{d} (a/x_{nm}) J'_n (x_{nm} \rho/a) \cos n\phi \sin (p\pi z/d) \tag{7.26}$$

$$H_z = 0 \tag{7.27}$$

$$H_\phi = -j E_0 \omega\varepsilon (a/x_{nm}) J'_n (x_{nm}\rho/a) \cos n\phi \cos (p\pi z/d) \tag{7.28}$$

$$H_\rho = -j \omega\varepsilon E_0 (a/x_{nm})^2 J_n (x_{nm}\rho/a) \sin n\phi \cos (p\pi z/d) \tag{7.29}$$

where, $n = 0, 1, 2, \cdots$, $m = 1, 2, 3, \cdots$, and $p = 0, 1, 2, \cdots$ and $k_\rho = x_{nm}/a$. x_{nm} is the mth root of $J_n(k_\rho a) = 0$, so that $x_{01} = 2.405$, $x_{11} = 3.832$, $x_{21} = 5.135$, $x_{02} = 5.520$, $x_{12} = 7.016$, etc. The resonant frequencies for these modes are given by

$$TE_{nmp}: f_r = \frac{1}{2\pi \sqrt{(\mu\varepsilon)}} \sqrt{\left[(x'_{nm}/a)^2 + (p\pi/d)^2\right]}; H_z \tag{7.30}$$

$$TM_{nmp}: f_r = \frac{1}{2\pi \sqrt{(\mu\varepsilon)}} \sqrt{\left[(x'_{nm}/a)^2 + (p\pi/d)^2\right]}; H_z \tag{7.31}$$

where x'_{nm} and x_{nm} are the mth roots of the equation $J'_n (x) = 0$ and $J_n(x) = 0$, respectively. The dominant mode in circular cavity will depend on the dimensions of the cavity. The smallest root out of x_{01} and x'_{11} generates the dominant mode. We can write,

$$F = \frac{f_r(TM_{011})}{f_r(TE_{111})} = \frac{x_{01}}{\sqrt{\left[x'^2_{11} + (\pi a/d)^2\right]}} \tag{7.32}$$

For $d < 2a$, $F < 1$, the TM_{010} mode is dominant. For $d \geq 2a$, $F \geq 1$, the TE_{111} mode is dominant. Another mode TE_{011} is important for its high Q. The field configuration for these modes are shown in Fig. 7.4.

Here

$$k_\rho^2 + k_z^2 = \left(\frac{x_{nm}}{a}\right)^2 + \left(\frac{p\pi}{d}\right)^2 = \omega_r\sqrt{\mu\varepsilon} = \omega_r/c \tag{7.33}$$

Equations 7.30 and 7.31 show that TM and TE modes of the same order n, m, p have identical frequencies but have different field patterns. Such modes are known as degenerate modes. Equations 7.30 and 7.31 can also be written as

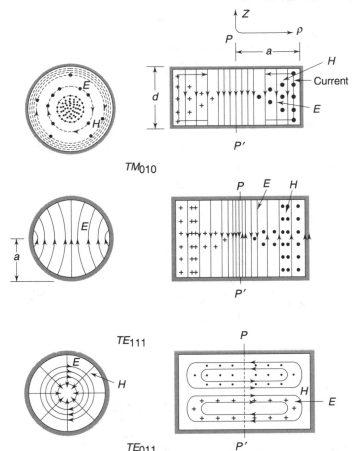

TM_{010}

TE_{111}

TE_{011}

Fig. 7.4 *Circular cavity field configurations (Left hand side is the cross-section through PP')*

$$(2af_r)^2 = \left[\left(\frac{p\pi}{\sqrt{\mu\varepsilon}}\right)^2 \left(\frac{a}{d}\right)^2\right] + \left(\frac{x_{nm}}{\sqrt{\mu\varepsilon}}\right)^2 \tag{7.34}$$

where x_{nm} is for TM_{nmp} mode and x'_{nm} for TE_{nmp} mode. The above equation is used to draw a straight line mode chart as shown in Fig. 7.5 which is very useful in cavity design. It shows the range of frequencies and the ratio of diameter/length, in which only a single mode can be resonated if there is no degeneracy.

Example 7.2 Find the resonant frequencies of the five lowest modes of an air-filled cylindrical cavity of radius 1.905 cm and length 2.54 cm. List them in ascending order.

Solution

Given $a = 1.905$ cm, $d = 2.54$ cm the resonant frequencies are calculated as follows

$$TE_{nmp} : f_r = \frac{1}{2\pi\sqrt{(\mu\varepsilon)}} \sqrt{\left[\left(x'_{nm}/a\right)^2 + \left(p\pi/d\right)^2\right]} ; H_z$$

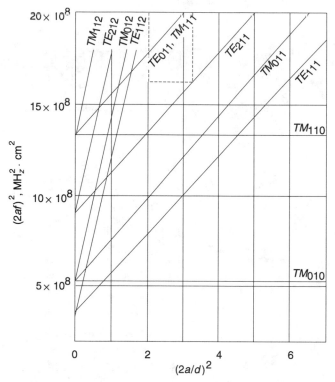

Fig. 7.5 *Circular cavity mode chart [Source: Ref.1]*

$$TM_{nmp}: f_r = \frac{1}{2\pi\sqrt{(\mu\varepsilon)}}\sqrt{\left[(x_{nm}/a)^2 + (p\pi/d)^2\right]}\,;\, H_z$$

Modes	TM_{010}	TE_{111}	TM_{011}	TE_{211}	TM_{111} TE_{011}
f_r (GHz)	6.03	7.49	8.44	9.67	11.27

7.4 CAVITY EXCITATION AND TUNING

The power between the waveguide cavity resonator or a coaxial resonator, the signal source and the load can be coupled by means of a coaxial line whose centre conductor is extended inside the cavity in the form of a probe or a loop. The resonator cavity may also be coupled to a waveguide by means of a small aperture in the common wall. Various coupling configurations are shown in Fig. 7.6.

A coaxial line probe parallel to **E**, placed at the position of maximum **E**-field should be used to couple power between the cavity and the external circuit. The loop should be used for power coupling at the position of maximum magnetic field with plane of the loop perpendicular to **H**-field.

Microwave cavity resonators are used in a transmission system as reflection type or transmission type cavities. In reflection type, the cavity is used as load

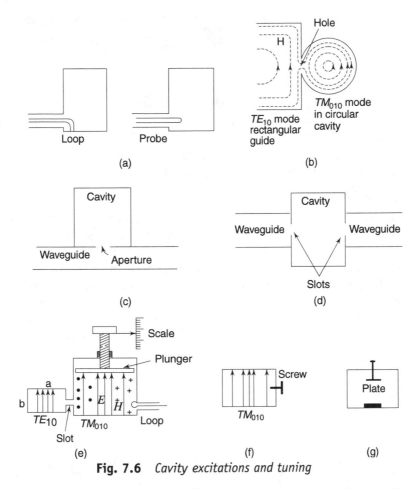

Fig. 7.6 *Cavity excitations and tuning*

producing maximum impedance, while transmission type cavity is placed in series with the line producing maximum absorption of power.

There are a number of methods by which the cavity may be mechanically tuned to resonance:

1. A short circuit plunger can be moved parallel to the axis to change the length of the cavity and thereby f_r as shown in Fig. 7.6(e).
2. A metallic screw may be inserted into the curved wall of the cavity (klystron tubes) to lower the cavity's distributed inductance and thereby increasing f_r as shown in Fig. 7.6(f).
3. A small plate may be placed at the centre of the cavity and its position is varied to control the variation in capacitance at the centre and thereby the f_r as shown in Fig. 7.6 (g).

7.5 Q-FACTOR OF MICROWAVE CAVITIES

The quality factor Q is a measure of the frequency selectivity of a cavity resonator and is defined by

$$Q = 2\,\pi\,\frac{\text{Maximum energy stored }(W)\text{ during a cycle}}{\text{Average energy dissipated per cycle}}$$

$$= \frac{\omega_r W}{P} \qquad (7.35)$$

where P is the average power loss and $\omega_r = 2\pi/$oscillation time period, T.

To a very good approximation, the bandwidth Δf around the resonant frequency f_r is given by $Q = f_r/\Delta f$. When a cavity is coupled to a circuit, several different Q factors are defined: unloaded $Q = Q_0$, external $Q = Q_e$ and the loaded $Q = Q_L$. These are described in the following paragraphs.

7.5.1   Unloaded Q_0

When a cavity is assumed to be not connected to any external circuit or load, Q accounts for the internal losses and is called the unloaded quality factor Q_0. The Q_0 of a cavity is reduced when for all practical purposes the cavity is coupled to an external circuit or load. The equivalent circuits of an unloaded cavity, cavity coupled to a generator and cavity coupled to a load are shown in Fig. 7.7.

The total energy stored in a cavity resonator is a constant and equal to the sum of the electric and magnetic energy stored. At resonant frequency, the electric and magnetic energies are equal and when the electric energy is maximum, the magnetic energy is zero and vice versa. The total maximum energy stored, therefore, can be written as

$$W = 2W_e = \frac{E}{2}\int_V |E|^2\,dV = 2W_m = \mu/2\int_V |H|^2\,dV \qquad (7.36)$$

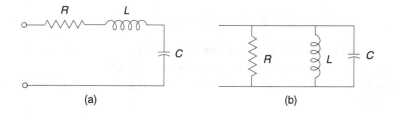

(a)   (b)

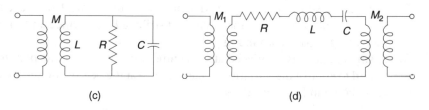

(c)   (d)

Fig. 7.7   *Cavity equivalent circuits (a) Series equivalent of an unloaded cavity (b) Parallel connection of an unloaded cavity (c) Cavity coupled to a generator (d) Cavity coupled to a load and a generator*

where V is the resonant cavity volume. For a lossless dielectric inside the cavity the energy dissipation is only due to joule loss in the cavity walls S and can be expressed by

$$P = R_s/2 \int_S |\hat{n} \times \mathbf{H}|^2 \ ds \tag{7.37}$$

where n is the unit outward normal to the inside surface S having current density $\mathbf{J} = \hat{n} \times \mathbf{H}$.

Therefore, the unloaded Q factor can be computed from the field components using the following equation:

$$Q_0 = \frac{\omega \varepsilon \int_V |\mathbf{E}_t|^2 \ dV}{R_s \int_S |\mathbf{H}_t|^2 \ dS} \tag{7.38}$$

where suffix t stands for the tangential component. For highly conducting cavity walls (copper, $\sigma = 5.8 \times 10^9$ mhos/m) at normal temperature, Q_0 has values of the order of 10^4–10^5. Even higher Q_0 up to 10^8–10^{10} can be obtained by using super conductive materials such as lead cooled to $2 - 5$ K.

In practice, power loss by conductors P_c, dielectric fills P_d and radiation P_r from the openings can contribute to unloaded Q_0,

$$Q_0 = \frac{\omega_r W}{P_c + P_d + P_r} \tag{7.39}$$

or,

$$\frac{1}{Q_0} = \frac{1}{Q_c} + \frac{1}{Q_d} + \frac{1}{Q_r} \tag{7.40}$$

Here conductor quality factor is defined by

$$Q_c = \omega_r W/P_c \tag{7.41}$$

The dielectric quality factor

$$Q_d = \omega_r W/P_d = \frac{\omega_r \varepsilon \int_V |E|^2 \ dV}{\sigma_d \int_V |E|^2 \ dV} = \frac{\omega_r \varepsilon}{\sigma_d}$$

$$= 1/\tan \delta \tag{7.42}$$

where the loss tangent for a dielectric of conductivity σ_d is given by

$$\tan \delta = \sigma_d/\omega_r \varepsilon \tag{7.43}$$

The radiation Q factor

$$Q_r = \omega_r W/P_r \tag{7.44}$$

For a completely closed cavity with highly conducting walls, radiation loss approaches zero and Q_r approaches infinity. For a cavity with apertures or transmission line or microstrip configuration, non-zero finite Q_r can be calculated. For the design of resonators unloaded Q is required to be determined as described in the following paragraphs.

7.5.2 Q-Factor of Transmission Line Resonator

The Q-factor of a transmission line resonator can be computed by taking into account the distributed nature of currents and voltages. The instantaneous voltage and current on the low loss short circuited line are given by

$$v(x, t) = V_0 \sin \beta x \sin \omega t \tag{7.45}$$

$$i(x, t) = (V_0/Z_0) \cos \beta x \cos \omega t \tag{7.46}$$

where $x = 0$ at the right hand side end (load). The stored electric and magnetic energies along the line of length l are given by

$$W_e(t) = (1/2) CV_0^2 \sin^2 \omega t \int_0^l \sin^2 \beta x \, dx$$

$$= (CV_0^2/4) \sin^2 \omega t \, [\,(\sin 2\beta \, l/2\beta) - l\,] \tag{7.47}$$

$$W_m(t) = \frac{L V_0^2}{2 Z_0^2} \cos^2 \omega t \left[\frac{\sin 2\beta l}{2\beta} + l \right] \tag{7.48}$$

Total stored energy

$$W = W_e(t) + W_m(t) = \frac{1}{4} ClV_0^2 \left[1 + \frac{\sin 2\beta l}{2\beta l} \cos 2\omega t \right] \tag{7.49}$$

The maximum stored energy ($wt = 0$)

$$W_{max} = \frac{1}{4} ClV_0^2 \left[1 + \frac{\sin 2\beta l}{2\beta l} \right] \tag{7.50}$$

For both $\lambda/4$ and $\lambda/2$ lines, $\sin 2\beta l/2\beta l = 0$.
Therefore,

$$W_{max} = 1/4 \, (ClV_0^2) \tag{7.51}$$

The energy dissipation per cycle is given by P = average power T. Due to the conductor heating, average power loss is

$$P_c = (R_s/2) \int i^2 \, dx$$

$$= \frac{1}{4} \frac{RlV_0^2}{Z_0^2} \left[1 + \frac{\sin 2\beta l}{2\beta l} \right] \tag{7.52}$$

The insulation dielectric heating loss is

$$P_d = G/2 \int V^2 \, dx$$

$$= \frac{GlV_0^2}{4} \left(1 - \frac{\sin 2\beta l}{2\beta l} \right) \tag{7.53}$$

For resonant lengths $l = \lambda/4$ or $\lambda/2$,

$$P_c = RlV_0^2/4Z_0^2 \tag{7.54}$$

$$P_d = \frac{GlV_0^2}{4} \tag{7.55}$$

The Q-factor of the short circuited line cavity of resonant length is

$$Q = \frac{2\pi}{T} \frac{W}{P_c + P_d} = \frac{\omega_r LC}{RC + GL} \tag{7.56}$$

Thus for small R and G, Q is very high and approaches infinity for lossless line ($R = 0 = G$). In a low loss case,

$$Z_0 = \sqrt{L/C} \tag{7.57}$$

$$\alpha = \frac{1}{2} \left(\frac{R}{Z_0} + GZ_0 \right) \tag{7.58}$$

$$\beta_r = w_r \sqrt{LC} \tag{7.59}$$

Then for quarter and halfwave transmission line resonators.

$$Q = \beta_r/2\alpha \tag{7.60}$$

Because of the open structure, the skin effect loss and radiation loss at high frequencies lower the predicted Q.

7.5.3 Unloaded Q-Factor of Coaxial Cavity

Let us assume that a coaxial resonator is formed by short circuiting two ends of an air-filled coaxial line of length d. The radii of the conductors, a and b are such that only *TEM* mode exists at resonance. The standing wave fields inside the resonator are

$$E_\rho = \frac{E_0}{\rho \ln b/a} \left(e^{-jk_z z} - e^{+jk_z z} \right)$$

$$= \frac{2jE_0}{\rho \ln b/a} \sin k_z z \tag{7.61}$$

$$H_\phi = \frac{2E_0}{\eta \rho \ln b/a} \left(e^{-jk_z z} + e^{+jk_z z} \right)$$

$$= \frac{2 E_0}{\eta \rho \ln b/a} \cos k_z z \tag{7.62}$$

where $k_z = p\pi/d = \omega\sqrt{\mu\varepsilon}$, $p = 1, 2, 3,...$ The resonant frequencies are given by

$$f_r = \frac{p}{2\sqrt{(\mu\varepsilon)d}} \tag{7.63}$$

It can be seen that both the electric and magnetic fields undergo p half cycles of variation along the length of the resonator.

The time-average total energy stored is

$$W = 2W_e = \frac{\varepsilon}{2} \int_a^b \int_0^d \int_0^{2\pi} |E_\rho|^2 \, \rho d \, \rho d\phi \, dz$$

$$= \frac{\varepsilon}{2} \frac{4 E_0^2}{\left(\ln \dfrac{b}{a}\right)^2} \int_0^{2\pi} d\phi \int_0^d \sin^2 (p\pi z/d) dz \int_a^b \frac{d\rho}{\rho}$$

$$= \frac{\varepsilon}{2} \left[\frac{2 E_0}{\ln (b/a)}\right]^2 .2\pi \frac{d}{2} . \ln \frac{b}{a};$$

$$= \frac{2\pi d \varepsilon E_0^2}{\ln b/a} \tag{7.64}$$

The time-average power loss on the cavity walls is given by

$$P = R_s/2 \int_S |H_\phi|^2 \, ds$$

$$= (R_s/2) \int_0^{2\pi} \left[\int_0^d \left(\frac{2 E_0}{\eta a \ln(b/a)}\right)^2 \cos^2 \frac{p\pi z}{d} \, dz \cdot a \right.$$

$$+ \int_0^d \left(\frac{2 E_0}{\eta b \ln (b/a)}\right)^2 \cos^2 \frac{p\pi z}{d} \, dz . b$$

$$\left. + 2 \int_a^b \left(\frac{2 E_0}{\eta \rho \ln (b/a)}\right)^2 s\cos^2 \left[\frac{p\pi z}{d}\right]_{z=0} \rho \, d\rho \right] d\phi$$

$$= \frac{R_s}{2} \left(\frac{2 E_0}{\eta \ln (b/a)}\right)^2 \left[2\pi a \frac{1}{a^2} \int_0^d \cos^2 \frac{p\pi z}{d} \, dz \right.$$

$$\left. + 2\pi b \frac{1}{b^2} \int_0^d \cos^2 \frac{p\pi z}{d} \, dz + 4\pi \int_a^b \frac{d\rho}{\rho} \right]$$

$$= \frac{\pi R_s}{2} \, [2E_0/(\, \eta \ln b/a)]^2 \left[\frac{d}{a} + \frac{d}{b} + 4 \ln b/a \right] \quad (7.65)$$

Therefore,

$$Q = \omega_r W/P$$

or,

$$Q_{TEM} = \frac{p \pi \eta \ln (b/a)}{R_s \left[d \left(\frac{1}{a} + \frac{1}{b} \right) + 4 \ln \left(\frac{b}{a} \right) \right]} \quad (7.66)$$

where

$$\omega_r \varepsilon = \frac{p \pi}{d} \sqrt{(\varepsilon/\mu)}$$

$$= \frac{p \pi}{\eta d}$$

$$\eta = \sqrt{(\mu/\varepsilon)} \quad (7.67)$$

7.5.4 Unloaded Q of Rectangular Cavity

Dominant TE_{101} mode: The dominant mode in a rectangular cavity of dimensions $(a \times b \times d)$, $b < a < d$, is TE_{101}. The field distribution in such a cavity can be obtained from Eqns 7.4 – 7.9 and after normalising by the factor

$$E_0 = \frac{j\omega \, \mu_0 \, H_0}{k_c^2} \, (\pi/a); \, \mu = \mu_0, \, \varepsilon = \varepsilon_0 \quad (7.68)$$

the field components becomes

$$E_y = E_0 \sin\left(\frac{\pi x}{a} \right) \sin \left(\frac{\pi z}{d} \right) \quad (7.69)$$

$$H_x = -j \, (E_0/\omega\mu_0) \, (\pi/d) \sin\left(\frac{\pi x}{a} \right) \cos\left(\frac{\pi z}{d} \right) \quad (7.70)$$

$$H_z = j \, (E_0/\omega\mu_0) \, (\pi/a) \cos\left(\frac{\pi x}{a} \right) \sin\left(\frac{\pi z}{d} \right) \quad (7.71)$$

where,

$$k_c = \pi/a \quad (7.72)$$

Since at resonance, total energy is equal to the maximum energy stored in the E-field, or that in the magnetic field, the resonant energy stored inside the cavity can be calculated from the maximum electric energy

$$W = (\varepsilon_0/2) \int_0^a \int_0^b \int_0^d |E_y|^2 \, dx \, dy \, dz$$

$$= (\varepsilon_0 \, E_0^2/2) \int\limits_0^a \int\limits_0^b \int\limits_0^d \, \sin^2 \, (\pi \, x/a) \, \sin^2 \, (\pi \, z/d) \, dx \, dy \, dz$$

$$= \varepsilon_0/8 \; abd \; E_0^2$$

$$= \varepsilon_0 \, E_0{}^2/8 \; x \; \text{cavity volume} \tag{7.73}$$

For a lossless dielectric the total power loss P in the cavity can be obtained from the ohmic losses in the six walls,

Therefore,

$$P = R_s/2 \left\{ 2 \int\limits_0^a \int\limits_0^b |H_x|^2_{z=0} \; dx \, dy + 2 \int\limits_0^b \int\limits_0^d |H_z|^2_{x=0} \, dy \, dz \right.$$

$$\left. + 2 \int\limits_0^a \int\limits_0^d \left[|H_x|^2 + |H_z|^2 \right]_{y=0} \, dx \, dz \right\} \tag{7.74}$$

Substituting the expressions for the field components and evaluating the integrals,

$$P = \frac{R_s \, \lambda^2 \, E_0^2}{8 \eta^2} \left[\frac{ab}{d^2} + \frac{bd}{a^2} + \frac{1}{2} \left(\frac{a}{d} + \frac{d}{a} \right) \right]; \tag{7.75}$$

where $\eta = \sqrt{\mu_0/\varepsilon_0}$, the intrinsic impedance of free space.

The Q-factor is then obtained from Eqns 7.73 and 7.75 :

$$Q_0 = \frac{\omega_r \, W}{P}$$

$$= \frac{\pi \eta}{4 \, R_s} \; \frac{2b \left(a^2 + d^2 \right)^{3/2}}{ad \left(a^2 + d^2 \right) + 2b \left(a^3 + d^3 \right)} \tag{7.76}$$

From the symmetry of Q_0 in a and d, it is seen that Q_0 is maximum for a square base cavity $a = d$, given by

$$Q_{0 \, \text{max}} = \frac{1.11 \sqrt{(\mu/\varepsilon)}}{R_s \, (1 + a/2 \, b)} \tag{7.77}$$

where $Q_{0 \, \text{max}}$ increases with decrease in a/b.

For an air-filled cubic cavity, $a = b = d$, so that

$$Q_0 = \frac{0.74 \, \eta}{R_s} = \frac{279}{R_s} \tag{7.78}$$

For an air dielectric, $\eta = 120 \, \pi$ ohm, and for copper, $R_s = \sqrt{(w\mu_0/2\sigma)} = 0.0261$ ohm at 10 GHz. Then $Q \approx 10{,}690$ for a cubic cavity of dimensions $a = b = d$. The half power frequency bandwidth Δf is $\Delta f = f_r/Q = 10{,}000/10{,}690$ MHz $= 935.5$ kHz.

TE$_{mnp}$ **and TM**$_{mnp}$ **mode:** The general expressions for Q for the TE_{mnp} and TM_{mnp} modes can be obtained as

$$\text{TE: } Q_{mnp} = \frac{\eta\, abdk_c^2\, k^3}{4R_s\left[bd\left(k_c^4 + k_y^2\, k_z^2\right) + ad\left(k_c^4 + k_x^2\, k_z^2\right) + abk_c^2\, k_z^2\right]} \tag{7.79}$$

$$\text{TM: } Q_{mnp} = \frac{\eta\, abdk_c^2\, k}{4R_s\left[b\left(a+d\right)k_x^2 + a\left(b+d\right)k_y^2\right]} \tag{7.80}$$

where,

$$k_x = m\pi/a,\ k_y = n\pi/b,\ k_z = p\pi/d,$$

$$k = \sqrt{\left(k_x^2 + k_y^2 + k_z^2\right)}\ k_c = \sqrt{\left(k_x^2 + k_y^2\right)} \tag{7.81}$$

Closed rectangular cavities are used in resonant wavemeters, gun diode oscillators, etc. The cavities can be tuned with the aid of a short circuited plunger at one end wall.

7.5.5 Unloaded Q of a Circular Cavity

For a circular cavity of radius a and length d, the unloaded Q-factor can be computed from the expressions of field components of a given mode. The three very important modes of practical interest are TM_{010}, TE_{111} and TE_{011}. For $d/2a < 1$ the TM_{010} is dominant, while for $d/2a \geq 1$ the TE_{111} mode is dominant. The ϕ symmetric TE_{011} mode is of particular interest because its Q is two to three times that of the dominant modes and there are no axial currents ($H_\phi = 0$). Thus the cavity tuning by changing the length d with a short circuit plunger does not obstruct the electric currents on the wall. However, since TE_{011} mode is higher than the dominant mode, care must be taken to excite the cavity without the generation of other possible modes.

The resonant frequencies for various ratios of $d/2a$ are tabulated in Table 7.1 for different modes.

Table 7.1 $(f_r)_{nmp}/(f_r)_{dominant}$

$d/2a$	TM_{010}	TE_{111}	TM_{110}	TM_{011}	TE_{211}	TM_{111} TE_{011}
0.5	1.0	1.5	1.59	1.63	1.80	2.05
1.0	1.0	1.0	1.59	1.19	1.42	1.72
1.5	1.13	1.0	1.80	1.24	1.52	1.87

Dominant TM$_{010}$ **mode** $Q(d<2a)$

Here $n = 0$, $m = 1$, $p = 0$. Therefore, non zero cavity field components can be obtained from Eqns 7.24-7.29,

$$E_z = E_0\, J_0\, (x_{01}\, \rho/a) \tag{7.82}$$

$$H_\phi = -E_0 \, j\omega\varepsilon_0 \left(\frac{a}{x_{01}}\right) J'_0 \left(\frac{x_{01}\,\rho}{a}\right)$$

$$= jE_0\omega\varepsilon_0 \left(\frac{a}{x_{01}}\right) J_1 \left(\frac{x_{01}\,\rho}{a}\right)$$

$$= j\frac{E_0}{\eta} J_1 \left(\frac{x_{01}\,\rho}{a}\right) \tag{7.83}$$

where

$$k_\rho = x_{01}/a, \; x_{01} = 2.405,$$

$$\omega\varepsilon_0 \left(\frac{a}{x_{01}}\right) = \frac{\omega\varepsilon_0}{k_\rho} = \sqrt{\varepsilon_0/\mu_0} = 1/\eta$$

and

$$J'_0 = -J_1 \tag{7.84}$$

The resonant frequency from Eqn. 7.31 can be obtained as

$$f_r = \frac{x_{01}}{2\pi a\sqrt{(\varepsilon_0\,\mu_0)}} \tag{7.85}$$

The quality factor

$$Q = \omega \frac{\text{Average stored energy}}{\text{Power dissipated}}$$

can be determined from the average stored energy and power dissipated in the cavity walls. At resonance the electric energy stored W_e = average magnetic energy stored W_m. Therefore, average stored energy is $W = 2W_e$ or,

$$2\,W_e = \varepsilon_0/2 \int_0^a \int_0^{2\pi} \int_0^d \mathbf{E.E^*} \, \rho \, d\rho \, d\phi \, dz \tag{7.86}$$

Assuming no dielectric loss inside the cavity, the total power dissipation is due to ohmic loss in the conducting walls:

$$P = R_s/2 \int_{\text{walls}} |H_{\text{tan}}|^2 \, ds \tag{7.87}$$

Now,

$$\mathbf{E.E^*} = |E_z|^2 = E_0^2 \, J_0^2 \, (x_{01}\rho/a)$$

Therefore,

$$W_e = (\varepsilon_0/4)\int_0^a \int_0^{2\pi} \int_0^d E_0^2 \, J_0^2 \, (x_{01}\,\rho/a) \, \rho d \, \rho d\phi \, dz$$

$$= (\varepsilon_0 \, E_0^2/4) \, 2\pi.d. \int_0^a \rho \, J_0^2 \, (x_{01}\rho/a) \, d\rho$$

$$= (\varepsilon_0 E_0^2/4)\, 2\pi.d\, a^2/2\, J_1^2\,(x_{01})$$

$$= (\varepsilon_0 E_0^2/4)\, \pi\, da^2\, J_1^2\,(x_{01}) \tag{7.88}$$

$$\int_{\text{walls}} |H_{\text{tan}}|^2\, ds = \int_{\substack{\text{curved}\\ \text{walls}}} + \int_{\substack{\text{end}\\ \text{walls}}}$$

$$= \int_0^{2\pi}\int_0^{d} |H_\phi|^2\big|_{\rho=a}\, ad\phi dz + 2\int_0^{a}\int_0^{2\pi} |H_\phi|^2\big|_{z=0,d}\, \rho d\rho\, d\phi \tag{7.89}$$

First integral $= \dfrac{E_0^2}{\eta^2}\displaystyle\int_0^{2\pi}\int_0^{d} \left[J_1^2\,(x_{01}\,\rho/a)\right]_{\rho=a}\, a\, d\phi\, dz$

$$= (E_0/\,\eta)^2\, 2\pi ad\, J_1^2\,(x_{01}) \tag{7.90}$$

Second integral $= 2\dfrac{E_0^2}{\eta^2}\displaystyle\int_0^{a}\int_0^{2\pi} J_1^2\,(x_{01}\,\rho/a)\, \rho d\rho\, d\phi$

$$= 2\left(\frac{E_0}{\eta}\right)^2 .2\pi \int_0^{a} \rho J_1^2\!\left(\frac{x_{01}\,\rho}{a}\right) d\rho$$

$$= 4\pi\left(\frac{E_0}{\eta}\right)^2 .\left(\frac{a^2}{2}\right) J_1^2\,(x_{01}) \tag{7.91}$$

From Eqns 7.86, 7.89 and 7.90

$$P = R_s \frac{E_0^2}{\eta^2}\, \pi a J_1^2\,(x_{01})(a+d) \tag{7.92}$$

Therefore, $\quad Q_0 = \omega_r\, 2W_e/P$

$$= \omega_r\, \frac{\varepsilon_0\, da\, \eta^2}{2\, R_s\,(a+d)}$$

$$= \frac{x_{01}\, \eta}{2\, R_s\,(1+a/d)} \tag{7.93}$$

Since $x_{01}/a = \omega_r\sqrt{(\mu\varepsilon)}$ for $p=0$, $x_{01} = 2.405$.

$$Q_{TM_{010}} = \frac{1.202\, \eta}{R_s\,(1+a/d)} \tag{7.94}$$

Dominant TE$_{111}$ mode Q (d/2a ≥ 1)

Substituting the expressions for the field components from Eqns 7.18–7.23 for $n=1$, $m=1$ and $p=1$ in integrals Eqns 7.86 and 7.87, the final result obtained for the Q is

$$Q_0 = \frac{\lambda_0}{2\pi\delta_s}\, \frac{\left[1-(1/x_{11}')^2\right]\left[(x_{11}')^2+(\pi a/d)^2\right]^{3/2}}{\left[(x_{11}')^2+2a/d\,(\pi a/d)^2+(1-2a/d)\,(\pi a/x_{11}'\,d)^2\right]} \tag{7.95}$$

This shows that Q decreases as $1/\sqrt{f}$ for dominant TE_{111} mode.

TE_{011} *mode Q*

Field distribution in circular cavity for TE_{011} modes is ϕ independent ($n = 0$, $m = 1$, $p = 1$) and can be written as

$$H_z = H_0 \, J_0 \, (x'_{01} l/a) \, \sin \, (\pi z/d) \tag{7.96}$$

$$H_\rho = H_0 \, (\pi a/dx'_{01}) \, J'_0 \, (x'_{01} \, \rho/a) \, \cos \, (\pi z/d); \tag{7.97}$$

$$E_\phi = j\omega\mu \, H_0 \, (a/x'_{01}) \, J'_0 \, (x'_{01} \, \rho/a) \, \sin(\pi z/d); \tag{7.98}$$

where,

$$k'_\rho = x'_{01}/a, \; x'_{01} = 3.832. \tag{7.99}$$

The unloaded Q_0 can be obtained in a similar way

$$Q_0 = \frac{\eta}{2 R_s} \frac{\left[\left(\dfrac{x'_{01}}{a} \right)^2 + \left(\dfrac{\pi}{d} \right)^2 \right]^{3/2}}{\dfrac{1}{a} \left(\dfrac{x'_{01}}{a} \right)^2 + \dfrac{2}{d} \left(\dfrac{\pi}{d} \right)^2} \tag{7.100}$$

where

$$\eta = \sqrt{(\mu/\varepsilon)} \, , R_s = \sqrt{(\omega\mu/2\sigma)} \; = 1/\sigma \, \delta \tag{7.101}$$

For a cavity filled with a homogeneous dielectric of conductivity σ_d, the dielectric Q-factor is given by

$$Q_d = \omega\varepsilon/\sigma_d = 1/\text{loss tangent} \tag{7.102}$$

Example 7.3 A circular cylindrical air-filled cavity with radius 3 cm and length 10 cm is excited in TE_{111} mode. The 3dB bandwidth is 2.5 MHz. Calculate the resonant frequency and the Q.

Solution

Here $n = m = p = 1$.
Therefore,

$$f_r = \frac{1}{2\sqrt{(\mu_0 \, \varepsilon_0)}} \sqrt{\left[(x'_{11}/a)^2 + (\pi/d)^2 \right]}$$

$$f_r = \frac{1}{2\sqrt{\left(4\pi \times 10^{-7} \times 8.854 \times 10^{-12}\right)}} \left[(1.841/0.03)^2 + (\pi/0.1)^2 \right]^{1/2}$$

$$= 10.42 \text{ GHz}$$

$$Q_0 = f_r/\text{bandwidth}$$

$$= \frac{10.42 \times 10^9}{2.5 \times 10^6} = 4168$$

Example 7.4 A circular air-filled copper cavity is excited in the TM_{010} mode at 9.375 GHz. The cavity has ratio length: radius = 1.5. Find the Q-factor.

Solution

Surface resistance of Cu is $R_s = \sqrt{(w\mu/2\sigma)}$, $f = 9.375 \times 10^9$ H_z. $\eta = 377$ ohm. $d/a = 1.5$, $1 + a/d = 1 + 1/1.5 = 2.5/1.5 = 5/3$; $\sigma = 5.8 \times 10^7$ mho/m. Therefore,

$$Q = 1.202 \times 377 \times 3/5/R_s;$$

$$R_s = \sqrt{[(6.28 \times 9.375 \times 10^9 \times 4\pi \times 10^{-7})/(2 \times 5.8 \times 10^7)]}$$

$$= 2.5 \times 10^{-2}$$

$$Q \approx 10875$$

Q-factors for TE_{nmp} and TM_{nmp} modes

The expression for the unloaded Q-factor for general TE_{nmp} and TM_{nmp} modes are

$$TE_{nmp} \cdot Q_0 = \frac{\lambda_0}{2\pi\delta} \frac{\left[1 - (n/x'_{nm})^2\right]\left[(x'_{nm})^2 + (p\pi a/d)^2\right]^{3/2}}{\left[(x'_{nm})^2 + 2a/d\,(p\pi a/d)^2 + (1 - 2a/d)(np\pi a/x'_{nm}\,d)^2\right]} \tag{7.103}$$

$$TM_{nmp} \cdot Q_0 = \frac{\lambda_0}{2\pi\delta} \frac{\left[(x_{nm})^2 + (p\pi a/d)^2\right]^{1/2}}{(1 + 2a/d)} \; ; \text{ for } p > 0 \tag{7.104}$$

$$= \frac{\lambda_0}{2\pi\delta} \frac{x_{nm}}{(1 + a/d)} \; ; \text{ for } p = 0 \tag{7.105}$$

The parameter $Q\,\delta/\lambda_0$ is independent of frequency but varies with the ratio diameter/length $(= 2a/d.)$ Therefore,

$$Q \propto \lambda_0/\delta \propto \lambda_0 \sqrt{f}$$

$$\propto 1/\sqrt{f} \tag{7.106}$$

Figure 7.8 shows the variation of normalised value $Q_0\delta/\lambda_0$ with $2a/d$ for several modes. It is seen that Q_0 is higher for the TE_{011} mode as compared to the dominant mode TM_{010} for $2a/d > 1$ or TE_{111} for $2a/d \leq 1$.

Circular cavity provides a very high Q over a wide frequency range. For this reason a circular cavity is preferably used for frequency meters when excited in the TE_{011} mode. The advantages of TE_{011} mode are: (1) Q_0 is 2 to 3 times that of TE_{111} mode, and (2) Since $H_\phi = 0$, there is no axial current on the walls and consequently a short circuit plunger at one end can move freely for tuning purpose without intersecting the current path. Thus no significant leakage of signal takes place through the gap between the plunger plate and the circular walls.

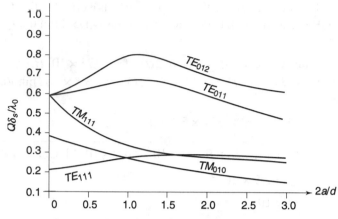

Fig. 7.8 Q_0 for circular cavity [Source: Ref. 1]

Example 7.5 Show that Q of a circular cavity operating in TM_{010} mode is greater than that of the square based rectangular cavity with TE_{101} mode by 8.26 %. The dimension of each cavity is such that the circular cavity is circumscribed by the square based rectangular cavity.

Solution

For TM_{010} mode in a circular air-filled cavity of radius a and length d,

$$Q_1 = \frac{1.2025\,\eta}{R_s\,(1 + a/d)}$$

For TE_{101} mode in a square based ($a' \times a'$) rectangular cavity ($a' \times b' \times a'$), $Q_2 = \eta\,1.1107/[R_s(1+a'/2\,b')]$. For a square based rectangular cavity to circumscribe the circular cavity, the height and one of the sides of its base must be equal to the diameter $2a$, i.e., height $b' = d$ and base $a' = 2a$. Therefore,

$$\frac{Q_1 - Q_2}{Q_2} = \left[\frac{1.2025\,\eta}{R_s\,(1 + a/d)} - \frac{1.1107\,\eta}{R_s\,(1 + a'/2b')}\right]\Big/Q_2$$

$$= \left[\frac{1.2025\,\eta}{R_s\,(1 + a/d)} - \frac{1.1107\,\eta}{R_s\,(1 + 2a/2d)}\right]\Big/Q_2$$

$$= \frac{1.2025 - 1.1107}{1.1107} = 0.0826$$

Therefore, Q of the circular cavity is greater than that of the square-based cavity by 8.26%. This is expected since $Q\,\alpha$ volume / surface area.

7.6 LOADED AND EXTERNAL Q

Since a cavity is always coupled to a generator and load through either a coaxial line probe or a slot in a common wall of a waveguide, energy dissipation should be calculated taking the dissipation in external load along with that in the cavity.

The power loss P_e due to the presence of an external load in a cavity system results in an external quality factor $Q_e = \omega_r W/P_e$. We define loaded Q of the cavity as the total Q for the system expressed by

$$Q_L = 2\pi \frac{\text{Maximum energy stored } (W)}{\left(\begin{array}{c}\text{Energy dissipation in cavity per cycle, } P_0 + \text{energy} \\ \text{dissipation due to the presence of external load, } P_e\end{array}\right)}$$

(7.107)

or,

$$Q_L = \omega_r W/(P_0 + P_e)$$ (7.108)

or,

$$1/Q_L = 1/Q_0 + 1/Q_e$$ (7.109)

where,

$$Q_0 = \omega_r W/P_0 = \text{unloaded } Q$$ (7.110)
$$Q_e = \omega_r W/P_e = \text{external } Q$$ (7.111)

Q_L and Q_0 are related by the coupling coefficient

$$\beta_e = P_e/P_0.$$ (7.112)

Therefore,

$$Q_L = Q_0/(1+\beta_e)$$ (7.113)

or

$$\beta_e = Q_0/Q_L - 1$$ (7.114)

This shows that $Q_L < Q_0$ always. The following special cases are important for cavity resonators:

(1) When $Q_e = Q_0$, the energy dissipation in the resonant cavity and that in the load are equal. Under this condition the coupling is called critical and $Q_L = Q_0/2$.

(2) When $Q_e < Q_0$, energy dissipation in the cavity is less than that in load, and the coupling is called over coupling.

(3) When $Q_e > Q_0$, energy dissipated in the cavity is greater than that in the load and the coupling is called under coupling.

(4) When $Q_e = \text{infinite}$, coupling $= 0$.

Q increases at a given frequency for higher order modes, because of the fact that to support higher modes at a given frequency, the dimensions of the cavity increase and the ratio, volume / surface increase.

7.7 COUPLED CAVITIES

Different type of coupled cavities are used in microwave filter circuits and wavemeters. These are described below.

7.7.1 Reflection Cavity

When a cavity resonator is excited by a generator matched to the feeder line impedance Z_0, by means of a small centred hole in the transverse wall, the equivalent circuit is as shown in Fig. 7.9.

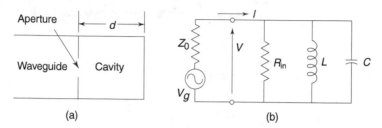

(a)   (b)

Fig. 7.9 *Aperture coupled cavity and its lumped equivalent circuit*

The input impedance can be expressed by

$$Z_{in}(\omega) = \frac{P + 2j\,(W_m - W_e)}{1/2\,\Pi\,*} = \frac{R_{in}}{1 + jQ_0\,(\Delta f / f_r)}; \qquad (7.115)$$

where $\omega_r = 1/\sqrt{LC}$. At half power point $Z_{in} = 0.707$ times the maximum value R_{in}. Therefore, $Q_0\,\Delta f/f_r = 1$, or the bandwidth

$$\Delta f = f_r/Q_0. \qquad (7.116)$$

For a rectangular cavity under dominant TE_{101} mode input impedance at any frequency ω can be written as [7.1]

$$Z_{in}(\omega) = -jZ_0 \frac{\omega^2\,r^L}{2(\omega - \omega_r - j\omega_r/2Q_0)} \qquad (7.117)$$

At resonance ($\omega = \omega_r$) input impedance is pure resistance which accounts for cavity losses. Therefore,

$$Z_{in}(\omega_r) = +jZ_0 \frac{\omega_r^2\,L}{2\,j\omega_r/2Q_0} \qquad (7.118)$$

$$= \omega_r\,L\,Z_0\,Q_0 \qquad (7.119)$$

$$= R_{in} \qquad (7.120)$$

Therefore,

$$Q_0 = R_{in}/(\omega_r L Z_0) \qquad (7.121)$$

The external Q is given by

$$Q_e = 1/\omega_r L \qquad (7.122)$$

Loaded Q is given by

$$Q_L = (1/Q_0 + 1/Q_e)^{-1} \qquad (7.123)$$

From Eqns 7.121 – 7.123

$$Q_L = [\omega_r L Z_0/R_{in} + \omega_r L]^{-1}$$
$$= R_{in}/(\omega_r L Z_0 + \omega_r L R_{in})$$
$$= R_{in}/[\omega_r L (Z_0 + R_{in})] \tag{7.124}$$

Therefore,

$$1/Q_L = 1/Q_0 + 1/Q_e = (1+\beta_e)/Q_0$$

or,

$$Q_0 = (1+\beta_e) Q_L \tag{7.125}$$

where,

$$\beta_e = Q_0/Q_e = \text{coupling parameter} = R_{in} \omega_r L/(\omega_r L Z_0) \tag{7.126}$$

$$= R_{in}/Z_0 \tag{7.127}$$

β_e is a measure of the degree of coupling between the cavity and the input waveguide. Similarly for a series resonant circuit

$$Q_0 = \omega_r L Z_0/R_{in} \tag{7.128}$$
$$Q_e = \omega_r L \tag{7.129}$$
$$\beta_e = Z_0/R_{in} \tag{7.130}$$
$$= Q_0/Q_e \tag{7.131}$$

Different degrees of coupling are described below:

Coupling	Parallel resonance	Series resonance
Over	$\beta_e > 1$	$\beta_e < 1$
Critical	$\beta_e = 1$	$\beta_e = 1$
Under	$\beta_e < 1$	$\beta_e > 1$

7.7.2 Transmission Cavity

A transmission type cavity is coupled to both a generator and load. Therefore, two apertures are involved, one on each side as shown in Fig. 7.10. The equivalent circuit permits one to devise many ways of measuring Q. The coupling of a cavity mode to a waveguide is represented by ideal transformers of turn ratios n_1 : 1 and n_2 : 1 at the generator and the load sides, respectively. Finally a combined equivalent circuit transforming load and generator sides into a single mesh at resonance facilitates the mesh analysis.

Following main parameters are important
1. Resonant frequency f_r and corresponding VSWR, S_r.
2. Input and output coupling parameters β_1 and β_2, which measure the coupling efficiency (= external power loss/Internal powerloss).
3. Transmission loss $T(\omega_r)$ at resonance (= Power output/Power input).
4. Q factors: Q_0, Q_L and Q_e.

By definition

$$Q_0 = \frac{1/2 \, \omega_r \, LI^2}{1/2 \, RI^2} = \frac{\omega_r \, L}{R} \tag{7.132}$$

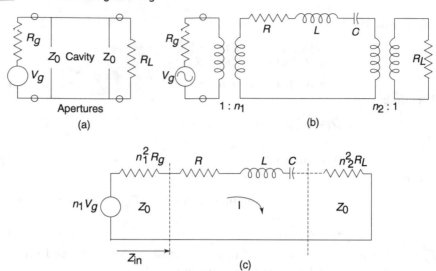

Fig. 7.10 *Transmission type cavity resonator and its equivalent circuits*

$$Q_L = \frac{1/2\, \omega_r\, LI^2}{1/2\left(R + n_1^2\, R_g + n_2^2\, R_L\right)I^2} = \frac{\omega_r\, L}{\left(R + n_1^2\, R_g + n_2^2\, R_L\right)}; \quad (7.133)$$

$$Q_e = \frac{1/2\, \omega_r\, LI^2}{1/2\, Z_0\, I^2} = \frac{\omega_r\, L}{Z_0} \quad (7.134)$$

where $\quad 1/Q_L = 1/Q_0 + 1/Q_e$ $\hfill (7.135)$

$$\beta_1 = \frac{1/2\, n_1^2\, R_g\, I^2}{1/2\, RI^2} = \frac{n_1^2\, R_g}{R} \quad (7.136)$$

$$\beta_2 = n_2^2 R_L / R \quad (7.137)$$

$\beta = 1$ critical coupling

$\beta < 1$ under coupling

$\beta > 1$ over coupling

From Eqns 7.132–7.137

$$Q_0 = Q_L\,(1 + \beta_1 + \beta_2) \quad (7.138)$$

The impedance of equivalent circuit can be expressed by

$$Z = R + n_1^2\, R_g + n_2^2\, R_L + j\,(\omega L - 1/\omega C)$$
$$= R\,[1 + \beta_1 + \beta_2 + jQ_0\,(\omega/\omega_r - \omega_r/\omega)] \quad (7.139)$$

and $\qquad I = n_1^2\, V_g^2/Z^2 \quad (7.140)$

In terms of matched generator and load, $R_g = Z_0 = R_L$. The load power

$$P_L = n_2^2 Z_0 |I|^2 = \beta_2 R |I|^2$$

$$= \frac{\beta_1 \, \beta_2 \, V_g^2 / Z_0}{(1 + \beta_1 + \beta_2)^2 + Q_0^2 \, (\omega/\omega_r - \omega_r/\omega)^2} \qquad (7.141)$$

Generator power available under matched condition

$$P_0 = V_g^2 / 4Z_0 \qquad (7.142)$$

Therefore, the transmission loss of the cavity

$$T(\omega) = P_L/P_0 = \frac{4\beta_1 \, \beta_2}{(1 + \beta_1 + \beta_2)^2 + Q_0^2 \, (\omega/\omega_r - \omega_r/\omega)^2} \qquad (7.143)$$

At resonance $\omega = \omega_r$

$$T(\omega_r) = \frac{4\beta_1 \, \beta_2}{(1 + \beta_1 + \beta_2)^2} \qquad (7.144)$$

Therefore,

$$T(\omega) = \frac{T(\omega_r)}{1 + Q_L^2 (\Delta\omega/\omega_r)^2} \; ; \; \text{for } \omega = \omega_r + \Delta\,\omega/2 \qquad (7.145)$$

If $\Delta\,\omega =$ half power bandwidth

$$T(\omega) = 1/2\, T\,(\omega_r)$$

or,

$$T(\omega)/T(\omega_r) = 1/2 = \frac{1}{1 + Q_L^2 (\Delta\omega/\omega_r)^2} \qquad (7.146)$$

Therefore,

$$Q_L = \omega_r / \Delta\,\omega \qquad (7.147)$$

An equivalent resistance, seen by the generator is R_{eg} for maximum power transfer and can be expressed by

$$n_1^2 \, R_{eg} = R + n_2^2 \, R_L$$

or,

$$R_{eg} = \frac{R + n_2^2 \, R_L}{n_1^2} \qquad (7.148)$$

The normalised value

$$R_{eg}/Z_0 = \frac{R + n_2^2 \, R_L}{n_1^2 \, Z_0} \qquad (7.149)$$

At resonance, for resistive impedance, the cavity VSWR

$$S_r \approx R_{eg}/Z_0$$

$$= \frac{R + n_2^2 R_L}{n_1^2 Z_0} = \frac{1 + \beta_2}{\beta_1} \qquad (7.150)$$

Therefore,

$$\beta_1 = \frac{4}{4 S_r - T(w_r)(1 + S_r)^2} \qquad (7.151)$$

$$\beta_2 = \beta_1 S_r - 1 \qquad (7.152)$$

The above analysis provides working formulae for experimental measurements of Q:

$$Q_L = f/\Delta f \qquad (7.153)$$

$$Q_0 = Q_L (1 + \beta_1 + \beta_2) \qquad (7.154)$$

$$\beta_1 = \frac{4}{4 S_r - T(\omega_r)(1 + S_r)^2} \qquad (7.155)$$

$$\beta_2 = \beta_1 S_r - 1 \qquad (7.156)$$

The measurement parameters at resonance are f_r, S_r, $T(w_r)$ and 3 db bandwidth Δf.

7.7.3 Loop-coupled Cavity

In loop-coupled cavity, since the loop is very small, the current in the loop can be considered to be constant. The plane of the loop is placed perpendicular to the magnetic flux lines of a given resonant mode. The equivalent circuit of the loop coupled cavity is shown in Fig. 7.11.

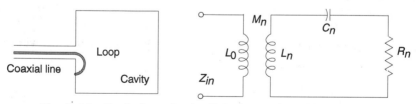

Fig. 7.11 *Equivalent circuit of single-port loop-coupled cavity*

The input impedance of the circuit in the vicinity of the nth resonance can be expressed by

$$Z_{in} = j\omega L_0 + \frac{\omega_n^2 M_n^2}{2 j L_n (\omega - \omega_n - j\omega_n/2Q_n)} \qquad (7.157)$$

where L_0 is the self-inductance of the coupling loop, $L_n = Q_n R_n/\omega_n$, $C_n = Q_n/\omega_n R_n$. For maximum coupling, the loop is placed at a position of maximum magnetic field of a given mode.

7.8 RE-ENTRANT CAVITY

In microwave tubes it is essential that for efficient energy transfer from an electron beam to high Q cavity resonators, the electron transit time across the cavity field region must be very small. Consequently the cavity grids need to be spaced very closely to form re-entrant structure as shown in Fig. 7.12. The E-field is concentrated in the small gap g. This capacitance region can be used to control flow of electrons passing through this gap when it is grided (for a klystron tube). The tuning of the cavity can be accomplished by means of short-circuit plungers.

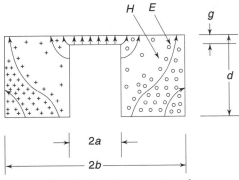

Fig. 7.12 *Re-entrant cavity*

The re-entrant cavity of length d with a gap thickness $g \ll d$ may be considered as a coaxial line with radii of the inner and outer conductors as a and b, respectively. The gap capacitance C and the coaxial line below the gap provide equal and opposite reactances at the plane of the capacitance at resonance such that

$$1/\omega_r C = Z_0 \tan(2\pi d/\lambda_g)$$

or, $$d = (\lambda_g/2\pi)\tan^{-1}(1/Z_0\omega_r C) \qquad (7.158)$$

where $$C = \varepsilon_0 \pi a^2/g, \qquad (7.159)$$

and $$Z_0 = 60 \ln b/a \qquad (7.160)$$

for air dielectric. For a small capacitance C, $Z_0 \, \omega_r \, C \ll 1$, $\beta d \approx \pi/2$, i.e., the line is practically a quarter wave long. For large C or small gap, the length d should be shortened. Thus for a given gap width or C, the resonant lengths d of the cavity can be varied from multiple of quarter wavelength to a smaller value satisfying the resonant condition in Eq. 7.158.

By increasing C the electric energy stored in the cavity increases. A corresponding increase in the magnetic energy ($W_e = W_m$ at resonance) has to be provided by a larger microwave current in the cavity walls, which results in higher dissipative losses due to finite conductivity of the cavity walls. Consequently the unloaded Q_0 of coaxial cavity decreases.

The cavity mode in these configurations is *TEM* type when $d > (b-a)$; $b < \lambda_g/4$. If $d < (b-a)$ and $b \approx \lambda_g/4$, the cavity mode is *TM* type with the electric field directed mainly along the cavity axis. The *TM* mode cavity can be easily tuned by adjusting the capacitive gap g, provided that the resonant mode is TM_{010}.

7.9 HOLE-AND-SLOT CAVITY

Microwave resonant cavity of the hole-and-slot type, shown in Fig.7.13, is used in multicavity magnetron oscillators. The resonant frequency of the cavities can be determined from the relation

$$\omega_r = 1/\sqrt{LC} \qquad (7.161)$$

The function of the lumped inductance L is carried out by the surface of the hole which is equivalent to a loop of metallic band. The lumped capacitance C is formed by the slot cut through the copper block.

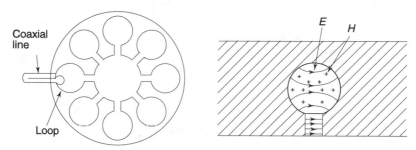

Fig. 7.13 *Hole-and-slot cavity*

7.10 MICROSTRIP RESONATORS

In microwave integrated circuits, resonators of planar configurations are used for fixed tuned circuits, determining the effective microstrip permittivity ε_{eff} at a given frequency and various discontinuity parameters. A microstrip resonator can be formed by a closed loop on a substance, or a line left open or short-circuited at both ends or simply by a metallic patch of rectangular or circular geometry. By loosely coupling an r.f. test signal to such a resonator, frequency pulling effects are avoided, and the reflected energy and resonator characteristics can be analysed.

There are three phenomena associated with an open circuit
(a) The fringing fields extending beyond the physical end of the strip.
(b) Launching of surface waves from the end
(c) Radiation loss from the open end.

Phenomena (b) and (c) require equivalent shunt conductance at the open end of the line. In most cases this will not be taken into consideration. But their minimisation may be carried out. The phenomenon (a) dominates and is accounted for by an equivalent fringe capacitance C_f to be connected at the open end. A variety of resonator configurations are described below.

7.10.1 Microstrip Line Gap-Coupled Resonators

The microstrip line gap-coupled resonator is formed by sections of microstrip lines coupled through gaps as shown in Fig. 7.14. The resonator consists of a half wavelength open circuited microstrip line that is capacitively coupled to an input microstrip line along the length or from the side as shown in Figs. 7.14 (a) and (b), respectively. The open circuited end contains fringing electric field, resulting in accumulation of electric charge. This in turns results in an equivalent fringing capacitance C_{f0}. The capacitance C_{f0} is compensated by an additional inductance which arises due to increase in the length d beyond one wavelength by Δd_0. Δd_0 is the equivalent extra length of microstrip line, having all the propagation characteristics of those applicable to the main line i.e., having same Z_0 and hence same W/h and ε_{eff}. Equating the reactances of C_{f0} and that of the electrically open circuited extra length Δd_0,

$$\frac{1}{\omega C_{f0}} = Z_0 \cot\beta \ \Delta d_0 \ \frac{Z_0}{\tan \beta \Delta d_0} \tag{7.162}$$

Since

$$\Delta d_0 \ll \lambda g, \ \tan\beta \Delta d_0 = \beta \Delta d_0 \ \text{and}$$

$$\Delta d_0 = cZ_0 \ c_{f0}/\sqrt{\varepsilon_{\text{eff}}} \ ; c = \frac{1}{\sqrt{\mu_0 \ \varepsilon_0}} = 3 \times 10^8 \ \text{m/s}$$

or,

$$\Delta d_0 = 0.412 \ h \left[\frac{\varepsilon_{\text{eff}} + 0.3}{\varepsilon_{\text{eff}} - 0.258} \right] \left[\frac{W/h + 0.262}{W/h + 0.813} \right] \tag{7.163}$$

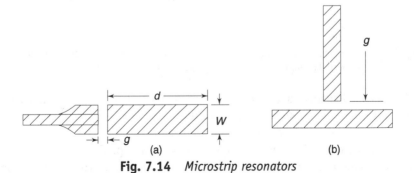

(a)   (b)

Fig. 7.14 *Microstrip resonators*

Similarly, at the gap region also there is end effect length Δd_g. The resonant condition can be expressed by

$$d + \Delta d_g + \Delta d_0 = \lambda_g/2 \tag{7.164}$$

The amount of fringing capacitance can be calculated as a function of line width W and the substrate thickness h using the assumption of static potential fields [1, 22, 23]. Typical values of the open-circuited fringing capacitance are shown in Fig. 7.15. The quasi -*TEM* mode analysis of microstrip lines can be used with reasonable accuracy for frequencies up to 4 GHz and for substrate thickness of 1 mm. At higher frequencies the electric field is more confined between the patch and ground plane resulting in an increase in the effective dielectric constant as well as increased dielectric loss and conductor loss. This frequency dispersion of the effective dielectric constant helps to determine the effective dielectric constant of a substrate. Taking a pair of resonators of lengths d_1 and d_2 having identical widths, gaps and substrate, the dispersion results can be expressed for closely resonant frequencies f_1 and f_2 as

$$\varepsilon_{\text{eff}} = \left[\frac{hc \, (2f_1 - f_2)}{2 f_1 f_2 \, (d_2 - d_1)} \right]^2 \tag{7.165}$$

Total end effect length can be determined as

$$\Delta d = \Delta d_g + \Delta d_0 = \frac{f_2 d_2 - 2 f_1 d_1}{2 f_1 - f_2} \tag{7.166}$$

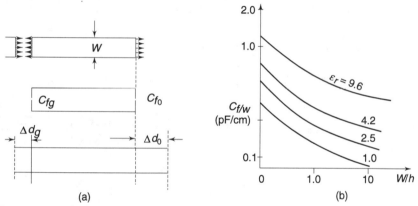

(a)   (b)

Fig. 7.15 *Fringing capacitance as a function of the line width W and the substrate thickness h (Source: Ref. 1)*

The series gap capacitance can be represented by an equivalent π-network as shown in Fig. 7.16. The significant capacitance that exists across the gap is denoted by C_2, whereas, C_1 represents the fringing capacitance between the line and the ground at each end of the line at a symmetrical gap. The values of the gap capacitance can be determined from the graphs of Fig. 7.16 (b) as a function of the gap spacing / substrate thickness. Resonators of this type are commonly used in microstrip filters. Garg and Bahl [21] have given expressions for the gap capacitances as

$$C_1 = \frac{C_e(\varepsilon_r)}{2} \quad ; 0.5 \leq W/h \leq 2.0 \tag{7.167}$$

$$C_2 = \frac{C_0(\varepsilon_r) - C_1(\varepsilon_r)}{2} \quad ; 2.5 \leq \varepsilon_r \leq 15 \tag{7.168}$$

where,

$$C_0 \, (\varepsilon_r) = C_0(9.6)(\varepsilon_r/9.6)^{0.8}$$
$$C_e \, (\varepsilon_r) = C_e(9.6)(\varepsilon_r/9.6)^{0.9}$$
$$C_0(9.6)/W = (g/W)_0^m \exp(k_0) \text{ pF/m}$$
$$C_e(9.6)/W = (g/W)_e^m \exp(k_e) \text{ pF/m}$$

$$
\begin{aligned}
m_0 &= W/h(0.619 \log(W/h) - 0.3855) \\
k_0 &= 4.26 - 1.453 \log(W/h)
\end{aligned}
\quad \Bigg] \quad 0.1 \le g/W \le 1.0
$$

$$
\begin{aligned}
m_e &= 0.8675 \\
k_e &= 2.043 \, (W/h)^{0.12}
\end{aligned}
\quad \Bigg] \quad 0.1 \le g/W \le 0.3
$$

$$
\begin{aligned}
m_e &= 1.565/(W/h)^{0.16} - 1 \\
k_e &= 1.97 - 0.03/(W/h)
\end{aligned}
\quad \Bigg] \quad 0.3 \le g/W_- \le 1.0
$$

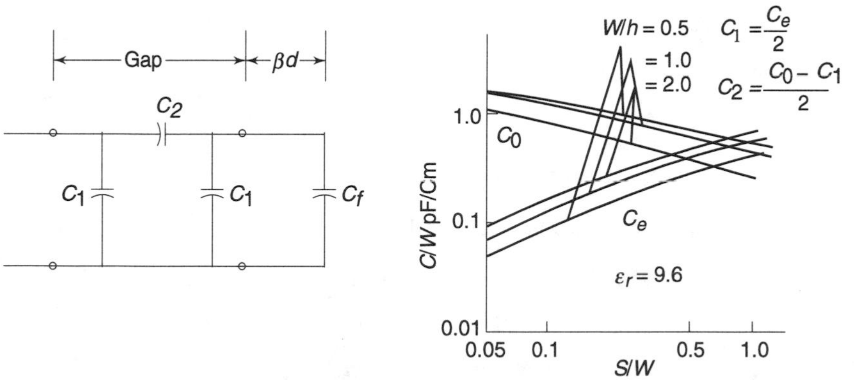

Fig. 7.16 *Gap capacitance*

7.10.2 Rectangular Microstrip Disk Resonator

A microstrip rectangular patch on a dielectric substrate over a ground plane (Fig. 7.17) forms a dielectric loaded cavity with the ground plane and the patch as electric walls and magnetic side walls at the periphery where $\mathbf{H}_{\text{tan}} = 0$ (simulates an open circuit). The cavity exhibits higher order mode resonances. The fields inside the cavity can be found accurately by assuming that the substrate thickness $h \ll \lambda$, so that only one **E**-field component E_z (constant in z) exists which satisfies the scalar *Helmholtz equation*

$$(\nabla^2 + k^2) \, E_z \, (x, y) = 0 \tag{7.169}$$

where $k = \omega\sqrt{(\mu_0 \, \varepsilon)}$. If the length and width of the patch are l and W, respectively, and the origin of the coordinates is at one corner such that $0 \le z \le h$, $0 \le x \le l$, $0 \le y \le W$, the following boundary conditions must be satisfied.

$$H_x(y = 0) = H_x(y = W) = 0$$

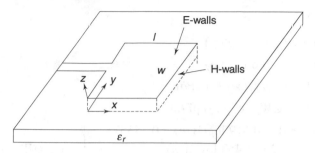

Fig. 7.17 *Microstrip rectangular disk resonator*

or
$$\left.\frac{\partial E_z}{\partial y}\right|_{y=0} = \left.\frac{\partial E_z}{\partial y}\right|_{y=W} = 0$$

$$H_y(x=0) = H_y(x=l) = 0$$

or,

$$\left.\frac{\partial E_z}{\partial x}\right|_{x=0} = \left.\frac{\partial E_z}{\partial x}\right|_{x=l} = 0 \qquad (7.170)$$

to yield the field solutions of TM_{mn0} modes

$$E_z = A_{mn} \cos(m\pi x/l) \cos (n\pi y/W) \qquad (7.171)$$

$$H_x = \frac{j}{\omega\mu}\frac{\partial E_z}{\partial y} = \frac{n\pi A_{mn}}{W\,\omega\mu} \cos\left(\frac{m\pi x}{l}\right) \sin\left(\frac{n\pi y}{W}\right) \qquad (7.172)$$

$$H_y = \frac{j}{\omega\mu}\frac{\partial E_z}{\partial x} = \frac{-m\pi A_{mn}}{\omega\mu l} \sin\left(\frac{m\pi x}{l}\right) \cos\left(\frac{n\pi y}{W}\right) \qquad (7.173)$$

where $k_x^2 + k_y^2 = (m\pi/l)^2 + (n\pi/W)^2 = k_r^2 = \omega_r^2\mu_0\varepsilon$ and A_{mn} is an *amplitude constant* that depends on the excitation. Thus the resonant frequency of the cavity in TM_{mn0} mode to z is

$$(f_r)_{mn} = \frac{1}{2\pi\sqrt{(\mu_0\,\varepsilon)}} [(m\pi/l)^2 + (n\pi/W)^2]^{1/2} \; ; m = 0,1,2,\ldots$$

$$n = 0,1,2,\ldots \qquad (7.174)$$

Resonant frequencies occur when l or W is equal to an integral multiple of half wavelength in the substrate. For $l > W$, the lowest order resonant frequency occurs when $m = 1$ and $n = 0$, i.e., for TM_{100} mode:

$$(f_r)_{100} = \frac{1}{2l\sqrt{(\mu_0\,\varepsilon)}} \qquad (7.175)$$

The Q-factor of the resonator is determined in the following manner. The total stored energy in the resonator for dominant mode TM_{100} is

$$W_s = 2 W_e = \frac{\varepsilon'}{2} \int\limits_0^h \int\limits_0^W \int\limits_0^l |E_z|^2 \, dx \, dy \, dz$$

$$= \frac{\varepsilon'}{2} \int\limits_0^h \int\limits_0^W \int\limits_0^l A_{10}^2 \cos^2(\pi x/l) \, dx \, dy \, dz$$

$$\frac{\varepsilon' h A_{10}^2 W}{2} \int\limits_0^l \cos^2\left(\frac{m \pi x}{l}\right) dx$$

$$= \frac{\varepsilon' h A_{10}^2 \cdot h W l}{4} \tag{7.176}$$

The power loss in the dielectric

$$P_d = \frac{\omega \varepsilon_r''}{2} \int\limits_0^h \int\limits_0^W \int\limits_0^l |E_z|^2 \, dx \, dy \, dz$$

$$= \frac{\omega \varepsilon_r''}{2} \frac{2 W}{\varepsilon'} = \frac{\omega \varepsilon_r'' W}{\varepsilon'}$$

$$= \frac{\omega A_{10}^2 \varepsilon_r'' h W l}{4} \tag{7.177}$$

The conductor power loss for the bottom and top conducting surfaces :

$$P_c = 2 \frac{R_s}{2} \int\limits_0^l \int\limits_0^W \left(|H_x|^2 + |H_y|^2\right) dx \, dy$$

$$= R_s \int\limits_0^l \int\limits_0^W |H_y|^2 \, dx \, dy \; ; H_x = 0 \text{ for } n = 0$$

$$= R_s \int\limits_0^l \int\limits_0^W \left(\frac{\pi A_{10}}{\omega \mu l}\right) \sin^2\left(\frac{\pi x}{l}\right) dx \, dy$$

$$= R_s \left(\frac{\pi A_{10}}{w \mu l}\right)^2 \frac{W l}{2} \tag{7.178}$$

The Q-factor

$$Q_0 = \frac{\omega W_s}{P_c + P_d} = \frac{\omega \cdot \varepsilon' A_{10}^2 h \omega l / 4}{\omega A_{10}^2 \varepsilon_r'' (W l h / 4) + R_s (\pi A_{10} / \omega \mu l)^2 \cdot W l / 2}$$

$$= \frac{\omega \varepsilon' h}{\omega \varepsilon_r'' h + 2 R_s (\pi/\omega \mu l)^2} \tag{7.179}$$

The above expression for Q neglects the radiation loss due to fringing fields at the edges of the patch.

7.10.3 Microstrip Circular Disk Resonator

A circular disk on the substrate of a high dielectric constant forms a cavity with the patch and the ground plane as electric walls on which $\hat{n} \times \mathbf{E} = 0$ and cylindrical magnetic walls surrounding the patch on which $\hat{n} \times \mathbf{H} = 0$ as shown in Fig. 7.18. The patch can be excited by a coaxial line probe from the ground plane or a input microstrip line on the plane of the patch. It will be assumed that the height (h) of the substrate of relative dielectric constant (ε_r) is small compared to the free space wavelength (λ_0). The resonator is excited by a line current (I_0) along z-axis through a probe of negligible diameter which forms the centre conductor of a coaxial line fed from the ground plane at $\rho = \rho_0$, $\phi = 0°$, where $0 < \rho_0 < a$.

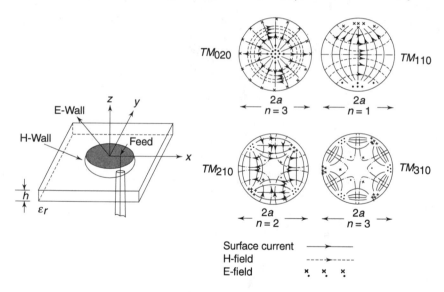

Fig. 7.18 *Circular disk microstrip resonator*

The cavity model of the disk structure will be based on the following assumptions
  (a) Because of the close proximity between the disk and the ground plane the source current will be assumed uniform along z and therefore the modal fields will have no z variations in the frequency band of interest, and there is no charge accumulation in the region between the disk and the ground plane.
  (b) Above implies that only non-zero field components are E_z, H_ρ and H_ϕ in the region bounded by the microstrip and the ground plane.

(c) The electric current in the microstrip must have no components normal to the edge at any point on the edge, implying zero tangential component of **H** along the edges: $H_\phi = 0$ at $\rho = a$.

The region between the disk and the ground plane will therefore be considered as a cavity with magnetic walls at the edges and electric walls at the top and bottom. It will be assumed that such a cavity would radiate no power and have a purely reactive input impedance of either zero or infinite value at resonance. The resonant modes are denoted by $TM_{nm\,0}$, where n and m denote variations in the azimuthal and radial directions, respectively, with zero variation in the z-direction.

The components of modal fields in the resonator can be obtained by solving the wave equation in the cylindrical coordinate system with delta function excitation. However, without much loss of generality, the cavity fields can be obtained from the source free wave equation

$$\Delta_t^2 E_z + k_\rho^2 E_z = 0 \tag{7.180}$$

Following the method of separation of variables, we seek to find solutions of the form

$$E_z = R(\rho)\, \Phi(\phi) \tag{7.181}$$

The ϕ-symmetric ring structure yields harmonic solution

$$\Phi(\phi) = \cos n\phi \tag{7.182}$$

The solution $R(\rho)$ corresponds to Bessel's equation of order n. k_ρ represents the wave number in the radial direction and is always real. Therefore, in the disk structure there will be formation of cylindrical standing waves due to two magnetic side walls. This implies that the solution of R should be represented by the Bessel function of the first kind $J_n(k_\rho \rho)$,

$$R(\rho) = J_n(k_\rho \rho) \tag{7.183}$$

Since the fringing fields in the radial direction extend into free space, wave number k_ρ is determined by the effective dielectric constant ε_{eff}: $k_\rho = k_0 \sqrt{\varepsilon_{\text{eff}}}$. For a very thin substrate ($h \ll a$), $\varepsilon_{\text{eff}} \approx \varepsilon_r$ and $k_\rho = k_0 \sqrt{\varepsilon_r}$. The components of the modal fields for $TM_{nm\,0}$ modes are therefore

$$E_z = J_n(k_\rho \rho) \cos n\phi \tag{7.184}$$

$$H_\rho = \frac{j}{\omega \mu_0} \frac{\partial E_z}{\partial \phi}$$

$$= (-jn/\omega \mu_0)\, J_n(k_\rho \rho)\, \sin n\phi \tag{7.185}$$

$$H_\phi = \frac{j}{\omega \mu_0} \frac{\partial E_z}{\partial \rho}$$

$$= (-jk_\rho/\omega \mu_0)\, J_n'(k_\rho \rho)\, \cos n\phi \tag{7.186}$$

From the boundary conditions

$$H_\phi = 0 \text{ at } \rho = a \qquad (7.187)$$

or,

$$\frac{\partial E_z}{\partial \rho} = 0 = k_\rho J'_n (k_\rho a) \qquad (7.188)$$

where prime indicates differentiation with respect to the argument. The resonant frequencies are obtained, from the roots x'_{nm} of

$$J'_n(k_\rho a) = 0 \text{ or } k_\rho = x'_{nm}/a \text{ or } f_r = x'_{nm}/2\pi a \sqrt{\mu \varepsilon a} \qquad (7.189)$$

The dominant mode is designated as the TM_{110} mode for which x'_{nm} is minimum $= x'_{11} = 1.841$.

The total stored energy in the resonator for the dominant mode is given by

$$W_s = 2W_e = \frac{\varepsilon'_r \varepsilon_0}{2} \int_0^h \int_0^{2\pi} \int_0^a J_1^2 (k_\rho \rho) \cos^2 \phi \, \rho \, d\rho \, d\phi \, dz$$

$$= \frac{\pi \varepsilon'_r \varepsilon_0 h a^2}{4} \left(1 - \frac{1}{k_\rho^2 a^2} \right) J_1^2(k_\rho a) \qquad (7.190)$$

since

$$J'_1 (k_\rho a) = 0.$$

The power loss in the dielectric

$$P_d = \frac{\omega \varepsilon''_r}{2} \int_0^h \int_0^{2\pi} \int_0^a |E_z|^2 \, \rho \, d\rho \, d\phi \, dz$$

$$= \frac{2\omega \varepsilon''_r}{\varepsilon'_r} W_e \qquad (7.191)$$

The conductor power loss

$$P_c = 2 \frac{R_s}{2} \int_0^{2\pi} \int_0^a (|H_\phi|^2 + |H_\rho|^2) \, \rho d \rho \, d\phi$$

$$= \frac{\pi R_s}{k_0^2 \eta^2} \int_0^a \left[k_\rho^2 J_1^2 (k_\rho \rho) + \frac{J^2 (k_\rho \rho)}{\rho^2} \right] \rho \, d\rho$$

$$= \frac{\pi R_s k_\rho^2 a^2}{2 k_0^2 \eta^2} \left(1 - \frac{1}{k_\rho^2 a^2} \right) J_1^2 (k_\rho a) \qquad (7.192)$$

The Q-factor

$$Q = \frac{\omega W}{P_d + P_c} = \frac{k_0 h}{2 R_s/\eta + \varepsilon''_r k_0 h/\varepsilon'_r} \qquad (7.193)$$

This expression neglects the radiation loss due to fringing field and conductor loss for currents on the top of the patch. These reduce the practical Q achievable.

Example 7.6   Find Q of a microstrip disk resonator at dominant TM_{110} mode operating at 5 GHz. The copper patch is etched on alumina substrate with $\varepsilon'_r = 9.6$, $\varepsilon''_r = 0.0002$ and $h = 1/16$ inch.

Solution

$$Q = \frac{k_0 h}{2 R_s / \eta + \varepsilon''_r k_0 h / \varepsilon'_r}$$

Here $\varepsilon'_r = 9.6$, $\varepsilon''_r = 0.0002$, $h = 1/16$ inch $= 1.5875$ mm

$$\lambda_0 = 30/5 = 6 \text{ cm}$$

$$k_0 = 2\pi/6 = \pi/3$$

$$k_0 h = \pi/3 \times 0.15875 = 0.1663$$

$$R_s = \sqrt{(\omega \mu_0 / 2\sigma)}$$

$$= \left[\frac{2\pi \times 10^9 \times 4\pi \times 10^{-7}}{2 \times 5.8 \times 10^7} \right]^{1/2} = 1.8 \times 10^{-2} \qquad \text{(position)}$$

$$Q = \frac{0.1663}{2 \times 1.8 \times 10^{-2}/120\pi + 0.0002 \times 0.1663/9.6} = 1680.50$$

7.10.4 Microstrip Ring Resonator

Figure 7.19 shows a circular microstrip ring resonator with inner and outer radii a and b respectively. Ring resonators have found applications in circulators, hybrid junctions and filters. They can also be used to measure propagation constants in microstrips, based on the principle that the ring is resonant when its mean circumference equals an integral number of half wavelengths, i.e., $(a + b)/2 = nc/(2 f_r \sqrt{\varepsilon_{\text{eff}}})$. A simple approach to find the resonant frequencies will be described from a cavity model similar to the case of a disk. The region between the ring and the ground plane will therefore be considered as a cavity with magnetic walls at the edges and electric walls at the top and bottom. The resonant modes are denoted by TM_{nm0}, where n and m denote variations in the azimuthal and radial directions respectively, with zero variation in the z-direction.

The components of modal fields in the ring structure can be obtained by solving the wave equation in the cylindrical coordinate system in a similar way as in the case of disk. Solution R should be represented by the Bessel function of the first kind $J_n (k_\rho \rho)$ and Bessel function of the second kind $N_n (k_\rho \rho)$

$$R(\rho) = J_n (k_\rho \rho) - A_n N_n (k_\rho \rho) \qquad (7.194)$$

to exclude origin, where the modal field is zero. Here A_n is a constant dependent on frequencies and to be determined from the boundary conditions. Since the

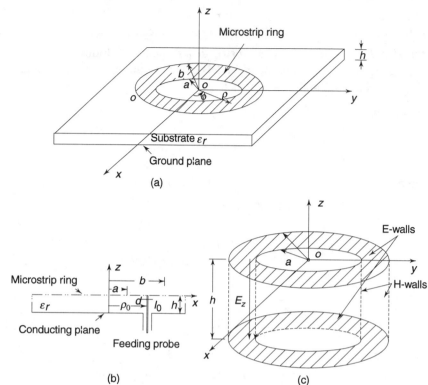

Fig. 7.19 *Microstrip ring resonator(a) Isometric view (b) Excitation by probe (c) Cavity model*

fringing fields in the radial direction extend into free space, wave number k_ρ is determined by the effective dielectric constant ε_{eff}, $k_\rho = k_0 \sqrt{\varepsilon_{\text{eff}}}$. For a very thin substrate ($b - a \gg d$), $\varepsilon_{\text{eff}} \approx \varepsilon_r$ and $k_\rho = k_0\sqrt{\varepsilon_r}$. The components of the modal fields for TM_{nm0} modes are therefore

$$E_z = R(\rho) \cos n\phi \tag{7.195}$$

$$H_\rho = \frac{j}{\omega\mu_0} \frac{\partial E_z}{\partial \phi} \tag{7.196}$$

$$H_\phi = \frac{j}{\omega\mu_0} \frac{\partial E_z}{\partial \rho} \tag{7.197}$$

From the boundary conditions

$$H_\phi = 0 \text{ at } \rho = a, b \tag{7.198}$$

or,

$$\frac{\partial E_z}{\partial \rho} = 0 = k_\rho [\, J'_n (k_\rho a) - A_n N'_n(k_\rho a) \,]$$

$$= k_\rho \left[J'_n (k_\rho\, b) - A_n\, N'_n(k_\rho\, b) \right] \tag{7.199}$$

where prime indicates differentiation with respect to the argument. This leads to the eigen value equation of the ring resonator for the resonances

$$\frac{J'_n(k_\rho\, a)}{N'_n(k_\rho\, a)} = \frac{J'_n(k_\rho\, b)}{N'_n(k_\rho\, b)} = A_n \tag{7.200}$$

The eigen value solutions k_ρ give the resonant frequency f_r, with

$$k_\rho = \frac{2\pi f_r}{c}\, \sqrt{\varepsilon_r}\, ,\ c = \text{speed of light in vacuum} \tag{7.201}$$

The field components are

$$E_z = \left[J_n\left(k_\rho\, \rho\right) - \frac{J'_n(k_\rho\, a)}{N'_n(k_\rho\, a)}\, N_n\left(k_\rho\, \rho\right) \right] \cos n\ \phi \tag{7.202}$$

$$H_\rho = (-jn/\omega\mu_0) \left[J_n\left(k_\rho\, \rho\right) - \frac{J'_n(k_\rho\, a)}{N'_n(k_\rho\, a)}\, N_n\left(k_\rho\, \rho\right) \right] \sin n\phi \tag{7.203}$$

$$H_\phi = (-jk/\omega\mu_0) \left[J_n\left(k_\rho\, \rho\right) - \frac{J'_n(k_\rho\, a)}{N'_n(k_\rho\, a)}\, N_n\left(k_\rho\, \rho\right) \right] \cos n\phi \tag{7.204}$$

These equations determine the modal field patterns of TM_{nm0} modes inside microstrip ring resonator.

The surface electric current on the inner surface of the patch is given by

$$\mathbf{J_S} = -\hat{z} \times \mathbf{H} = \hat{\rho} H\phi - \hat{\phi} H_\rho$$

$$= -\ \hat{\rho}\ (jk/\omega\mu_0) \left[J_n\left(k_\rho\, \rho\right) - \frac{J'_n(k_\rho\, a)}{N'_n(k_\rho\, a)}\, N_n\left(k_\rho\, \rho\right) \right] \cos n\phi$$

$$+\ \hat{\phi}\ (jn/\omega\mu_0) \left[J_n\left(k_\rho\, \rho\right) - \frac{J'_n(k_\rho\, a)}{N'_n(k_\rho\, a)}\, N_n\left(k_\rho\, \rho\right) \right] \sin n\phi \tag{7.205}$$

The mode field patterns and current distributions on the microstrip ring can be obtained from Eqns 7.202–7.205 and are shown in Fig. 7.20.

The Q-factor of the microstrip ring resonator can be computed in the same way as in the case of a disk.

EXERCISES

. 7.1 A short-circuited two-wire line is made of copper. The conductor diameter is 1 cm, the spacing is 3 cm, and the length is 40 cm. Find the antiresonant frequency, the Q, and the input resistance at resonance.

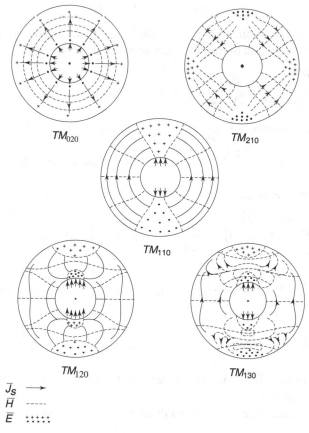

Fig. 7.20 *Mode field patterns and current distributions in a microstrip ring resonator*

7.2 An air-filled 50 ohm transmission line is shorted at one end and terminated at the other end in a lumped capacitor of 4 pF. Calculate the length of the transmission line for resonance at 1 GHz.

7.3 Find the smallest possible size of a cubical cavity and the modes which resonate at 5 GHz.

7.4 A rectangular waveguide resonator of cross-sectional dimensions 2.3 cm × 1 cm is filled with air. What should be the length of the resonator for TE_{101} mode resonance at 10 GHz? What is the next higher mode of resonance and the corresponding resonant frequency?

7.5 Calculate the resonant frequency for the TE_{101} mode of a rectangular cavity formed by shorting the ends of an X-band waveguide with $a = 0.9''$, $b = 0.4''$ and $d = 2.4''$.

7.6 Design a rectangular cavity to have a resonant frequency of $f_r = 9.8$ GHz, having dimension $a = d$ and $b = a/2$.

7.7 Find the resonant frequency and Q of a copper cubic rectangular cavity of dimension 1000 cubic cm for the TE_{101} mode.

7.8   A rectangular copper cavity of height $b = 1.2$ cm, width $a = 2.5$ cm resonates at 10 GHz. The cavity is critically coupled to a rectangular guide of dimensions $a \times b$. Determine the cavity length d and the radius of the centred circular aperture. Find the unloaded and loaded Q's for lossless dielectric inside the cavity.

7.9   An air-filled circular cylindrical cavity resonates at 3 GHz in TM_{010} mode. The resonator is now filled with a lossless dielectric material with dielectric constant 2.2. Calculate the new resonance frequency.

7.10   An air-filled cylindrical cavity has a radius of 3 cm. The cavity is tuned by means of a plunger that allows the length to be varied from 5 to 7 cm. Determine the resonant modes and the range of resonant frequencies.

7.11   The resonant frequency of a cavity is 8.8 GHz. It is critically coupled to an external circuit. The measured bandwidth with loading is 4.0 MHz. Calculate the loaded and unloaded Q's.

7.12   A cylindrical cavity of radius $a = 2$ cm and length 5 cm is filled with a dielectric with permittivity $\varepsilon = (2.5 - j0.0001)\varepsilon_0$. The cavity is made of copper. Find the resonant frequency and Q for the dominant mode.

7.13   Design a microstrip circular disk resonator operating in the TM_{110} mode at 2 GHz with the substrate of relative dielectric constant $6 - j0.005$ and is $1/16''$ thick. The disk is made of copper. Determine the required radius a of the disk and Q of the resonator.

REFERENCES

1.   Collin, R E *Foundation of microwave engineering*, Second edition, McGraw-Hill, Inc. 1992.

2.   Montgomery, CG, R H Dicke and E M Purcell, *Principles of microwave circuits*, McGraw-Hill Book Company, New York, 1948.

3.   Ragan, G.L, (ed.) *Microwave transmission circuits*, McGraw-Hill Book Company, New York, 1948.

4.   Slater, JC, *Microwave Electronics*, D. Van Nostrand Company, Inc., Princeton, N.J., 1950.

5.   Goubau G, *Electromagnetic waveguides and cavities*, Chap. 2, Pergamon Press, New York, 1961.

6.   Kurokawa, K, "The expansions of electromagnetics fields in cavities", *IRE Trans.*, Vol. MTT-6, pp. 178–187, April 1958.

7.   Kaifez, D and P Guilon, *Electron resonators*, Artech House Books, Dedham, Mass., 1986.

8.   Van Bladel, J, "The Excitation of Dielectric Resonators of Very High Permittivity", *IEEE Trans.*, Vol. MTT-23, pp. 208-217, 1975.

9.   Chow, K K, "On the solution and field pattern of cylindrical dielectric resonators", *IEEE Trans.*, Vol. MTT-14, 439, 1966.

10.   Kraus, J D, *Electromagnetics*, McGraw-Hill Book Company, Inc., 1953.

11.   Slater, PC, and N H Frank, *Electromagnetism*, McGraw-Hill Book Company, 1947.

12. Plonsey, R and RE Collin, *Electromagnetic fields*, McGraw-Hill Book Company, Inc., 1961.

13. Javod, M and P M Brown, *Field analysis and electromagnetics*, McGraw-Hill Book Company, 1963.

14. Argence, E and T Kahan, *Theory of waveguides and cavity resonators*, Blackie and Son Limited, London, 1967.

15. Marcavitz, N, *Waveguide handbook*, M.I.T. Radiation laboratory series, vol.10, McGraw-Hill, New York, 1951.

16. Ramo, S, J R Whinnery and T Van Duzer, *Fields and waves in communication electronics*, 2nd ed. New York, John Wiley, 1984.

17. Waldron, R A, *Theory of guided electromagnetic waves*, London, Van Nostrand Reinhold, 1970.

18. Wu,Y S and Rosenboum, "Mode chart for microstrip ring resonators", *IEEE Trans. on microwave theory and techniques*, Vol. MTT-21, July 1973, pp. 487– 489.

19. Ali, SM, W C Chew, and JA Kong, "Vector hankel transform analysis of annular ring microstrip antenna", *IEEE Trans. on antennas propagation*, Vol. AP-30, July 1982, pp 637-644.

20. Terry Edwards, *Foundations for microstrip circuit design*, Second Edition, John Wiley and Sons, 1992.

21. Garg, Ramesh and I J Bahl, "Microstrip discontinuities", *Int. J. Electronics*, 45, No. 1, 1978, pp. 81–87.

22. Silvester P and Benedek, "Equivalent Capacitances of Microstrip Open Circuits", *IEEE Trans*. Vol. MTT-20, pp. 511–516, 1972.

23. Benedek, P and P Silvester, "Equivalent Capacitances for Microstrip Gaps and Steps", *IEEE Trans.*, Vol. MTT-20, pp. 729–733, 1972.

24. Das, A, S K Das and SP Mathur, "Radiation characteristics of higher order modes in annular ring microstrip antenna", *Proc. IEE, Pt. H.*, October, 1984.

Chapter 8

Microwave Filters

8.1   INTRODUCTION

Microwave filters are two-port, reciprocal, passive, linear devices which attenuate heavily the unwanted signal frequencies while permitting transmission of wanted frequencies. There are three types of construction of filters—a *reflective filter* which consists of capacitive and inductive elements producing ideally zero reflection loss in the passband and very high attenuation in the stop band, *absorptive filters* which dissipate the unwanted signal internally and pass the wanted signal, and a *lossy filter* which uses lossy material in the filter to produce heavy loss in the rejected signal but low loss to the wanted signal. This chapter describes mainly the reflective microwave filters constructed from purely reactive elements. The practical filters have a small non-zero attenuation in the pass band and a small signal output in the attenuation or stop band due to the presence of resistive losses in reactive elements and propagating medium. A microwave filter is designed to operate between resistive source and load impedances of 50 ohm in most microwave systems.

8.2   FILTER PARAMETERS

In designing a filter, the following important parameters are generally considered.
1. Pass bandwidth
2. Stop band attenuation and frequencies
3. Input and output impedances
4. Return loss
5. Insertion loss
6. Group delay

The most important parameters among the above is the amplitude response given in terms of the insertion loss vs frequency characteristics. Figure 8.1 shows a basic block diagram of a filter fed by a generator and terminated by a load.

Let P_i be the incident power at the filter input, P_r is the reflected power, P_L is the power passed on to the load. The insertion loss of the filter is then defined by

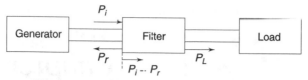

Fig. 8.1 *A filter between a generator and a load*

$$IL(\text{dB}) = 10 \log \frac{P_i}{P_L} = 10 \log \frac{P_i}{P_i - P_r}$$

$$= 10 \log \frac{1}{1 - |\Gamma|^2} \tag{8.1}$$

where $P_L = P_i - P_r$ if the filter is lossless and Γ is the voltage reflection coefficient given by $|\Gamma|^2 = P_r/P_i$.

The return loss of the filter is defined by

$$RL(\text{dB}) = 10 \log \frac{P_i}{P_r} = 10 \log \frac{1}{|\Gamma|^2} \tag{8.2}$$

which quantifies the amount of impedance matching at the input port.

The group delay is important for the multi-frequency or pulsed signals to determine the frequency dispersion or deviation from constant group delay over a given frequency band and is defined by

$$T_d = \frac{1}{2\pi} \times \frac{d\phi_t}{df} \tag{8.3}$$

where ϕ_t is the *transmission phase*.

The most commonly used microwave filters are the low-pass, high-pass, band-pass and band-stop filters. The electrical equivalent circuits of these filters along with typical insertion loss response characteristics are shown in Fig. 8.2.

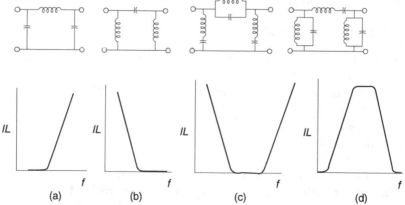

Fig. 8.2 *Electrical equivalent circuits of filters and their responses (a) low-pass, (b) high-pass, (c) band-pass (d) band-stop*

8.3 MISMATCH EFFECTS

A filter has a nominal characteristic resistive impedance R_o such that when it is terminated by this impedance at both ends there is no reflection at either of the ports. If the filter is not matched terminated there are mismatched effects as described below.

Consider that a lossless filter network with characteristic resistive impedance R_o is terminated by a resistive load R_L and fed by a generator of voltage V_g and resistive impedance R_g as shown in Fig. 8.3, where $R_g = R_L \neq R_o$. The maximum power P_1 delivered to the load without the presence of filter is given by

$$P_1 = \frac{|V_g|^2}{4R_g} \tag{8.4}$$

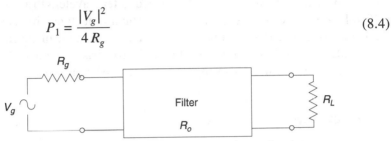

Fig. 8.3 *Filter network with mismatch terminations*

The power delivered to the load when the filter is inserted is given by

$$P_2 = \frac{|V_L|^2}{R_L} \tag{8.5}$$

The insertion loss of the filter is given by

$$IL \text{ (dB)} = 10 \log \frac{P_1}{P_2} = 20 \log \left[\frac{1}{2} \sqrt{\left(\frac{R_L}{R_g} \right)} \left| \frac{V_g}{V_L} \right| \right] \tag{8.6}$$

If $$R_g = R_L, \tag{8.7}$$

$$IL\text{(dB)} = 20 \log \left(\frac{1}{2} \frac{V_g}{V_L} \right) = \alpha_0, \text{ say} \tag{8.8}$$

If the filter is matched to the line, $R_g = R_o = R_L$, $V_L = V_g/2$, and

$$IL \text{ (dB)} = 20 \log \left(\frac{1}{2} \left| \frac{V_g}{V_L} \right| \right) = 0 \tag{8.9}$$

Therefore, the mismatched insertion loss is, in general,

$$IL \text{ (dB)} = \alpha = 20 \log \sqrt{(R_L / R_g)} + 20 \log \left(\frac{1}{2} \left| \frac{V_g}{V_L} \right| \right)$$

$$= \alpha_o + 10 \log (R_L/R_g) \tag{8.10}$$

This shows that, if $R_L > R_g$, $\alpha > \alpha_o$. If $R_L < R_g$, $\alpha < \alpha_o$ and insertion loss becomes insertion gain, i.e., the output level becomes higher than input level in the pass-band.

8.4 MICROWAVE REALISATION OF THE FILTER ELEMENTS

In microwave filters, lumped elements of the filter circuit are simulated by means of sections of waveguides, coaxial lines, strip or microstrip lines, cavity resonators, and resonant irises, etc. A few examples of these circuits are shown in Figs 8.4(a)–(d). Figure 8.4 (a) shows that a series inductance is realised by means of a short circuit stub of length less than the quarter wavelength formed in the narrow wall of a rectangular waveguide. A series capacitance can be realized by a similar stub of length greater than quarter wavelength but less than half wavelength or by a coaxial line gap formed by choke of length l less than quarter wavelength as shown in Fig. 8.4(b). Figure 8.4(c) shows simulation of a shunt inductance by a short circuit waveguide stub of length less than quarter wavelength in the plane parallel to the waveguide broadwall, or by means of an inductive iris, or a coaxial-T. A shunt capacitance can be simulated by means of capacitive iris as shown in Fig. 8.4(d).

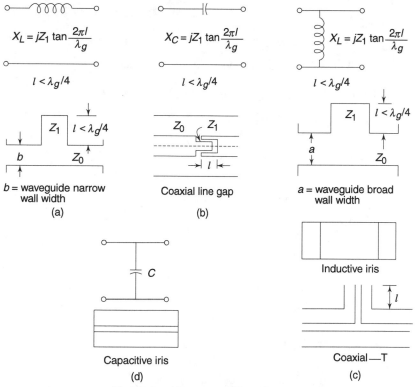

$$X_L = jZ_1 \tan \frac{2\pi l}{\lambda_g}$$

$$l < \lambda_g/4$$

$$X_C = jZ_1 \tan \frac{2\pi l}{\lambda_g}$$

$$l < \lambda_g/4$$

$$X_L = jZ_1 \tan \frac{2\pi l}{\lambda_g}$$

$$l < \lambda_g/4$$

Z_1 $l < \lambda_g/4$

b Z_0

b = waveguide narrow wall width

(a)

Z_0 Z_1

l

Coaxial line gap

(b)

Z_1 $l < \lambda_g/4$

a Z_0

a = waveguide broad wall width

(c)

C

Capacitive iris

(d)

Inductive iris

l

Coaxial — T

(c)

Fig. 8.4 *Microwave filter elements*

The equivalent lumped element values of the microwave components are themselves functions of the frequency and have in general, an infinite number of poles or zeroes, or both. Whereas, according to the Foster's Reactance Theorem, input impedance of lumped element filter circuits possess finite number of poles or zeroes or both. Hence the established synthesis procedure for the design of lumped elements filters can be applied to design microwave filters only for a narrow frequency band. For a physical realisability of a passive filter network, the magnitude of reflection coefficient $|\Gamma(\omega)| \leq 1$ and can be expressed as the ratio of two polynomials to result in the insertion loss expression in terms of even polynomials $M(\omega^2)$ and $N^2(\omega)$:

$$\text{Insertion loss} = 1 + \frac{M(\omega^2)}{N^2(\omega)} \tag{8.11}$$

Suitable forms for the polynomials M and N are chosen to obtain desired response characteristics in the filter design which are described below.

8.5   FILTER DESIGN

There are two-filter synthesis techniques popularly used. These are the *image parameter method*, and the *insertion loss method*. Out of these two methods, only insertion loss method gives complete specifications of a physically realisable frequency characteristic over the entire pass and the stop bands from which the microwave filters are synthesized or designed most preferably.

8.5.1   Prototype Low-pass Design by Insertion Loss Method

Basic design of microwave filters of types low-pass, high-pass, band-pass and band-stop, operating at arbitrary frequency bands and between arbitrary resistive loads, are made from a prototype low-pass design through some frequency transformation, element normalisation and the simulation of these elements by means of sections of microwave transmission line. In this method a physically realizable network is synthesized that will give the desired insertion loss vs. frequency characteristics. This method consists of the following steps

1. Design of a prototype low-pass filter with the desired pass band characteristics.
2. Transformation of this prototype network to the required type (low-pass, high-pass, band-pass, or band-stop) filter with the specified centre and band edge frequencies.
3. Realisation of the network in microwave form by using sections of microwave transmission lines whose reactances correspond to those of distributed circuit elements.

Figure 8.5 shows a prototype low pass circuit where the element values are designed from standard low-pass response approximations such as *Butterworth* (or Maximally flat or Binomial) *pass band response* and *Chebyshev* (or equal ripple) *pass band response*. In the basic prototype filter derived from these approximations, the angular cut-off frequency ω'_c and termination resistances r are both normalized to unity. The Butterworth and Chebyshev response characteristics are shown in Fig. 8.6 and are described in the following paragraphs.

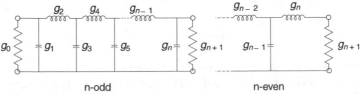

n-odd   n-even
Fig. 8.5   *Prototype low pass filter*

Butterworth response Insertion loss approximation for a low-pass Butterworth filter prototype is expressed by

$$IL = 1 + a_m^2 \omega'^{2n} \; ; \; \omega' = \omega'/\omega'_c \tag{8.12}$$

where the pass band ranges from $\omega = 0$ to $\omega = \omega_c$. Butterworth approximation exhibits a flat response in the pass-band and a monotonically increasing attenuation in the stop band. The maximum insertion loss in the pass band is 3 dB at ω_c so that $a_m^2 = 1$. The rate of increase of the insertion loss for $\omega > \omega_c$ depends on the exponent $2n$, which in turn is related to the number of filter sections used in a filter network of n reactive elements. L_x represents an insertion loss at a given frequency $\omega'_x = \omega_x/\omega_c$ in the stop band.

Chebyshev response The approximation for a low-pass Chebyshev prototype is expressed by

$$IL = 1 + a_m^2 T_n^2(\omega') \; ; \; \omega' = \omega/\omega_c \tag{8.13}$$

where n denotes the degree of approximation (i.e., number of reactive elements) and a_m is the ripple factor. $T_n(\omega')$ is the Chebyshev polynomial of degree n given by

$$T_1(x) = x, \; T_2(x) = 2x^2 - 1, \; T_3(x) = 4x^3 - 3x,$$
$$T_4(x) = 8x^4 - 8x^2 + 1,$$
$$T_n(x) = 2xT_{n-1}(x) - T_{n-2}(x) = \cos(n \cos^{-1}x) \quad \text{for } x \le 1$$
$$= \cosh(n \cosh^{-1}x) \quad \text{for } x > 1 \tag{8.14}$$

where $x = \omega'$. The insertion loss oscillates between 1 and $1+a_m^2$ in the pass band, becomes $1+a_m^2$ at the cutoff frequency ω_c and increases monotonically beyond cut-off (stop band $\omega' > 1$) at a much faster rate as compared with the Butterworth filter. L_x represents insertion loss at a frequency $\omega'_x = \omega_x/\omega_c$ in the stop band.

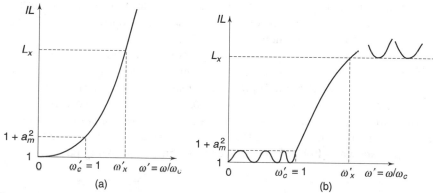

Fig. 8.6 *Butterworth and Chebyshev IL responses (a) Butterworth (b) Chebyshev*

The element values of low-pass ladder network derived from Butterworth and Chebyshev responses for the normalised values of cut-off angular frequency $\omega'_c = 1$, and load resistance $r =$ unity, can be calculated from the following equations

Butterworth prototype element values

$$g_0 = 1$$
$$g_k = 2 \sin [(2k - 1) \, \pi/2n], \, k = 1,2,...,n$$
$$g_{n+1} = 1, \text{ for all } n \qquad (8.15)$$

Chebyshev prototype element values

$$g_0 = 1$$
$$g_1 = 2p_1/\sinh (\beta/2n)$$
$$g_k = 4p_{k-1} p_k / (q_{k-1} g_{k-1}) \, ; \, k = 2, 3, 4, ..., n$$
$$g_{n+1} = 1 \text{ for } n \text{ odd}$$
$$= \tanh^2 \beta/4 \text{ for even } n \qquad (8.16)$$

where,
$$p_k = \sin \frac{(2k - 1) \, \pi}{2n}, \, k = 1,2,3,...,n$$

$$q_k = \sinh^2(\beta/2n) + \sin^2(k\pi/n), \, k = 1,2,...,n$$
$$\beta = \ln (\coth A_m/17.37) \qquad (8.17)$$

Here A_m is the ripple height = a_m^2. The values of g_k can be obtained from tables of g_k as given below.

Table 8.1 *Values of g_k for Butterworth filter (IL = 3 dB at $\omega' = 1$)*

k	n						
	1	2	3	4	5	6	7
1	2.0	1.414	1. 0	0.7654	0.618	0.5176	0.445
2	1.0	1.414	2.0	1.8480	1.618	1.4140	1.247
3		1.000	1.0	1.8480	2.000	1.9320	1.802
4			1.0	0.7654	1.618	1.9320	2.000
5				1.0000	0.618	1.4140	1.802
6					1.000	0.5176	1.247
7						1.0000	0.445

Table 8.2 *Values of g_k for Chebyshev filter*

0.1 dB ripple

k	n						
	1	2	3	4	5	6	7
1	0.3052	0.8430	1.0315	1.1088	1.1468	1.1681	1.1811
2	1.0000	0.6220	1.1474	1.3061	1.3712	1.4039	1.4228
3		1.3554	1.0315	1.7703	1.9750	2.0562	2.0966
4			1.0000	0.8180	1.3712	1.5170	1.5733
5				1.3554	1.1468	1.9029	2.0966
6					1.0000	0.8618	1.4228
7						1.3554	1.1811

0.2 dB ripple

k	1	2	3	4	5	6	7
				n			
1	0.4342	1.0378	1.2275	1.3028	1.3394	1.3598	1.3722
2	1.0000	0.6745	1.1525	1.2844	1.3370	1.3632	1.3781
3		1.5386	1.2275	1.9761	2.1660	2.2934	2.2756
4			1.0000	0.8468	1.3370	1.4555	1.5001
5				1.5386	1.3394	2.0974	2.2756
6					1.0000	0.8838	1.3781
7						1.5386	1.3722

0.5 dB ripple

k	1	2	3	4	5	6	7
				n			
1	0.6986	1.4029	1.5963	1.6703	1.7058	1.7254	1.7372
2	1.0000	0.7071	1.0969	1.1926	1.2296	1.2479	1.2583
3		1.9841	1.5963	2.3661	2.5408	2.6064	2.6381
4			1.0000	0.8419	1.2296	1.3137	1.3444
5				1.9841	1.7058	2.4758	2.6381
6					1.0000	0.8696	1.2583
7						1.9841	1.7372

8.5.2 Filter Transformations from Prototype

In order to design actual low-pass, high-pass, band-pass and band-stop filters, the transformations of the low-pass prototype filters with normalised cut-off frequency $\omega'_c = 1$ and having the source and load resistances of 1 ohm are made into the desired type with required source and load impedances using frequency and impedance transformations. This procedure is described below.

Low-pass filters For the design of loss-pass filter from the low-pass prototype response parameters, frequency scaling is required to change the normalised cut-off frequency to absolute cut-off frequency ω_c by dividing all inductances and capacitances by ω_c without altering the resistances. Impedance scaling is done by changing the source and load resistances from 1 to R_L (with $R_L = R_G$), multiplying all resistances and inductances by R_L, and dividing all capacitances by R_L. Above two scalings are done simultaneously. Thus, the actual low-pass filter element values are

$$C_k = \frac{g_k}{2 \pi f_c Z_L} \text{ farad} \tag{8.18}$$

$$L_k = \frac{g_k Z_L}{2 \pi f_c} \text{ henry} \tag{8.19}$$

High-pass filters The low-pass prototype network is transformed into a high-pass filter by transforming series inductances into series capacitances and shunt capacitances into shunt inductances using the frequency transformation

$$\omega' = \frac{\omega_C}{\omega} \tag{8.20}$$

where ω_c and ω are the bandedge and variable angular frequencies of the high-pass filter. The element values are

$$C_k = \left(\frac{1}{g_k \, \omega_C \, Z_L}\right) \tag{8.21}$$

$$L_k = g_k \left(\frac{Z_L}{g_k \, \omega_C}\right) \tag{8.22}$$

Band-pass filters To map the low-pass prototype to a band-pass filter, the following frequency transformation is used :

$$\omega' = \frac{f_0}{f_2 - f_1}(f/f_0 - f_0/f); \quad f_0 = \sqrt{f_1 \, f_2}, \tag{8.23}$$

where f_0, f and $f_2 - f_1$ are the centre frequency, variable frequency, and bandwidth, respectively, and f_1 and f_2 are the frequency band limits.

Applying the frequency transformation to series inductances and shunt capacitances of the low-pass prototype gives,

Series-tuned series elements

$$L_k = g_k \frac{Z_L}{2\pi (f_2 - f_1)}; \quad C_k = \frac{2\pi (f_2 - f_1)}{g_k \, Z_L \, f_0^2} \tag{8.24}$$

Shunt-tuned shunt elements

$$L_k = \frac{2\pi (f_2 - f_1) \, Z_L}{g_k \, f_0^2}; \quad C_k = \frac{g_k}{2\pi (f_2 - f_1) \, Z_L} \tag{8.25}$$

where,

$$\omega_0^2 = \frac{1}{L_k \, C_k} \tag{8.26}$$

Band-stop filters The transformation from low-pass prototype to band-stop is given by

$$\frac{1}{\omega'} = \frac{f_0}{(f_2 - f_1)}(f/f_0 - f_0/f) \tag{8.27}$$

where all the quantities used in Eq. 8.27 are defined as in a band-pass filter. Here series inductance is mapped into a shunt-tuned circuit with element values

$$\omega_0 C_k = \frac{1}{\omega_0 \, L_k} = \frac{f_0}{2\pi(f_2 - f_1) \, Z_L \, g_k} \tag{8.28}$$

and shunt capacitance into a series-tuned circuit with element values

$$\omega_0 L_k = \frac{1}{\omega_0 C_k} = \frac{f_0 Z_L}{2\pi(f_2 - f_1) g_k} \tag{8.29}$$

The above results are summarized in Table 8.3.

Table 8.3 *Filter element values*

Proto- type element	Low-pass filter elements	High-pass filter elements	Band-pass filter elements	Band-stop filter elements
Series arm g_k	$L_k = \dfrac{g_k Z_L}{\omega_c}$	$C_k = \dfrac{1}{g_k Z_L \omega_c}$	$L_k = \dfrac{g_k Z_L}{(\omega_2 - \omega_1)}$	$L_k = \dfrac{g_k Z_L (\omega_2 - \omega_1)}{\omega_0^2}$
			$C_k = \dfrac{\omega_2 - \omega_1}{\omega_0^2 g_k Z_L}$	$C_k = \dfrac{1}{g_k Z_L (\omega_2 - \omega_1)}$
Shunt arm g_k	$C_k = \dfrac{g_k}{Z_L \omega_c}$	$L_k = \dfrac{Z_L}{g_k \omega_c}$	$L_k = \dfrac{Z_L}{g_k (\omega_2 - \omega_1)}$	$C_k = \dfrac{g_k (\omega_2 - \omega_1)}{Z_L \omega_0^2}$
			$C_k = \dfrac{g_k (\omega_2 - \omega_1)}{\omega_0^2 Z_L}$	$L_k = \dfrac{Z_L}{g_k (\omega_2 - \omega_1)}$

8.6 MICROWAVE LOW-PASS FILTERS

Low-pass filter elements can be designed from the low-pass prototype by using the scaled Eqns. 8.18 and 8.19. Low pass filters are conveniently constructed using *TEM* structures such as coaxial lines, strip lines and microstrip lines. Microstrip filters are used in satellite, airborne communication, and EW systems for their small size, light weight, and low cost. Coaxial line, strip line and microstrip filters can be realised using such line sections having appropriate impedances to simulate reactive elements. The design is a good approximation of an idealised lumped-element circuit. In coaxial line and strip line low-pass filters, the shunt capacitors are realised by very short sections ($<< \lambda_g/4$) of relatively low characteristic impedance, and the series inductors by short sections ($< \lambda_g/4$) of relatively high characteristic impedance. Figure 8.7 shows realisation of some filter elements.

From the basic transmission line theory, the length of the predominantly inductive line and the predominantly capacitive line are obtained from the input reactance and susceptance formulae of the filter section, respectively.

Inductive length For inductive reactance X_L of a short length l_L of a lossless line with short circuit termination,

$$\omega_c L = Z_{0L} \tan \beta l_L$$

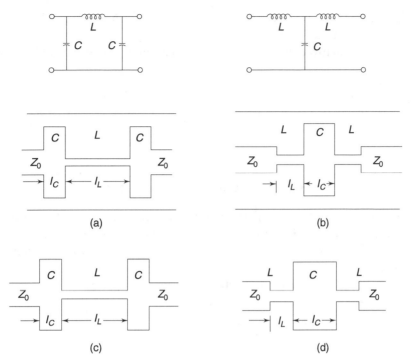

Fig. 8.7 *Microwave low-pass filters (a) Coaxial π section (b) Microstrip π section (c) Coaxial T section (d) Microstrip T section*

$$\approx Z_{0L}\frac{\sin \beta l_L}{\beta l_L}; \ l_L << \lambda_{gL}/4 \tag{8.30}$$

or
$$l_L = \frac{\lambda_{gL}}{2\pi} \sin^{-1}\left(\frac{w_C L}{Z_{0L}}\right) \tag{8.31}$$

Capacitive length For capacitive reactance X_c of a short length l_c of lossless line with open circuit termination.

$$\frac{1}{\omega_c C} = Z_{0c} \cot \beta l_c \approx Z_{0c}/\sin \beta l_c \tag{8.32}$$

or
$$l_C = \frac{\lambda_{gC}}{2\pi} \sin^{-1}(\omega_c Z_{0c} C) \tag{8.33}$$

where λ_g represents the corresponding guide wavelength, L and C are the inductance and capacitance required, and Z_0 represents the characteristic impedance of the corresponding section.

In practice, however, fringe capacitance C_f at the ends of the inductive line must be taken into account, and must be added to C where C_f is given by

$$C_f = \frac{1}{\omega Z_{0L}}\left(\tan \frac{\pi l_L}{\lambda_{gL}}\right) \approx \frac{l_L}{2 f Z_{0L} \lambda_{gL}} \tag{8.34}$$

Waveguide form of low-pass filters cannot be realised since waveguides are basically high-pass lines.

A five-section low pass microstrip filter configuration is shown in Fig. 8.8.

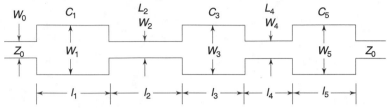

Fig. 8.8 *Five-section low-pass microstrip filter*

Example   8.1   Design a microstrip low-pass filter with cut-off frequency 2 GHz, 30 dB attenuation at frequency 3.5 GHz for Chebyshev attenuation response with 0.2 dB ripple. Use alumina substrate of thickness 0.63 mm.

Solution

Given, $f_c = 2$ GHz, 30 dB attenuation frequency $f_x = 3.5$ GHz and $1 + a_m^2 = 0.2$ dB. For low-pass prototype attenuation,

$$A = 1 + a_m^2 T_n^2(\omega') \; ; \; \omega' = \omega/\omega_C$$

For the given ripple magnitude

$$0.2 \text{ dB} = 10 \log (a_m^2 + 1)$$

or,

$$a_m^2 = 10^{0.2/10} - 1 = 0.047$$

Stop-band attenuation at f'_x.

$$30 \text{ dB} = 10 \log [1 + a_m^2 \cosh^2(n \cosh^{-1} \omega'_x)]$$

or,

$$\omega'_x = \omega_x/\omega_c = 3.5/2 = 1.75$$

Therefore,

$$1 + 0.047 \cosh^2 [n \cosh^{-1}(1.75)] = 10^{30/10} = 1000$$

or,

$$\cosh^2 [n \cosh^{-1}(1.75)] = (1000 - 1)/0.047$$

or,

$$n \cong 5$$

Therefore a five-section LP filter is required.

The prototype elements for $n = 5$, ripple = 0.2 dB, are $g_0 = 1 = g_6$, $g_1 = g_5 = 1.3394$, $g_2 = g_4 = 1.337$, $g_3 = 2.166$.

The lumped element values for cascaded π sections for 50 ohm terminations are obtained as

$$C_k = \frac{g_k}{Z_L \, \omega_C}$$

$$L_k = \frac{g_k \, Z_L}{\omega_C}$$

or,

$$C_1 = C_5 = \frac{1.3394}{50 \times 2\pi \times 2 \times 10^9} \text{ F}$$

$$= 2.132 \text{ pF}$$

$$C_3 = \frac{2.166}{50 \times 2\pi \times 2 \times 10^9} \text{ F}$$

$$= 3.447 \text{ pF}$$

$$L_2 = L_4 = \frac{1.337 \times 50}{2\pi \times 2 \times 10^9} = 5.32 \text{ nH}$$

These lumped elements can be realised by sections of microstrip lines having length and width with corresponding suffices as shown in Fig. 8.8.

These elements are realised as follows.

Capacitive lines of 20 ohm impedance (say)

$$l_{c1} = l_{c5} = f_c \, \lambda g_C \, Z_{oc} C_1$$
$$l_{c3} = f_c \, \lambda_{gC} \, Z_{0c} C_3$$

For alumina substrate at 2 GHz, $\varepsilon_r = 9.9$. The effective permittivity

$$\varepsilon_{\text{eff}} = \frac{\varepsilon_r + 1}{2} + \frac{\varepsilon_r - 1}{2} \left[\left(1 + \frac{12h}{w} \right)^{-1/2} + 0.04(1 - W/h)^2 \right]; \frac{W}{h} \leq 1$$

$$= \frac{\varepsilon_r^2 + 1}{2} + \frac{\varepsilon_n - 1}{2} \, (1 + 12h/W)^{-1/2}; \, W/h \geq 1$$

$$= \frac{9.9 + 1}{2} + \frac{9.9 - 1}{2} \left(1 + \frac{12h}{W} \right)^{-1/2}$$

$$= 5.45 + 4.45 \left(1 + \frac{12h}{W} \right)^{-1/2}$$

$$Z_0 = \frac{120\pi}{2\pi \sqrt{\epsilon_{\text{eff}}}} \ln \left[\frac{8h}{W} + \frac{0.25h}{W} \right]; \, W/h \leq 1$$

$$= \frac{120\,\pi}{\sqrt{\varepsilon_{\text{eff}}}} \left[\frac{W}{h} + 1.393 + 0.667 \ln\left(\frac{W}{h} + 1.444\right) \right]^{-1} \quad W/h \geq 1$$

$$\lambda_g = \frac{\lambda_0}{\sqrt{\varepsilon_{\text{eff}}}}$$

Design parameters at 2 GHz

$$\lambda_0 = 30/2 = 15 \text{ cm}, \qquad h = 0.63 \text{ mm}$$

From the above equations for Z_0, ε_{eff} and λ_{g0}, the following are found.

For $\qquad Z_0 = 50$ ohm, $\quad W_0 = 0.6$ mm, $\quad \varepsilon_{\text{eff}} = 6.66$, $\quad \lambda_{g0} = 5.81$ cm

For $\qquad Z_{0c} = 20$ ohm, $\quad W_{0c} = 2.64$ mm, $\quad \varepsilon_{\text{eff}} = 7.71$, $\quad \lambda_{gc} = 5.40$ cm

For $\qquad Z_{0L} = 100$ ohm, $\quad W_{0L} = 0.075$ mm, $\varepsilon_{\text{eff}} = 6.03$, $\quad \lambda_{gL} = 6.11$ cm

and $\qquad l_{c1} = l_{c5} = 4.59$ mm

$$l_{L1} = l_{L4} = 7.25 \text{ mm}$$

$$l_{c3} = 7.43 \text{ mm}$$

8.7 MICROWAVE HIGH-PASS FILTERS

High-pass microwave filters can be designed from the low-pass prototype by using the transformation (8.20)-(8.22) and the response characterstics of Fig. 8.9. In coaxial and microstrip forms, short ($< \lambda_g/4$) lengths of relatively high characteristic impedance tee-connected to the main line, approximates the shunt inductors. The series capacitors are obtained by very small ($<< \lambda_g/4$) gaps in the line as shown in Fig. 8.9.

In waveguide form, the change in broadwall dimension may be utilised to simulate the lump equivalent circuit. The shunt inductor can be realised by an inductive iris or by a H-planc T-junction.

8.8 MICROWAVE BAND-PASS FILTERS

Band-pass microwave filters are narrow band filters and can be designed from the low-pass prototype using the transformations (8.23)-(8.26).The low-pass prototype circuit in Fig. 8.5 can be transformed to band-pass filter circuit of Fig. 8.10(a). The response characteristics of low-pass and its transformation to band-pass is shown in Fig. 8.10 (b). The resonator circuits of these filters are realised by the use of a cascaded strip line, coaxial line or cavity resonators of suitable configurations as described below.

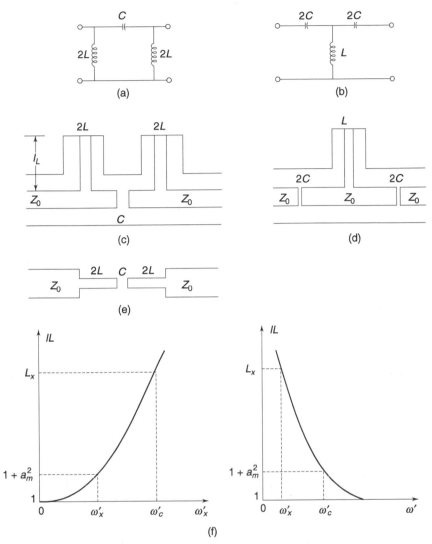

Fig. 8.9 *High pass filters (a) π-section (b) T-section (c) Coaxial π-section (d) Coaxial T-section (e) Microstrip π-section (f) Butterworth response of protoype low-pass and high-pass*

8.8.1 Quarterwave Coupled Cavity Bandpass Filters

These filters are realised by waveguide cavities coupled through irises as shown in Fig. 8.11(a). The equivalent circuit for any kth section loaded with two identical inductive irises with normalised susceptance $-jb_k$ is also shown in Fig. 8.11 (b). Each pair of irises separated by distance l_k forms a cavity k. Mumford and Collin have given detailed analysis of such filters. For high Q narrow band filter $b_k \gg 1$ and is given by

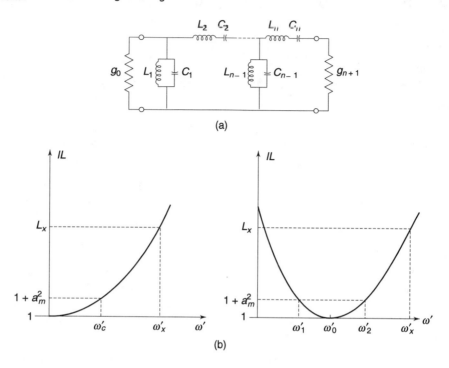

(a)

(b)

Fig. 8.10 *(a) A bandpass filter circuit derved from the low prototype (b) the bandpass response transformed from a low pass prototype*

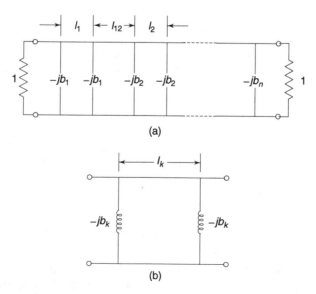

Fig. 8.11 *(a) Quarterwave coupled waveguide cavity filter (b) Single section equivalent circuit*

$$b_k = 2(Q_k^2 - 1)^{1/2} \tag{8.35}$$

whereas for the kth resonator Q

$$Q_k = \frac{g_k}{2} \frac{\beta_0}{\beta_2 - \beta_1} \tag{8.36}$$

and the length of the cavity

$$l_k = \frac{1}{\beta_0} \left[\tan^{-1} \left(\frac{-2}{b_k} \right) \right] \tag{8.37}$$

where g_k is the normalised value of the element of prototype filter. The physical length of the quarterwave coupling line between the kth and $k + 1$th cavities will be

$$l_{k, k+1} = \frac{l_k + l_{k+1}}{2} - \frac{\lambda_{g0}}{4} \tag{8.38}$$

The required iris dimensions can be found from Chapter 5. The insertion loss of the Chebyshev band-pass filter is given by

$$IL = 1 + a_m^2 \, T_n^2 \left[\frac{\beta_0}{\beta_2 - \beta_1} \left(\frac{\beta}{\beta_0} - \frac{\beta_0}{\beta} \right) \right] \tag{8.39}$$

where $\qquad \beta_0 = \sqrt{(\beta_1 \beta_2)} \tag{8.40}$

Example 8.2 Design a quarterwave coupled three-cavity filter having Chebyshev response with maximum passband ripple of 0.1 dB using a waveguide of dimensions $a = 2.286$ cm and $b = 1.016$ cm. The passband extends from $f_1 = 10$ GHz to $f_2 = 10.4$ GHz. Inductive diaphragm with circular holes are to be used.

Solution

For narrow band and large susceptance values of the irises,

$$b_k = 2 \, (Q_k^2 - 1)^{1/2}$$

$$Q_k = \frac{\beta_0}{\beta_2 - \beta_1} \times \frac{g_k}{2}$$

At centre frequency

$$\beta_0 = 2\pi / \lambda_{g0}, \ \tan \beta_0 l_k = -\, 2/b_k$$

Actual distance between cavity k and $k + 1$ is

$$l_{k, k+1} = \frac{l_k + l_{k+1}}{2} - \frac{\lambda_{g0}}{4}$$

For a three-cavity Chebyshev filter with $n = 3$, and ripple 0.1 dB, $a_m^2 = 0.0233$; $g_1 = 1.0315$, $g_2 = 1.1474$, $g_3 = 1.0315$.

Now, at band edges the propagation constants k_0 in free space and β in a rectangular waveguide ($a \times b$) excited in dominant TE_{10} mode are given by

$$k_{01} = \frac{2\pi \times 10 \times 10^9}{3 \times 10^{10}} = 2.0944$$

$$k_{02} = \frac{2\pi \times 10 \cdot 4 \times 10^9}{3 \times 10^{10}} = 2.1782$$

$$\beta_1 = [k_{01}^2 - (\pi/a)^2]^{1/2} = 1.5805$$
$$\beta_2 = [k_{02}^2 - (\pi/a)^2]^{1/2} = 1.6899$$

Centre of the band occurs at

$$\beta_0 = \sqrt{(\beta_1 \beta_2)} = 1.6343$$
$$f_0 = \sqrt{(f_1 f_2)} = 10.198 \text{ GHz}$$

$$\lambda_0 = \frac{30}{f_0} = 2.942 \text{ cm}, \quad \lambda_{g0} = \frac{2\pi}{\beta_0} = \frac{2 \times 3.1415}{1.6343} = 3.845 \text{ cm}$$

Now, $\qquad \dfrac{\beta_0}{\beta_2 - \beta_1} = \dfrac{1.6343}{1.6899 - 1.58011} = 14.9386$

Therefore, $\qquad Q_1 = 14.9386 \times \dfrac{1.0315}{2} = 7.7046$

$$Q_2 = 14.9386 \times \frac{1.1474}{2} = 8.5703$$

$$Q_3 = 14.9386 \times \frac{1.0315}{2} = 7.7046$$

and, $\qquad b_1 = 2 \times (7.7046^2 - 1)^{1/2} = 15.2788$

$$b_2 = 2 \times (8.5703^2 - 1)^{1/2} = 17.0235$$
$$b_3 = b_1 = 15.2788$$

Calculation of cavity length l_k

$$\tan \beta_0 l_k = \frac{-2}{b_k}$$

or, $\qquad \tan\left(\beta_0 \times \dfrac{180°}{\pi}\right) l_k = -\dfrac{2}{b_k}$

or,
$$\tan\left(\frac{1.6343 \times 180^\circ}{3.1415}\right) l_k = -\frac{2}{b_k} \ ;$$

For $k = 1$,
$$\tan 93.64 \ l_k = -2/b_k = -2/15.2788$$
$$= -0.1309$$
$$= -\tan 7.4576$$
$$= \tan(180^\circ - 7.4576^\circ)$$
$$= \tan 172.5424$$

Therefore,
$$l_1 = 172.5424/93.64 = 1.843 \text{ cm}$$

$$\tan 93.64 \ l_2 = -2/b_2 = -2/17.0235 = -\tan 6.7006$$
$$= \tan(180^\circ - 6.7006)$$
$$= \tan 173.299$$

Therefore,
$$l_2 = \frac{173.299}{93.64} = 1.851 \text{ cm}$$

$$l_3 = l_1 = 1.843 \text{ cm}$$

Distance between cavities

$$l_{k,\,k+1} = \frac{l_k + l_{k+1}}{2} - \frac{\lambda_{g0}}{4}$$

$$= \frac{l_k + l_{k+1}}{2} - \frac{3.845}{4}$$

$$= \frac{l_k + l_{k+1}}{2} - \frac{1}{4} \times \frac{2\pi}{\beta_0} = \frac{l_k + l_{k+1}}{2} - 0.96114$$

$$l_{1,2} = \frac{l_1 + l_2}{2} - 0.96114$$

$$= \frac{1.843 + 1.851}{2} - 0.96114 = 0.885 \text{ cm}$$

$$l_{2,3} = \frac{l_2 + l_3}{2} - 0.96114$$

$$= \frac{1.843 + 1.483}{2} - 0.96114 = 0.882 \text{ cm}$$

Radii of diaphragm holes Assuming TE_{10} mode of excitation, radius of the kth hole is given by

$$r_k^3 = 3ab/8\beta_0 b_k$$

Therefore,

$$r_1 = \left[\frac{3 \times 2.286 \times 1.016}{8 \times 1.6343 \times 15.2788} \right]^{1/3}$$

$$= 0.3538 \text{ cm}$$

$$r_2 = \left[\frac{3 \times 2.286 \times 1.016}{8 \times 17.0235 \times 1.6343} \right]^{1/3}$$

$$= 0.3152 \text{ cm}$$

$$r_3 = 0.3267 \text{ cm}$$

Example 8.3 Design a symmetrical three-section, maximally flat BP quarterwave coupled filter such that the centre frequency in the passband and the bandwidth are 10,000 and 100 MHz, respectively.

Compare the insertion loss of this filter with that of a four-section filter in the same pass and attenuation bands in the frequency range 9800 to 10,200 MHz.

Solution

Fractional bandwidth of the filter $= \dfrac{\text{bandwidth}}{\text{centre frequency}}$

$$= \frac{100}{10,000} = \frac{1}{100} = 1\%$$

In our case $n = 3$ (odd) and we assume maximally flat response with 3 dB insertion loss at the band edges. Hence elements of LP prototype are

$$g_0 = 1, g_1 = 1, g_2 = 2, g_3 = 1$$

Given, $f_0 = 10,000 \text{ MHz} = \sqrt{(f_1 f_2)} = 10^4$, and

$$f_2 - f_1 = 100 \text{ MHz}$$

Therefore,

$$f_2 = 10.05 \text{ GHz}$$

$$f_1 = 9.95 \text{ GHz}$$

X-band waveguide dimensions

$$a = 0.9" = 2.286 \text{ cm}$$

$$b = 0.4" = 1.016 \text{ cm}$$

$$\beta_0 = 2\pi \sqrt{\left[(f_0/c)^2 - (1/2a)^2 \right]}$$

$$= 2\pi \sqrt{\left[\left(\frac{10}{30} \right)^2 - \left(\frac{1}{2 \times 2.286} \right)^2 \right]}$$

$$= 1.5805 \text{ rad/cm}$$

$$\lambda_{g0} = \frac{2\pi}{\beta_0} = \frac{2\pi}{1.5805} = 3.975 \text{ cm}$$

$$\beta_1 = 2\pi \sqrt{\left[(f_1/c)^2 - (1/2a)^2 \right]}$$

$$= 2\pi \sqrt{\left[\left(\frac{9.950}{30} \right)^2 - \left(\frac{1}{2 \times 2.286} \right)^2 \right]}$$

$$= 1.5665 \text{ rad/cm}$$

$$\beta_2 = 2\pi \sqrt{\left[(f_2/c)^2 - (1/2a)^2 \right]}$$

$$= 2\pi \sqrt{\left[\left(\frac{10.05}{30} \right)^2 - \left(\frac{1}{2 \times 2.286} \right)^2 \right]}$$

$$= 1.5943 \text{ rad/cm}$$

$$\frac{\beta_0}{\beta_2 - \beta_1} = \frac{1.5805}{1.5943 - 1.5665} = 56.8525$$

Loaded Q of the sections are

$$Q_1 = \frac{\beta_0}{\beta_2 - \beta_1} \frac{g_1}{2} = 28.4263$$

$$Q_2 = \frac{\beta_0}{\beta_2 - \beta_1} \frac{g_2}{2} = 56.8525$$

$$Q_3 = \frac{\beta_0}{\beta_2 - \beta_1} \frac{g_3}{2} = 28.4263$$

Therefore, for hole sincefitances

$$b_1 = 2 (Q_1^2 - 1)^{1/2} = 56.8173$$
$$b_2 = 2 (Q_2^2 - 1)^{1/2} = 113.6874$$
$$b_3 = 2 (Q_3^2 - 1)^{1/2} = 56.8173$$

Calculation of l_k at $f = f_0$

$$\tan \beta_0 l_k = - 2/b_k$$

or, $\qquad \tan (90.555\ l_1)° = -\ 2/b_1 = \tan (177.98°)$

or, $\qquad (90.555\ l_1)° = 177.98°$

or, $\qquad l_1 = 177.98°/90.555 = 1.965$ cm

$\qquad \tan (90.555\ l_2)° = -\ 2/b_2 = \tan (178.992°)$

or, $\qquad (90.555\ l_2)° = 178.992°$

$\qquad l_2 = 178.992/90.555 = 1.976$ cm

Since $\qquad b_1 = b_3,\ l_3 = l_1 = 1.965$ cm

Calculation of $l_{k,k+1}$ at f_0:

$$l_{1,2} = \frac{l_1 + l_2}{2} - \frac{\lambda_{g0}}{4} = 0.9767 \text{ cm}$$

$$l_{2,3} = \frac{l_2 + l_3}{2} - \frac{\lambda_{g0}}{4} = 0.9767 \text{ cm}$$

Radius 'r' of diaphragm

Assuming TE_{10} mode is excited:

$$b_k = \frac{3\ ab}{8\,\beta_0\ r_k^3}$$

Therefore,

$$r_k = \left[\frac{3\ ab}{8\,\beta_0\ b_k}\right]^{1/3}$$

$$r_3 = r_1 = \left[\frac{3 \times 2.286 \times 1.016}{8 \times 1.5805 \times 56.8173}\right]^{1/3} = 0.2136 \text{ cm}$$

$$r_2 = \left[\frac{3 \times 2.286 \times 1.016}{8 \times 1.5805 \times 113.6874}\right]^{1/3} = 0.1693 \text{ cm}$$

Insertion loss vs frequency

$$IL = 1 + (\omega')^{2n}$$

$$= 1 + \left[\frac{f_0}{f_2 - f_1}\left(\frac{f}{f_0} - \frac{f_0}{f}\right)\right]^{2n}$$

$$= \left[1 + 10^{12}\left(\frac{f}{10} - \frac{10}{f}\right)^6\right] ; \text{ Three-cavity filter}$$

$$= \left[1 + 10^{16}\left(\frac{f}{10} - \frac{10}{f}\right)^8\right] ; \text{ Four-cavity filter}$$

The results are tabulated below.

f(GHz)	Band	IL(dB) (3 cavity)	IL(dB) (4 cavity)
9.8	Attenuation	36.4	48.5
9.9	Attenuation	18.2	24.3
9.95	Bandedge	3.0	3.05
10.0	Centre	0.0	0.0
10.05	Bandedge	·3.0	3.0
10.1	Attenuation	17.0	24.0
10.2	Attenuation	35.9	47.8

Therefore, insertion loss characteristic is more sharp in a four-cavity filter.

8.8.2 Direct Coupled Cavity Band-pass Filters

Direct coupled cavity filters consist of a number of cavity resonators directly coupled by inductive irises as shown in Fig. 8.12. Due to zero separation between the cavities these filters are more compact than the corresponding quarterwave coupled cavity filters. The design formulae are given below based on *S.B. Cohn's [12] design method.*

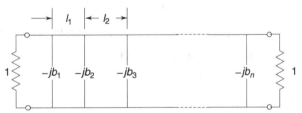

Fig. 8.12 *Direct coupled cavity band-pass filter*

The normalised susceptances of the irises are

$$b_1 = \frac{1 - \omega/g_1}{\sqrt{(\omega/g_1)}} \tag{8.41}$$

$$b_k = \frac{1}{\omega}\left(1 - \frac{\omega^2}{g_k\, g_{k-1}}\right)\sqrt{(g_k\, g_{k-1})} \tag{8.42}$$

$$b_n = \frac{1 - \omega r/g_n}{\sqrt{(\omega r/g_n)}} \tag{8.43}$$

where

$$\omega = \frac{\pi}{2}\frac{\beta_2 - \beta_1}{\beta_0} \tag{8.44}$$

Here, $r = 1$ for odd n in Chebyshev filter and for all integer values of n for Butterworth filter. For Chebyshev filter with even n

$$r = 2a_m^2 + 1 - 2a_m\sqrt{(1 + a_m^2)} \tag{8.45}$$

$$\omega' = \frac{\beta_0}{\beta_2 - \beta_1} \left(\frac{\beta}{\beta_0} - \frac{\beta_0}{\beta} \right) \tag{8.46}$$

and g_k are the element values from the low-pass prototype filter. The length of the kth cavity at centre frequency f_0 is

$$l_k = \frac{\lambda_{g0}}{2} - \frac{\lambda_{g0}}{4\pi} \left[\tan^{-1} \frac{2}{b_{k+1}} + \tan^{-1} \frac{2}{b_k} \right] \tag{8.47}$$

Example 8.4 Design a four-cavity direct coupled microwave filter of Chebyshev response where the passband is to extend from $f_1 = 9500$ MHz to $f_2 = 10,500$ MHz with 0.1 dB ripple.

Solution

The values of $k_0 = \omega/c$ at band edge frequencies f_1 and f_2 are

$$k_{01} = \left[\frac{2 \times 3.1416 \times 0.95 \times 10^{10}}{3 \times 10^{10}} \right] = 1.9897$$

$$k_{02} = \left[\frac{2 \times 3.1416 \times 1.05 \times 10^{10}}{3 \times 10^{10}} \right] = 2.1991$$

The corresponding values of β_1 and β_2 for a rectangular waveguide $(a \times b)$ are

$$\beta_1 = [k_{01}^2 - (\pi/a)^2]^{1/2} = 1.4468$$
$$\beta_2 = [k_{02}^2 - (\pi/a)^2]^{1/2} = 1.7235$$

where $a \approx 2.3$ cm and $b \approx 1$ cm.

The centre of the band occurs at

$$\beta_0 = (\beta_1 \beta_2)^{1/2} = 1.5791$$

or,

$$f_0 = \sqrt{(f_1 f_2)} = 9987.5 \text{ MHz}$$

From the table of values of g_k for the Chebyshev filter with 0.1 dB ripple and $n = 4$,

$$g_1 = 1.1088$$
$$g_2 = 1.3061$$
$$g_3 = 1.7703$$
$$g_4 = 0.8180$$

The value of ω is calculated from the formula

$$\omega = \frac{\pi}{2} \cdot \frac{\beta_2 \sim \beta_1}{\beta_0} = 0.2752$$

Ripple 0.1 dB = 10 log $(1 + a_m^2)$,

or, $\qquad a_m^2 = 0.0233$.

Because n is even and the response required is Chebyshev, the source impedance $r \neq 1$ but is given by the relation

$$r = 2a_m^2 + 1 - 2a_m\sqrt{(1 + a_m^2)}$$

$$= 2 \times 0.0233 + 1 - 2\sqrt{(0.0233)}\ \sqrt{(1 + 0.0233)}$$

$$= 0.7377$$

Calculation of b_k

$$b_1 = \frac{1 - \omega/g_1}{\sqrt{(\omega/g_1)}} = \frac{1 - 0.45036/1.1088}{\sqrt{0.4536/1{\cdot}1088}}$$

$$= 0.9318$$

$$b_2 = \frac{1}{\omega}\left(1 - \frac{\omega^2}{g_1 g_2}\right)\sqrt{(g_2\, g_1)}$$

$$= \frac{1}{0.45036}\left(1 - \frac{(0.45036)^2}{1.1088 \times 1.3061}\right)\sqrt{(1.1088 \times 1.3061)}$$

$$= 2.298$$

$$b_3 = \frac{1}{\omega}\left(1 - \frac{\omega^2}{g_3 g_2}\right)\sqrt{(g_3\, g_2)}$$

$$= 3.080209$$

$$b_4 = \frac{1 - \omega r/g_4}{\sqrt{(\omega r/g_4)}} = \frac{1 - 0.45036 \times 0.7377/0.8180}{\sqrt{[0.45036 \times 0.7377/0.8180]}}$$

$$= 0.9318$$

The length of the kth cavity at $\beta = \beta_0$ is

$$l_k = \frac{\lambda_{g0}}{2} - \frac{\lambda_{g0}}{4\pi}\left(\tan^{-1}\frac{2}{b_{k+1}} + \tan^{-1}\frac{2}{b_k}\right)$$

$$\lambda_0 = \frac{2\pi}{\left[\beta_0^2 + (\pi/a)^2\right]^{1/2}} = 3.001 \text{ cm}$$

and $$f_0 = 9.6721 \times 10^9 \text{ Hz}$$

Now $$\lambda_{g0} = \frac{\lambda_0}{\sqrt{\left[1 - (\lambda_0/2a)^2\right]}}$$

$$= \frac{3.001}{\sqrt{\left[1 - (3.001/2 \times 2.3)^2\right]}}$$

$$= 4.2218 \text{ cm}$$

$$l_1 = \frac{\lambda_{g0}}{2} - \frac{\lambda_{g0}}{4\pi} \left[\tan^{-1} \frac{2}{2.2988} + \tan^{-1} \frac{2}{0.9318} \right] = 0.1575 \text{ cm}$$

$$l_2 = \frac{\lambda_{g0}}{2} - \frac{\lambda_{g0}}{4\pi} \left[\tan^{-1} \frac{2}{3.08021} + \tan^{-1} \frac{2}{2.298} \right] = 0.7471 \text{ cm}$$

$$l_3 = \frac{\lambda_{g0}}{2} - \frac{\lambda_{g0}}{4\pi} \left[\tan^{-1} \frac{2}{0.9318} + \tan^{-1} \frac{2}{3.08021} \right] = 0.8718 \text{ cm}$$

To calculate diaphragm dimensions we use circular irises for which

$$r_0^3 = 3ab/8b_k\beta_0$$

For $a = 0.9''$ and $b = 0.4''$

$$r_{01} = \left[\frac{3 \times 0.9 \times 0.4 \times (2.54)^2}{8 \times 0.9318 \times 1.48826} \right]^{1/3}$$

$$= 0.85635 \text{ cm.}$$

$$r_{02} = \left[\frac{3 \times 0.9 \times 0.4 \times (2.54)^2}{8 \times 2.2988 \times 1.48826} \right]^{1/3}$$

$$= 0.63378 \text{ cm.}$$

$$r_{03} = \left[\frac{3 \times 0.9 \times 0.4 \times (2.54)^2}{8 \times 3.08021 \times 1.48826} \right]^{1/3}$$

$$= 0.57489 \text{ cm.}$$

$$r_{04} = 0.70341$$

Thus the radii and the spacings are

$r_{01} = 0.85635$ cm	$l_1 = 0.1575$ cm
$r_{02} = 0.63378$ cm	$l_2 = 0.7471$ cm
$r_{03} = 0.57489$ cm	$l_3 = 0.8718$ cm
$r_{04} = 0.70341$ cm	

This is the final form of the filter.

8.8.3 Microstrip Band-pass Filters

Band-pass filters can be constructed in microstrip forms by using end coupled, parallel coupled, and interdigital configurations as shown in Figs 8.13-8.15.

Parallel coupling is much stronger than end coupling so that the realisable band-widths are greater. The most compact configuration is the interdigital one. The design procedure of these filters are described below.

End coupled microstrip band-pass filters Figure 8.13 shows the general layout of an end coupled microstrip band-pass filter.

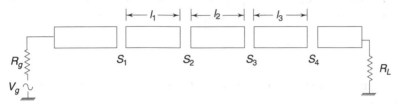

Fig. 8.13 *End coupled microstrip band-pass filter*

Here the tuned circuits are realised from open or short circuited transmission line sections of one quarter or half wavelength long. The resonators are coupled by means of gap capacitances between the resonator sections. The resonator lengths l and the coupling gaps S between successive resonators are important design parameters. To reduce insertion loss in the pass-band, the gaps are usually much smaller than the substrate height to enable tight coupling. The resonator lengths depend on the guide wavelength, coupling reactance and the gap capacitance. Since this structure is large, it is not much preferred configuration.

Parallel coupled band-pass filter To achieve a more compact structure than end coupled filter, parallel coupled microstrip configuration as shown in Fig. 8.14 is preferred. As discussed in the theory of couplers in Chapter 6, since the maximum coupling occurs over a quarterwave long coupling region, to achieve resonance, each resonator element is nearly half wavelength long at the centre frequency and kept open at both ends.

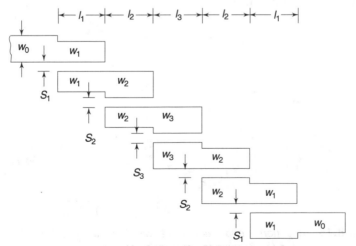

Fig. 8.14 *Parallel coupled band-pass filter*

Detail derivation of the coupling and impedance of parallel coupled filters have been given by Matthaei [6] using even and odd mode excitations which are summarised below for Chebyshev response of n section bandpass filter.

Let the filter be terminated in input and output lines with characteristic impedance Z_0. Each resonator has an electrical length of π at the centre frequency f_0. The maximum effective length l_e of each resonator is calculated after correcting for the capacitive end loading at each open circuited end, such that

$$l_e = \frac{\lambda_g}{4} - l_{e0} \tag{8.48}$$

where l_{e0} is the hepothetical extension of the line to make up the end fringing given by [7]

$$l_{e0} = 4.12 \ln \left(\frac{\varepsilon_{\text{eff}} + 0.3}{\varepsilon_{\text{eff}} - 0.258} \right) \left(\frac{W/h + 0.262}{W/h + 0.813} \right) \tag{8.49}$$

and due to Cohn [15]

$$\left(\frac{l_{e0}}{h} \right)_{\text{max}} = \frac{2}{\pi} \ln 2 \approx 0.441 \tag{8.50}$$

The band-pass filter circuit consisting of parallel resonant circuit in the shunt arm and series resonant circuit in the series arm can be converted to either cascade of all parallel resonator networks together with impedance inverters K or to cascade of all series resonator networks together with admittance inverters J as shown in Fig. 8.15, where the inverters are basically quarterwave transformers to transform any load impedance by the quantity Z'^2, where Z' is the characteristic impedance of the quarterwave transforming line. In this process (1) the resonant frequency must remain constant : $w_0^2 = 1/LC$ and, (2) impedance at similar planes in each network must be equal.

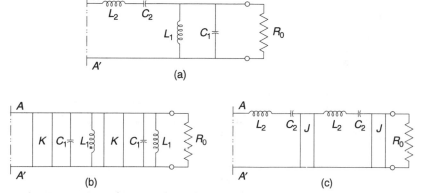

Fig. 8.15 *Impedance and admittance inverters (a) A section of band-pass filter (b) K-inverter (c) J-inverter*

The inverter concept can be explained by considering a quarterwave lossless line of characteristic impedance Z_0 terminated in an impedance Z_L. The input impedance in the line

$$Z_i = \frac{Z'^2}{Z_L} \tag{8.51}$$

Normalising by Z',

$$\frac{Z_i}{Z'} = \frac{1}{Z_L/Z'}$$

or,

$$z_i = 1/z_L \tag{8.52}$$

where normalised impedances are $z_i = Z_i/Z'$, $z_L = Z_L/Z'$. This concept is represented by the following equivalent circuit of a simple L-section as shown in Fig. 8.16, where the impedances shown are normalised with respect to characteristic impedance of the quarterwave section.

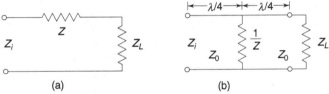

(a)   (b)

Fig. 8.16 *Quarterwave coupling network (a) Lumped element circuit
(b) Quarter wave coupling equivalent*

The details of impedance and admittance inverter formulae are given by Matthaei et al [6] and Collin [2] which are summarized below without any derivation.

For all series network configurations with admittance inverters J, the admittances of the inverters are as follows.

For the first coupling section

$$\frac{J_{10}}{Y_0} = \left(\frac{\pi F}{2 g_0 \, g_1} \right)^{1/2} \tag{8.53}$$

For the final coupling section

$$\frac{J_{n+1,n}}{Y_0} = \left(\frac{\pi F}{2 g_n \, g_{n+1}} \right)^{1/2} \tag{8.54}$$

For the intermediate coupling section

$$\frac{J_{k+1,k}}{Y_0} = \left(\frac{\pi F}{2 \sqrt{(g_k \, g_{k+1})}} \right), \quad k = 1 \text{ to } n - 1 \tag{8.55}$$

where $F = (f_2 - f_1)/f_0$, the fractional bandwidth.

In the microstrip filter, the line dimensions are determined from the knowledge of even and odd mode coupled line impedances which are given in term of admittances of the inverters

$$(Z_{0e})_{k+1,\,k} = Z_0(1+J_{k+1,\,k}Z_0 + J^2_{k+1,\,k}Z^2_0) \tag{8.56}$$

and

$$(Z_{00})_{k+1,\,k} = Z_0(1-J_{k+1,\,k}Z_0 + J^2_{k+1,\,k}Z^2_0) \tag{8.57}$$

where Z_0 is the characteristic impedance of the input/output lines of the filter.

The design steps for the parallel coupled band-pass filters are

1. Determine the number of sections from the specified attenuation characteristics.
2. Determine the values of the prototype elements to realise the specifications.
3. Determine the one-type resonator network (cascaded series resonators, say) and determine the inverter values.
4. Obtain the even and odd mode coupled line characteristic impedances Z_{0e} and Z_{00}.
5. Determine the microstrip widths and separations (W, s) of the parallel coupled resonant lines from the graphs of Z_{0e}, Z_{00} vs W/h and S/h.
6. Calculate the coupled section length l_k which is slightly less than quarter wavelength at centre frequency to account for the end fringing.

Example 8.5 Design a parallel coupled microstrip band-pass filter having Chebyshev response characteristics with passband ripple of 0.1 dB between the bandedge frequencies 9.98 GHz and 11.03 GHz. The 20 dB minimum attenuation occurs at 11.33 GHz.

Solution

Centre frequency $f_0 = \sqrt{(9.98 \times 11.03)} = 10.49$ GHz

For ripple 0.1 dB $= 10 \log (a^2_m + 1)$

$$a^2_m = (10^{0.1/10}) - 1 = 0.0233$$

Number of resonators   Fractional bandwidth

$$F = (f_2 - f_1)/f_0 = 0.1001$$

For n resonator filter stopband attenuation

$$20 \text{ dB} = 10 \log [1 + 0.0233 \cosh^2(n \cosh^{-1}\omega'_x)]$$

where

$$\omega'_x = \frac{\omega_0}{\omega_2 - \omega_1} \left(\frac{\omega_x}{\omega_0} - \frac{\omega_0}{\omega_x} \right)$$

$$= \frac{10.49}{11.03 - 9.98} \left(\frac{11.33}{10.49} - \frac{10.49}{11.33} \right)$$

$$= 1.541$$

Therefore,

$$\cosh(n \cosh^{-1} 1.541) = [(10^2-1)/0.0233)]^{1/2}$$
$$= 65.18$$
$$= \cosh(4.87°)$$

or, $$n = \frac{4.87°}{0.998} = 4.88 \approx 5$$

Therefore, five resonator band-pass filter is needed.
Prototype values
For $n = 5$, and ripple of 0.1 dB, the low pass prototype elements are $g_0 = g_6 = 1.0$, $g_1 = g_5 = 1.1468$, $g_2 = g_4 = 1.3712$ and $g_3 = 1.9750$. Thus the filter is symmetrical structure.
Admittance inverter admittances For symmetry for the first and final coupling

$$\frac{J_{10}}{Y_0} = \frac{J_{6,5}}{Y_0} = \sqrt{\left(\frac{\pi F}{2 g_0 g_1}\right)} = \sqrt{\left(\frac{\pi F}{2 g_5 g_6}\right)}$$

$$= \sqrt{\left(\frac{\pi \times 0.1001}{2 \times 1.1468}\right)} = 0.3703$$

For the intermediate coupling section ($k = 1$ to 2)

$$\frac{J_{2,1}}{Y_0} = \frac{\pi F}{2\sqrt{(g_1 g_2)}} = \frac{J_{5,4}}{Y_0} = \frac{\pi \times 0.1001}{2\sqrt{(1.1468 \times 1.3712)}}$$
$$= 0.1254$$

$$\frac{J_{3,2}}{Y_0} = \frac{\pi F}{2\sqrt{(g_2 g_3)}} = \frac{J_{4,3}}{Y_0} = \frac{\pi \times 0.1001}{2\sqrt{(1.3712 \times 1.9750)}}$$
$$= 0.09555$$

Even and odd mode coupled line impedances For symmetry only $k = 0$ to 2 are required to be computed :

$k = 0$ $(Z_{oe})_{10} = 50 (1 + 0.3703 + 0.1371); (Z_{oo})_{10} = 38.54$ ohm
             $= 75.37$ ohm
$k = 1$ $(Z_{oe})_{21} = 50 (1 + 0.1254 + 0.0157); (Z_{oo})_{21} = 44.52$ ohm
             $= 57.06$ ohm
$k = 2$ $(Z_{oe})_{32} = 50 (1 + 0.09555 + 0.00913) (Z_{oo})_{32} = 45.68$ ohm
             $= 55.23$ ohm

Microstrip line dimensions for $\varepsilon_r = 9.0$
Line width and spacing using Bryant and Weiss curves [17].

	ohm	ohm	W/h	s/h
$k = 0$,	$Z_{oe} = 75.3$	$Z_{00} = 38.34$	0.76	0.3
$k = 1$,	$Z_{oe} = 57.1$	$Z_{00} = 44.52$	1.05	1.0
$k = 2$,	$Z_{oe} = 55.23$	$Z_{00} = 45.68$	1.1	1.5

For 50 ohm input and output lines the characteristic impedance is given by the analytical formula [18] for narrow strips ($W/h < 3.3$.)

$$50 = \frac{119.9}{\sqrt{2}\,(\varepsilon_{r+1})} \left\{ \ln\!\left(\frac{4h}{W}\right) + \sqrt{\left[16\!\left(\frac{h}{W}\right)^2 + 2\right]} \right\}$$

For $\varepsilon_r = 9$,

$$\ln\left\{ \left(\frac{4h}{W}\right) + \sqrt{\left[16\!\left(\frac{h}{W}\right)^2 + 2\right]} \right\} = \frac{50\sqrt{[2\,(9+1)]}}{119.9} = 1.865$$

or,

$$\left(\frac{4h}{W}\right) + \sqrt{\left[16\!\left(\frac{h}{W}\right)^2 + 2\right]} = e^{1.865} = 6.46$$

$$W/h = 1.3$$

Resonator length   The physical lengths of the resonators and the coupled regions are half wavelengths and quarter wavelengths, respectively.

Coupled region length $l = \dfrac{\lambda_{gm}}{4}$

where $\quad \lambda_{gm} = \dfrac{\lambda_{g0} + \lambda_{ge}}{2} =$ midband wavelength

$$\lambda_{go} = \text{odd-mode guide wavelength} = \frac{300}{f\,(\text{GHz})}\,\frac{Z_{oo}}{Z_{oo\,\text{air}}}\ (\text{mm})$$

$$\lambda_{ge} = \text{even-mode guide wavelength} = \frac{300}{f\,(\text{GHz})}\,\frac{Z_{oe}}{Z_{oe\,\text{air}}}\ (\text{mm})$$

$Z_{oe\text{air}}$, $Z_{oo\text{air}}=$ even and odd mode characteristics impedances of the air-spaced lines.

From the Bryant and Weiss curves [17].

	Z_{oeair} (ohm)	Z_{ooair} (ohm)	λ_{ge} (mm)	λ_{go} (mm)	λ_{gm} (mm)
$k = 0$	185	90	11.60	12.2	11.9
$k = 1$	150	100	10.88	12.73	11.8
$k = 2$	140	110	11.3	11.87	11.6
$\lambda_{gm/4}$ (mm)					
$k = 0$	2.98				
$k = 1$	2.95				
$k = 2$	2.90				

Final design values for $\varepsilon_r = 9.0$, 50 ohm input/output lines. $W/h = 1.3$

Coupled lines	W/h	s/h	$l \approx \lambda_{gm}/4$ (mm)
$k = 0$	0.76	0.3	2.98
$k = 1$	1.05	1.0	2.95

$k = 2$	1.1	1.5	2.90
$k = 3$	1.05	1.0	2.95
$k = 4$	0.76	0.3	2.98

The filter configuration is shown below.

Interdigital band-pass filters The interdigital configuration is the most compact filter where the resonators are placed side by side with one end short circuited and other end open circuited alternatively as shown in Fig. 8.17.

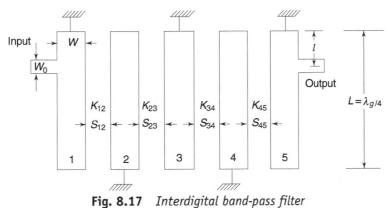

Fig. 8.17 *Interdigital band-pass filter*

The normalised coupling coefficient of a pair of resonators is given by [5]

$$K_{n,\,n+1} = \frac{f_2 - f_1}{f_0 \sqrt{(g_n\, g_{n+1})}} \tag{8.58}$$

where $f_0 = (f_1 + f_2)/2$, the centre frequency, g'_ns are low-pass prototype element values normalised to $\omega'_c = 1$ and $r = 1$. The singly loaded Q for the filter is given by

$$Q_L = \frac{f_0\, g_1}{(f_2 - f_1)} = \frac{f_0\, g_{n+1}}{(f_2 - f_1)} \tag{8.59}$$

The position of the input and output line points l can be calculated from

$$\frac{Q_L}{Z_0 / Z_{01}} = \frac{\pi}{\left[4\sin^2\left(\pi l/2L\right)\right]} \tag{8.60}$$

The dimensions of the filter are found from K vs s/h graph and $Q/(Z_0/Z_{01})$ vs l/L graph of Fig. 8.18.

Example 8.6 Design a microstrip band-pass interdigital filter having centre frequency $f_0 = 4$ GHz, bandwidth $\Delta f = 0.4$ GHz, 35 dB attenuation points at $f_x = 4 \pm 0.4$ GHz. Attenuation response is Chebyshev type with 0.2 dB ripple. Use substrate dielectric constant $\varepsilon_r = 9.8$ and height $h = 1.27$ mm.

Solution

For the low pass prototype Chebyshev response, the attenuation for normalised cut-off frequency $\omega'_c = 1$ is given by

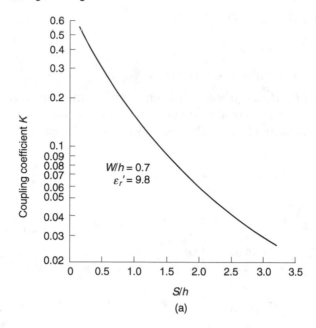

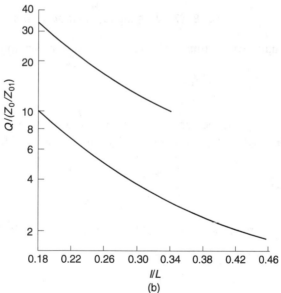

Fig. 8.18 *Coupling coefficient and Q of interdigital filter (J.S. Wong [19])
(a) K vs S/h (b) $Q_L/(Z_0/Z_{01})$ vs l/L*

$$A = 1 + a_m^2 T_n^2(\omega') \; ; \; \omega' = \omega/\omega_c$$

where a_m^2 is the ripple height, n is the order of the filter, ω'_c = bandwidth over which attenuation has maximum ripple a_{mr}^2

Given ripple magnitude

$$0.2 \text{ dB} = 10 \log (a_m^2 + 1)$$

or, $\qquad a_m^2 = 10^{0.2/10} - 1$

Hence Chebyshev attenuation response is

Pass band: $A = 1 + (10^{0.2/10} - 1) \cos^2 (n \cos^{-1} \omega')$; $\omega' \le 1$

Stop band: $A = 1 + (10^{0.2/10} - 1) \cosh^2 (n \cosh^{-1} \omega')$; $\omega' \ge 1$

For low-pass to band-pass frequency transformation

$$\omega' = \frac{\omega_0}{\omega_2 - \omega_1} \left(\frac{\omega}{\omega_0} - \frac{\omega_0}{\omega} \right)$$

or,

$$\omega'_x = \frac{f_0}{f_2 - f_1} \left(\frac{f_x}{f_0} - \frac{f_0}{f_x} \right)$$

$$= \frac{4}{0.4} \left(\frac{4.4}{4} - \frac{4}{4.4} \right)$$

$$= 1.9091$$

For stop band 35 dB attenuation point at $\omega'_x = 4 \pm 0.4$ GHz.
Therefore,

$$35 = 10 \log [1 + (10^{0.2/10} - 1) \cosh^2 (n \cosh^{-1} \omega'_x)]$$

or, $\qquad \dfrac{35}{10} = \log [1 + 0.047 \cosh^2 (n \cosh^{-1}(1.9091))]$

or, $\qquad [1 + 0.047 \cosh^2 (n \cosh^{-1}(1.9091))] = 10^{3.5} = 3162.3$

or, $\qquad [1 + 0.047 \cosh^2 (n \times 1.26281)] = 3162.3$

or, $\qquad \cosh^2 (1.26281\ n) = 3161.3/0.047 = 67261.7$

or, $\qquad \cosh (1.26281\ n) = 259.35$

or, $\qquad n = \dfrac{\cosh^{-1}(259.35)}{1.26281} = 4.95$

$$n \approx 5$$

Therefore, five resonators are required (odd number).

The prototype low-pass values obtained from Table 8.2 for $n = 5$ and ripple $= 0.2$ dB are

$$g_1 = 1.0 = g_6$$
$$g_1 = g_5 = 1.3394$$
$$g_2 = g_4 = 1.337$$
$$g_3 = 2.166$$

In microwave form, the interdigital band-pass filter can be realised by selecting input and output line characteristic impedance $Z_0 = 50$ ohm and the filter internal impedance $Z_{01} = 58$ ohm (say).

The loaded Q of the filter resonator

$$Q_L = f_0 g_1 / \Delta f = 4 \times 1.3394 / 0.4 = 13.394$$

Normalised coupling coefficients between n and $n+1$th resonator is

$$K_{n, n+1} = \frac{f}{f_0 \sqrt{(g_n \, g_{n+1})}}$$

Therefore,

$$K_{12} = K_{45} = \frac{0.4}{4\sqrt{1.3394 \times 1.337}}$$

$$= 0.0747$$

$$K_{23} = K_{34} = \frac{0.4}{4\sqrt{1.337 \times 2.166}}$$

$$= 0.0588$$

Taking $W/h = 0.7$,

$$W = 0.889 \text{ mm}, \; h = 1.27 \text{ mm}, \; W/h < 1$$

$$W_0 = 1.265 \text{ mm}$$

$$S_{12} = S_{45} = 2.2 \text{ mm}$$

$$S_{23} = S_{34} = 2.52 \text{ mm}$$

$$l/L = 0.145,$$

$$L = \lambda_0 / \left(4 \sqrt{\varepsilon_{\text{eff}}} \right); \; \lambda_0 = c / f_0 = 30/4 = 7.5 \text{ cm}$$

$$\varepsilon_{\text{eff}} = \frac{\varepsilon_r + 1}{2} + \frac{\varepsilon_r - 1}{2} \left(1 + \frac{12h}{W} \right)^{-1/2} + 0.04 \left(1 - \frac{W}{h} \right)^2$$

for $\quad W/h \leq 1$

or, $\quad \varepsilon_{\text{eff}} = \frac{9.8 + 1}{2} + \frac{9.8 - 1}{2} \left(1 + \frac{12 \times 1.27}{0.889} \right)^{-1/2} + 0.04 \left(1 - \frac{0.889}{1.27} \right)^2$

$$= 5.4 + 4.4 \times 0.2348 + 3.6 \times 10^{-3}$$

$$= 6.44$$

$$L = 7.4 \text{ mm}, \; l = 1.07 \text{ mm}$$

To find input/output line width W_0

$$Z_0 = 50 \text{ ohm} = \frac{120 \pi}{2 \pi \sqrt{\varepsilon_{\text{eff}}}} \ln \left[\frac{8h}{W_0} + 0.25 \frac{W_0}{h} \right]; \; W_0/h < 1$$

$$= 23.65 \ln \left[\frac{8h}{W_0} + 0.25 \frac{W_0}{h} \right]$$

or, $$\ln \left[\frac{8 \times 1.27}{W_0} + 0.25 \frac{W_0}{1.27} \right] = \frac{50}{23.65} = 2.1142$$

or, $$\ln \left[\frac{10.167}{W_0} + 0.1969 \, W_0 \right] = 2.1142$$

or, $$\left[\frac{10.167}{W_0} + 0.1969 \, W_0 \right] = e^{2.1169} = 8.283$$

or, $$\frac{10.16 + 0.1969 \, W_0^2}{W_0} = 8.283$$

or, $$W_0 = 1.265 \text{ mm}$$

8.9 MICROWAVE BAND-STOP FILTERS

The band-stop microwave filters are narrow band filters and can be designed from the low-pass prototype by using the frequency transformation

$$\frac{1}{\omega'} = \frac{\omega_0}{\omega_2 - \omega_1} \left(\frac{\omega}{\omega_0} - \frac{\omega_0}{\omega} \right) \tag{8.61}$$

The low-pass prototype circuit, band-stop filter circuit and their attenuation responses are shown in Fig. 8.19, where ω_x represents a frequency for high attenuation in the stop band.

8.9.1 Realisation by Quarterwave Couplings

To realise a band-stop filter in transmission line form, it is more convenient to use only resonant series or shunt branches linked by quarterwave sections of transmission lines i.e., by the impedance or admittance inverters. Therefore, a lumped element ladder network consisting of reactive elements can be replaced by the quarterwave coupling network of Fig. 8.20.

The characteristic impedances of the quarterwave coupling sections are Z' for n even and Z_0 for n odd, where n is the number of resonators. Therefore,

For n odd, $\qquad Z'/Z_0 = 1$ $\qquad\qquad\qquad\qquad$ (8.62)

(Butterworth Chebyshev filter)

For n even, $\qquad Z'/Z_0 = [(1/g_0 \, g_{n+1})^{1/2}]$ $\qquad\qquad$ (8.63)

The filter can be designed by defining reactance slope parameters of the resonators as follows

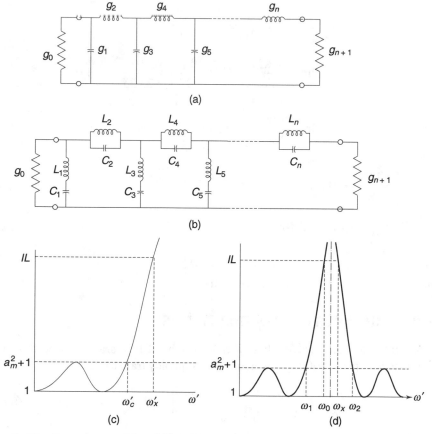

Fig. 8.19 *Prototype and band-stop filters (a) Prototype circuit (b) Equivalent band-stop circuit(c) Prototype low-pass response (d) Band-stop response*

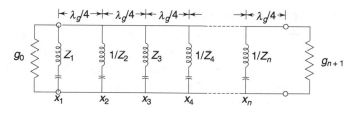

Fig. 8.20 *Microwave equivalent circuit of a ladder band-stop network*

$$\frac{x_1}{Z_0} = \frac{\omega_0}{g_0 \, g_1 \, (\omega_2 - \omega_1)} \tag{8.64}$$

$$\frac{x_k}{Z_0} = \left(\frac{Z'}{Z_0}\right)^2 \frac{g_0 \, \omega_0}{g_k \, (\omega_2 - \omega_1)} \; ; k = \text{even} \tag{8.65}$$

$$\frac{x_1}{Z_k} = \frac{\omega_0}{g_0 \, g_k \, (\omega_2 - \omega_1)} \; ; k = \text{odd} \neq 1 \tag{8.66}$$

The dual realisation circuit is a quarterwave coupled network consisting of parallel resonant circuit in series as shown in Fig. 8.21. The slope parameters are obtained by replacing Z_0, Z' by Y_0, Y' and x_k by b_k keeping all other conditions same.

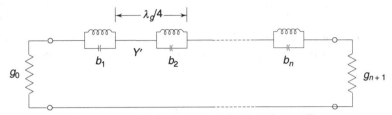

Fig. 8.21   *Quarterwave coupled band-stop filter*

8.9.2   Practical Realisation in Microwave Form

The series and parallel resonant circuits can be approximately realised by short-circuited or open-circuited, resonant stubs with a series gap capacitance in coaxial, strip and microstrip forms and with shunt inductive irises in waveguide form. These are summarized in the Table 8.4 along with the approximate design formulae. The details are described by Mathaei [6].

Table 8.4   *Band-stop filter element realisation*

Resonator	Approximate configuration		Design formulae
$\circ\!-\!|\!(\!-\!\text{mmm}\!-\!\circ$ x	Gap C_g Z_1 sc $\leftarrow \lambda_g/4 \rightarrow$	Coaxial strip Microtrip	$F(\phi_0) = 2x/Z_1$ $\omega_0 C_g = 1/(Z_1\tan\phi_0)$
$\circ\!-\!|\!(\!-\!\text{mmm}\!-\!\circ$ x	gap C_g Z_1 oc $\lambda_g/2$	Coaxial strip Microstrip	$F(\phi_0) \approx x/Z_1$ $\omega_0 C_g = 1/(Z_1\tan\phi_0)$
b	Iris L_g Y_1 oc $\lambda_g/4$	Waveguide	$F(\phi_0) = 2b/Y_1$ $\omega_0 C_g = 1/(Y_1\tan\phi_0)$
b	Iris L_g Y_1 sc $\lambda_g/2$	Waveguide	$F(\phi_0) = b/Y_1$ $\omega_0 L_g = 1/(Y_1\tan\phi_0)$
	$F(\phi_0) = \phi_0 \sec^2\phi_0 + \tan\phi_0$		

The length of the stub is determined by

$$l_k = \frac{\lambda_g \, \phi_0}{2\pi} \tag{8.67}$$

The line width can be determined from the knowledge of the characteristic impedance. The gaped length for the capacitance C_g can be approximated by

$$d_k = \frac{\lambda_g}{4} \text{ or } \frac{\lambda_g}{2} - l_k \tag{8.68}$$

as the case may be depending on the short circuit and open circuit conditions. The inductive irises can be designed by calculating the value of L_g from Table 8.4. Values of $F(\phi_0)$ vs ϕ_0 is shown in Table 8.5.

Table 8.5 $F(\phi)$ Vs. $\phi°$

ϕ	$F(\phi)$	ϕ	$F(\phi)$	ϕ	$F(\phi)$	ϕ	$F(\phi)$	ϕ	$F(\phi)$	ϕ	$F(\phi)$	ϕ	$F(\phi)$
20.00	0.76	68.00	10.96	75.50	24.99	80.55	58.56	83.30	116.48	86.00	328.42	88.70	3226.52
21.00	0.80	68.50	11.47	75.60	25.34	80.60	59.18	83.35	118.24	86.05	336.85	88.75	3497.78
22.00	0.85	69.00	12.01	75.70	25.69	80.65	59.81	83.40	120.40	86.10	345.61	88.80	3804.74
23.00	0.90	69.50	12.60	75.80	26.80	80.70	60.46	83.45	121.89	86.15	354.72	88.85	4153.96
24.00	0.95	70.00	13.23	75.90	26.42	80.75	61.11	83.50	123.77	86.20	364.20	88.90	4553.58
25.00	1.00	70.50	13.91	76.00	26.79	80.80	61.78	83.55	125.70	86.25	374.06	88.95	5013.77
26.00	1.06	70.60	14.06	76.10	27.17	80.86	62.45	83.60	127.68	86.30	384.33	89.00	5547.42
27.00	1.10	70.70	14.19	76.20	27.57	80.90	63.14	83.65	129.70	86.35	395.02	89.05	6171.05
28.00	1.16	70.80	14.34	76.30	27.97	80.95	63.84	83.70	131.77	86.40	406.17	89.10	6906.09
29.00	1.22	70.90	14.49	76.40	28.38	81.00	64.55	83.75	133.90	86.45	417.80	89.15	7780.72
30.00	1.28	71.00	14.64	76.50	28.80	81.05	65.27	83.80	136.07	86.50	429.93	89.20	8832.70
31.00	1.34	71.10	14.79	76.60	29.22	81.10	66.00	83.85	138.30	86.55	442.61	89.25	10113.40
32.00	1.40	71.20	14.95	76.70	29.66	81.15	66.75	83.90	140.58	86.60	455.85	89.30	11694.29
33.00	1.47	71.30	15.11	76.80	30.11	81.20	67.51	83.95	142.92	86.65	469.69	89.35	13677.07
34.00	1.54	71.40	15.27	76.90	30.57	81.25	68.28	84.00	145.32	86.70	484.18	89.40	16210.46
35.00	1.61	71.50	15.43	77.00	31.04	81.30	69.07	84.05	147.78	86.75	499.35	89.45	19518.95
36.00	1.69	71.60	15.60	77.10	31.52	81.35	69.87	84.10	150.31	86.80	515.24	89.50	23954.27
37.00	1.77	71.70	15.77	77.20	32.01	81.40	70.68	84.15	152.90	86.85	531.91	89.55	30092.89
38.00	1.85	71.80	15.94	77.30	32.51	81.45	71.51	84.20	155.55	86.90	549.39	89.60	38933.26
39.00	1.94	71.90	16.11	77.40	33.02	81.50	72.35	84.25	158.28	86.95	567.76	89.65	52329.04
40.00	2.03	72.00	16.29	77.50	33.55	81.55	73.21	84.30	161.08	87.00	587.06	89.70	74046.28
41.00	2.13	72.10	16.47	77.60	34.09	81.60	74.08	84.35	163.95	87.05	607.36	89.75	112731.26
42.00	2.23	72.20	16.65	77.70	36.64	81.65	74.97	84.40	166.91	87.10	628.74	89.80	192063.13
43.00	2.34	72.30	16.84	77.80	35.21	81.70	75.88	84.45	169.94	87.15	651.27	89.85	397933.81
44.00	2.45	72.40	17.03	77.90	35.79	81.75	76.80	84.50	173.06	87.20	675.03	89.90	1266238.98
45.00	2.57	72.50	17.22	78.00	36.39	81.80	77.74	84.55	176.26	87.25	700.11	89.95	27096342.23
46.00	2.70	72.60	17.42	78.10	37.00	81.85	78.70	84.60	179.55	87.30	726.63		
47.00	2.84	72.70	17.62	78.20	37.62	81.90	79.67	84.65	182.94	87.35	754.67		
48.00	2.98	72.80	17.82	78.30	38.27	81.95	80.66	84.70	186.42	87.40	784.38		
49.00	3.14	72.90	18.03	78.40	38.92	82.00	81.67	84.75	190.01	87.45	815.87		
50.00	3.31	73.00	18.24	78.50	39.60	82.05	82.70	84.80	193.70	87.50	849.30		
51.00	3.49	73.10	18.45	78.60	40.29	82.10	83.75	84.85	197.49	87.55	884.83		
52.00	3.68	73.20	18.67	78.70	41.01	82.15	84.82	84.90	201.40	87.60	922.64		
53.00	3.89	73.30	18.80	78.80	41.74	82.20	85.92	84.95	205.43	87.65	962.92		
54.00	4.11	73.40	19.12	78.90	42.49	82.25	87.03	85.00	209.58	87.70	1005.91		

(Contd)

φ	F(φ)	φ	F(φ)	φ	F(φ)	φ	F(φ)	φ	F(φ)	φ	F(φ)	φ	F(φ)
55.00	4.35	73.50	19.35	79.00	43.26	82.30	88.16	85.05	213.86	87.75	1051.84		
56.00	4.61	73.60	19.58	79.10	44.06	82.35	89.32	85.10	218.27	87.80	1100.99		
57.00	4.90	73.70	19.82	79.20	44.87	82.40	90.50	85.15	222.82	87.85	1153.67		
58.00	5.21	73.80	20.06	79.30	45.71	82.45	91.70	85.20	227.51	87.90	1210.22		
59.00	5.55	73.90	20.31	79.40	46.58	82.50	92.93	85.25	232.36	97.95	1271.04		
60.00	5.93	74.00	20.56	79.50	47.46	82.55	94.18	85.30	237.36	88.00	1336.56		
61.00	6.34	74.10	20.82	79.60	48.38	82.60	95.46	85.35	242.52	88.05	1407.28		
61.50	6.57	74.20	21.08	79.70	49.32	82.65	96.76	85.40	247.85	88.10	1483.78		
62.00	6.80	74.30	21.35	79.80	50.29	82.70	98.09	85.45	253.36	88.15	1566.68		
62.50	7.05	74.40	21.62	79.90	51.28	82.75	99.45	85.50	259.06	88.20	1656.74		
63.00	7.31	74.50	21.90	80.00	52.31	82.80	100.84	85.55	264.95	88.25	1754.79		
63.50	7.59	74.60	22.18	80.10	53.37	82.85	102.26	85.60	271.05	88.30	1861.81		
64.00	7.88	74.70	22.47	80.15	53.91	82.90	103.70	85.65	277.36	88.35	1978.94		
64.50	8.19	74.80	22.76	80.20	54.46	82.95	105.18	85.70	283.89	88.40	2107.48		
65.00	8.51	74.90	23.06	80.25	55.02	83.00	106.69	85.75	290.66	88.45	2248.96		
65.50	8.86	75.00	23.37	80.30	55.59	83.05	108.24	85.80	297.67	88.50	2405.19		
66.00	9.23	75.10	23.68	80.35	56.16	83.10	109.81	85.85	304.94	88.55	2578.29		
66.50	9.62	75.20	24.00	80.40	56.75	83.15	111.42	85.90	312.48	88.60	2770.77		
67.00	10.04	75.30	24.32	80.45	57.34	83.20	113.07	85.95	320.30	88.65	2985.64		
67.50	10.48	75.40	24.65	80.50	57.94	83.25	114.76						

The design procedure follows the following steps.

1. Number of resonators n is determined from the maximum attenuation at band edge and the value of given attenuation in the stop band at a given frequency ω_x using the frequency transformation.
2. From n and band edge attenuation, the prototype element values are determined.
3. Slope parameters are determined for a selected realisable network.
4. Selecting the stub impedances, the gap capacitances or iris inductances and the electrical length of the stub are determined at resonance frequency.
5. Stub lengths and the dimensions of the capacitance gaps or the irises are found.

A few examples of the microwave form of band-stop filters are shown in Fig. 8.22.

Example 8.7   Design a stripline Chebyshev narrow bandstop filter having frequency of infinite attenuation at 4 GHz, fractional bandwidth 0.05, passband ripple 0.5 dB and 26 dB minimum attenuation occurs at the 2% of the centre frequency. Use dielectric substrate of $\varepsilon_r = 2.56$ and height $h = 1/8"$. Filter's nominal characteristics impedance is 50 ohm.

Solution

(i) Number of resonators = n

The 2 % of $f_0 = 4 \times 0.02 = 80$ MHz.

The frequency at 26 dB attenuation is

$$f_x = 4 \text{ GHz} + \frac{80\,\text{MHz}}{2} = 4.04 \text{ GHz}.$$

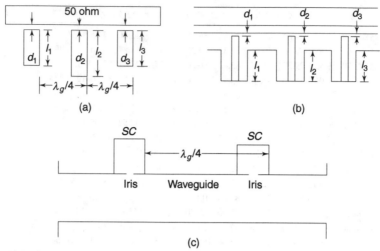

Fig. 8.22 *Band-stop filter (a) Strip line(b) Coaxial line (c) Waveguide*

Therefore, $f_x/f_0 = 4.04/4 = 1.01$

$$\Delta f/f_0 = 0.05$$

From frequency transformation at 26 dB attenuation point

$$\frac{1}{\omega'_x} = \frac{f_0}{\Delta f}\left(\frac{f_x}{f_0} - \frac{f_0}{f_x}\right)$$

$$= \frac{1}{0.05}(1.01 - 0.99)$$

or, $\omega'_x = 2.5$

Ripple $0.5 \text{ dB} = 10 \log(a_m^2 + 1)$

or, $a_m^2 = 0.122$

At 26 dB attenuation point

$$26 = 10 \log[1 + a_m^2 \cosh^2(n \cosh^{-1}\omega'_x)]$$

or, $[1 + 0.122 \cosh^2(n \cosh^{-1} 2.5)] = 10^{2.6}$

$$[\cosh^2(n \cosh^{-1} 2.5)] = \frac{10^{2.6}}{0.122}$$

or, $[\cosh(n \cosh^{-1} 2.5)] = 57.1 = \cosh 4.74°$

Therefore,

$$n = \frac{4.74}{\cosh^{-1} 2.5} = 3$$

(ii) LP prototype element values
    For $n = 3$, ripple = 0.5 dB ripple, from Table 8.2

$$g_0 = g_4 = 1.0$$
$$g_1 = g_3 = 1.5963$$
$$g_2 = 1.0969$$
$$\omega'_c = 1$$

(iii) Microwave realisation
The three-resonator band-stop filter, its transformed configuration using quarter wavelength coupling lines and strip line realisation with open ended resonant half wavelength stubs and gap capacitance are shown below.
    Since $n = 3$ (odd), the impedance of the quarter wavelength coupling lines are $Z'_1 = Z_0$. The reactance slope parameters of the series resonators are

$$\frac{x_1}{R_0} = \frac{\omega_0}{g_0 g_1 (\omega_2 - \omega_1)} = \frac{x_3}{R_0} = \frac{1}{1.5963 \times 0.05} = 12.53$$

$$\frac{x_2}{R_0} = \left(\frac{Z_1}{R_0}\right)^2 \frac{g_0 \omega_0}{g_2 (\omega_2 - \omega_1)} = \frac{1}{1.0969 \times 0.05} = 18.23$$

Therefore,

$$x_1 = x_3 = 12.53 \times 50 = 626.5 \text{ ohm}$$

$$x_2 = 18.23 \times 50 = 911.6 \text{ ohm}$$

The gap capacitance reactance at resonant

$$\omega_0 C_{gk} = \frac{1}{Z_k \tan \phi_0}$$

where
$$F(\phi_0) = \phi_0 \sec^2 \phi_0 + \tan \phi_0 = \frac{x_k}{Z_k}$$

Guide wavelength:
Strip line substrate height $h = 1/8" = 3.175$ cm,

$$\varepsilon_r = 2.56$$

$$\lambda_{g0} = \frac{\lambda_0}{\sqrt{\varepsilon_r}} = \frac{c/f_0}{\sqrt{\varepsilon_r}} = \frac{30/4}{\sqrt{2.56}} = 4.68 \text{ cm}$$

$$\lambda_{g0}/4 = \frac{4.68}{4} = 1.17 \text{ cm}$$

Line widths   Assuming resonator line impedances $Z_{b1} = Z_{b2} = Z_{b3} = 60 \ \Omega$. $(t = 0)$ line widths are calculated as follows.
For $Z_0 = 50$ ohm line, $\sqrt{\varepsilon_r} \ Z_0 = \sqrt{2.56} \times 50 = 80$ ohm
For $Z_0 = 60$ ohm line, $\sqrt{\varepsilon_r} \ Z_0 = \sqrt{2.56} \times 60 = 96$ ohm

From $\sqrt{\varepsilon_r}$, Z_0 vs W/h graphs in Chapter 4,

$$Z_0 = 50 \text{ ohm}, W_0/h = 0.74, W_0 = 0.74 \times 3.175 = 2.35 \text{ mm}$$
$$Z_0 = 60 \text{ ohm}, W_1/h = 0.55, W_1 = 0.55 \times 3.175 = 1.75 \text{ mm}$$
$$W_2 = W_3$$

Stub length

$$F(\phi_{01}) = F(\phi_{03}) = \phi_{01}\sec^2\phi_{01} + \tan\phi_{01} = \frac{x_1}{Z_{b_1}}$$

$$= \frac{626.5}{60} = 10.442$$

$$F(\phi_{03}) = \frac{911.6}{60} = 15.19$$

From the table of ϕ_0 vs $F(\phi_0)$, we find

$$\phi_{01} = \phi_{03} = 67.5° = 1.1781 \text{ radians}$$
$$\phi_{02} = 71.35° = 1.2453 \text{ radians}$$

The stub lengths are given by

$$l_1 = l_3 = \frac{\lambda_{g0}\,\phi_{01}}{2\pi} = \frac{4.68 \times 1.1781}{2\pi} = 0.8775 \text{ cm}$$

$$l_2 = \frac{\lambda_{g0}\,\phi_{01}}{2\pi} = \frac{4.68 \times 1.2453}{2\pi} = 0.9275 \text{ cm}$$

Capacitance gap

$$d_k = \frac{\lambda_{g0}}{4} - l_k$$
$$d_1 = 1.17 - 0.8775 = 0.2925 \text{ cm} = d_3$$
$$d_2 = 1.17 - 0.9275 = 0.2425 \text{ cm}$$

EXERCISES

8.1 Design a maximally flat low pass 50 ohm microstrip filter having cut-off frequency 2 GHz, and insertion loss of 30 dB at 4 GHz. Use lossless dielectric substrate of $\varepsilon_r = 9.6$ and thickness 0.635 mm.

8.2 Design a low-pass 50 ohm Chebyshev coaxial line filter with 0.1 dB pass-band ripple and number of sections is 3. The band edge frequency is 2 GHz.

8.3 Design a low-pass stripline Chebyshev filter having a ripple of 0.1dB and stopband attenuation of 30 dB at 3 GHz. The filter may be realised using high impedance and low impedance sections. The input and output imped-ances are 50 ohm.

8.4 Design coaxial line maximally flat 50 ohm high pass filter when the number of sections is 3 and the band edge frequency is 1 GHz.

8.5 It is required to design a symmetrical three-section, band-pass, quarter-wave coupled filter such that the central frequencies in the pass band and the bandwidth are 9375 and 125 MHz, respectively.
   (a) Determine the loaded Q_L per section.
   (b) Plot the estimated insertion loss due to the filter in the frequency range of 9000 and 10,000 MHz.

8.6 Design a three-cavity quarterwave coupled Chebyshev filter with following specifications.
   Waveguide width $a = 0.9''$
   Band edges at $f_1 = 10,400$ MHz
   Pass band tolerance $a_m^2 = 0.0233$
   Inductive diaphragms with circular holes are to be used. Determine the hole radii and diaphragms spacings.

8.7 Design a 6-section parallel coupled Chebyshev microstrip band-pass filter at 4 GHz with 100 MHz bandwidth. The microstrip parameters are $\varepsilon_r = 9.8$ and $h = 0.5$ mm. The input and output impedances are 50 ohm.

REFERENCES

1. Mumford, WW, "Maximally flat filters in waveguides", *BSTJ*, Vol. 27, pp. 684–714, Oct. 1948.

2. Collin, RE, *Foundations for microwave engineering*, McGraw-Hill Inc., International Editions, 1992.

3. Altman, JL, *Microwave circuits*, D.Van Nostrand Company, Inc., 1964.

4. Ghose, RN, *Microwave circuit theory and analysis*, McGraw-Hill Book Company, Inc., 1963.

5. Bahl, IJ and P Bhartia, *Microwave solid state circuit design* , John Wiley and Sons, 1988.

6. Matthaei, GL, L. Young and EMT Johns, *Microwave filters, impedance matching networks and coupling structures*, McGraw-Hill, New York, 1964.

7. Zverev, AL, *Handbook of filter synthesis*, Wiley, New York, 1967.

8. Howe, H, Jr, "Nomographs aid and filter designer ", *Microwave and RF*, Vol. 24, Oct.1985, pp.103–107.

9. Fox, AG, "Waveguide filters and transformers", *Bell telephone system lab*, Rept. MM-41-160-25, 1941.

10. Rhodes, JD, *Theory of electrical filter*, John Wiley and Sons, Inc., New York, 1976.

11. Malherbe, JAG, *Microwave transmission line filters*, Aptech house books, Dedham, Mass, 1979.

12. Cohn, S B, "Microwave bandpass filter containing high-Q dielectric resonators, "*IEEE trans.*, vol. MTT -16, pp. 218–227, 1968.

13. Cohn, S B "Direct coupled resonator filters" *Proc. IRE*, Vol. 45, pp. 187–196, February 1957.

14. Matthaei, G L "Design of wide-band (and narrow-band) band-pass microwave filters on the insertion loss basis, *IRE, Trans. vol. MTT* - 8, pp. 580–593, 1960.

15. Cohn. S B, "Problems in strip transmission lines", *IRE, Trans.*, PG MTT-3, No. 2, March 1955, pp. 119–126.

16. Hammerstad, E O, and F Bekkadal, "A microstrip handbook", *ELAB Report*, S T F 44 A74169, N7034, University of Trondheim-NTH, Norway, 1975.

17. Bryant, T G and J A Weiss, "Parameter of microstrip transmission lines and of coupled pairs of microstrip lines", *IEEE Trans. MTT-16*, No.12, Dec. 1968, pp. 1021–1027.

18. Terry Edwards, *Foundations for microstrip circuit design* , Sec. Edition, John Wiley and Sons, 1981.

19. Wong, J S, "Microstrip tapped-line filter design", *IEEE Trans. MTT.,* Vol. MTT – 27, Jan. 1979, pp. 44–50.

Chapter **9**

Microwave Vacuum Tube Devices

9.1 INTRODUCTION

The principle of operation of a vacuum tube at microwave frequencies (above 1 GHz) is different from electronic vacuum tubes, such as triodes, tetrodes, and pentodes. These conventional electronic vacuum tubes fail to operate above 1 GHz because of two reasons, viz., the electron transit time from the cathode to the grid becomes comparable to the time period of the sinusoidal signal, and also appearance of stray reactances due to the lead wire inductances and the inter-electrode capacitances. Due to appreciable transit time of the electrons in the cathode-grid space, the grid potential may attain a negative half cycle by the time the electrons, which started from the cathode during the positive half cycle of the grid, reach the grid. This makes the electrons oscillate back and forth in the cathode-grid space or return to the cathode resulting in reduction of efficiency. On the other hand the stray reactances cause an increase in the real part of the input admittance, which overloads the input circuit and reduces the operating efficiency of the tube. Because of these effects concepts of microwave tube design are different.

Microwave tubes are commonly known as *klystrons*, *magnetrons* and *travelling wave tubes* (TWT), which differ from all conventional electronic vacuum tubes, in that transit time is utilised for microwave oscillations or amplification. The principle uses an electron beam on which space-charge waves interact with electromagnetic fields in the microwave cavities to transfer energy to the output circuit of the cavity (klystrons and magnetrons) or interact with the electromagnetic fields in a slow-wave structure to give amplification through transfer of energy (travelling wave tubes).

Klystrons and TWTs are linear beam or 'O'-type tubes in which the accelerating electric field is in the same direction as the static magnetic field used to focus the electron beam. Magnetrons are crossed field devices (M-type) where the static magnetic field is perpendicular to the electric field.

This chapter describes the operating principles of four well known microwave tubes—klystrons, reflex klystrons, travelling wave tubes (TWT) and magnetrons.

9.2 KLYSTRONS

There are two basic configurations of klystron tubes, one is called *reflex klystron*, used as a low power microwave oscillator and another is called *multicavity klystron*, used as a low power microwave amplifier.

9.2.1 Reflex Klystron Oscillator

The schematic configuration of a reflex klystron tube is shown in Fig. 9.1, which uses only a single re-entrant microwave cavity as resonator. The electron beam emitted from the cathode K is accelerated by the grid G and passes through the cavity anode A to the repeller space between the cavity anode and the repeller electrode.

9.2.1.1 Mechanism of Oscillation

Due to dc voltage in the cavity circuit, RF noise is generated in the cavity. This electromagnetic noise field in the cavity becomes pronounced at cavity resonant frequency. The electrons passing through the cavity gap d experience this RF field and are velocity modulated in the following manner.

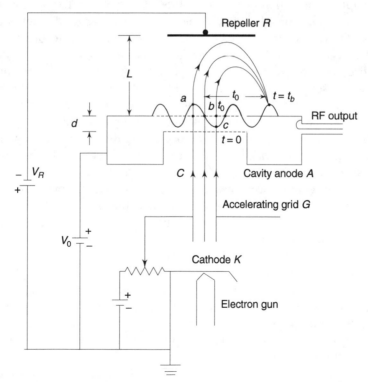

Fig. 9.1 *Reflex klystron*

The electrons a shown in Fig. 9.2, which encountered the positive half cycle of the RF field in the cavity gap d will be accelerated, those (reference electrons) b, which encountered zero RF field will pass with unchanged original velocity, and the electrons c, which encountered the negative half cycle will be retarded on entering the repeller space.

All these velocity modulated electrons will be repelled back to the cavity by the repeller due to its negative potential. The repeller distance L and the voltages can be adjusted to receive all the velocity modulated electrons at a same time on the positive peak of the cavity RF voltage cycle. Thus the velocity modulated electrons are bunched together and lose their kinetic energy when they encounter the positive cycle of the cavity RF field. This loss of energy is thus transferred to the cavity to conserve the total power. If the power delivered by the bunched electrons to the cavity is greater than the power loss in the cavity, the electromagnetic field amplitude at the resonant frequency of the cavity will increase to produce microwave oscillations. The RF power is coupled to the output load by means of a small loop which forms the centre conductor of the coaxial line. When the power delivered by the electrons becomes equal to the total power loss in the cavity system, a steady microwave oscillation is generated at resonant frequency of the cavity.

9.2.1.2 Mode of Oscillation

The bunched electrons in a reflex klystron can deliver maximum power to the cavity at any instant which corresponds to the positive peak of the RF cycle of the cavity oscillation. If T is the time period at the resonant frequency, t_0 is the time taken by the reference electron to travel in the repeller space between entering the repeller space at b and the returning to the cavity at positive peak voltage on formation of the bunch, then

$$t_0 = (n + 3/4)\, T = NT \tag{9.1}$$

where $N = n + 3/4$, $n = 0, 1, 2, 3, \ldots$, as shown in Fig. 9.2.

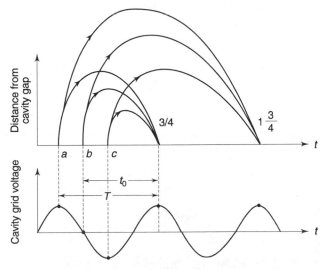

Fig. 9.2 *Reflex klystron modes*

Thus by adjusting repeller voltage for a given dimension of the reflex klystron, the bunching can be made to occur at $N = n + 3/4$ positive half cycle. Accordingly the mode of oscillation is named as $N = \dfrac{3}{4}, 1\dfrac{3}{4}, 2\dfrac{3}{4}$, etc, for modes $n = 0, 1, 2, 3$, ..., respectively. It is obvious that the lowest order mode 3/4 occurs for a maximum value of repeller voltage when the transit time t_0 of the electrons in the repeller space is minimum. Higher modes occurs at lower repeller voltages. Since at the highest repeller voltage the acceleration of the bunched electrons on return is maximum, the power output of the lowest mode is maximum.

9.2.1.3 Power Output and Efficiency

The power output of a reflex klystron is maximum if the bunched electrons on return cross the cavity gap when the gap field is positive maximum. This power can be calculated from various parameters as described below with reference to Figs 9.1 and 9.2.

Velocity modulation For calculation of RF power, it will be assumed that
1. Cavity grids and repeller are plane parallel and very large in extent.
2. No RF field is excited in the repeller space.
3. Electrons are not intercepted by the cavity anode grid.
4. No debunching takes place in the repeller space.
5. The cavity RF gap voltage amplitude V_1 is small compared to the dc beam voltage V_0: $V_1 \ll V_0$.

The electron velocity u attained due to the dc beam voltage V_0 while entering the cavity gap at $t = 0$ is uniform, given by

$$u = u_0 = \sqrt{(2eV_0/m)}$$

$$= 5.93 \times 10^5 \sqrt{V_0} \ \text{m/sec.} \tag{9.2}$$

where V_0 is in volt and $u = 0$ at cathode. The instantaneous cavity RF voltage can be written as

$$V(t) = V_1 \sin \omega t \tag{9.3}$$

where $V_1 \ll V_0$. The average transit time through the cavity gap d and the transit angle are given by, respectively,

$$t_g = d/u_0, \ \theta_g = \omega t_g \tag{9.4}$$

The average microwave voltage in the cavity gap can be written as

$$V_{av} = 1/t_g \int_0^{t_g} V_1 \sin \omega t \, dt$$

$$= \frac{V_1[1 - \cos \omega t_g]}{\omega t_g}$$

$$= \frac{V_1 \sin^2 (\theta_g/2)}{\theta_g/2}$$

$$= V_1 \beta_1 \sin \theta_g/2 \tag{9.5}$$

where
$$\beta_1 = \frac{\sin (\theta_g/2)}{\theta_g/2} \tag{9.6}$$

is called the *beam coupling coefficient* of the cavity gap. Therefore, the coupling between the electron beam and the cavity varies with the cavity gap d in sin X/X form. Here $\beta_1 = 0$ for $d \to 0$.

The exit velocity from the cavity gap after velocity modulation is given by

$$u(t_g) = \sqrt{[2e(V_0 + V_{av})/m]}$$

$$= \sqrt{\left[\frac{2eV_0}{m}(1 + V_1\beta_1/V_0 \cdot \sin (\theta_g/2))\right]} \tag{9.7}$$

where $\beta_1 V_1/V_0$ is called the *depth of modulation*. If the modulation amplitude is small ($<<1$), Eqn. 9.7 for velocity modulation becomes,

$$u(t_g) \approx u_0 \left[1 + \frac{\beta_1 V_1}{2V_0} \sin (\theta_g/2)\right]$$

$$= u_0 \left[1 + \frac{\beta_1 V_1}{2V_0} \sin (\omega t_g - \theta_g/2)\right] \tag{9.8}$$

Transit time   The round trip transit time in the repeller space is given by

$$t_r = \frac{2 \text{ velocity } (u)}{\text{acceleration } (a)} \tag{9.9}$$

The factor 2 in the numerator arises because of the to and fro journey. The electron acceleration is given by

$$a = \frac{eE}{m} = \frac{e}{m} \frac{V_0 \mid V_R + V_1 \sin \omega t}{L}$$

$$\approx \frac{e}{m} \frac{V_0 + V_R}{L} \tag{9.10}$$

Therefore,

$$t_r = \frac{2u(t_g)}{a} = \frac{2u_0 \, m L \left[1 + \dfrac{\beta_1 V_1}{2V_0} \sin (\omega t_g - \theta_g/2)\right]}{e (V_0 + V_R)} \tag{9.11}$$

Since the reference electron does not undergo any velocity modulation, its transit time in repeller space

$$t_0 = \frac{2u_0}{a} = \frac{2u_0 \, m L}{e(V_0 + V_R)} = NT \tag{9.12}$$

$$= \frac{2\pi N}{\omega} \tag{9.13}$$

From Eqns. 9.11 and 9.13

$$t_r = t_0 \left[1 + \frac{\beta_1 V_1}{2 V_0} \sin\left(\omega t_g - \frac{\theta_g}{2} \right) \right] \tag{9.14}$$

Density modulation and beam current The time of arrival of electron to the cavity gap can be expressed by

$$t_b = t_g + t_r = t_g + t_0 \left[1 + \frac{\beta_1 V_1}{2 V_0} \sin\left(\omega t_g - \frac{\theta_g}{2} \right) \right] \tag{9.15}$$

$$= t_g + \frac{2 \pi N}{\omega} + \frac{\pi N}{\omega} \frac{\beta_1 V_1}{V_0} \sin\left(\omega t_g - \frac{\theta_g}{2} \right)$$

$$= t_g + \frac{2 \pi N}{\omega} + \frac{X}{\omega} \sin\left(\omega t_g - \frac{\theta_g}{2} \right) \tag{9.15}$$

where

$$X = \pi N \beta_1 V_1 / V_0, \tag{9.16}$$

called the *bunching parameter* of the reflex klystron.
From the 9.15

$$\frac{dt_b}{dt_g} = 1 + X \cos\left(\omega t_g - \theta_g / 2 \right) \tag{9.17}$$

The bunched electrons on return constitute the bunched beam current i_b such that the conservation of charge gives,

$$I_0 |dt_g| = i_b |dt_b| \tag{9.18}$$

where I_0 is dc the beam current. From Eqns. 9.17 and 9.18, we get

$$i_b = \frac{I_0}{\dfrac{dt_b}{dt_g}} = I_0 [1 + X \cos\left(\omega t_g - \theta_g / 2 \right)]^{-1} \tag{9.19}$$

From Eqn. 9.15, since $V_1 \ll V_0$, $X \ll 1$, $t_b = t_g + 2 \pi N / \omega$, so that

$$i_b = I_0 [1 + X \cos\left(\omega t_b - 2 \pi N - \theta_g / 2 \right)]^{-1} \tag{9.20}$$

By Fourier expansion, the beam current of a reflex klystron oscillator is

$$i_b = I_0 + 2 I_0 \sum_{n=1}^{\infty} J_n(nX) \cos n \left(\omega t_b - 2 \pi N - \frac{\theta_g}{2} \right)$$

$$= I_0 + 2 I_0 J_1(X) \cos \omega \left(t_b - t_0 - \frac{t_g}{2} \right)$$

$$+ \sum_{n=2}^{\infty} 2 I_0 J_n(nX) \cos n \omega \left(t_b - t_0 - \frac{t_g}{2} \right); \tag{9.21}$$

The fundamental component of the RF induced current in the cavity is, therefore,

$$i_{RF} = \beta_1 \, 2I_0 \, J_1 \, (X) \, \cos \omega \left(t_b - t_0 - \frac{t_g}{2} \right)$$

or, $i_{RF} \, (t_b) \approx 2I_0 \, \beta_1 \, J_1 \, (X) \, \cos \, (\omega t_b - 2\pi N)$ (9.22)

where $t_g/2 \ll 2\pi N$, is neglected for smallness.

Power output The magnitude of the fundamental RF current in the cavity is given by

$$|i_{RF}| = 2I_0 \, \beta_1 J_1(X)$$ (9.23)

The rms RF power delivered to the cavity is

$$P_{RF} = V_1 \, i_{RF}/2 = V_1 \, I_0 \, \beta_1 \, J_1(X)$$ (9.24)

Since, from Eqn. 9.16

$$\frac{V_1}{V_0} = \frac{2X}{\beta_1 \, 2\pi N} = \frac{X}{\beta_1 \pi N}$$

Therefore, Eqn. 9.24 reduces to

$$P_{RF} = \frac{V_0 I_0 \, X J_1 \, (X)}{\pi N}$$ (9.25)

From Eqns. 9.2 and 9.13

$$\frac{2\sqrt{(2eV_0/m)} \, mL}{e(V_0 + V_R)} = \frac{2\pi N}{\omega}$$

or, $\pi N = \sqrt{(2V_0 m/e)} \cdot \omega L/(V_0 + V_R)$

Therefore, $P_{RF} = \dfrac{V_0 I_0 \, X J_1(X) \, (V_0 + V_R)}{2\pi f L} \cdot \sqrt{(e/2mV_0)}$ (9.26)

Efficiency The dc power supplied by the beam voltage V_0 is

$$P_{dc} = V_0 I_0$$ (9.27)

Therefore, the electronic efficiency of a reflex klystron oscillator is

$$\eta = \frac{P_{RF}}{P_{dc}} = \frac{X J_1(X)}{\pi N}$$ (9.28)

where $X = \pi N \beta_1 V_1/V_0$, $N = n + 3/4$, the mode number ($n = 0, 1, 2, 3, ...$), $\beta_1 = \sin (\theta_g/2)/(\theta_g/2)$ and $u_0 = \sqrt{(2eV_0/m)}$. It can be shown from Bessel's function table, $X J_1 \, (X)$ attains a maximum value of 1.252 at $X = 2.408$. Thus the optimum values are

$$P_{RF} = 0.3986 \, I_0 \, V_0/N$$ (9.29)

$$\eta = 0.3986/N$$ (9.30)

Although these equations show that the lowest is N, the highest are P_{RF} and η, it has been observed that it is not possible to get 3/4 mode in reflex klystron, so that $N = 1\ 3/4$ mode leads maximum RF power output and efficiency:

$$P_{RFmax} = 0.227\ V_0\ I_0 \tag{9.31}$$

$$\eta_{max} = 22.7\ \% \tag{9.32}$$

9.2.1.4 Mode Curve

Since the output power and frequency can be electronically controlled by varying the repeller voltage, expansions for these parameters in terms of repeller voltage are important to draw mode curves. From Eqns. 9.26 and 9.29.

$$P_{RF} = \frac{0.3986\ V_0\ I_0\ (V_0 + V_R)}{2\ f\ L} \sqrt{\frac{e}{2mV_0}} \tag{9.33}$$

Operation frequency

$$f_{MHz} = \frac{(V_0 + V_R)\ N}{L_{cm}\ \sqrt{V_0} \times 6.74 \times 10^{-2}} \tag{9.34}$$

and

$$|V_R| = \sqrt{(8\ m/e)} \cdot (fL/N) \cdot \sqrt{V_0} - V_0 \tag{9.35}$$

$$= 6.74375 \times 10^{-6}\ f\ L/N\ \sqrt{V_0} - V_0 \tag{9.36}$$

where f is in Hz and L in metres.

The nature of variation of output power and frequency with repeller voltage for different modes are shown in Fig. 9.3.

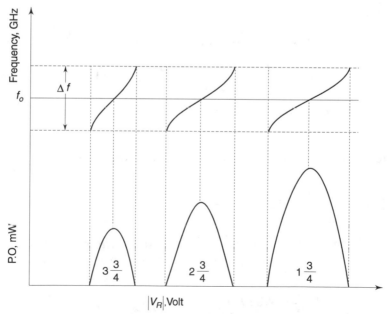

Fig. 9.3 *Reflex klystron mode curves*

From Eqn. 9.36,

$$d\,|V_R|/df_{Hz} = 6.74375 \times 10^{-6}\,L\,\sqrt{V_0}\,/N\,;$$
$$N = n + 3/4,\ n = 0,\ 1,\ 2,\ 3,\ \ldots \qquad (9.37)$$

From Eqns. 9.35 and 9.37 it is seen that once the cavity of a reflex klystron has been tuned to the correct frequency f_0, both the beam voltage V_0 and repeller voltage V_R are adjusted to get maximum power output for a given mode N. The tuning frequency f_0 can be varied mechanically by a short circuit plunger as shown in Fig. 9.4. and adjusting the beam voltage V_0 with klystron power supply.

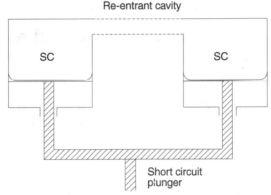

Fig. 9.4 *Reflex klystron tuning*

The performance of a reflex klystron is specified as
Frequency range : 2 – 200 GHz
Bandwidth : ± 30 MHz for $\Delta V_R = \pm 10$ Volt.
Power output : 10 mW – 2.5 W
Reflex klystron are widely used as microwave source in the laboratory and in microwave transmitter.

Example 9.1 A reflex klystron is to be operated at frequency of 10 GHz, with dc beam voltage 300 V, repeller space 0.1 cm for $1\dfrac{3}{4}$ mode. Calculate P_{RFmax} and corresponding repeller voltage for a beam current of 20 mA.
Solution

$$P_{RFmax} = \frac{0.398\,V_0\,I_0}{N} = \frac{0.398 \times 300 \times 20 \times 10^{-3}}{1\dfrac{3}{4}}$$

$$= 1.365 \text{ Watts}$$

$$|V_R| = 6.74 \times 10^{-6} f_{(Hz)}\,L_{(m)}\,\sqrt{V_0}\,/N - V_0$$
$$L_{(m)} = 0.1 \times 10^{-2}\,m = 10^{-3}\,m$$

$$N = 1\frac{3}{4} = 1.75$$

$$|V_R| = 6.74 \times 10^{-6} \times 10 \times 10^9 \times 10^{-3} \times \sqrt{300} / 1.75 - 300$$

Therefore, $\quad V_R = -367.08$ volts

Example 9.2 A reflex klystron is operated at 5 GHz with dc beam voltage 350 V, repeller spacing 0.5 cms for $N = 3\dfrac{3}{4}$ mode. Calculate bandwidth over $\Delta V_R = 1$ V.

Solution

$$N = 3 \ 3/4 = 15/4$$

$$\Delta V_R = 6.7438 \times 10^{-6} \times L_m \times \Delta f_{Hz} \sqrt{V_0} / N$$

or, $\quad 1 = 6.74 \times 10^{-6} \times 0.5/100 \times \Delta f_{Hz} \times \sqrt{350} \times 4/15)$

or, $\quad \Delta f = 5.948$ MHz

9.2.1.5 Electronic Admittance

The electronic admittance Y_e of the reflex klystron is defined by the ratio of induced bunch beam current i_b of Eqn. 9.21 and cavity gap voltage at the time of bunching t_b of Eqn. 9.3.

$$Y_e = i_{RF} \ (t_b)/V(t_b)$$

$$= \frac{2 I_0 \ \beta_1 J_1(X) \cos (\omega t_b - \omega t_0)}{V_1 \sin \omega t_b}$$

$$= \frac{2 I_0 \ \beta_1 J_1(X)}{V_1} \ \frac{e^{j\omega(t_b - t_0)}}{e^{j(\omega t_b - \pi/2)}}$$

$$= \frac{2 I_0 \ \beta_1 \ J_1(X)}{V_1} \ e^{j(\pi/2 - \omega t0)} \tag{9.38}$$

From Eqns. 9.11 and 9.14, substituting $V_1 = X V_0 / \pi N \beta_1$ and $2 \pi N = \omega t_0$ in the above Eqn. 9.38

$$Y_e = \frac{\pi N \beta_1 2 I_0 \beta_1 J_1(X)}{X V_0} \ e^{j(\pi/2 - \omega t_0)}$$

or, $\quad Y_e = \dfrac{\pi N I_0 (\beta_1)^2 J_1(X)}{V_0 X} \ e^{j(\pi/2 + \omega t_0)} \tag{9.39}$

$$= G_e + j B_e$$

Thus the electronic admittance is a function of the dc beam admittance I_0/V_0, the dc transit angle $2\pi N$, the transit angle through the cavity gap θ_g and the signal voltage V_1.

The equivalent circuit for the reflex klystron is shown in Fig. 9.5, where L and C represent the magnetic and electric storage of energy in the cavity, G_c the cavity loss conductance, G_L the load conductance. If the bunched electrons return to the cavity gap a little before reference transit time $t_0 = (n + 3/4)T$, the ac beam current lags behind the field and an inductive reactance appears in the circuit for

B_e and if the bunched electrons return to the gap a little after t_0, the ac current leads the field and a capacitive reactance is presented to the circuit for B_e. The condition for oscillations is satisfied when G_e becomes negative and the total conductance in the circuit is negative.

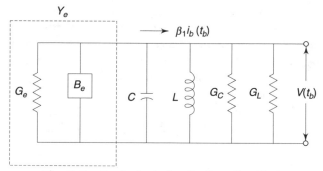

Fig. 9.5 *Equivalent circuit of a reflex klystron*

$$|-G_e| \geq G_c + G_L = \frac{1}{R_{sh}}, \; R_{sh} = \text{effective shunt resistance} \qquad (9.40)$$

Variation of G_e and B_e in the complex Y_e plane is schematically shown in Fig. 9.6 which forms a spiral. $|V_R| = \infty$ corresponds to the origin. A decrease in $|V_R|$ moves outwards along the spiral. The oscillation will be there for the value of $2\pi N$ for which the spiral lies in the area to the left of the line $-(G_c + G_L)$ when Eqn. 9.40 is satisfied for

$$\omega t_0 = 2\pi(n + 3/4) = 2\pi N, \; n = 0, 1, 2, \ldots$$

If $Y_e = \pm jB_e$, oscillations take place at a frequency lower or higher than the cavity resonance frequency as shown in Fig. 9.6.

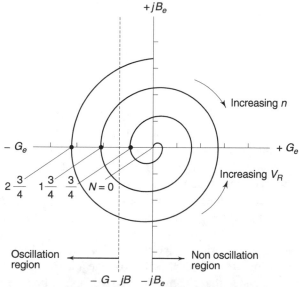

Fig. 9.6 *Electronic admittance of a reflex klystron*

9.2.1.6 Modulation of a reflex klystron

A reflex klystron is widely used as a microwave source for laboratory experiments. To find the characteristics of the line or a load, the most common measurable parameters are VSWR and the position of a voltage standing wave minimum. To avoid the use of the costly microwave receivers or power meters, microwave signals are low frequency modulated and this modulated signal is probed and detected by a crystal detector. The detected microwave signals that carry the original amplitude and phase variation information of the microwave signal are measured using low frequency receiver to obtain the desired parameters.

Two basic modulations are used for this purpose (1) amplitude modulation by square waves and (2) frequency modulation by saw tooth waves, as shown in Figs 9.7 and 9.8.

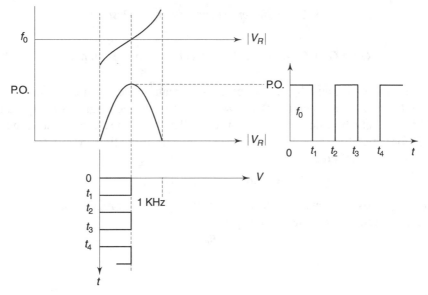

Fig. 9.7 *Amplitude modulation*

In amplitude modulation, the dc repeller voltage is adjusted at the left edge of the mode power curve and a low frequency (usually 1 KHz) square voltage wave is superimposed on the dc repeller voltage. The amplitude of the square pulse is adjusted to attain the power maximum point. Due to square wave variation of effective repeller voltage, the output is a pulsed signal at a constant frequency f_0. Here care must be taken such that the negative half cycle of the modulating amplitude should not enter into the higher mode or part of the same mode to avoid oscillation at two or more frequencies simultaneously.

Frequency modulation is useful for frequency sweeping over the band of frequency for a given mode by using a superimposed saw tooth voltage at the repeller plate. Here the positioning of the dc repeller voltage and the amplitude of the saw tooth wave are so adjusted that the frequency sweeping taken place over nearly

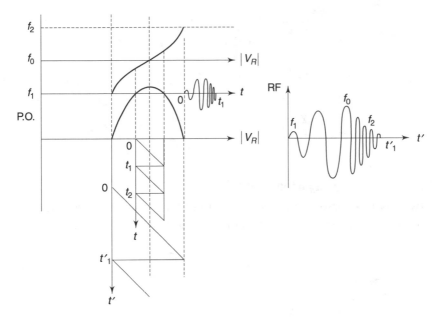

Fig. 9.8 *Frequency modulation*

linear part of the frequency variation and at the same time over a small portion of the power output curve to maintain almost constant power output and linear frequency sweeping for the response characteristics obtainable in a oscilloscope, where the detected output is fed to the *y* plate and the modulating signal to the *x* plate for triggering.

Example 9.3   A reflex klystron is operated at 9 GHz with dc beam voltage 600 V for $1\dfrac{3}{4}$ mode, repeller space length 1 mm, dc beam current 10 mA. The beam coupling coefficient is assumed to be 1. Calculate, the repeller voltage, electronic efficiency and output power.

Solution

Given,

$$f_0 = 9 \text{ GHz}, \quad V_R = ?$$
$$N = 1\ 3/4, \quad P_{RF} = ?$$
$$V_0 = 600 \text{ Volt}, \quad \eta = ?$$
$$L = 1 \text{ mm}$$
$$\beta_1 = 1, \quad I_0 = 10 \text{ mA}.$$

From Eqn. 9.36

$$|V_R| = 6.74 \times 10^{-6} \times 1 \times 10^{-3} \times 9 \times 10^9 \times \frac{\sqrt{600}}{1\dfrac{3}{4}} - 600$$

$$= 249 \text{ V}$$

$$P_{RFmax} = \frac{0.398\ V_0 I_0}{N}$$

$$= \frac{0.398 \times 600 \times 10 \times 10^{-3}}{1\frac{3}{4}} \text{ watts}$$

$$= 0.2274 \times 600 \times 10 \times 10^{-3}$$

$$= 1.3644 \text{ Watts}$$

$$\eta_{max} = \frac{XJ_1(X)}{\pi N} = \frac{0.398}{N} = \frac{0.398}{1\frac{3}{4}} = 22.74 \%$$

9.2.2 Two-cavity Klystron Amplifier

The two cavity klystron tubes are widely used for microwave amplification as shown by the schematic circuit diagram in Fig. 9.9.

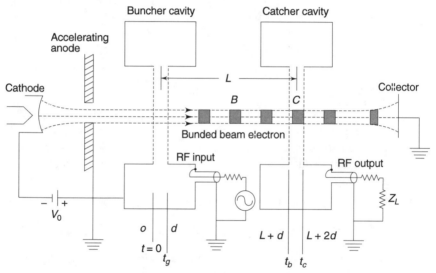

Fig. 9.9 *Two-cavity klystron amplifier*

9.2.2.1 *Mechanism of operation*

The basic principles of two-cavity klystron operation can be described in following manner. A high velocity electron beam produced by the accelerating anode is successively passed through an input re-entrant cavity resonator (buncher cavity) grid, a field free drift space of length L, an output re-entrant cavity resonator (catcher cavity) grid and finally collected by a collector electrode. The electron beam is focussed to travel axially without spreading during transit by applying an axial magnetic field produced by an external coil current. The input RF signal to be amplified excites the buncher cavity with a coupling loop. The combination of the anode voltage V_0 and the cavity gap width d are such that the transit of electrons through each cavity gap is less than the quarter of the time period T of the input signal cycle. The electrons passing through the buncher grids are accelerated/retarded/passed through with unchanged initial dc velocity depending upon when they encounter the RF signal field at the buncher cavity gap at positive/ negative/zero crossing phase of the cycle, respectively, as shown by distance— time plot in Fig. 9.10. This is called the *applegate diagram*. Thus the electron beam is velocity modulated to form bunches or undergoes density modulation in accordance with the input RF signal cycle. While passing through the catcher cavity grid, this density modulated electron beam induces RF current in the output cavity and thereby excites the RF field in the output cavity at input signal cycle. The phase of field in the output cavity is opposite to that of the input cavity so that the bunched electrons are retarded by the output gap voltage. The loss of kinetic energy of the electrons on retardation process transfers RF energy to the output cavity continuously at signal cycle. The amplitude of the signal at output cavity attains a steady large value when the loss of kinetic energy of the bunched elec- trons compensates the output cavity circuit losses. The amplified signal is cou- pled out from the catcher cavity through a current loop to the load.

The current induced in the catcher cavity is rich in harmonics up to say, 15. The cavity is tuned to the fundamental or any harmonic as desired.

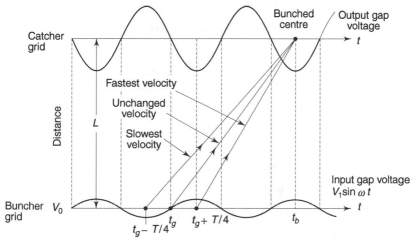

Fig. 9.10 *Applegate diagram*

9.2.2.2 Analysis of two-cavity klystron amplifier

The analysis for RF amplification by a two-cavity klystron amplifier is based on the following assumptions:

1. The transit time in the cavity gap is very small compared to the period of the input RF signal cycle.
2. The input RF signal amplitude V_1 is very small compared to the dc beam voltage $V_0 =$ the anode potential with respect to the cathode potential.
3. The cathode, anode, cavity grids and collector are all parallel and the cavity grids do not intercept any electron while passing.
4. No space charge or debunching take place at the bunch point.
5. The RF fields are totally confined in the cavity gaps so that field is zero in the drift space L.

6. The electrons leave the cathode with zero initial velocity.

Velocity modulation Due to potential difference V_0 between the anode and cathode, the electrons form a high current density beam with the axial velocity u_0 which is obtained by equating kinetic energy and potential energy:

$$u_0 = \sqrt{(2eV_0/m)} = 5.93 \times 10^5 \sqrt{V_0} \ \text{m/s} \qquad (9.41)$$

Let the signal voltage across the gap be $V_1 \sin \omega t$. The electron beam enters the buncher cavity at a time $t = t_1$ with velocity u_0 and passes out at $t = t_2$ as shown in Fig. 9.11.

The transit time and transit angle through the buncher gap d is

$$t_2 - t_1 = t_g = d/u_0, \ \theta_g = \omega t_g \qquad (9.42)$$

respectively. Due to input RF signal in the buncher cavity, the average RF voltage in the buncher gap can be obtained as

$$V_{av} = (1/t_g) \int_{t_1}^{t_2} V_1 \sin \omega t \, dt$$

$$= (V_1/\omega t_g) (\cos \omega t_1 - \cos \omega t_2)$$

or, $$V_{av} = V_1 \beta_1 \sin (\omega t_1 + \theta_g/2) \qquad (9.43)$$

where, $\beta_1 = (\sin \theta_g/2)/(\theta_g/2)$, is called the buncher cavity beam coupling coefficient. As electrons pass through the buncher gap, their velocities are increased, decreased or unchanged, depending on the positive, negative, or zero RF voltage across the grids when they pass through. At time $t_{av} = (t_1 + t_2)/2$ the electron is midway across the gap with velocity say u_{av} say. Then

$$u_{av}/u_0 = \sqrt{[(V_0 + V_{av})/V_0]}$$

or, $$u_{av} = u_0 \left[1 + \frac{V_1 \beta_1}{V_0} \sin (\omega t_1 + \theta_g/2 \right]^{1/2} \qquad (9.44)$$

Since $V_1 \ll V_0$, the binomial expansion of Eqn. 9.44 gives

$$u_{av} \approx u_0 [1 + m/2 \sin (\omega t_1 + \theta_g/2)] \qquad (9.45)$$

Thus the electrons in the beam are velocity modulated by the input RF signal with depth of velocity modulation $m = \beta_1 V_1/V_0$. These electrons exit from the buncher gap at time $t_1 + t_2$ to the field free drift space between the two cavities with velocity

$$u_a (t_2) = u_0 [1 + m/2 \sin (\omega t_2 - \theta_g/2)] \qquad (9.46)$$

Transit time in the drift space If t_3 is the time when the bunched electrons are at the catcher grid after travelling through the field free drift space L

$$t_3 = t_2 + \frac{L}{u(t_2)}$$

$$= t_2 + \frac{L}{u_0[1 + m/2 \sin (\omega t_2 - \theta_g/2)]}$$

$$\approx t_2 + L/u_0 [1 - m/2 \sin (\omega t_2 - \theta_g/2)] \qquad (9.47)$$

Therefore, the transit time t_d in the drift space is given by

$$t_d = t_3 - t_2 = L/u_0 [1 - (m/2) \sin (wt_2 - \theta_g/2) \qquad (9.48)$$

Density modulation   Because of the difference in velocities of the electrons in the velocity modulated beam, the electrons will form bunches i.e., becomes density modulated, in accordance with the input signal cycle.

This bunching action can be illustrated graphically by plotting distance as a function of time for the drift space, what is known as *Applegate diagram* as shown in Fig. 9.10 for the velocities $u_{min} = u_0 (1 - m/2)$, $u_{max} = u_0 (1 + m/2)$ and $u_{unchange} = u_0$. A maximum degree of bunching takes place when the buncher and catcher cavities are spaced to satisfy the condition

$$t_d = t_3 - t_2 = L/u(t_2) = t_0[1 - (m/2) \sin (\omega t_2 - \theta_g/2)] \quad (9.49)$$

where $t_0 = L/u_0$, is the dc transit time in drift space. The corresponding transit angle in the drift space L is

$$\omega t_d = \omega (t_3 - t_2) = \omega t_0 - (m\omega t_0/2) \sin (\omega t_2 - \theta_g/2)$$

$$= \theta_0 - (m/2) \theta_0 \sin (\omega t_2 - \theta_g/2) \qquad (9.50)$$

where $\theta_0 = \omega t_0$, the dc transit angle. If N is the number of RF cycles elapsed during the transit time of reference electron with velocity u_0, at the point of bunching ωt_0's

$$t_0 = \frac{L}{u_0} = NT = \frac{N}{f} \quad \frac{2\pi N}{w}$$

Therefore,

$$\theta_0 = \omega t_0 = 2\pi N \qquad (9.51)$$

The catcher cavity is placed so that the electron bunching occurs within the catcher gap. During traversal through the gap the bunched electrons induce positive charges in the catcher grids by repelling conduction electrons in it as shown in Fig. 9.11.

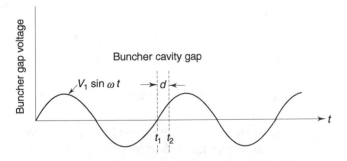

Fig. 9.11   *The buncher gap voltage*

This process results in a series of positive current pulses in the external circuit at the rate equal to the periodic time of the input signal and width equal to the transit time across the gap. The Fourier transform of this pulse waveform gives a large number of frequency components. The output cavity is tuned to the fundamental component equal to the frequency of the input cycle and provides a sinusoidal signal at the output port of the catcher cavity.

Let I_0 = dc beam current in the buncher, and i_b = bunched beam current at catcher = dQ/dt_3

Assuming there is no loss of electrons, total charge

$$Q = \Sigma \, dQ = \Sigma I_0 \, dt_2 = \Sigma i_b \, dt_3$$

or,

$$i_b = \frac{I_0}{\Sigma (dt_3/dt_2)} \tag{9.52}$$

From Eqn. 9.48

$$t_3 = t_2 + \frac{2\pi N}{\omega} \left[1 - \frac{m}{2} \sin \left(\omega t_2 - \frac{\theta_g}{2} \right) \right]$$

$$= t_2 + \frac{2\pi N}{\omega} - \frac{m\pi N}{\omega} \sin \left(\omega t_2 - \frac{\theta_g}{2} \right)$$

$$\frac{dt_3}{dt_2} = 1 - m\pi N \cos \left(\omega t_2 - \frac{\theta_g}{2} \right)$$

$$= 1 - X \cos \left(\omega t_2 - \frac{\theta_g}{2} \right) \tag{9.53}$$

where

$$X = \pi m N = \frac{\pi N \beta_1 V_1}{V_0}, \tag{9.54}$$

and is defined as the *bunching parameter* of a two-cavity klystron.

From Eqns. 9.52 and 9.53

$$i_b = \frac{I_0}{\Sigma [1 - X \cos (\omega t_2 - \theta_{g/2}]} \tag{9.55}$$

Equation 9.55 is the expression of the bunched beam current in the output cavity. Substituting $t_2 = t_3 - t_0 = t_3 - 2\pi N/\omega$ in Eqn. 9.55.

$$i_b = I_0 \, \Sigma \, [1 - X \cos (\omega t_3 - 2\pi N - \theta_{g/2})]^{-1} \tag{9.56}$$

Expanding in Fourier series and finding the coefficients

$$i_b = I_0 + 2I_0 \sum_{n=1}^{\infty} J_n \, (nX) \cos n(\omega t_3 - \theta_g - \theta_0) \tag{9.57}$$

This is the generalised expression of a bunched beam current consisting of a dc component I_0 plus the fundamental ac component of amplitude $2 \, I_0 \, J_1(X)$ and harmonics of amplitude $2I_0 \, J_n \, (nX)$, $n = 2, 3, 4, \ldots$.

The klystron is generally tuned to fundamental ac component of current given by

$$i_f(t_3) = 2I_0 J_1(X) \cos(\omega t_3 - \theta_g - \theta_0) \tag{9.58}$$

The current can be made maximum when $J_1(X)$ is max $= 0.582$ at $X = 1.841$ by adjusting the dc beam voltage V_0.

From Eqns. 9.51, 9.53 and 9.58, it can be shown that the fundamental component of current is maximum for an optimum drift space length,

$$L_{op} = \frac{3.682 \, u_0 \, V_0}{\omega \, \beta_1 \, V_1} \tag{9.59}$$

Beam spreading/debunching   At the point of bunching the electrostatic force of repulsion between the electrons does not allow electronic collision, but may cause beam spreading or undesirable debunching. This reduces the efficiency of the klystrons.

Power output   The fundamental component of RF beam current passing through the output cavity gap induces a current in the catcher cavity

$$i_c = i_f \beta_2 \tag{9.60}$$

where β_2 is the beam coupling coefficient of the catcher cavity gap. $\beta_2 = \beta_1$ when both buncher and catcher cavities are identical.

Therefore, from Eqns. 9.58 and 9.60

$$\begin{aligned} i_c &= 2I_0 \beta_2 J_1(X) \cos(\omega t_3 - \theta_g - \theta_0) \\ &= I_2 \beta_2 \cos(\omega t_3 - \theta_g - \theta_0); \; I_2 = 2I_0 J_1(X) \end{aligned} \tag{9.61}$$

The corresponding RF voltage across the catcher cavity is given by

$$V_c = V_2 \cos(\omega t_3 - \theta_g - \theta_0 - \phi) = \beta_2 I_2 R_{sh} \tag{9.62}$$

where ϕ is the phase angle between i_c and V_c. The average power delivered to the output cavity is

$$\begin{aligned} P_0 &= \frac{1}{2\pi} \int_0^{2\pi} i_c V_c \, d(\omega t_3); \\ &= \beta_2 I_0 V_2 J_1(X) \cos\phi \end{aligned} \tag{9.63}$$

The coupling between the beam and RF field in the gap is better for higher values of $\beta \leq 1$.

For maximum power output $\cos\phi = 1$, $\phi = 0°$, $X = 1.841$ and $\beta_2 = 1$. Therefore,

$$P_{0\,max} = I_0 V_2 J_1(X) = 0.582 \, I_0 \, V_2 \tag{9.64}$$

Since power output is a function of $X = \pi N \beta_1 V_1/V_0$, it can be varied by changing V_0 as shown in Fig. 9.12. From Eqn. 9.51.

$$u_0 = \frac{Lf}{N}$$

or, $\qquad \sqrt{(2e\,V_0/m)} - \dfrac{Lf}{N}$

or, $\qquad\qquad V_0 = \dfrac{m}{2e}\,(Lf/N)^2$ $\qquad\qquad$ (9.65)

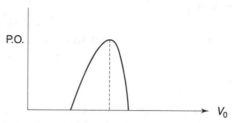

Fig. 9.12 *Power output vs beam voltage*

Efficiency Efficiency η of a two-cavity klystron can be defined as the ratio of RF output power to the dc beam power

$$\eta = \frac{P_{ac}}{P_{dc}} = \frac{\beta_2\,I_0\,V_2 J_1(X)}{I_0 V_0}$$

$$= \frac{\beta_2\,V_2\,J_1(X)}{V_0} \qquad\qquad (9.66)$$

The efficiency becomes maximum when $X = 1.841$

$$\eta_{max} = 0.582\,\beta_2\,V_2/V_0 \qquad\qquad (9.67)$$

If the coupling is perfect $\beta_2 = 1$

$$i_{c\,max} = 2\,I_0 \times 0.582 \qquad\qquad (9.68)$$

and $\quad V_2 = V_0$. Then the $\eta_{max} = 58.2\%$

9.2.2.3 *Equivalent circuit*

The output cavity can be represented by an equivalent circuit as shown in Fig. 9.13 where G_{sh} represents the effective shunt conductance which accounts for the catcher cavity wall resistance, beam loading resistance and external load resistance. The electric and magnetic stored energies are represented by C and L, respectively, so that the resonant frequency of the cavity is

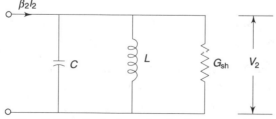

Fig. 9.13 *Equivalent circuit of output cavity*

$$f_0 = 1/(2\pi\sqrt{LC}) \tag{9.69}$$

The power output is given by

$$P = |i_c|^2/(2G_{sh}) = 4 I_0^2 \beta_2^2 J_1^2(X)/(2G_{sh}) \tag{9.70}$$

9.2.2.4 Voltage gain and mutual conductance

The voltage gain for the klystron is defined by

$$A_V = |V_2/V_1|$$

From the equivalent circuit

$$V_2 = \frac{\beta_2 I_2}{G_{sh}}$$

$$X = \frac{\pi N V_1 \beta_1}{V_0}$$

Therefore,

$$V_1 = \frac{X V_0}{N \pi \beta_1} = \frac{2 V_0}{\beta_2 \theta_0}$$

and

$$A_V = \frac{\beta^2 \theta_0 I_0 J_1(X)}{X V_0 G_{sh}} \tag{9.71}$$

where $\beta_1 = \beta_2$ for identical cavities.

The dc beam conductance $G_0 = I_0/V_0$

The mutual conductance of the klystron amplifier is defined as

$$|G_m| = |i_c|/V_1 = 2\beta_2 I_0 J_1(X)/V_1$$

$$= \beta^2 \theta_0 G_0 J_1(X)/X \tag{9.72}$$

Therefore,

$$A_V = G_m/G_{sh} \tag{9.73}$$

G_m decreases with increase in X.

For maximum power output, $X = 1.841$, so that

$$|G_m| = 0.316 \, G_0 \, \beta^2 \, \theta_0 \tag{9.74}$$

Two-cavity klystron amplifier is not a low noise device. Therefore, it is normally used in the transmitter, not in the receiver. Some typical applications are, troposcatter transmitter, satellite communication ground stations and UHF TV transmitter power amplifier. The typical frequency range of applications is from C-band to about 60 GHz. The average cw power output is in the range 100 kW–250 kW for high power applications. A 30–60 dB power gain is possible with a bandwidth of 10–60 MHz and 30–40% efficiency in practice.

Example 9.4 A two-cavity klystron amplifier is tuned at 3 GHz. The drift space length is 2 cm, beam current 25 mA. The catcher voltage is 0.3 times the beam voltage. It is assumed that the gap length of the cavity << the drift space so that the input and output voltages are in phase ($\beta = 1$). Compute (a) power output and efficiency for $N = 5\ 1/4$ (b) beam voltage, input voltage and output voltage for maximum power output for $5\frac{1}{4}$ mode.

Solution

Beam voltage

$$V_0 = \frac{m}{2e} (Lf/N)^2$$

$$= \frac{9.1 \times 10^{-31}}{2 \times 1.6 \times 10^{-19}} \left[\frac{2 \times 10^{-2} \times 3 \times 10^9}{5 + 1/4} \right]^2 \text{ volts}$$

$$= 371.4 \text{ volts}$$

For maximum power output, the bunching parameter

$$X = 1.84 = \pi N V_1 / V_0$$

Therefore the buncher voltage magnitude

$$V_1 = \frac{X V_0}{\pi N} = \frac{1.84 \times 371.4}{3.1415 \times 5 \frac{1}{4}} = 41.43 \text{ volts}$$

Catcher voltage $V_2 = 0.3 \ V_0 = 111.4$ volts
Output power

$$P_0 = \beta_2 \ I_0 \ V_2 \ J_1(X) \cos \phi$$

$$P_{0\,max} = I_0 \ V_2 \ J_1(X)$$

or, $\qquad P_{0max} = 0.582 \ I_0 \ V_2$

$$= 0.582 \times 25 \times 10^{-3} \times 111.4 \text{ watts} = 1.621 \text{ watts}$$

$$P_{dc} = I_0 V_0 = 25 \times 10^{-3} \times 371.4 \text{ watts}$$

$$= 9.3 \text{ watts}.$$

Efficiency $\eta_{max} = 0.582 \ V_2/V_0$

$$= 0.582 \times 111.4/371.4$$

$$= 17.46 \ \%$$

Example 9.5 A two-cavity klystron operates at 5 GHz with dc beam voltage 10 kV, cavity gap 2 mm. For a given input RF voltage, the magnitude of the gap voltage is 100 volts. Calculate the transit time at the cavity gap, the transit angle, and the velocity of the electrons leaving the gap.

Solution

DC beam velocity $u_0 = 0.593 \times 10^6 \sqrt{V_0}$

$$= 0.593 \times 10^6 \sqrt{(10 \times 10^3)} \text{ m/s}$$

$$= 0.593 \times 10^8 \text{ m/s}$$

Gap transmit time $t_g = \dfrac{d}{u_0} = \dfrac{2 \times 10^{-3}}{0.593 \times 10^8} = 33.7 \times 10^{-12}$ sec.

The gap transit angle $\theta_g = \omega t_g = 2\pi \times 5 \times 10^9 \times 33.7 \times 10^{-12}$ rad

$$= 1.059 \text{ rad} = 60.7 \text{ deg.}$$

The beam coupliing coefficient

$$\beta_1 = \frac{\sin \theta_g/2}{\theta_g/2}$$

$$= 0.505/0.5295 = 0.9537$$

The velocity of electron leaving the input cavity gap

$$u(t) = u_0 \left[1 + \frac{\beta_1 V_1}{2 V_0} \sin (\omega t + \theta_g/2) \right]$$

$$= 0.593 \times 10^8 \, [1 + (0.954 \times 100)/(2 \times 10 \times 10^3) \sin(\omega t + \theta_g/2)]$$

$$= 0.593 \times 10^8 \, [1 + 0.00477 \sin (\omega t + 0.5295)]$$

is varying sinusoidally at the input cycle.
The maximum velocity

$$u(t)_{max} = u_0 \, (1 + m/2) = 0.593 \times 10^8 \, (1 + 0.00477)$$

$$= 0.5958 \times 10^8 \text{ m/s}$$

The minimum velocity

$$u(t)_{min} = u_0 \, (1 - m/2) = 0.593 \times 10^8 \, (1 - 0.00477)$$

$$= 0.5902 \times 10^8 \text{ m/s}$$

Example 9.6 A two-cavity klystron operates at 10 GHz with $I_0 = 3.6$ mA, V_0 = 10 kV. The drift space length is 2 cms and the output cavity total shunt conductance is $G_{sh} = 20 \, \mu$ mho and beam coupling coefficient $\beta_2 = 0.92$. Find the maximum voltagc and power gain.

Solution

Maximum voltage gain

$$A = \frac{\beta^2 \theta_0 I_0 J_1(X)_{max}}{X V_0 G_{sh}}$$

dc beam velocity $u_0 = 0.593 \times 10^6 \, \sqrt{V_0}$

$$= 0.593 \times 10^6 \times \sqrt{10} \times 10^3$$

$$= 0.593 \times 10^8 \text{ m/s}$$

Transit angle in drift space

$$\theta_0 = \frac{\omega L}{u_0}$$

$$= \frac{2\pi \times 10 \times 10^9 \times 2 \times 10^{-2}}{0.593 \times 10^8}$$

$$= 21.19 \text{ rad.}$$

$$A_{max} = \frac{0.92 \times 0.92 \times 21.19 \times 3.6 \times 0.582}{1.841 \times 10 \times 10^3 \times 20 \times 10^{-6}}$$

$$= 102.1$$

Example 9.7 An identical two-cavity klystron amplifier operates at 4 GHz with $V_0 = 1$ kV, $I_0 = 22$ mA, cavity gap 1 mm, drift space 3 cms. If dc beam conductance and catcher cavity total equivalent conductance are 0.25×10^{-4} mhos and 0.3×10^{-4} mhos, respectively, calculate,

(a) the beam coupling coefficient, dc transit angle in the drift space and the input cavity voltage magnitude for maximum output voltage.

(b) voltage gain and efficiency, neglecting the beam loading.

Solution

(a) DC beam velocity $u_0 = 0.593 \times 10^6 \sqrt{V_0} = 0.59 \times 10^6 \sqrt{1} \times 10^3$

$$= 1.88 \times 10^7 \text{ m/s}$$

Gap transit angle $\theta_g = \dfrac{\omega d}{u_0} = \dfrac{2\pi \times 4 \times 10^9 \times 1 \times 10^{-3}}{1.88 \times 10^7}$ rad

$$= 1.337 \text{ rad} = 76.6 \text{ deg.}$$

The beam coupling coefficient

$$\beta_1 = \beta_2 = \frac{\sin \dfrac{\theta_g}{2}}{\dfrac{\theta_g}{2}} = \sin 38.3°/0.6685 = 0.927$$

DC transit angle in the drift space

$$\theta_0 = \frac{\omega L}{u_0} = \frac{2\pi \times 4 \times 10^9 \times 3 \times 10^{-2}}{1.88 \times 10^7} \text{ rad}$$

$$= 40.11 \text{ rad}$$

For maximum output voltage $X = 1.84$, $J_1(X) = 0.582$, so that the input cavity gap voltage magnitude

$$V_1 = \frac{2V_0 X}{\beta_1 \theta_0} = \frac{2 \times 10^3 \times 1.841}{0.927 \times 40.11} \text{ volts}$$

$$= 99 \text{ V}$$

(b) The voltage gain

$$A_v = \frac{\beta^2 \theta_0 J_1(X) I_0}{X V_0 G_{sh}}$$

$$= \frac{0.927 \times 0.927 \times 40.11 \times 0.582 \times 22 \times 10^{-3}}{1.841 \times 0.55 \times 10^{-4} \times 10^3}$$

$$= 4.36$$

$$= 12.8 \text{ dB}$$

Catcher voltage $V_2 = A_V \times V_1 = 4.36 \times 99 = 431.64$ V
Power efficiency

$$\eta = \frac{P_{RF}}{P_{dc}} = \frac{\beta_2 I_0 J_1(X) V_2}{I_0 V_0} = \frac{\beta_2 J_1(X) V_2}{2 V_0}$$

$$= \frac{0.927 \times 0.582 \times 431.64}{10^3}$$

$$= 23.28\%$$

9.2.2.5 Beam loading

It can be shown that when the buncher cavity gap is negligibly small, the average energy of the electron in the beam leaving the cavity over a cycle, is nearly equal to the energy with which they enter the cavity. However, when the buncher cavity gap in appreciable, the average energy of the electrons leaving the buncher gap is larger than that when they enter the gap because the electrons encounter the RF field for a longer duration. This excess energy must be supplied by the buncher cavity to the electron beam for bunching. Thus the electron beam is loaded by the cavity energy. This phenomenon is called "beam loading."

9.2.3 Multicavity Klystron Amplifier

In klystron the bandwidth is limited due to its resonant structure. The bandwidth of the klystron amplifier can be significantly improved by using stagger turned multiple (more than two) cavities as shown in Fig. 9.14. The increase in bandwidth, however, is offset by a decrease in gain.

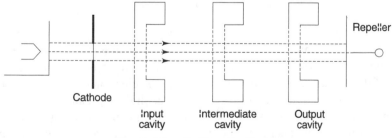

Fig. 9.14 *Multicavity klystron*

9.3 TRAVELLING WAVE TUBE AMPLIFIER

A travelling wave tube amplifier (TWTA) circuit uses a helix slow-wave non-resonant microwave guiding structure and thus a broad band microwave amplifier. The electron beam from the cathode continuously interacts with an axial RF input field over a long distance inside the helix where both velocity and density modulations of the electron beam occur as shown by the schematic diagram in Fig. 9.15.

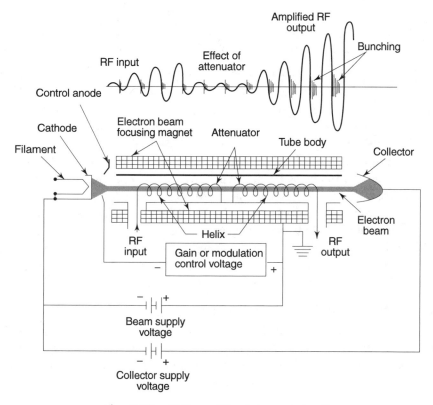

Fig. 9.15 *TWT amplifier tube and circuit*

The electron beam is focussed axially by a static axial magnetic field and collected in a collector circuit. The microwave input signal is injected on the helix slow-wave circuit surrounding the electron beam, which provides an axial component of the signal wave that can interact with the electron beam. The dc beam voltage is adjusted so that the beam velocity is slightly greater than the axial component of field on the slow-wave structure. During transit along the axis, the electron beam transfers energy to the travelling signal wave and thus the signal field amplitude increases. When the loss in the system is compensated by this energy transfer a steady amplification of the microwave signal appears at the output end of the helix.

An attenuator is placed over a part of the helix near the output end to attenuate any reflected waves due to impedance mismatch that can be fed back to the input to cause oscillations. The attenuator is placed after sufficient length of the interaction region so that the attenuation of the amplified signal is insignificant compared to the amplification. TWTAs are commercially available for application at frequencies above 1 GHz , with a power gain up to 60 dB.

9.3.1 Analysis of TWTA

Wave propagation in helix In TWTA, a microwave field is being created inside a helix slow-wave guide by injecting the input signal into it. This field will decelerate the electrons in the axial beam at some sections, and at other sections it will accelerate them resulting in density modulation in the beam in accordance with the input signal. The helix reduces the phase velocity v_p of the wave propagation from the velocity of light to that of the beam velocity so that the continuous interaction between specific regions of wave and beam can be maintained. The signal, however, travels along the helix conductor profile at approximately the velocity of light c.

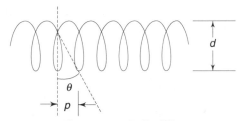

Fig. 9.16 *Helical line*

If d is the diameter of the helix and p is the helix pitch, the time taken by the signal along the wire must be equal to that taken by the axial wave, so that

$$\frac{p}{v_p} = \frac{\sqrt{[p^2 + (\pi d)^2}}{c}$$

or,
$$v_p = \frac{cp}{\sqrt{[p^2 + (\pi d)^2]}} \approx \frac{cp}{\pi d} = \frac{\omega}{\beta} \tag{9.75}$$

when $p \ll \pi d$. Equation 9.75 is useful in designing a helix slow wave structure.
Beam and field interaction   The electromagnetic wave travelling along the helix axis, has a longitudinal component of electric field as shown in Fig. 9.17. These fields exert force $\mathbf{F} = -e\mathbf{E}$ on the electrons.

An electric field directed against the electron flow accelerates the electrons and that directed along the electron flow decelerates the electrons. Thus a density modulation in accordance with the input signal frequency will occur to form bunches.

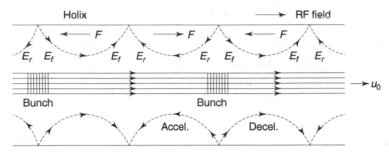

Fig. 9.17 *Bunch formation in a TWT due to beam and field interaction*

As the dc velocity of the beam is maintained slightly greater than the phase velocity of the travelling wave, more electrons face the retarding field than the accelerating field, and a great amount of kinetic energy is transferred from the beam to the electromagnetic field. Thus the field amplitude increases forming more compact bunch and a larger amplification of the signal voltage appears at the output end of the helix.

Focussing   The entire electron beam moves axially without spreading by the influence of an axial magnetic field generated by means of a magnet around the helix using a permanent magnet in low power tubes and a current carrying solenoid for high power tubes. The disadvantages of the solenoid are that it is relatively bulky and it consumes power. These arrangements are suitable for high power tubes where RF power output is more than a few kW. For satellite communications and other low power requirements where weight and power consumption must be minimised, permanent magnet is used for focussing.

Reflection of waves   An attenuator placed midway along the helix attenuates the reflected waves propagating from any mismatch load at the output end to prevent them from reaching the input and causing oscillations.

The attenuator will attenuate both the forward and reflected waves on the helix without affecting the electron beam. Thus the bunched electrons after exit from the attenuator, reinduces the forward wave on the helix with the same frequency and the corresponding field induces a new amplified microwave signal on the helix.

9.3.1.1 Gain Characteristics
The analysis of TWT amplifier involves complicated mathematical techniques. A detailed description of theory and design was developed by Pierce. The output power gain is defined as

$$A_p = 10 \log \left| \frac{\text{output voltage}}{\text{input voltage}} \right|^2 = -9.54 + 47.3 \, NC \text{ dB} \quad (9.76)$$

The first term -9.54 dB represents a loss due to the fact that the input wave divides into three waves of equal magnitude and only one of these waves is amplified. N is the length of the interaction region in wavelengths,

$$N = l/\lambda_e \tag{9.77}$$

where l is the length of the slow wave structure in metres and

$$\lambda_e = 2\pi u_0/\omega, \; u_0 = \sqrt{[(2\,e/m)\,V_0]} \tag{9.78}$$

The factor C is the gain parameter of the circuit defined by

$$C = (I_0\,Z_0/4V_0)^{1/3} \tag{9.79}$$

where I_0 is the dc beam current, V_0 is the dc beam voltage, and Z_0 the characteristic impedance of the helix.

The peak power output of a single-helix-type tube is limited to about 3 kW because of the difficulty in removing heat due to ohmic loss from the helix conductor. For low inputs, the small signal (or linear) gain is almost constant. As the RF power input is increased, the RF output power does not increase in proportion, but instead attains a maximum and finally starts to decrease. The point at which the output power is maximum is termed as the saturation point, and the gain at this point is the saturation gain as shown in Fig. 9.18. There will be a range of input for which the output remains in saturation, and this range defines the overdrive capability of the TWT.

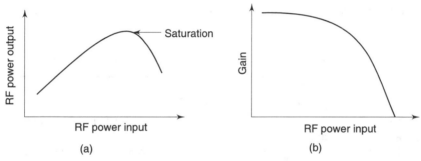

Fig. 9.18 *RF power output and gain characteristics*

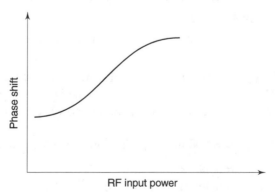

Fig. 9.19 *Phase shift characteristic*

Distortion As the operating level approaches saturation, the nonlinearity of the transfer characteristic gives rise to intermodulation distortion.

Another form of disortion occurs in a TWTA as a result of phase nonlinearities. The absolute time delay between input and output of a TWT at a fixed input level is generally not significant. However, at higher input levels, where more of the energy in the beam is converted to output power, the average beam velocity is reduced, and therefore, the delay time is increased. Since phase delay is directly proportional to time delay, this results in a phase shift at the output relative to the phase shift at saturation as shown in Fig. 9.19. Such phase shift converts phase modulation at the output for an amplitude modulation on the input. This distortion can be troublesome where unwanted amplitude variations at the input could appear as an angle modulated signal.

When TWT is operated near or at saturation, harmonic distortion occurs due to more intense electron bunching, which produces sharp current peaks rich in harmonics. These harmonics must be filtered at the output.

Fluctuations in cathode emission and in electron velocity contribute to the current noise in TWT. For low noise operation the TWTs have noise factors of less than 10 dB. Medium power TWTs have noise factors in the range 18 to 30 dB, and high power TWTs have noise factors in the range 24 to 27 dB.

9.3.2 Comparison between TWTA and Klystron Amplifier

The performance characteristics of TWTA and klystron amplifiers are compared in Table 9.1.

Table 9.1 *Comparison of TWTA and klystron amplifier*

Klystron amplifier	TWTA
1. Linear beam or 'O' type device	Linear beam or 'O' type device
2. Uses cavities for input and output circuits	Uses nonresonant wave circuit
3. Narrow band device due to use of resonant cavities	Wideband device because use of non-resonant wave circuit

9.3.3 Applications of TWT

TWT amplifiers are used in medium power satellite and higher power satellite transponder output.

Example 9.8 A helix travelling-wave tube operates at 4 GHz under a beam voltage 10 kV and beam current 500 mA. If the helix impedance is 25 ohm and the interaction length is 20 cm, find the output power gain in dB.

Solution

Given $V_0 = 10$ kV, $I_0 = 500$ mA, $Z_0 = 25$ ohm, $f = 4$ GHz and $l = 20$ cm.

$$u_0 = 0.593 \times 10^6 \sqrt{V_0}$$
$$= 0.593 \times 10^6 \, (10 \times 10^3)^{1/2}$$
$$= 0.593 \times 10^8 \text{ m/sec.}$$
$$N = l/\lambda_e = l\omega/2\pi u_0$$
$$= \frac{0.2 \times 2\pi \times 4 \times 10^9}{2\pi \times 0.593 \times 10^8}$$

$$= 13.49$$
$$C = (I_0 Z_0/4V_0)^{1/3}$$
$$= \left[\frac{500 \times 10^{-3} \times 25}{4 \times 10 \times 10^3} \right]^{1/3}$$
$$= 0.068$$

Therefore, $\quad A_p = -9.54 + 47.3 \times 13.49 \times 0.068$
$$= 33.85 \text{ dB}$$

9.4 MAGNETRON OSCILLATOR

9.4.1 Introduction

A magnetron oscillator is used to generate high microwave power. Magnetrons are crossed field tubes (M-type) in which the dc magnetic field and the dc electric field are perpendicular to each other. A high power microwave oscillator uses travelling-wave cylindrical magnetron tube as shown in the schematic cross-sectional diagram in Fig. 9.20.

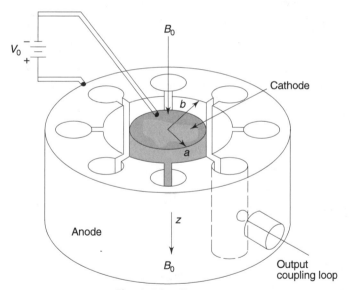

Fig. 9.20 *Magnetron*

It consists of a cylindrical cathode of finite length and radius a at the centre surrounded by a cylindrical anode of radius b. The anode is a slow wave structure consisting of several re-entrant cavities equi-spaced around the circumstance and coupled together through the anode cathode space by means of slots. Radial electric field is established by dc voltage V_0 in between the cathode and the anode and a dc magnetic flux denoted by B_0 is maintained in the positive z-direction by means of a permanent magnet or an electromagnet. The electrons emitted from the cathode try to travel to anode. But with the influence of crossed fields **E** and **H** in the space between the anode and the cathode, it experiences force $\mathbf{F} = -e\mathbf{E} - e(\mathbf{v} \times \mathbf{B})$, where $\mathbf{v}$ is the velocity vector of the electron considered and take curved trajectory. Due to excitation of the anode cavities by RF noise voltage in the biasing circuit, the RF fieldlines are fringed out of the slot to the space between the anode and cathode. The accelerated electrons in the trajectory, when retarded by this RF field, transfer energy from the electron to the cavities to grow RF oscillations till the system RF losses balance the RF oscillations for stability.

9.4.2 Equations of Electron Trajectory

After emergence from the cathode with zero velocity (say), the electrons will acquire a velocity $\mathbf{V}$ having a tangential as well as a radial velocity component due to force $\mathbf{F}$ exerted by the crossed fields $\mathbf{E}$ and $\mathbf{H}$;

$$\mathbf{F} = -e\mathbf{E} - e(\mathbf{v} \times \mathbf{H}) \tag{9.80}$$

In Fig. 9.21, trajectories a', b', c', and d' of the electrons are shown for different magnetic field strengths. At zero magnetic field, the electrons take the straight path a', by the influence of electric field only. For a given V_0 if the magnetic field is increased, the electrons take curved path b' to reach the anode. At a critical value of magnetic field B_c, say, the electrons just graze the anode surface at radius b and take the path c' to return to the cathode for a given voltage V_0. This value B_c is called the cut-off magnetic flux density. If the magnetic field is greater than B_c all the electrons return to the cathode as shown by a typical path d' without reaching the anode. The average velocity of the electron in the z-direction is constant, and is given by

$$v_z = E_0/B_0 \tag{9.81}$$

where $E_0 = V_0/(b - a)$ is the dc electric field.

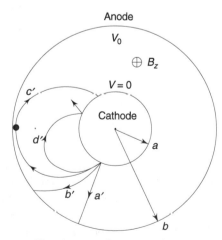

Fig. 9.21 *Electron trajectory*

In general, an electron in the trajectory of radius of curvature r and at velocity v experiences three radial forces: $-e\mathbf{E}$, $-e(\mathbf{v} \times \mathbf{B})$ and the centrifugal force mv^2/r such that for equilibrium

$$\frac{mv^2}{r} + eE = evB \tag{9.82}$$

where the electric field is in the radial direction only and is given by

$$E\,(r) = -\frac{V_0}{r\ln\dfrac{b}{a}} \tag{9.83}$$

In the absence of an electric field the electrons move in a circular path and return to the cathode, when

$$mv^2/r = evB \tag{9.84}$$

or, $\qquad\qquad v/r = eB/m = \omega$, called the *cyclotron angular frequency*.

The equations of motion for electrons in a crossed electric and magnetic fields can be given as

$$m\,(dV/dt) = \mathbf{F} = -\,e\,\mathbf{E} - ev \times \mathbf{B} \tag{9.85}$$

Since the electron emitted from the cathode in the direction opposite to $\mathbf{E}$ the equation of motion for electrons in cylindrical coordinates, are

$$\frac{d^2r}{dt^2} - r\,(d\phi/dt)^2 = +\,(e/m)\left[E_r - rB_z\,\frac{d\phi}{dt}\right] \tag{9.86}$$

and $\quad \dfrac{1}{r}\dfrac{d}{dt}\left(r^2\,\dfrac{d\phi}{dt}\right) = \dfrac{eB_z}{m}\dfrac{dr}{dt}$

where $B_0 = B_z$ is assumed in the +z-direction.

9.4.2.1 Cut-off magnetic field and voltage

From the second equation,

$$\frac{d}{dt}(r^2\,d\phi/dt) = \frac{eB_z}{m}\frac{rdr}{dt}$$

$$= \frac{\omega}{2}\frac{d}{dt}\,(r^2)$$

or, $\qquad\qquad r^2\,d\phi/dt = \omega r^2/2 + K \tag{9.87}$

where K is an integration constant. At $r = a$, $d\phi/dt = 0$, and $K = -\,\omega a^2/2$. Therefore, the angular velocity of the electrons is

$$d\phi/dt = \omega/2\,(1 - a^2/r^2) \tag{9.88}$$

Since the electrons move in direction perpendicular to the magnetic field, the kinetic energy of the electrons is given by the electric field only

$$eV = 1/2\,m\,[(dr/dt)^2 + (r\,d\phi/dt)^2] \tag{9.89}$$

At $r = b$, $V = V_0$, and $dr/dt = 0$ for the electrons to just graze the anode, so that

$$(b \, d\phi/dt)^2 = 2e/m \, V_0$$

or, $\quad b^2[\omega/2.(1 - a^2/b^2)]^2 = 2e/m \, V_0 \qquad\qquad (9.90)$

Substituting $\omega = eB_c/m$ at grazing.

$$b^2 \, [eB_c/2m \, (1 - a^2/b^2)]^2 = 2eV_0/m$$

or, $\qquad\qquad\qquad B_c = \dfrac{(8 V_0 \, m/e)^{1/2}}{b(1 - a^2/b^2)} \qquad\qquad (9.91)$

Thus if $B_0 > B_c$ for a given V_0, the electrons will not reach the anode. For a given B_0, the cut-off voltage is given by

$$V_c = e/8 \, m \cdot b^2 \, (1 - a^2/b^2)^2 \, B_0^2 \qquad\qquad (9.92)$$

Here, if $V_0 < V_c$, for a given B_0, the electrons will not reach the anode. Equations 9.91 and 9.92 for B_c and V_c are called the *Hull cut-off* magnetic and voltage equations, respectively.

9.4.3 Resonant Modes in a Magnetron

The nature of field distribution in the magnetron cavities is such that the alternating RF magnetic flux lines pass through the cavities parallel to the cathode axis, and the alternating RF electric fields are concentrated across the slot and fringe out to the interaction space between the anode and the cathode, in the transverse direction. For N resonant coupled cavities of the anode, there exist N resonant frequencies or modes. The equivalent resonant circuit is shown in Fig. 9.22.

Since the slow-wave structure is closed on itself, the total phase shift around the internal periphery must be an integral multiple of 2π for possible oscillations. The phase shift between two adjacent cavities is given by

$$\phi_n = \frac{2\pi n}{N} \qquad\qquad (9.93)$$

where $n = 0, \pm 1, \pm 2, \pm 3, \ldots, \pm N/2$, indicates the nth mode of oscillation; $n \neq 0$ because this would indicate zero RF fringing fields in the interaction region.

For continuous interaction between the electrons and the RF fields for transfer of energy, the anode dc voltage V_0 is adjusted to coincide the average rotational velocity of the electrons with the phase velocity of the RF field in the interaction space.

Since having opposite phase in successive cavities, excitation is maximum in the cavities, $\phi_n = \pi$ or π-mode is commonly used for magnetron oscillators, where $n = N/2$.

9.4.3.1 Mode separation

Since each mode corresponds to a different frequency, the various modes are detuned differently from the fundamental resonant frequency of the cavities but differ very little in frequency from each other as shown in Fig. 9.22. This makes it difficult to separate the π-mode from the next immediate mode for maximum excitation of the cavities. Hence it is possible for the frequency to jump from one mode to another, which is highly undesirable. The separation of π-mode fre-

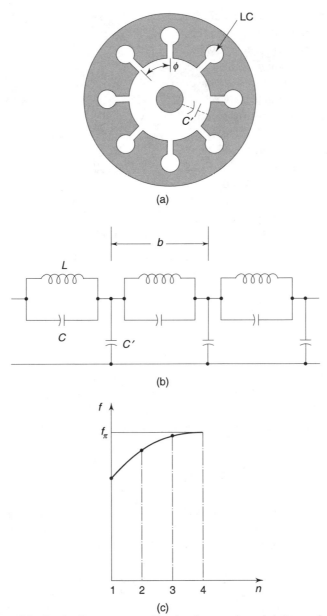

Fig. 9.22 *Equivalent resonant circuit of magnetron (a) Approximate equivalent (b) Band-pass equivalent circuit schematic (c) Mode frequencies*

quency from other modes is commonly done by the strapping method as shown in Fig. 9.23.

In this method two metallic rings are arranged with one ring connected to the even numbered anode and the others to the odd numbered anode poles. Thus for π-mode each ring is at same potential and no π-mode current flows in the straps and the straps inductance has no effect. But the two rings having opposite potentials provide a capacitive loading in parallel to the capacitance C at each slot of the resonant cavities and lowers the frequency of π-mode. For the other modes each ring experiences a phase difference between the successive connection points and the resulting current flow gives rise to an inductive field with reduced capacitive effect. The inductance is in parallel with the slot, thus raising the unwanted mode frequency. Therefore, the strapping method increases the frequency separation between the π-mode and the higher adjacent modes.

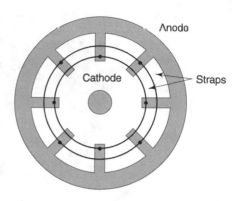

Fig. 9.23 *Strapping method*

Average drift velocity Although the electrons move in both ϕ and r directions, they possess average drift velocity in the ϕ direction given by

$$v_\phi = \frac{Er}{Bz} \tag{9.94}$$

for $v_\phi < Er/Bz$, the electrons will tend to be deflected towards the anode and be collected. For $v_\phi > Er/Bz$, the electrons will be deflected towards the cathode.

Hartree voltage Magnetrons are usually designed to operate in the π-mode where the phase difference between adjacent resonators is 180°. For strong interaction between the wave on the anode structure and the electron beam, the phase velocity of the wave should be nearly equal to the drift velocity v_ϕ and the oscillations for π-mode start at beam voltage

$$V_{0h} = \frac{2\pi f}{N}(b^2 - a^2)B_0 \tag{9.95}$$

which is known as the *Hartree* voltage. Here f is the operating frequency and N is number of the resonators. A plot of Hull cut-off voltage and the Hartree voltage vs B_0 is shown in Fig. 9.24 where the region of oscillations is indicated, by shaded area.

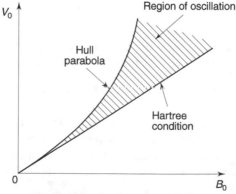

Fig. 9.24 *Region of oscillation*

9.4.4 Mechanism of Oscillations

The electrons emitted from the cathode try to travel towards the anode with the influence of dc electric field **E** but take a curved trajectory with the influence of dc magnetic field **B**. The total force on an electron is $\mathbf{F} = -e\,(\mathbf{E} + \mathbf{v} \times \mathbf{B})$, where **v** is the velocity of the electron in the space between the cathode and the anode. One can postulate the existence of RF oscillations in the resonant structure due to noise voltages in the dc biasing circuit. The RF electric field at the resonant frequency of the π-mode structure in the slot of the cavities and fringes out in the cathode anode space as shown in Fig. 9.25.

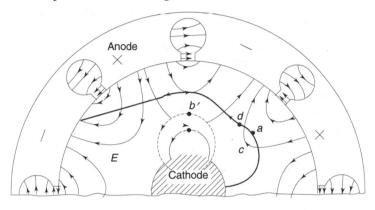

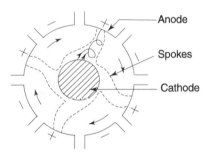

Fig. 9.25 *Mechanism of oscillation*

When the dc magnetic field exceeds the cut-off value, the electrons try to return back to the cathode in absence of the RF field. Due to fringing RF electric field the electron *a* experiences a retarding electric field and electron *b* experiences an accelerating electric field. The retarded electron *a* experiences reduced magnetic force $(-e\mathbf{v} \times \mathbf{B})$ due to its reduced velocity and moves towards the anode. By adjusting the dc anode voltage and dc magnetic field, the circumferential velocity component of the electron can be made such that the electron *a* takes approximately one half cycle of the RF oscillation to travel from one slot position to the next. This makes electron *a* experience a retarding field and ultimately reaches the anode surface after continuously delivering energy to the RF oscillations.

On the other hand the electron *b* is accelerated by the RF field to return quickly to the cathode. Thus since electron *b* remains in the interaction space for a much shorter duration compared to the electron *a*, the energy absorbed by *b* is much smaller than the energy delivered by *a* to the system to sustain oscillations.

The electron *b* on bombardment to the cathode causes heating loss in the cathode (~ 5% of the anode power). The oscillation amplitude grows due to retarded electrons and a steady state is reached when the losses in the system are compensated by the RF oscillation continuously.

9.4.4.1 *Phase focussing*

The electrons around *a*, such as *c* and *d* are acted upon both radial and tangential component of the RF field in such a way that the electron *c* moves faster than *a* and that *d* moves slower than *a* to form a bunch around *a*. Then these electrons from *c* to *d* are confined to spokes or focussed and terminated to the alternate anode. This is called phase focussing. For π-mode these spokes have angular velocity equal to two anode poles per cycle and the electrons within the spokes deliver energy to the oscillations before they are collected by the anode.

9.4.4.2 *Tuning*

The fine tuning of the magnetron oscillator can be done by changing (adding) the capacitance between the ring-strap by placing a tuning ring (C-ring) as shown in Fig. 9.26. This reduces the resonant frequency of the π-mode by an amount depending upon the position of the C-ring.

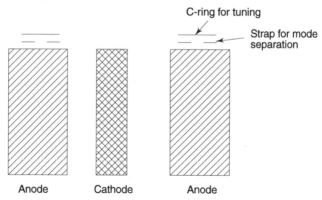

Fig. 9.26 *Tuning of a magnetron*

9.4.5 Power Output and Efficiency

A magnetron can deliver a peak power output of up to 40 MW with the dc voltage of 50 kV at 10 GHz. The average power output is of the order of 800 kW. The magnetron possesses a very high efficiency ranging from 40 to 70 %. Magnetrons are commercially available for peak power output from 3 kW and higher.

9.4.6 Applications

The magnetrons are widely used on radar transmitters, industrial heating, and microwave ovens. A microwave oven requires a standard power of 600 W and frequencies 915 MHz or 2450 MHz. For industrial heating magnetrons generate power of the order of kW at MHz frequencies. For radar applications a magnetron generates peak power of the order of MW in GHz frequency range.

Example 9.9 A pulsed cylindrical magnetron is operated with the following parameters

Anode voltage = 25 kV
Beam current = 25 A
Magnetic density = 0.34 Wb/m^2
Radius of cathode cylinder = 5 cm
Radius of anode cylinder = 10 cm

Calculate (a) the angular frequency, (b) the cut-off voltage, (c) the cut-off magnetic flux density.

Solution

(a) Angular frequency $= eB_0/m = 1.759 \times 10^{11} \times 134$

$$= 0.5981 \times 10^{11} \text{ radian}$$

(b) The cut-off voltage $= (e\ B_0^2 b^2\ /8m)\ (1 - a^2/b^2)^2$

$$= 1/8 \times 1.759\ 10^{11} \times 0.34^2 \times (10 \times 10^{-2})^2 \times (1 - 5^2/10^2)^2$$

$$= 142.97 \text{ kV.}$$

(c) The cut-off magnetic flux density $= \dfrac{(8\ V_0 m/e)^{1/2}}{b(1 - a^2/b^2)}$

$$= \left(\frac{8 \times 25 \times 10^3 \times 1}{1.759 \times 10^{11}} \right)^{1/2} [10 \times 10^{-2}\ (1 - 5^2/10^2)]^{-1}$$

$$= 142.2 \text{ mWb/m}^2.$$

EXERCISES

9.1 A reflex klystron is operated at 5 GHz with an anode voltage 1000 V and the cavity gap 2 mm. Calculate the gap transit angle. Find the optimum length of the drift region. Assume $N = 1\frac{3}{4}$, $V_R = -500$ V.

9.2 A reflex klyston operates at the peak mode of $n = 2$ with $V_0 = 280$ V, $I_0 = 22$ mA, and $V_1 = 30$ V. Determine (a) the input power, (b) the output power, (c) the efficiency.

9.3 A reflex klystron operates at 8 GHz at the peak of $n = 2$ mode with $V_0 = 300$ V, $R_{sh} = 20$ k-ohm, and $L = 1$ mm. If the gap transit time and beam loading are neglected, find the (a) repeller voltage, (b) beam current necessary to obtain a RF gap voltage of 200 V, and (c) the electronic efficiency.

9.4 A reflex klystron operates at the peak of $n = 1$ mode with dc power input of 30 mW and $V_1/V_0 = 0.3$. Determine (a) the efficiency and (b) total output power.

9.5 A two cavity klystron amplifier is operated with a beam voltage of 3 kV. If the coupling coefficient is 0.9 and the magnitude of the signal voltage at the input cavity gap is 100 V, find the velocities of the electrons leaving the input gap.

9.6 A two-cavity klystron amplifier has beam current 400 mA, an output cavity coupling coefficient 0.95 and the magnitude of the voltage across the output cavity gap is 75 V. Calculate the maximum magnitude of the fundamental induced current and the maximum power output.

9.7 A two-cavity klystron is operated at 10 GHz with $V_0 = 1200$ V, $I_0 = 30$ mA, $d = 1$ mm, $L = 4$ cm and $R_{sh} = 40$ k-ohm. Neglecting beam loading calculate (a) input RF voltage V_1 for a maximum output voltage, (b) voltage gain, (c) efficiency.

9.8 A helix travelling wave tube is operated with a beam current of 300 mA, beam voltage of 5 kV, and characteristic impedance of 20 ohm. What length of the helix will be selected to give an output power gain of 50 dB at 10 GHz?

9.9 A TWT has $V_0 = 3$ kV, $I_0 = 3$ mA, $f = 10$ GHz, $Z_0 = 25$ ohm and normalised circuit length N = 50. Calculate (a) the gain parameter C, (b) dB power gain.

9.10 A cylindrical magnetron is operated at 5 GHz with $a = 3$ cm, $b = 5$ cm, $N = 16$, $V_0 = 20$ kV, $B_0 = 0.05$ T. Calculate the Hull cut-off voltage and cut-off magnetic flux density, and Hartree voltage. How do the cut-off voltage and Hartree voltage vary with B_0? Indicate the operating range.

9.11 An X-band pulsed cylindrical magnetron has $V_0 = 30$ kV, $I_0 = 80$ A, $B_0 = 0.01$ Wb/sq.m, $a = 4$ cm, and $b = 8$ cm. Calculate the cyclotron angular frequency, cut-off voltage and cut-off magnetic flux density.

9.12 For a magnetron $a = 0.6$ m, $b = 0.8$ m, N = 16, $B = 0.06$ T, $f = 3$ GHz and $V_0 = 1.6$ kV. Calculate the average drift velocity for electrons in the region between the cathode and anode.

REFERENCES

9.1 Terman, FE, *Electronic and Radio Engineering,* McGraw-Hill Book Company, Inc. 1955.

9.2 Collin, RE, *Foundations for Microwave Engineering*, McGraw-Hill Book Company, 1966.

9.3 Atwater, HA, *Introduction to Microwave Theory*, McGraw-Hill Book Company, 1962.

9.4 Reich, HJ, *Microwave Principles*, Van Nostrand Reinhold Company, 1957 and Affiliated East-West Press Pvt, Ltd., New Delhi, 1972.

9.5 Sims, GD and IM Stephenson, *Microwave Tubes and Semiconductor Devices*, Interscience Publishers New York, 1963.

9.6 Gewartowski, JB and HA Watson, *Principles of Electron Tubes*, Princeton, N.J, Van Nostrand D, 1966.

9.7 Gittins, JF, *Power Travelling-wave Tubes*, New York, American Elsevier, 1965.

9.8 Kleen, WJ, *Electronics of Microwave Tubes*, Academic Press, New York, 1958.

9.9 Pierce, JR, *Travelling wave tubes*, Princeton, N.J., Van Nostrand D, 1950.

9.10 Mendel, JT, *Helix and Coupled-cavity Travelling-wave Tubes*, *Proc. IEEE*, 61, No. 3, 280–298, March 1973.

9.11 Liao, SY, *Microwave electron tubes*, Prentice-Hall, Inc., Englewood Cliffs, N.J., 1988.

9.12 Hansen, W James, *Eliminate Confusion in TWTA Specifications*, *Microwaves and RF*, July, 1984.

9.13 *Hughes TWT and TWTA Designer's Handbook*, Torrance, Calif., Hughes electron dynamics division.

Microwave Solid State Devices and Circuits

10.1 INTRODUCTION

A wide range of microwave semiconductor devices have been developed since 1960s for detection, mixing, frequency multiplication, phase-shifting, attenuating, switching, limiting, amplification, and oscillation. In most of the low power applications, solid-state devices have replaced electron beam devices because of the advantages of their small size, light weight, high reliability, low cost and capability of being incorporated into microwave integrated circuits. Some of the widely used devices are described in this chapter.

10.2 DIODES

Microwave diodes are classified as crystal diodes and Schottky diodes for mixing and detection, PIN diode for attenuation, modulation, switching, phase shifting and limiting, Varactor diode for frequency multiplication, parametric amplification and tuning, Tunnel diode and Gunn diode for oscillation, and Read (IMPATT, TRAPATT and BARITT) diodes for amplification and oscillation. An important consideration of frequency and power limitations of these diodes is the requirement of reduced thickness of the active layer to minimise the transit time effect.

10.2.1 Crystal Diode

A typical silicon crystal diode and the equivalent electrical circuit are illustrated in Fig. 10.1. The diode essentially consists of a pointed tungsten wire (~0.08 mm dia) made in the form of a spring that pressed against the surface of a silicon (p-type) wafer (~1.6 mm square) suitably "doped" with impurities making a rectifying contact. The equivalent circuit parameters are described below.

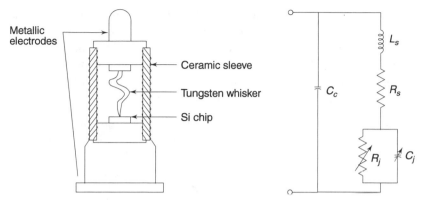

Fig. 10.1 *Crystal diode and equivalent circuit*

Here R_s and L_s are the series lead resistance and inductance, respectively, C_c is the case capacitance, and R_j and C_j the effective resistance and capacitance for the junction. Value of R_j is small for forward bias and large for reverse bias. L_s and C_c can be tuned out by matching elements.

10.2.2 Schottky Diode

Schottky diodes are metal-semiconductor barrier diodes as shown in Fig. 10.2. The diode is constructed on a thin silicon (n^+-type) substrate by growing epitaxially on n-type active layer of about 2 micron thickness. A thin SiO_2 layer is grown thermally over this active layer. Metal-semiconductor junction is formed

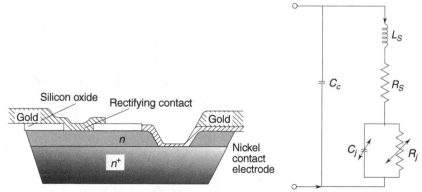

Fig. 10.2 *Schottky diode and its equivalent circuit*

R_j = resistance of metallic junction
C_j = barrier capacitance (0.3–0.5 pF)
R_s = bulk resistance of heavily doped Si substrate (4–6 ohm)
L_s = inductance of gold whisker wire (0.4–0.9 nH)
C_c = Case capacitance

by depositing metal over SiO_2. Schottky diodes also exhibit a square-law characteristic and have a higher burn out rating, lower $1/f$ noise and better reliability than point contact diodes. When the device is forward biased, the major carriers (electrons) can be easily injected from the highly doped n-semiconductor material into the metal. When it is reverse-biased, the barrier height becomes too high for the electrons to cross and no conduction takes place.

RF power flow in the device is limited by power dissipation in R_s and is shorted across C_j . C_c and L_s produce RF-mismatch and can be matched by external circuit.

10.2.3 Diode Detector Circuit

The microwave diode can be used for detection of microwave signal. For input signal power (<10 W) the forward I-V characteristic is approximately parabolic and follows the square law : $I \alpha V^2$ as shown in Fig. 10.3(a). If the microwave signal voltage $v = V \cos \omega t$ is applied across the diode, the diode current is given by

$$i = I_0 (e^{aV} - 1) \tag{10.1}$$

where I_0 is the diode reverse saturation current, $a = 1/(nV_T)$, n is a constant, typically 1.1 for Schottky diodes and 1.4 for point contact diodes, V_T the thermal voltage whose value is 26 mV at room temperature. For a small signal amplitude, series expansion of this current gives

$$i = I_0[aV \cos \omega t + (a^2 V^2 /4) (1+ \cos 2 \omega t)] \tag{10.2}$$

Thus the diode direct current $I = I_0 a^2 V^2 /4$ is proportional to the microwave input power and hence the name—square law detector. The ac components are filtered out by the detector circuit. For a large power input (> 10 W) the V-I characteristic becomes more linear, so that input power needs to be attenuated for operation in the square-law region. Reverse biased output (current or power) of the detector is nearly zero.

The series resistance R_s and junction capacitance C_j limit the sensitivity of the detector circuit by reducing the signal power output. After tuning out L_s and C_c, if the rms signal voltage across the diode is V and signal current I, total power in the circuit is

$$P_t = I^2 R_s + V^2/R_j \tag{10.3}$$

Power absorbed in R_j is

$$P_a = V^2/R_j \tag{10.4}$$

Hence the power loss is the ratio of useful power input to the total power input

$$P/P_t = \frac{1}{1 + \left(I^2 / V^2\right) R_s R_s} = \frac{1}{1 + \left|Y_j\right|^2 R_s R_j}$$

$$= \frac{1}{1 + R_s / R_j + \omega^2 C_j^2 R_s R_j} \tag{10.5}$$

The microwave detector diodes are sensitive and operate with RF signal without any dc bias. The diode is mounted in a waveguide or a coaxial line (Fig. 10.3), which contains matching elements, so that VSWR < 1.3 and the microwave power

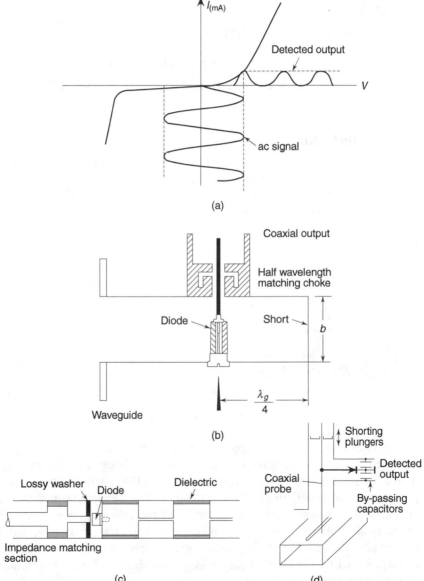

Fig. 10.3 *Microwave detector mount (a) Diode I-V and detection characteristics (b) Waveguide detector(c) Coaxial detector (d) Tunable probe detector*

is absorbed without appreciable reflection. An RF bypass capacitance in the output circuit is constructed so that the microwave signal is contained in the detector without coupling to the measuring instrument such as VSWR meter which is a high gain, low noise amplifier tuned to modulating frequency of 1 kHz. The output impedance of the detector circuit is designed either high (50–200 ohms) or low (2.5–10 ohm). The detector circuit is matched using short circuit stub.

Example 10.1   A low level point contact detector diode has $R_j = 2$ ohm, $R_s = 5$ ohm and $C_j = 0.5$ pf. Calculate the power loss in dB for operation at 5 GHz.

Solution

Power loss in dB = $10 \log (P/P_i)$

$$= 10 \log (1 + R_s/R_j + \omega^2 C_j^2 R_s R_j)^{-1}$$
$$= -10 \log(1 + 5/2 + (2\pi \times 5 \times 10^9 \times 0.5 \times 10^{-12})^2 \times 2 \times 5)$$
$$= -5.4 \text{ dB}$$

10.2.4   Diode Mixer

At frequencies above about 1 GHz a silicon crystal diode can be used satisfactorily as mixer to produce lower frequency IF signal. It has low conversion loss, low noise, and ability to stand momentary overloads. Figure 10.4 shows a schematic circuit of a coaxial line broadband mixer used up to frequencies of about 3 GHz. Single diode mixers are useful where signal levels are relatively large compared to noise.

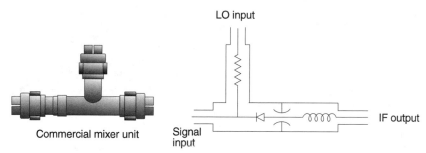

Fig. 10.4   *Coaxial line single diode mixer*

For most communication applications where signals are not always large, balanced mixers are employed using two or four diodes in a hybrid ring or a magic-T configuration, respectively, as shown in Fig. 10.5, where the RF signal port and the local oscillator port are decoupled. The RF signal splits at the input, appears at the remaining ports out of phase and recombines at these ports with in-phase local oscillator signals in the mixer diodes to produce an IF signal. Since the circuit is designed for 180° phased signals to produce IF, the oscillator noise close to signal frequency will cancel and the desired IF is coupled through the centre the tap coil.

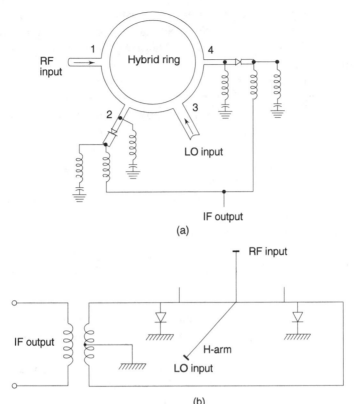

Fig. 10.5 *Balanced mixers (a) Microstrip hybrid ring (b) Wave guide magic-T*

10.2.5 Harmonic Mixer

Harmonic mixers are used in connection with frequency calibration set-ups where a standard low frequency signal causes harmonic generation in a non-linear device such as a crystal diode and then heterodyned with another microwave test signal to produce an IF. The harmonic power generated by a crystal decreases rapidly with the order of harmonic. Experience has shown that at about 10th and higher orders, the harmonic amplitude is inversely proportional to the harmonic number. The schematic circuit of a harmonic crystal mixer system is shown in Fig. 10.6. Crystals such as 1 N23 B, 1 N21 B are employed and mounted in waveguides or coaxial line. A standard signal level of 1 V is sufficient in order to compromise between a large value that gives low conversion loss and a small value that gives minimum noise.

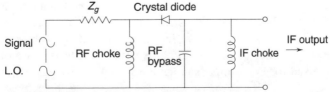

Fig. 10.6 *Schematic circuit of a harmonic crystal mixer*

The RF choke provides dc return for the crystal. The RF bypass prevents RF leakage through the IF terminals. The IF choke provides a dc return with negligible effect on the IF power. The local oscillator voltage $V_p \cos \omega_p t$ produces currents in the crystal that contain the fundamental frequency plus harmonics $n\omega_p$, $n = 1,2,3,...$ An incoming signal $V_s \cos \omega_s t$ will beat with these frequencies and produce sum and difference frequency components $n\omega_p \pm \omega_s$. The difference frequencies can be tuned out through an IF amplifier. Thus harmonic operation can be used with low frequency local oscillator to generate IF.

A reasonably good impedance match should exist between the crystal unit and the source of signal power, and load impedance Z_{IF} for optimum IF signal level. Since the input and output impedances will depend upon the amplitude of the local oscillations, proper amplitude of local oscillator is to be adjusted. The degree to which input RF power is converted to IF power (at the difference frequency) is expressed in a quantity known as *conversion loss*:

$$\text{Conversion loss (dB)} = 10 \log_{10} \frac{\text{IF output power available}}{\text{RF input power available}} \tag{10.6}$$

The available IF power is the IF power delivered to a load whose impedance is the conjugate of the IF output impedance Z_{IF} of the mixer.

The conversion loss in crystal mixers ordinarily decreases with increasing amplitude of the local oscillator signal. Part of the signal power which does not reappear as IF output power is dissipated within the crystal. Typical conversion loss values range from 4.5 to 7 dB for the fundamental frequency. Conversion loss is higher for mixing with higher harmonics.

The output noise of mixer crystals has a distinct effect on the receiver operation. This is measured in terms of the output noise ratio:

$$\text{Output noise ratio, } N_r = \frac{\text{available IF output noise power}}{\begin{array}{c}\text{available thermal or Johnson noise}\\\text{power of an equivalent resistance}\end{array}}$$

$$= \frac{N_0}{k\,T\,B_{IF}} \tag{10.7}$$

The corresponding noise temperature

$$T_n = N_r T = \frac{N_0}{k\,B_{IF}} \tag{10.8}$$

A design example of a harmonic mixture An X-band waveguide harmonic mixer is shown in Fig. 10.7. The RF and local oscillator signals are heterodyned in the crystal and the difference frequencies are obtained at a type BNC output jack. A reflecting choke (high series reactance) is used at the type N input to prevent loss of signal and the particular harmonic to be used in mixing. Waveguide is below cut-off for the fundamental and the lower harmonics. A tunable short is used for input impedance matching. A RF by-pass gap capacitor is used at the BNC port to prevent RF leakage through the IF terminal. The reflecting choke can be considered as a short circuited radial guide of axial width δ < $\lambda/2$. Under this condition only TM_{on} modes will be supported by this guide [6]. A high series reactance is obtained by keeping the depth A slightly less than the quarter wavelength at the signal or the harmonic frequency to be used.

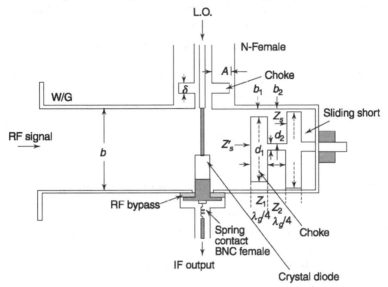

Fig. 10.7 *Waveguide harmonic crystal mixer*

Thus the designed values are f_0 = 10 GHz, λ_0 = 3 cm, δ = 5 mm, and A = 7.2 mm. For 50 Ω coaxial line with air dielectric (ε_r = 1), the characteristic impedance

$$Z_0 = 50 = \frac{138}{\sqrt{\varepsilon_r}} \log_{10} \text{(outer radius/inner radius)} \qquad (10.9)$$

or
outer radius/inner radius = 2.3.

The gap bypass capacitance C_g is selected to pass the desired IF and bypass the RF, local oscillator signal and its harmonics and also the higher frequency IF's. Therefore,

$$\frac{1}{\omega C_g} << 50 \ \Omega$$

is achieved by a half millimeter gap with teflon filling. A good contact between the BNC connector and diode is obtained by means of a spring contact.

A choke type short circuit plunger is used for impedance tuning. The impedance seen at the input of the plunger is

$$Z'_s = (Z_1/Z_2)^2 \ Z_s \qquad (10.10)$$

where Z_s is the impedance of the sliding short and Z_1 and Z_2 are the characteristic impedances of the coaxial lines formed between the waveguide walls and the plunger. If $Z_2 \gg Z_1$, Z_s' will approximate to a short circuit by a factor $(Z_1/Z_2)^2$ better than Z_s does. The width of the plunger is uniform and slightly less than the interior guide width. However, the height of the plunger is made non-uniform. The gap $b_1 \ll b_2$ is to make $Z_1 \ll Z_2$. The back section makes a sliding fit in the waveguide. The selection of the dimensions are made as

$$b_2 = 8b_1 \text{ with } b_1 = 0.5 \text{ mm, and } b_2 = 4 \text{ mm.}$$

For an X-band waveguide of $0.9'' \times 0.4''$ size,

$$b = 10.16 \text{ mm}$$
$$d_1 = b - 2b_1 = 9.16 \text{ mm}$$
$$d_2 = b - 2b_2 = 2.16 \text{ mm}$$
$$g/4 = 9.938 \text{ mm at 10 GHz}$$

A bandwidth of 10 to 20 % can be achieved with this design.

10.2.6 PIN Diode and Its Applications

A PIN diode consists of a high-resistivity intrinsic semiconductor layer between two highly doped p^+ and n^+ Si layers as shown in Fig. 10.8 along with its equivalent circuit. The device acts as electrically variable resistor related to the i layer thickness.

The intrinsic layer has a very large resistance in reverse bias and it decreases in forward bias. When mobile carriers from p and n regions are injected into the i layer, carriers take time such that the diode ceases to act as a rectifier at microwave frequency and appears as a linear resistance. This property makes it usable as a variable attenuator at microwave frequencies.

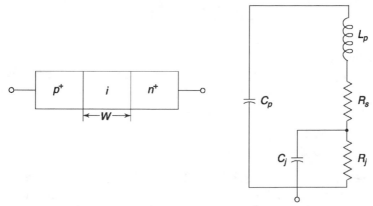

Fig. 10.8 *PIN diode and equivalent circuit*
$R_j, C_j =$ Junction resistance, capacitance of i layer
$R_s =$ Bulk semiconductor (p^+ and n^+)
layer and contact resistance
$L_p, C_p =$ package inductance, capacitance

R_j is a variable element, and C_j is constant (≈ 0.2 pf) at frequencies > 1 GHz. The resistivity of the i layer is of the order of 1000 ohm-cm, and $R_s \approx 1$ ohm. L_p and C_p may be omitted when the diode is directly mounted in chip form into microwave integrated circuit. Because of the large breakdown voltage (~ -500 V) compared to an ordinary pn junction diode, the PIN diode can be biased at a high negative region so that large ac signal superimposed on dc cannot make the device forward biased. For the device to remain in the high impedance state, in the presence of large microwave signal, the dc reverse voltage must be high. At high power applications the i region width W should be wide but less than diffusion length L_d, otherwise the centre layer voltage drop will increase and the forward bias impedance will become high. To keep reverse bias capacitance low, the cross-section area of the device must be so small such that forward bias impedance does not become high.

PIN switch

The PIN device is designed such that under reverse or zero biasing R_j is extremely large and C_j (0.02–2 pf) plays the dominant role, whereas C_p, L_p, R_s are negligibly small. For forward biasing R_j is very small ($<< 1/j\omega C_j$) so that R_f (0.1–2 ohm) $= R_j + R_s$ plays the dominant role. Therefore, for reverse bias, a high capacitive impedance is presented to the microwave signal and for forward bias, the diode represents a very low resistance. Thus by changing the bias it acts as a switch.

(a) Single switch

Figure 10.9 shows schematic circuits of a single PIN switch in shunt and series mounting configurations. AC blocking inductor is realised from a high impedance strip line section and dc blocking capacitor is realised from a gap in the line. For shunt configuration reverse biasing produces transmission ON due to high impedance shunt and forward biasing produces transmission OFF due to low impedance shunt. For series configuration, transmission is ON for forward bias and OFF for reverse bias. In practice for shunt mounting, due to non-zero forward bias resistance, isolation between input and output is not infinite. Similarly, for reverse bias shunt capacitor is not infinite and a non-zero insertion loss results. Reverse is the case for series mounting.

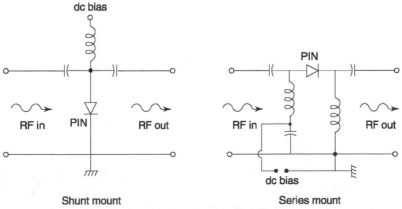

Fig. 10.9 *Single PIN switch*

Because of the large breakdown voltage (~ -500 V) compared to an ordinary diode, PIN diode can be biased at high -ve region so that large ac signal, superimposed on dc, cannot make the device forward biased. Referring to Fig. 10.8, where a microwave signal superimposed on the dc negative bias voltage, it is apparent that for the device to remain in the "high impedance" state, in the presence of large microwave voltages, the dc reverse bias must be high. A high power PIN microwave switch is therefore characterised by a wide i-region of width W. This width W must be much larger than the diffusion length L_d, otherwise, the centre layer voltage drop will increase and the forward bias impedance will become high. Also to keep reverse bias capacitance low, the cross-section area of the device must be small. But it must not be so small that the forward bias impedance increases.

Example 10.2   A shunt mounted PIN diode in a TEM transmission line having characteristic impedance $Z_0 = 50$ ohm can be represented by a shunt impedance $Z = R + jX$. Calculate the insertion loss and isolation at a frequency $f_0 = 2$ GHz. The forward resistance $R_f = 0.1$ Ω and capacitance $C_j = 0.02$ pf.

Solution

Let the line be terminated by characteristic impedance Z_0. Therefore, input impedance measured across Z at AA' towards the load is $Z_{in} = Z_0$.

Therefore, effective load at $AA' = Z_e = ZZ_0/(Z + Z_0)$

Reflection coefficient at AA' is

$$\Gamma = (Z_e - Z_0)/(Z_e + Z_0)$$

Therefore, the transmission coefficient

$$T = 1 + \Gamma = 1 + \frac{Z_e - Z_0}{Z_e + Z_0}$$

$$= \frac{2Z_e}{Z_e + Z_0}$$

$$= 2Z/(2Z + Z_0)$$

Now insertion loss,

$$\alpha = \text{Incident power/Power delivered to load}$$

$$= \frac{|V_i|^2}{|V_t|^2}$$

$$= \frac{|V_i|^2}{|TV_i|^2}$$

$$= \frac{1}{|T|^2}$$

$$= \left| \frac{2Z + Z_0}{2Z} \right|^2$$

$$= \left(1 + \frac{Z_0}{2Z}\right)^2$$

assuming that there is no loss in the line and in the diode,

$$\alpha\,(\text{dB}) = 20\,\log_{10}\left(1 + \frac{Z_0}{2Z}\right)\,\text{dB}$$

Now

$$Z \approx R_f = 0.1\ \Omega \text{ for forward bias}$$
$$\approx 1/\omega c$$
$$= 1/(2\pi f \times 0.02 \times 10^{-12})\ \Omega$$
$$= 1/[(2\pi \times 2 \times 10^9 \times 0.02 \times 10^{-12})]\text{ reverse bias}$$
$$= 3.98\text{ k}\Omega$$

Therefore,

$$\text{Isolation (forward bias)} = 20\,\log\left(1 + \frac{50}{0.2}\right) = 48\text{ dB}$$

$$\text{Insertion (Reverse bias)} = 20\,\log\left(1 + \frac{50}{3978.9}\right) = 0.108\text{ dB}$$

(b) Double switch

The PIN diode double switch circuit using two diodes called as single-pole dou-ble-throw (SPDT) is shown in Fig. 10.10. Diodes are biased through RF chokes to isolate ac component from dc source. The impedance matching between RF feeder line and the switch is obtained from the quarterwave lines which provide zero impedance when the switch is OFF (open circuit) and infinite impedance when the switch is ON (short circuit). Thus,

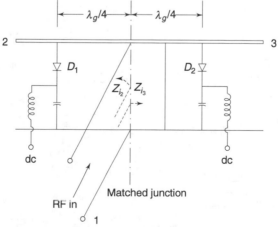

Fig. 10.10 *SPDT double switch*

(i) When D_1 is forward biased, Z_{i2} = infinite

(ii) When D_2 is reverse biased, $Z_{i3} = 0$,

Therefore, input RF power is directed to port (3), while port (2) remains isolated. Input power can be switched to port (2), keeping port (3) isolated by changing bias of D_1 and D_2, i.e., D_1 is reverse biased and D_2 forward biased.

Since in practice under forward bias condition a diode has non zero finite resistance ($R_f \neq 0$), some power is transmitted to port (2) in the first case giving rise to finite isolation. Similarly under reverse bias condition the diode reactance X becomes finite and hence some non-zero finite insertion loss to port (3) appears during transmission.

SPDT switch using 4 pin diodes   Insertion loss and isolation in double switch can be improved by putting two other diodes on both sides at a distance $\lambda/4$ from D_1 and D_2. The basic configuration of a SPDT switch using four PIN diodes is shown in Fig. 10.11, along with an MIC model. To have power in port (3), the diode states will be as follows: D_1 forward biased, D_2 reverse biased, D_3 reverse biased and D_4 forward biased. The quarter wave transformer transforms the short

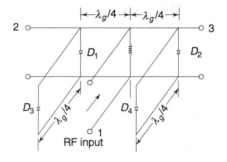

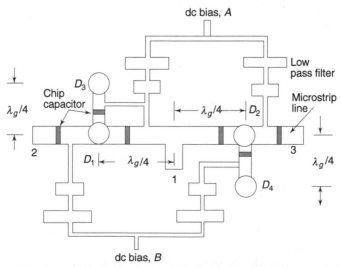

Fig. 10.11 *Four PIN diode SPDT switch and MIC model (D_1, D_2, D_3, D_4 are shunt mounted PIN diodes with grounded)*

circuit impedance of any forward biased diode to an open circuit at the RF port and eliminates any reactive loading on the RF port. Under this condition port (3) will be connected to port (1) with port (2) isolated. For power in port (2), the logic of the diode states will be just opposite as compared to above. In MIC model as shown in the figure, the diodes are shunt mounted with the cathodes grounded.

PIN duplexer   The basic PIN duplexer circuit is shown in Fig. 10.12, which uses two 3 dB, 90° hybrids and two PIN-diodes D_1 and D_2. For simplicity biasing circuits are not shown.

(a) Transmit mode   For transmission of RF power from T_x, both the diodes are forward biased to act as short circuit at AA' as shown in Fig. 10.13. Almost all RF power is transmitted through the antenna. Total reflected power at T_x is zero. Because of non-zero finite forward resistance $(R_f \neq 0)$ of the diode, some RF power is transmitted to matched load and got absorbed. Power in the R_x port is zero.

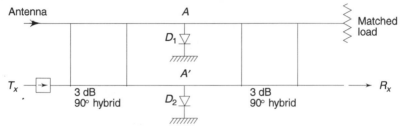

Fig. 10.12   *PIN duplexer*

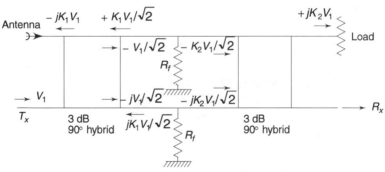

Fig. 10.13   *Transmit mode equivalent circuit of a PIN duplexer*

(b) Receive mode   For reception mode RF power enters R_x from the antenna and isolator isolates any power entering T_x as explained in Fig. 10.14.

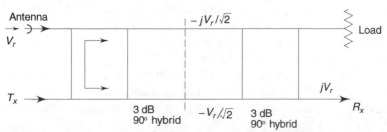

Fig. 10.14   *Receive mode equivalent circuit of PIN duplexer*

(c) PIN phase shifter   Electronic phase shifters designed using PIN diodes are extensively used in phased array in quantities of thousands. Three types of variable phase shifters using PIN diodes are described here

   (a)   Transmission type switched line ϕ-shifter
   (b)   Reflection type hybrid coupled ϕ-shifter
   (c)   Transmission type loaded line ϕ-shifter

(a) Transmission type switched line ϕ-shifter   Phase shift is obtained by perturbing the parameters of the transmission line. Here D_1, D_2, D_3 and D_4 are identical PIN diodes. Operation is based on switching the diodes from one type (FB) biasing to another (RB) so that a differential phase change occurs at the output. This is explained in Fig. 10.15. Four different states are described below.

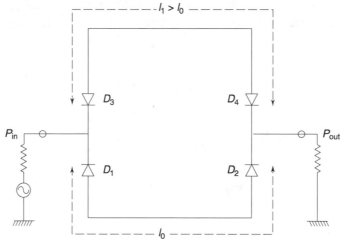

Fig. 10.15   *Transmission type switched line ϕ-shifter*

State 1

D_1 and D_2 are forward biased and offer short circuit.
D_3 and D_4 are reverse biased and offer open circuit.

    Therefore, the signal passes through l_0 line. If the input voltage is V_0, output voltage becomes $V_0 e^{j\phi_0}$, where $\phi_0 = \beta l_0$, the phase shift due to propagation through the line l_0.

State 2

D_1 and D_2 are reverse biased and offer open circuit.
D_3 and D_4 are forward biased and offer short circuit.

Therefore, the signal passes through l_1 line. If input voltage is V_0, output voltage becomes $V_0 e^{j\phi_1}$, where $\phi_1 = \beta l_1$ is the amount of phase shift. Therefore, differential ϕ-shift $\Delta \phi = \phi_1 - \phi_0 = \beta (l_1 - l_0)$ becomes a fixed ϕ-shift. Such ϕ-shifter is called the 1-bit ϕ-shifter. For 4-bit ϕ-shifter, a cascade of 22.5°, 45°, 90° and 180° ϕ-shifters gives all 360° ϕ-shift in step of 22.5°.

    If all upper line diodes are OFF, signal will pass through all lower lines with net ϕ-shift of 4 ϕ_0 which can be the reference. If lower diodes are not OFF, signal will pass through all upper diodes with net ϕ-shift $= 4\phi_1$. Therefore, the total phase shift is

$$\Delta \phi = 4 \, (\phi_1 - \phi_0)$$

By tapping from different sections we can get phase shifts of $\Delta \phi = \phi_1 - \phi_0$, $2\phi_1 - \phi_0$, etc. The disadvantage of this model is that, leakage through the reverse bias diode and non-zero forward resistance decreases isolation between the ports and increases the insertion loss. Moreover, the number of diodes per bit phase shift are large.

(b) Reflection type hybrid ϕ-shifter   As shown in Fig. 10.16, here ϕ-shift is obtained by changing the value or location of the terminating impedance. A hybrid coupler (or circulator) is used to separate input and output signals.

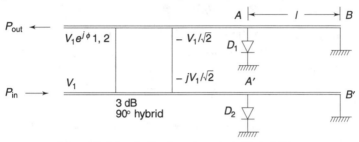

Fig. 10.16   *Reflection type hybrid ϕ-shifter*

Two states are described below for phase shift.

State 1

D_1 and D_2 are forward biased and offer short circuit at AA' (at shorter length) to produce a phase shift $\phi_1 = \phi_0 = \beta l_0$.

State 2

D_1 and D_2 are reverse biased and offer open circuit at AA'. The physical short at BB' produces an effective short at a longer length to produce a phase shift $\phi_2 = \phi_0 + 2\beta l$.

Therefore, phase change, $\Delta \phi = 2\beta l$. By changing the line length l or by changing the terminating impedance, the differential ϕ-shift can be obtained.

(c) Transmission type loaded line ϕ-shifter   This is a simplest model of a PIN phase shifter as shown in Fig. 10.17. Phase shift is obtained by loading the T_{x}-line between the input and output.

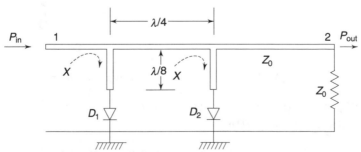

Fig. 10.17   *Transmission type loaded line ϕ-shifter*

Two states contributing the phase are as follows.

State 1

When D_1 and D_2 are forward biased, effective shorts are produced at the corresponding diode locations and the input reactance becomes inductive

$$X = j\omega L = jZ_0 \tan (\beta\lambda/8)$$

This produces a phase ϕ_0. Therefore, for input voltage V_1, output becomes $V_1 e^{j\phi_0}$.

State 2

When D_1 and D_2 are reverse biased, effective open circuits are produced at the corresponding diode locations and the input reactance becomes capacitive

$$X = 1/j\omega C$$

$$= -jZ_0 \cot (\beta\lambda/8)$$

This produces a phase ϕ_1. Therefore, for an input voltage V_1, output becomes $V_1 e^{j\phi_1}$. Therefore, phase change $\Delta\phi = \phi_1 - \phi_0$. This phase can be maximised by equating $1/\omega C = \omega L$. Here $\lambda/4$ line between the diodes provides $\lambda/2$ path for incidence and reflected signals to get cancelled at input side.

PIN attenuator

Since the resistance of diode decreases with increase of forward bias, the diode acts as a variable attenuator when put in series or in shunt on a line. At frequencies < 10 MHz the diode shows rectifying property, whereas at frequencies >> 10 MHz it behaves as variable resistor or attenuator. Figure 10.18 shows schematic circuits of series and shunt attenuator configurations . In series circuit, the attenuation decreases with increase of biasing current with consequent decrease of RF resistance. In shunt circuit attenuation increases with biasing current because most of the RF energy is absorbed in the diode.

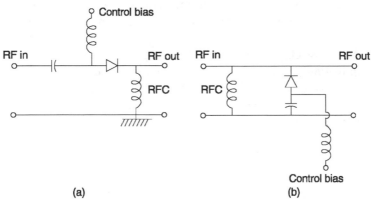

Fig. 10.18 *PIN attenuator(a) Series (b) Shunt*

10.3 TRANSFERRED ELECTRON DEVICES (TED)—GUNN DIODES

Gunn diodes are negative resistance devices which are normally used as low power oscillator at microwave frequencies in transmitter and also as local oscillator in receiver front ends. **J B Gunn** (1963) discovered microwave oscillation in Gallium arsenide (GaAs), Indium phosphide (InP) and cadmium telluride

(CdTe). These are semiconductors having a closely spaced energy valley in the conduction band as shown in Fig. 10.19 for GaAs. When a dc voltage is applied across the material, an electric field is established across it. At low **E**-field in the material, most of the electrons will be located in the lower energy central valley Γ. At higher **E**-field, most of the electrons will be transferred into the high-energy satellite L and X valleys where the effective electron mass is larger and hence electron mobility is lower than that in the low energy Γ valley. Since the conductivity is directly proportional to the mobility, the conductivity and hence the current decreases with an increase in **E**-field or voltage in an intermediate range, beyond a threshold value V_{th} as shown in Fig. 10.20. This is called the *transferred electron* effect and the device is also called 'Transfer Electron Device (TED) or Gunn diode.' Thus the material behaves as a negative resistance device over a range of applied voltages and can be used in microwave oscillators.

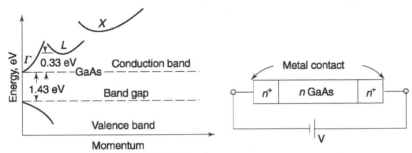

Fig. 10.19   *Multi-valley conduction band energies of GaAs*

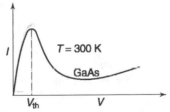

Fig. 10.20   *Current-voltage characteristics of GaAs*

The basic structure of a Gunn diode is shown in Fig. 10.21 (a), which consists of n-type GaAs semiconductor with regions of high doping (n^+). Although there is no junction this is called a diode with reference to the positive end (anode) and negative end (cathode) of the dc voltage applied across the device. If voltage or an electric field at low level is applied to the GaAs, initially the current will increase with a rise in the voltage. When the diode voltage exceeds a certain threshold value, V_{th}, a high electric field (3.2 KV/m for GaAs) is produced across the active region and electrons are excited from their initial lower valley to the higher valley, where they become virtually immobile. If the rate at which electrons are transferred is very high, the current will decrease with increase in voltage, resulting in equivalent negative resistance effect. Since GaAs is a poor conductor, considerable heat is generated in the diode. The diode should be well bonded into a heat sink (Cu-stud).

The electrical equivalent circuit of a Gunn diode is shown in Fig. 10.21(b), where C_j and $-R_j$ are the diode capacitance and resistance, respectively, R_s includes the total resistance of lead, ohmic contacts, and bulk resistance of the diode, C_p and L_p are the package capacitance and inductance, respectively. The negative resistance has a value that typically lies in the range −5 to −20 ohm.

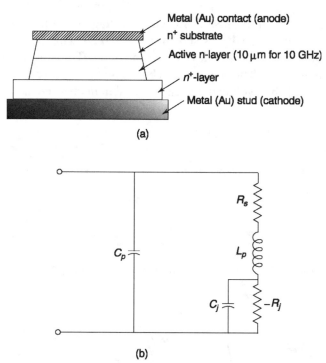

(a)

(b)

Fig. 10.21 *Constructional details and the electrical equivalent circuit of a Gunn diode*

10.3.1 Modes of Operation

There are two principal modes of operation that result in microwave oscillations in a Gunn diode. These are the Gunn mode or the transit-time (TT) mode and the limited-space-charge (LSA) mode. Under special circumstances two other modes—quenched domain mode and delayed mode are possible as described below.

Gunn or TT mode When the voltage applied across n^+ n n^+ GaAs crystal exceeds a threshold level, electrons are transferred from the low energy, high mobility conduction band to a higher energy, lower or nearly zero mobility sub conduction band, where these heavier electrons bunch together to form an electric field dipole domain near the cathode. Since the applied voltage remains constant, the electric field across the domain is greater than the average field. The consequent electric field remains below the threshold level across rest of the crystal. This prevents the formation of further domains. All the conduction band electrons drift across the crystal at the same velocity and the less mobile bunched electrons have reduced velocity. The current in the presence of domain also decreases. After the high field domain had travelled into the end contact, the current returns to its higher level and a high field domain is again formed. The movement of high field domain is explained in Fig. 10.22. Each domain results in a pulse of current at the output. These current fluctuations occur at microwave frequencies to produce output signal at the low impedance RF circuit with a period equal to the transit time. The high field domain is quenched before it reaches the anode. Therefore, the transit time is shortened and the frequency is increased. This mode of oscillation has a low efficiency of power generation and the frequency cannot be controlled by the external circuit.

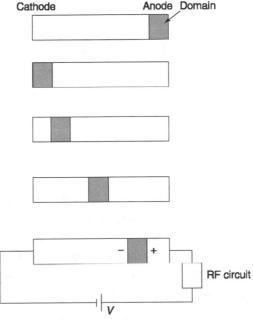

Fig. 10.22 *High field domain movement*

LSA mode LSA mode of operation can produce several watts of power with minimum efficiencies of 20%. The power output decreases with frequency, viz. 1W at 10 GHz and several mW at 100 GHz.

For LSA mode of operation, the Gunn diode works as a part of a resonant circuit as shown in Fig. 10.23. The resonant circuit is tuned to a frequency several times greater than that of the TT mode so that dipole domains do not have sufficient time to form and the circuit operates as a negative resistance oscillator when the dc voltage is adjusted to a value greater than the threshold voltage and nearly at the mid-point of the negative resistance region. The resistance load R_L is adjusted to a value of about 20% greater than the maximum negative resistance value of the device to enable oscillations to start and stay steady. The amplitude

of the oscillations builds up and becomes steady when the average negative resistance of the Gunn diode becomes equal to the load resistance R_L. The peak-to-
peak amplitude of the microwave oscillations is approximately equal to the voltage range in the negative resistance region.

Quenched domain mode   If the resonant circuit is tuned to a value slightly above
that of the TT mode, the dipole domain will be quenched before it arrives at the
anode by the negative swing of the oscillation voltage but the Gunn diode will
operate mostly like Gunn mode. This mode of operation is called a *quenched
domain mode*.

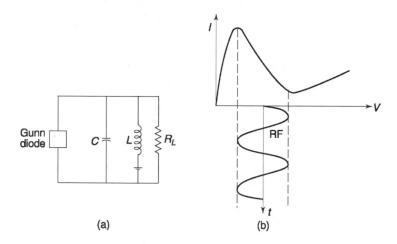

(a)   (b)

Fig. 10.23   *Gunn oscillator operating in LSA mode and RF oscillating voltage*

Delayed mode   If the resonator is tuned below that of the Gunn mode, the dipole
domains will arrive at the anode well in time but the formation of a new dipole
domain will be delayed until the oscillation voltage increases above the threshold
value. This type of operating mode is called the *delayed mode*.

10.3.2   Gunn Diode Oscillator

A Gunn diode oscillator can be designed by mounting the diode inside a
waveguide cavity formed by a short circuit termination at one end and by an iris
at the other end as shown in Fig. 10.24. The diode is mounted at the centre perpendicular to the broadwall where the electric field component is maximum under the
dominant TE_{10} mode. The intrinsic frequency f_0 of oscillation depends on the
electron drift velocity V_d due to high field domain through the effective length l.

$$f_0 = V_d/l \tag{10.11}$$

For GaAs, $V_d \approx 10^7$ cm/s. Normally, the cavity is tuned to resonate at the
intrinsic frequency f_0 by adjusting the short position. A tuning screw is inserted
perpendicularly at the centre of the broadwall for fine frequency tuning.

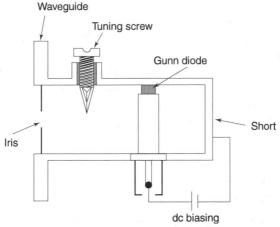

Fig. 10.24 *Gunn diode oscillator circuit*

The total resistive loading from the cavity and the external load should be around 20 per cent higher than the Gunn device resistance $-R_j$ so that the circuit resistance $-R_L R_j / (R_L - R_j)$ will be negative. The cavity has an impedance transforming property between the high impedance of the output waveguide and the required low value for the Gunn diode. The Gunn diode is placed on a metal post. The top of the post is insulated from the waveguide to form RF bypass capacitor and dc bias voltage is applied to the post. The degree of coupling to the external waveguide is adjusted by selecting the inductive iris dimension.

The power output of the Gunn diode oscillator is in the range of a few watts for CW operation at biasing values 10 V and 1A at $30 - 40$ GHz. A frequency tuning range of nearly 2% can be achieved. For pulsed operation peak powers are typically 100–200 W.

The power output of the Gunn diode is limited by the difficulty of heat dissipation from the small chip. The advantages are small size, ruggedness, and low cost.

Example 10.3 The drift velocity of electrons is 2×10^7 cm/s, through the active region of length 10×10^{-4} cm. Calculate the natural frequency of the diode and the critical voltage.
Solution

The natural frequency $f_0 = V_d/l = \dfrac{2 \times 10^7}{10 \times 10^{-4}} = 20$ GHz

The critical voltage $V = l \times$ critical field for GaAs.

$$= 10 \times 10^{-4} \times 3.2 \text{ kV/cm.}$$
$$= 10 \times 10^{-4} \times 3.2 \times 10^3 \text{ V}$$
$$= 3.2 \text{ volts}$$

10.4 AVALANCHE TRANSIT-TIME DEVICES (ATTD)

Avalanche transit-time devices (W.T. Read, 1958) are p - n junction diode with the highly doped p and n regions. They could produce a negative resistance at microwave frequencies by using a carrier impact ionization avalanche breakdown and carriers drift in the high field intensity region under reverse biased condition. There are three types of this device—(1) Impact Ionization Avalanche Transit Time (IMPATT) effect, (2) Trapped Plasma Avalanche Triggered Transit (TRAPATT) effect, and (3) Barrier Injected Transit Time (BARITT) effect. The IMPATT diodes have an efficiency of the order of 3% CW power and 60% pulsed power and can be operated from 500 MHz to 100 GHz. The power outputs lie between 1W (CW) and over 400 W (pulsed). TRAPATT is suitable for low frequency (1–3 GHz) applications with pulsed power output of several hundred watts and efficiency 20–60%. BARITT has the advantage of low-noise figures (< 15 dB) but with low power and smaller bandwidth.

10.4.1 IMPATT Diodes

IMPATT diodes have many forms, viz., $n^+ pip^+$ or $p^+ nin^+$ read device, $p^+ nn^+$ abrupt junction, and $p^+ in^+$ diode, which are shown in Fig. 10.25 together with their doping profiles. Such diodes can be manufactured from Ge, Si, GaAs, or InP. However, GaAs provides the highest efficiency, the highest operating frequency, and least noise figure. But the fabrication process is more difficult and is more expensive than Si. A typical construction and package are shown in Fig. 10.26. An n-type expitaxial layer is formed over the $n+$ substrate. On top of this is the diffused $p+$ layer. A metallised cathode and plated heat sink as anode are also included.

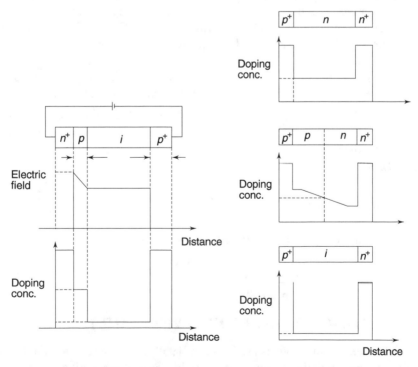

Fig. 10.25 *Types of IMPATT diodes and doping profile*

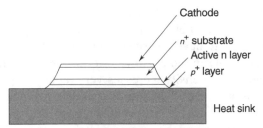

Fig. 10.26 *Construction and package of p^+ n n^+ IMPATT diode*

DC operating principles Under reverse biased condition, the characteristics of different regions of the diode, electric field distribution, the microwave voltage and the external carrier flow are shown in Fig. 10.27. The illustration is given with reference to $n^+ pip^+$ diode. When the reverse bias voltage exceeds the breakdown voltage V_B, a maximum electric field of very high value (MV/m) appears at the n^+p junction. The holes moving in this high field region acquire sufficient

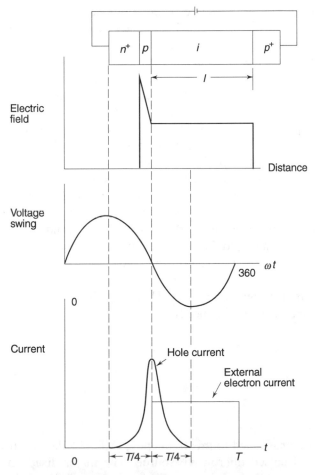

Fig. 10.27 *IMPATT diode operation*

energy to excite valence electrons of the atom into the conduction band resulting in avalanche multiplication of electron-hole pairs. By designing doping profile, the **E**-field can be made to have a very sharp peak, so that "impact avalanche" multiplication occurs only in the close vicinity of the junction. The process is cumulative and the carrier density increases very rapidly. To prevent the diode from burning, a constant current bias source is used to maintain the average current at safe limit I_0. The diode current is contributed by the conduction electrons, which move to the n^+ region and the associated holes, which drift through i space charge region to the p^+ region, under the influence of a lower but steady electric field. The drift time is given by

$$t_d = \frac{l}{V_d} \tag{10.12}$$

where V_d is the drift velocity of the holes ($\sim 10^5$ m/s for **E** ~ 0.5 MV/m), and l the length of the drift region. If $l \approx 2\ \mu m$, $t_d \approx 20$ ps.

Mechanism of oscillations When the diode is mounted in a resonator circuit, any noise voltage spike excites a resonant component of voltage across the circuit. Under the influence of reverse bias steady field and the ac field, the diode swings into and out of the avalanche condition. Since the hole drift time is very short, carriers drift to the end contacts before ac voltage swings the diode out of the avalanche. The ac field, therefore, takes energy from the carriers or, in effect, from the dc bias source. Thus the microwave oscillator voltage builds up across the diode.

It is also apparent that due to the ac field, the charge or hole current grows exponentially to a maximum and again decays exponentially to zero. The system can be designed such that the peak of the hole current lags the peak of the ac voltage by a quarter period or 90°. During hole drifting process, a constant electron current is induced into the external circuit equal to the average current in the space-charge region. Thus dc power is drawn from the reverse bias supply. This electron current starts flowing when hole current reaches its peak and continues for a half cycle (T/2) corresponding to the negative swing of the ac voltage. Thus a 180° phase shift between the external current and the ac microwave voltage provides a negative resistance for sustained oscillations. The maximum negative resistance occurs at drift transit angle $\theta = \omega t_d = \pi$. Therefore, the fundamental frequency of microwave oscillation f, is obtained from

$$\omega t_d = \frac{\omega l}{V_d} ; \left(f = \frac{1}{2t_d} \right)$$

or,

$$f = \frac{V_d}{2l} \tag{10.13}$$

If the resonator is tuned to this frequency, IMPATT diodes provide a high-power CW and pulsed microwave source. The major disadvantages of the IMPATT diodes are (1) since dc power is drawn due to induced electron current in the external circuit, IMPATT diodes have low efficiency (RF power output/dc input power), and (2) tend to be noisy due primarily to the avalanche process and to the high level of operating current. A typical noise figure is 30 dB which is worse than that of Gunn diodes.

Mounting and equivalent circuit   IMPATT diodes are mounted in coaxial lines, waveguides, or microstrip lines to form microwave circuits for oscillations and amplifications as shown in Fig. 10.28. A simplified equivalent circuit for IMPATT diode chip is shown in Fig. 10.29. Here R_d is the diode negative resistance consisting of the series lead resistance R_s and the negative resistance $-R_j$ due to impact avalanche process, C_j the junction capacitance at the breakdown volt-

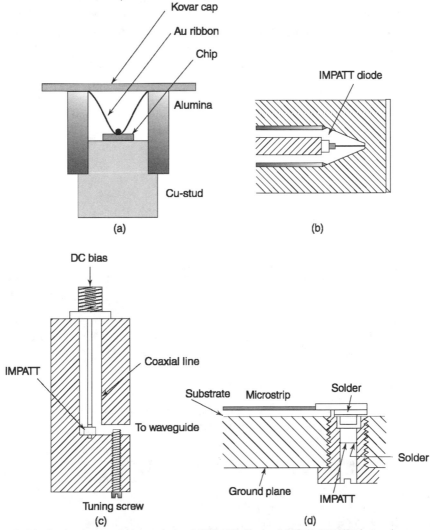

Fig. 10.28 *Packaging and mounting of IMPATT diode (a) IMPATT diode packages (b) Coaxial end mount (c) Coaxial to waveguide transition mount (d) Microstrip mount*

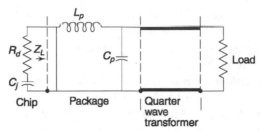

Fig. 10.29 *Equivalent circuit of IMPATT diode*

age, and L_p, C_p are the package lead inductance and capacitance, respectively. The diode mount is designed in such a way that the overall reactance of the circuit is tuned to zero at resonance by controlling L_p. Since the power extracted by the ac field from the dc field compensates for the power dissipation in positive resistance of the circuit, the total resistance R_d must also be zero. Since the diode chip impedance

$$Z_d = -|R_d| - j/(\omega C_j) \tag{10.14}$$

the load impedance must be $Z_L = -Z_d$, or

$$R_L = |R_d| \tag{10.15}$$

$$X_L = 1/(\omega C_j) \tag{10.16}$$

Since R_d varies with both bias and signal currents, for a given load and bias, stable oscillations are obtained when $|R_d| = R_L$. The corresponding peak RF current determines the load power.

Since total negative resistance in the circuit is low (≈ 10 ohm) for sustained oscillation, multisection quarterwave transformers must be incorporated for impedance matching between the diode circuit and the load as shown in Fig. 10.29.

Example 10.4 An IMPATT diode with nominal frequency 10 GHz, has $C_j = 0.5$ pf, $L_p = 0.5$ nH and $C_p = 0.3$ pF at breakdown bias of 80 V and bias current 80 mA. The RF peak current is 0.65 A for $R_d = -2\ \Omega$. Find (a) the resonant frequency of oscillation, (b) the efficiency.

Solution

(a) For sustained oscillation, physical load $R_L = |Z_d| = 2$ ohm

The effective load impedance (Ref. Fig. 10.29).

$$Z_L = j\omega L_p + \frac{R_L}{1 + j\omega C_p R_L} = j\omega L_p + \frac{R_L}{\sqrt{\left[1 + \left(\omega C_p R_{L,}\right)^2\right]}\ \omega C_p R_L}$$

At nominal operating frequency 10 GHz

$$\omega L_p = 2\pi \times 10 \times 10^9 \times 0.5 \times 10^{-9} \text{ ohm}$$

$$= 31.42 \text{ ohm}$$

$$\omega C_p R_L = 2\pi \times 10 \times 10^9 \times 0.3 \times 10^{-12} \times 2$$

$$= 0.038,$$

$$\tan^{-1}(\omega C_p R_L) = 2.18°$$

Therefore,

$$(\omega C_p R_L)^2 \ll 1 \text{ and } Z_L \approx R_L + j\omega L_p$$

This shows that the role of C_p in circuit resonance is negligible. Therefore, the resonant frequency

$$f = \frac{1}{2\pi\sqrt{(C_j L_p)}} = \frac{1}{2\pi\left(0.5 \times 10^{-12} \times 0.5 \times 10^{-9}\right)^{1/2}}$$

$$= \frac{\sqrt{10^3}}{2\pi \times 0.5}$$

$$= 10.06 \text{ GHz}$$

(b)   The output power

$$P_L = 1/2 \text{ (RF peak current)}^2 \times R_L$$

$$= \frac{1}{2} \times 0.65^2 \times 2$$

$$= 0.4225 \text{ W}$$

Average dc input power

$$P_{dc} = \text{Break down voltage} \times \text{dc bias current}$$
$$= 80 \times 0.08 = 6.4 \text{ W}$$

The efficiency

$$\eta \% = \frac{P_L}{P_{dc}} \times 100 = \frac{0.4225}{6.4} \times 100$$

$$\approx 6.6\%$$

10.4.2   IMPATT Diode Power Amplifier

The IMPATT diode can be used as amplifier with the same basic circuit arrangement as oscillator, provided $R_L > |R_d|$ and a circulator is incorporated as shown in Fig. 10.30.

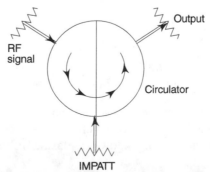

Fig. 10.30   *IMPATT circulator type amplifier*

The negative resistance is used to terminate one port of the circulator and actual load is connected to other port. The input RF power is fed from the remaining port. The negative resistance results in voltage reflection coefficient at the port greater than unity. Thus the average power from the source P_{av} circulates to the negative resistance port and the reflected power is greater than the incident power. The reflected power is delivered to the load. The power gain is

$$G_p = \frac{|\Gamma|^2 P_{av}}{P_{av}} = |\Gamma|^2 > 1$$

$$= \left|\frac{-|R_d| - R_L}{-|R_d| + R_L}\right|^2 \tag{10.17}$$

Therefore, if $R_d = -2$ ohm and $R_L = 3$ ohm, $G_p = |-5/1|^2 = 25$.

10.4.3 TRAPATT Diodes

The TRAPATT diodes are manufactured from Si, and have $p^+n\ n^+$ (or $n^+\ pp^+$) configurations as shown in Fig. 10.31. The p-n junction is reverse biased beyond the breakdown region, so that the current density is higher. This causes the electric field in the space charge region to be decreased and the carrier transit time is increased. Consequently the frequency of operation becomes lower and is limited to below 10 GHz, although efficiency of the diode increases due to low power dissipation.

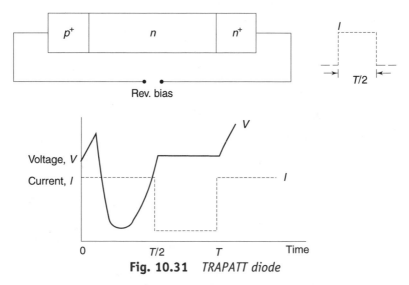

Fig. 10.31 *TRAPATT diode*

The TRAPATT diode is mounted inside a coaxial resonator at a position of maximum RF voltage swing. When avalanche occurs at dc reverse-bias plus RF swing beyond the threshold of breakdown, a plasma of holes and electrons are generated. This plasma density results in a high potential difference across the junction in opposition to the dc reverse bias voltage. At this relatively low voltage the plasma get trapped. But since the external circuit current flow, the voltage rises and the trapped plasma is released producing current pulse across the drift space. The total transit time is the sum of the delay time in releasing the trapped plasma and the drift time. Since the transit time is longer due to low voltage, the operating frequency is limited below 10 GHz. Moreover, the current pulse is associated with low voltage. Therefore, the power dissipation is low and efficiency is higher.   ,

The major disadvantages of TRAPATT are (1) high noise figure (60 dB) limits its use as an amplifier, (2) it generates strong harmonics due to the short duration current pulse.

10.4.4 BARITT Diodes

BARITT diodes are formed by forward biased p-n junction with p-n-p or p-n-i-p or p-n-metal, or metal-n-metal configurations. Minority charge carriers are injected into the drift region. The transit time through the drift region provides the required phase-shift between the current and voltage to give a negative resistance. When the diode is mounted in a resonator, a noise spike generates microwave voltage across the diode. During positive half cycle the total voltage produces a sharp pulse of minority carrier current in the drift region. During the drift time a constant external current delivers energy to the resonator from the dc bias source to maintain a continuous oscillation. BARITT diodes are low power (mw), low efficiency, and less noisy devices. They are used as local oscillator at microwave frequencies (4–8 GHz). Table 10.1 shows a comparison between various TEDs and ATTDs.

Table 10.1 *Comparison between Gunn, IMPATT, TRAPATT and BARITT*

	Gunn	*IMPATT*	*TRAPATT*	*BARITT*
Operating frequency	1–100 GHz	0.5–100 GHz	1–10 GHz	4–8 GHz
Bandwidth	2% of centre frequency.	One tenth of centre frequency	—	Narrow
Power output	A few watts (CW) 100-200W (pulsed)	1W(CW), 400 W pulsed	Several 100W(pulsed)	Low (mw)
Efficiency	—	3% CW, 60% pulsed	20–60%pulsed	Low (2%)
Noise figure	—	High, 30 dB	High, 60 dB	Less noisy than MPATT (< 15 dB)
Application	oscillator	oscillator, amplifier	oscillator	local oscillator
Construction	n^+nn^+ GaAs	n^+pip^+ reverse	p^+nn^+ or	p-n-p,or p-n-i-p,

(Contd)

single crystal	-bias *p–n* junction	n^+ *p* p^+ reverse-bias *p-n* junction	or *p-n*-metal or metal - *n* - metal, forward - bias *p - n* junction
Basic semi- GaAs	Si,Ge, GaAs or	Si	Si/metal
conductors InP	InP		
Harmonics —	Less	Strong	Less
Size Small	Small	Small	Small
Ruggedness Yes	Yes	Yes	Yes

10.5 TUNNEL DIODES

The tunnel diodes are heavily doped *p-n* junction diodes that have a negative resistance over a portion of its *I-V* characteristic as shown in Fig. 10.32. These diodes are used as microwave amplifiers or oscillators. Due to heavy doping, the width of the depletion region becomes very thin and an overlap occurs between the conduction-band level on the *n*-side and the valence-band level on the *p*-side. Initially under little forward bias condition the conduction-band electrons tunnel through the depletion layer resulting in increase of current with forward voltage. The current reaches a maximum of I_p at voltage V_p.

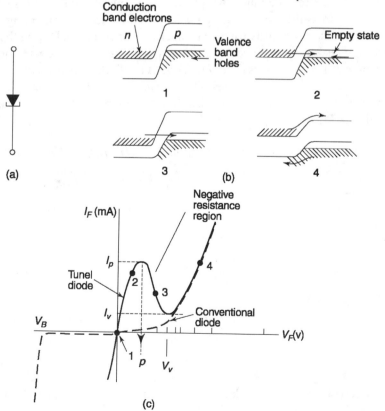

Fig. 10.32 *Tunnel diodes and I-V characteristics (a) Symbol, (b) Energy level diagram (c) I-V characteristics*

For further increase of forward bias, the conduction band electron energy levels are raised above the available energy levels in the valence band and become equal to levels in forbidden band. No direct tunnelling occurs and current decreases with increase in forward voltage till V_v showing a negative resistance characteristics for use in amplifiers or oscillators. As the forward voltage is increased beyond V_v, current increases with forward voltage in the same manner as a semiconductor diode.

The difference between tunnel diode and normal p-n junction diode is given in Table 10.2.

Table 10.2

Tunnel diode	*Normal p-n diode*
1. Doping levels at p and n sides are very high.	Doping is normal in both p and n sides.
2. Tunnelling current consists of majority carriers– electrons from n-side to the p-side.	Current consists of minority carriers—holes from p-side to the n-side.
3. Majority carrier current responds much faster to voltage changes—suitable to microwaves.	Majority carrier current does not respond so fast to voltage changes—suitable for low frequency applications only.
4. At a small values of reverse voltage a large current flows due to considerable overlap between conduction band and valence band—useful as frequency converter.	Current is extremely small (leakage current) up to considerable reverse bias voltage and then increases abruptly to extremely high at a particular voltage called breakdown voltage.
5. Low power device.	Lower power device.
6. Shows negative resistance characteristics—useful for reflection amplifiers and oscillators.	Does not show negative resistance –used as detector and mixers.
7. It is a low noise device.	Moderate noise characteristics.
8. Preferred semiconductors are Ge and GaAs.	Preferred semiconductors are Ge and Si.

10.5.1 Equivalent Circuit

The equivalent circuit of a tunnel diode is shown in Fig. 10.33. R_j and C_j are the diode junction resistance (100 ohm) and capacitance (pf), L_s(nH) is the lead inductance, R_s (few ohm) includes lead resistance and bulk resistance of the semiconductor. The impedance of the circuit is expressed by

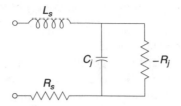

Fig. 10.33 *Equivalent circuit of tunnel diode*

$$Z = R_s - \frac{R_j}{1 + (\omega R_j C_j)^2} + j\omega R_j C_j \left[\frac{L_s}{R_j C_j} - \frac{R_j}{1 + (\omega R_j C_j)^2} \right] \quad (10.18)$$

From the condition of oscillation that

$$\text{Re } (Z) = 0 \quad (10.19)$$

$$\text{Im } (Z) = 0 \quad (10.20)$$

two important frequencies can be obtained as

Resistive cut-off frequency $f_r = \dfrac{1}{2\pi R_j C_j} \sqrt{\left[\dfrac{R_j}{R_s} - 1 \right]} \quad (10.21)$

Self-resonant frequency $f_0 = \dfrac{1}{2\pi} \sqrt{\left[\dfrac{1}{L_s C_j} - \dfrac{1}{(R_j C_j)^2} \right]} \quad (10.22)$

10.5.2 Tunnel Diode Amplifiers

Tunnel diode can be used as reflection amplifier (TDRA) incorporating a circulator to isolate the source from the load (Fig. 10.34).

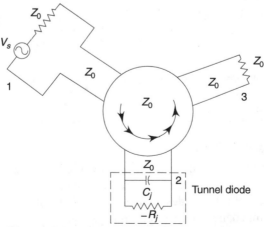

Fig. 10.34 *Tunnel diode reflection amplifier*

The characteristic impedance Z_0 of the circulator must be greater than that of the diode's negative resistance R_j. The voltage reflection coefficient at the diode port is

$$\Gamma = \frac{(Z_{in}/Z_0)-1}{(Z_{in}/Z_0)+1} \tag{10.23}$$

Neglecting C_j, for smallness,

$$\Gamma = \frac{[(-R_j/Z_0)-1]}{[(-R_j/Z_0)+1]}$$

or,

$$|\Gamma|^2 > 1 \text{ for } Z_0 > R_j \tag{10.24}$$

Therefore load power is

$$P_L = |\Gamma|^2 P_{in}$$

or,

Power gain $$G_p = |\Gamma|^2 = \left|\frac{R_j + Z_0}{R_j - Z_0}\right|^2 \tag{10.25}$$

Several advantages of tunnel diodes are
1. require very simple dc power supply
2. broadband operation possible
3. low noise figure (<5 dB at 10 GHz) due to low current levels.
4. immune to the natural radiation in the solar system and suitable for space communication.

10.5.3 Tunnel Diode Oscillator (TDO)

Tunnel diode oscillator circuit consists of a tank circuit coupled with the diode by means of capacitive divider (Fig. 10.35). When the power is switched on, a surge current produces oscillation in the tank circuit. The $R-C$ values make the dc bias at the centre of the negative resistance characteristic of the diode. Sustained oscillation occurs if the magnitude of the negative resistance of the diode is equal or greater than the resistance of the tank circuit. The oscillator circuit can generate microwave signal up to frequencies of about 100 GHz.

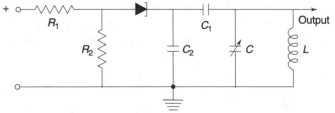

Fig. 10.35 *Tunnel diode oscillator*

10.6 VARACTOR DIODES

Varactor diodes are p-n junction diodes which provide a voltage variable junction capacitance in microwave circuits when reverse biased.

$$C_j = C_0 (1 + V_R/V_B)^{-n} \tag{10.26}$$

where C_0 is the junction capacitance for zero bias, V_R is the magnitude of reverse bias voltage below break down, V_B is the barrier potential and n is a constant whose values are 0.5 and 0.33 for abrupt and linear graded p - n junction, respectively. In this operation as the magnitude of reverse bias voltage is increased, the depletion layer width W_d will increase and the junction capacitance C_j will decrease in accordance with

$$C_j \, \alpha \, \frac{1}{W_d} \tag{10.27}$$

This non-linear C_j–V characteristic permits the varactor diode to be used for frequency multiplication and parametric amplification at microwave frequencies. Figure 10.36 shows the typical structure and equivalent circuit of a varactor diode. Here C_j is the junction capacitance, R_j is the junction reverse resistance, and R_s is the diode bulk resistance. Since R_j is very large, the simplified model is a series combination of C_j and R_s. The diode has a wide capacitance variation and the cut-off frequency of operation at a given reverse voltage is defined by

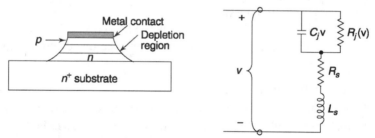

Fig. 10.36 *Varactor diode and its equivalent circuit*

$$f_c = \frac{1}{2 \pi R_s \, C_j \, (\text{min})} \tag{10.28}$$

Due to skin effect, R_s increases with increase in frequency, and therefore, the varactor is normally used at frequencies below $0.2 f_c$.

Varactor frequency multipliers When an ac voltage (pumping voltage) $v_p = V_p \sin \omega_p t$ is applied to a reverse biased diode, the diode capacitance varies as (Fig. 10.37),

$$C_j(t) = C_{j0} [1 + (V_R + V_p \sin \omega_p t)/V_B]^{-n}; \tag{10.29}$$

Expanding in a harmonic series

$$C_j(t) = C_0 - C_1 \sin \omega_p t + C_2 \sin 2\omega_p t - C_3 \sin 3\omega_p t + \dots \tag{10.30}$$

The diode current due to the above time varying capacitance can be expressed as

$$i(t) = I_1 \cos \omega_p(t) + I_2 \cos 2\omega_p t + I_3 \cos 3\omega_p t + \dots \qquad (10.31)$$

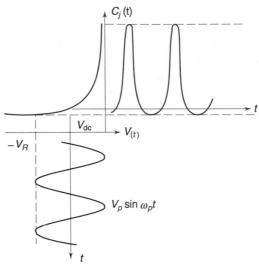

Fig. 10.37 *Variation of C_j with ac pump voltage*

Therefore, the varactor diode can be used as frequency multiplier or a harmonic generator where a resonant circuit tunes to the desired harmonic. Figure 10.38 shows a 1–2–3 tripler. The input signal at frequency f is fed from a stable crystal controlled low frequency generator to the reverse biased varactor diode through a buffer amplifier. The input resonant circuit 1 prevents unwanted frequencies reaching the diode from the input side and the input circuit from the diode circuit. The output resonator circuit 3 is tuned to the third harmonic to produce the final output of $3f$ frequency. The intermediate resonant circuit 2 called the idler, eliminates the heterodyning between the input and output frequencies. Since the harmonic current is capacitive, very little power loss occurs and the varactor multiplier operates at high efficiency without adding excessive noise.

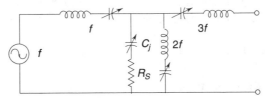

Fig. 10.38 *Varactor diode multiplier circuit*

10.7 THE PARAMETRIC AMPLIFIER

A parametric amplifier is named because of its operation due to the periodic variation of the device's parameters such as capacitance of the varactor diode, under the influence of a suitable pump signal. If a small input signal at a frequency f_s and the ac power source operated as a pumping signal at a frequency f_p, are applied together to the varactor diode, linear amplification of a small signal results in due to time-varying capacitance of the diode. Pump signal provides the power required for amplification and the power output is either at the input frequency f_s or at the idler frequency $f_i = f_p - f_s$.

Manley-Rowe Relations A set of power-conservation (Manley-Rowe) relations are described below which are useful in determining the maximum gain of the parametric amplifier. When two sinusoidal signals at frequencies f_s and f_p are applied across a lossless time varying non-linear capacitance $C_j(t)$, signals at the harmonics of f_s and f_p are generated which can be separated by band-pass filters and made their power dissipated in separate resistive loads (Fig. 10.39). From the conservation of power

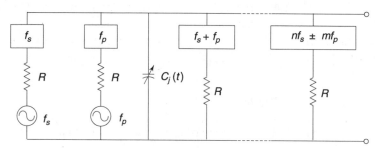

Fig. 10.39 *Manley-Rowe relations*

$$\sum_{n=0}^{\infty} \sum_{m=-\infty}^{\infty} \frac{n P_{nm}}{n f_s + m f_p} = 0 = \sum_{n=-\infty}^{\infty} \sum_{m=0}^{\infty} \frac{n P_{nm}}{n f_s + m f_p} \tag{10.32}$$

where P_{nm} represents the average power at the output frequencies $\pm |n f_s + m f_p|$. The above relations are called the *Manley-Rowe* or *power-conservation relations*. The sign convention for the power term P_{nm} is such that it is positive when the power is supplied by the two generators or power flowing into the non-linear capacitance, otherwise it is negative.

Example 10.5 A sinusoidal input signal at frequency f_s and pump signal at frequency f_p are applied across a time varying non-linear capacitance. If the output circuit is a band-pass filter with resistive series load at frequencies $f_s + f_p$, calculate the power gain.
Solution
From Manley-Rowe relations,

$$\sum_{n=0}^{\infty} \sum_{m=-\infty}^{\infty} \frac{n P_{nm}}{n f_s + m f_p} = 0 = \sum_{n=-\infty}^{\infty} \sum_{m=0}^{\infty} \frac{m P_{nm}}{n f_s + m f_p} \tag{10.33}$$

Here, m and n vary from -1 through zero to $+1$, respectively. Therefore, Manley-Rowe relations are

$$\frac{P_{10}}{f_s} + \frac{P_{11}}{f_s + f_P} = 0 = \frac{P_{01}}{f_p} + \frac{P_{11}}{f_s + f_p} \tag{10.34}$$

Since $f_s, f_p, f_s + f_p > 0$, P_{10} and $P_{01} > 0$, the average power at the frequency $f_s + f_p$ is $P_{11} < 0$ to satisfy above relations i.e., at the ouput frequency $f_0 = f_s + f_p$, the average power is negative in the output load. This means that power dissipation in the output load is negative. Therefore, power is delivered from the non-linear time varying capacitor $C_j(t)$ at the output frequency $f_0 = f_s + f_p$ to the load, where $f_0 > f_p > f_s$.

The power gain of signal is

$$\frac{P_{11}}{P_{10}} = \frac{f_0}{f_s} = \frac{f_s + f_p}{f_s} = 1 + \frac{f_p}{f_s} \qquad (10.34a)$$

The Manley-Rowe relations give the maximum signal gain possible. However, due to presence of losses in practical cases, the gain is less than $1 + f_p/f_s$. Such parametric amplifiers are called *up-converters*. If the output frequency $f_0 = f_s - f_p$, the parametric device is called the parametric *down-converter* having power loss in place of gain.

10.7.1 Analysis of a Parametric Amplifier

When a pump voltage $v_p = V_p \sin \omega_p t$ and a small signal voltage $v_s = V_s \sin \omega_s t$, where $V_s \ll V_p$, and $V_p \ll$ dc reverse bias voltage magnitude, $|V_R|$, are applied simultaneously to time varying non-linear capacitance $C_j(t)$ of a varactor diode, the current through the diode is

$$i = \frac{dQ(V)}{dt} = \frac{dQ(v_p + v_s)}{dt} \qquad (10.35)$$

Expanding $Q(v_p + v_s)$ in Taylor series about the small signal voltage $v_s = 0$,

$$Q(v_p + v_s) \approx Q(v_p) + \frac{\partial Q}{\partial v}\bigg|_{v_s = 0} \cdot v_s ; \text{ for } v_s \ll v_p \qquad (10.36)$$

Therefore,

$$i = \frac{dQ(v_p)}{dt} + \frac{d}{dt}\left[\frac{\partial Q}{\partial v}\bigg|_{v_s = 0} \cdot v_s\right] \qquad (10.37)$$

This shows that the diode current is given by

$$i = \frac{dQ(v_p)}{dt} + \frac{d}{dt}[C_j(t) \, v_s] \qquad (10.38)$$

and the junction capacitance is

$$C_j(t) = C_0\left[1 + \frac{V_R + V_p \sin \omega_p + V_s \sin \omega_s t}{V_B}\right]^{-n} \qquad (10.39)$$

For a linearly graded junction, $n = 1/3$ and since $|V_s| \ll |V_p| \ll V_R$, $C_j(t)$ can be expanded in a harmonic series

$$C_j(t) \approx C_0 - C_1 \sin \omega_p t \tag{10.40}$$

This time varying capacitance is utilised for parametric amplifications.

10.7.2 Negative Resistance Amplifier

Figure 10.40 shows a parametric amplifier equivalent circuit where the varactor diode represents a negative resistance at signal frequency to provide amplification of the signal. It is assured that the parasitic elements of the equivalent circuit of the varactor diode is tuned out by the external tuning elements L-C so that the diode can be represented by the series combination of $C_j(t)$ and R_s. The non-linear capacitance results in small-signal currents at the harmonics of the input frequency, and at the sum and difference components of the pumping and input frequencies. External tuned circuits and band-pass filters allow currents at desired frequencies, viz. i_s at f_s in the input loop, i_p at f_p in the pumping circuit and i_i at $f_i = f_p - f_s$ in the idler loop. The output is obtained across the load R_L in the idler circuit.

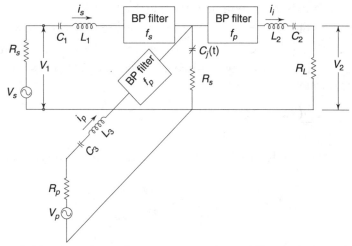

Fig. 10.40 *Negative resistance parametric amplifier equivalent circuit*

The presence of tuned circuits allow only the first harmonic of the signals to pass, so that

$$C_j(t) = C_0 - C_1 \sin \omega_p t \tag{10.41}$$

$$i_s = I_s \sin \omega_s t \tag{10.42}$$

$$i_i = I_i \sin \omega_i t \tag{10.43}$$

The diode current

$$i = i_s + i_i + i_p \tag{10.44}$$

The signal voltage across the diode is given by

$$V_d = Q/C = \int i \, dt / C \tag{10.45}$$

Since i_p involves only pumping frequency, it does not contribute to the signal current. Hence,

$$V_d = C_0 - C_1 \sin \omega_p t \quad -\left[\frac{I_s}{\omega_s} \cos \omega_s t + \frac{I_i}{\omega_i} \cos \omega_i t\right]$$

$$\approx \frac{1}{C_0} \left[\frac{I_s}{\omega_s} \cos \omega_s t + \frac{I_i}{\omega_i} \cos \omega_i t\right]\left(1 + \frac{C_1}{C_0} \sin \omega_p t\right) \quad (10.46)$$

This voltage contains terms containing cos $\omega_s t$, cos $\omega_i t$, sin $\omega_p t$, cos $\omega_s t$ and sin $\omega_p t$, cos $\omega_i t$. The product terms result in signal voltage components at frequencies $\omega_p \pm \omega_s$ and $\omega_p \pm \omega_i$. Due to resonant circuits only the difference frequency $\omega_p - \omega_i = \omega_s$ will contribute to the signal voltage across the diode.

$$V_{ds} = -\frac{1}{C_0}\left[\frac{I_s}{\omega_s} \cos \omega_s t + \frac{C_1}{C_0}\frac{I_i}{2\omega_i} \sin \omega_s t\right] \quad (10.47)$$

Factor 1/2 appears in the second term due to trigonometric identities sin $\omega_p t$ + cos $\omega_i t$ = 1/2 [sin $(\omega_p + \omega_i)t$ + sin $(\omega_p - \omega_i)t$]. The negative-resistance amplifier is an unstable device with narrow bandwidth and high gain. Since $|V_s| \ll |V_p|$, contribution of the first term is small compared to the second term and thus

$$V_{ds} \approx -\frac{1}{C_0^2}\frac{C_1}{2\omega_i} I_i \sin \omega_s t$$

$$= -\frac{1}{C_0^2}\frac{C_1}{2\omega_i}\frac{I_i}{I_s} I_s \sin \omega_s t$$

$$= -K I_s \sin \omega_s t \quad (10.48)$$

Thus the voltage component is out of phase to the current showing that the diode behaves like a negative resistance R^- to the signal frequency.

$$R^- = \frac{C_1 I_i}{C_0 2\omega_i I_s} \quad (10.49)$$

The total input resistance of the input loop is

$$R_{in} = R_s + R^- \quad (10.50)$$

Negative resistance circulator type amplifier can be designed if $|R^-| > R_s$ as shown in Fig. 10.41.

The signal power circulates around to the negative resistance port where the magnitude of the voltage reflection coefficient > 1, so that the reflected power is larger than the incident power which is obtained from the next port. The power gain is $G = |\Gamma|^2$ where,

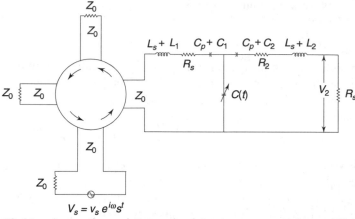

Fig. 10.41 *A negative resistance circulator type parametric amplifier*

$$\Gamma = \frac{R_{in}/Z_0 - 1}{R_{in}/Z_0 + 1} \tag{10.51}$$

Parametric amplifiers are very low noise amplifiers.

10.7.3 Parametric Up-Converter

Figure 10.42 shows an equivalent circuit of a parametric up-converter in which a pump voltage at frequency f_p and a signal voltage at frequency f_s are applied through respective tuned circuits to the varactor diode. The output power is taken from the idler circuit at either the lower side band frequency $f_i = f_p - f_s$ or at the upper sideband frequency $f_i = f_p + f_s$. Physically, power is transferred from the pumping source to the idler circuit to achieve power gain with low noise. The series tuned circuits allow only currents with respective frequencies, f_s, f_p and f_i in each loop. The circuit losses are small compared to the loss in the diode resistance

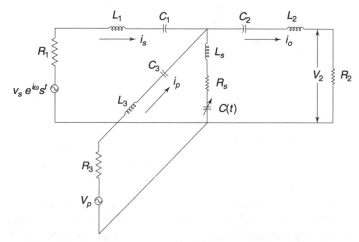

Fig. 10.42 *Parametric up-converter*

R_s and external loads R_g and R_L. For maximum gain, $R_L = R_g$ and maximum gain for a lossless ($R_s = 0$) diode is, $G_m = \omega_i/\omega_s$ as predicted by the Manley-Rowe relations.

For high gain, an up-converter requires a large ratio of output to input frequency, but limited to operation at $f_s = 1$ GHz. The up-converter is a stable device with a wide bandwidth and low gain. The advantages of the up-converter over the negative resistance device are as follows

1. Input impedance > 0
2. More stable
3. Power gain independent of source impedance
4. No circulator is required
5. Larger bandwidth (~5%)

10.7.4 Advantages and limitation of parametric amplifiers

Parametric amplifiers have the following advantages and limitations:

1. *Noise figure*   Because of minimum resistive elements, thermal noise in parametric amplifiers is lower compared to that of transistor amplifier. Typically, noise figure is in the range 1–2 dB.
2. *Frequency*   The upper frequency limit (40–200 GHz) is set by the difficulty of obtaining a source power at the pump frequency and by the frequency at which the varactor capacitance can be pumped. The lower frequency limit is set by the cut-off frequency of the microwave components used in the circuit.
3. *Bandwidth*   Bandwidth of the parametric amplifier is small due to the presence of tuned circuits. Bandwidth can be increased by stagger tuning.
4. *Gain*   The gain is limited (20–80 dB) by the stabilities of the pump source and the time varying capacitance.

Because of its low noise, parametric amplifiers are used in space communication systems, radio telescopes and tropo-receivers.

10.8 MICROWAVE TRANSISTORS

Microwave transistors are miniaturized designs to reduce device and package parasitic capacitances and inductances and to overcome the finite transit time of the charge carriers in the semiconductor materials. The most commonly used semiconductors are Si and GaAs. Since transit times are dependent on the electron mobility and saturation velocity in the semiconductor material, GaAs is significantly better than Si for high frequency devices. By means of molecular beam epitaxy techniques the High Electron Mobility Transistor (HEMT) are developed presently which can operate at frequencies of the order of 100 GHz.

Microwave transistors are used for amplifier and oscillator design. There are three categories of microwave transistors, (1) low noise transistor which is employed in first stage since this is the major contributor to the overall system noise, low level transistor which is used (2) to drive power stage, and (3) power transistor which amplifies final power output.

Microwave transistor amplifiers are constructed either as hybrid Microwave Integrated Circuits (MIC) where the transmission lines and matching networks are realised by microstrip circuit elements and the discrete components such as chip capacitors, resistors, and transistors are soldered in place, or as Monolithic Microwave Integrated Circuits (MMIC) where all active devices and passive circuit elements are fabricated on a single semiconductor (GaAs) crystal.

There are two basic types of construction of microwave transistors, bipolar and unipolar. Bipolar is three semiconductor (*pnp* or *npn*) region structure where charge carriers of both negative (electrons) and positive (holes) polarities are involved in transistor operation. Unipolar transistors are junction gate and insulated gate field-effect transistors (FETs). These are one or two semiconductor region structures where dominant carriers are of single polarity (electrons or holes). Most widely used transistors in microwave analog circuits are Si bipolar used in UHF-S band, and Si bipolar/ GaAs FET-used in S–C band.

Compared to GaAs devices Si bipolar transistors are inexpensive, durable, have higher gain and moderate noise figure. However, FETs are superior to bipolar transistors for the lower noise characteristics and the higher frequency of operation.

10.8.1 Microwave Bipolar Transistors

The microwave bipolar transistors are planar in form and mostly Si *n-p-n* type operating up to 5 GHz. GaAs also is used for improvements in operating frequency, in high temperatures, and in high radiation field. Geometries are (a) inter-digitated, (b) overlay, and (c) matrix forms. A bipolar construction is shown in Fig. 10.43. Epitaxial *n* layer is formed by condensing a single crystal film of semiconductor material upon a low resistivity Si wafer of substrate n^+. Above this a *p*-type diffused base and *n*+ type diffused emitter are formed. Typically the emitter width *W* is 1 micron, base thickness is 2 micron, and emitter length is 25 micron.

For power amplification class *C* operation is preferred to class A / AB mode to minimise collector-base transit time. In class C both emitter-base and collector-base junctions are reverse biased so that no current flows in absence of a signal. When an RF voltage of sufficient magnitude is applied to emitter-base junction, for a fraction of RF cycle the junction is forward biased, the electrons are injected to the base and the injected carriers transit the base by a combined diffusion and drift flow processes. They are accelerated in the collector-base depletion region.

High frequency limitations of bipolar devices   The microwave bipolar transistor is a non-linear *p-n-p* or *n-p-n* device, and its principle of operation is similar to that of the low-frequency device, but requirements for dimensions ($<< \lambda$), heat sinking and packaging are much more severe due to the following limitations.

1. At high frequencies the reactances due to junction capacitances limit the gain. The values of these capacitances depend on the depletion layer's width or the bias voltage and it provides feedback paths. However, the feedback effect is reduced in CB mode of operation. The junction capacitances must be reduced as much as possible.

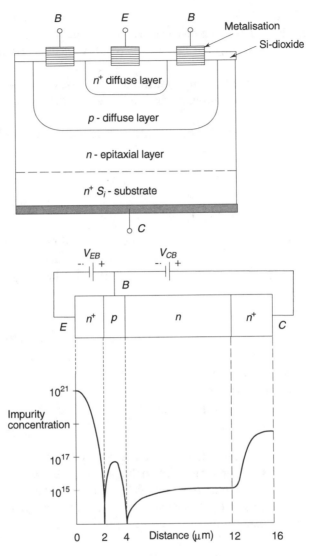

Fig. 10.43 *Geometries of a microwave bipolar transistor*

2. Signal loss in the lead inductances can be minimised by reducing the length in packaging.
3. The transit time taken by the majority charge carriers to cross from the emitter to collector limits the high frequency. The total transit time is contribution from

T_{jeb} =   emitter base junction capacitance charging time

T_b =   base region transit time

T_{bc} =   base collector-region depletion-layer transit time

T_{jbc}'=   base collector junction capacitance charging time.

Since a forward-biased emitter base junction is assumed, the transit time associated with this depletion region can be ignored. In general, the transit time can be reduced by using narrow p and n regions, especially the base. Transit time can also be reduced by reducing the depletion layer width at the reverse-biased collector-base junction. This needs the use of a higher collector voltage without avalanche breakdown. But corresponding increase in the power dissipation may damage the device.

The upper frequency limit of these devices is nearly 22 GHz. The cut-off frequency of the device is given by

$$f_T = 1/(2\pi T) \tag{10.52}$$

where

$$T = T_{jeb} + T_b + T_{bc} + T_{jbc}$$

$$\approx T_b + T_{bc} \tag{10.53}$$

since base and base-collector depletion layer transit time is much larger than the junction capacitance charging time. For $N_d = 5 \times 10^{15}$ /cc, or 1 ohm-cm resistivity, $T_{bc} = 1.7 \times 10^{-11}$ S and $T_b = 8.3 \times 10^{-10}\ W_B^2$ S, where W_B is the basewidth, Therefore,

$$f_T \approx 1/2\pi\,(T_b + T_{bc}) \tag{10.54}$$

The equivalent circuits of CB and CE modes are shown in Fig. 10.44, where r_e = emitter resistance, r_b = base resistance, C_c = collector depletion layer capacitance and C_e = emitter base junction capacitance.

Power frequency limitations

Let $\quad L\ =\quad$ emitter collector distance.

$\quad\ v_s\ =\quad$ maximum saturated drift velocity of carriers in semiconductor.

$\quad E_m =\quad$ maximum **E**-field that can be sustained in a semiconductor without having dielectric break down.

$\quad V_m =\quad E_m L$ = maximum allowable applied voltage.

$\quad X_c\ =\quad 1/(2\pi f_T C_0)$ = reactive impedance of collector base capacitance C_0.

$\quad I_m\ =\quad$ maximum current of the device.

Then, $\qquad V_m f_T = E_m L \cdot \dfrac{1}{2\pi T}$

$$= \frac{E_m L}{2\pi}\ \frac{1}{L/v_s}$$

$$= E_m v_s\,/2\pi \tag{10.55}$$

$$I_m f_T = \frac{V_m}{X_c}\ \frac{1}{2\pi T}$$

$$= \frac{E_m\,v_s}{2\pi X_c} \tag{10.56}$$

Therefore,

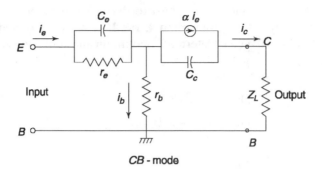

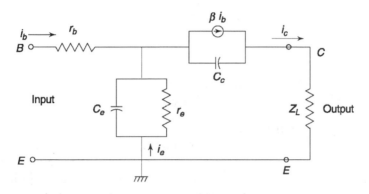

CB - mode

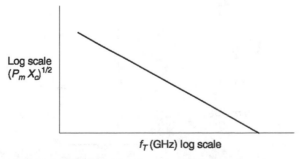

Fig. 10.44 *Microwave equivalent circuits of bipolar transistor*

$$V_m f_T \cdot I_m f_T = V_m I_m f_T^2 \tag{10.57}$$

or, $$P_m f\, T^2 = \frac{E_m\, V_s}{2\pi}\; \frac{E_m\, V_s}{2\pi X_c}$$

or, $$P_m X_c f_T^2 = \left(E_m V_s / 2\pi\right)^2$$

or, $$(P_m X_c)^{1/2} f_T = E_m V_s / 2\pi \tag{10.58}$$

For a given device impedance the power capacity decreases with increase device cut-off frequency as shown in Fig. 10.45.

Log scale
$(P_m X_c)^{1/2}$

f_T (GHz) log scale

Fig. 10.45 *Power vs frequency*

Example 10.6   A Si microwave transistor has reactance 1 ohm, transit time cut-off frequency 4 GHz, maximum **E**-field 1.6×10^5 V/m and saturation drift velocity 4×10^5 cm/s. Determine maximum allowable power.

Solution

Given

$$X_c = 1 \text{ ohm} , E_m = 1.6 \times 10^5 \text{ V/m}$$

$$f_T = 4 \text{ GHz} , v_s = 4 \times 10^5 \text{ cm/s}$$

$$P_m X_c = \left(\frac{E_m v_s}{2 \pi f_T} \right)^2$$

or,

$$P_m = \left(\frac{E_m v_s}{2 \pi f_T} \right)^2 \frac{1}{X_c}$$

$$= \left(\frac{1.6 \times 10^5 \times 4 \times 10^5}{2 \times 3.14 \times 4 \times 10^9} \right)^2$$

$$= 6.48 \text{ watts}$$

Biasing circuits   Microwave transistor biasing circuits are designed for (1) a stable operating point against variations in the device parameters and temperature, and (2) isolation of the biasing circuit from the high frequency circuit. The first requirement is fulfilled by designing a dc feedback and the second one by connecting RF chokes in series with bias resistances and capacitive shunt across the dc circuit elements as shown in Fig. 10.46.

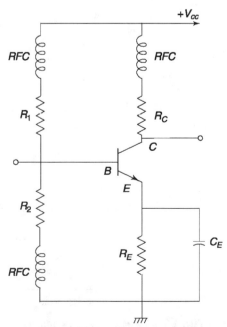

Fig. 10.46   *Biasing circuit*

10.8.2 Unipolar Field-Effect Transistors

Field-effect transistors are manufactured from GaAs for higher frequency of operation which is achieved due to the higher electron mobility compared to Si. The most commonly used FET for microwave frequencies is a Schottky barrier (Metal-semiconductor) gate and is called a MESFET. In recent years a hetero-junction is formed at the interface of an aluminum–GaAs (AlGaAs) doped alloy and an undoped GaAs layer which enables a channel with a very high electron mobility for higher frequency operations and lower noise. The latter construction is called HEMT. MESFET devices can give a single stage gain of 10–15 dB at 2 GHz with noise figures less than 1 dB and HEMT devices give a single stage gain of 15 dB at 8 GHz with noise figure of 0.4 dB and 6 dB gain at 50 GHz with noise figures of 1.8 dB.

The basic constructions and circuit symbol of MESFET and HEMT are shown in Fig. 10.47. The basic operations, equivalent circuit and biasing circuit of MESFET are described here. The three terminals of the transistors are named source, gate, and drain. The metal contacts of the terminals form a Schottky junction with the n type semiconductor region called the channel for charge carrier. The voltage on the gate controls the depletion region width and thus controls the current from the source to the drain. The bandwidth of the device is given by

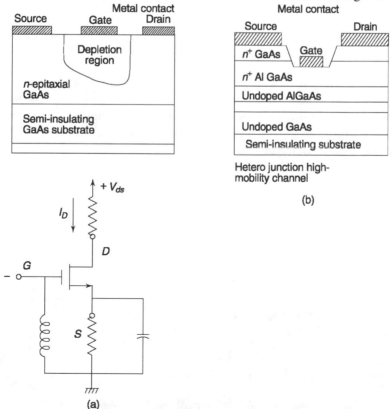

(a)

(b)

Fig. 10.47 *Constructions of MESFET and HEMT (a) MESFET (b) HEMT*

$$f_T = \frac{1}{2 \pi T} \tag{10.59}$$

Where T is the electron transit time through the channel. At microwave frequencies the channel length is extremely short so that the MESFET can operate in 5–20 GHz range.

The output drain current I_D versus drain-to-source voltage V_{ds} is shown in Fig. 10.48 as a function of gate-to-source voltage V_{gs}. For maximum dynamic range the dc voltage of the gate must be negative with respect to the source. This is achieved by grounding the gate through an RF choke.

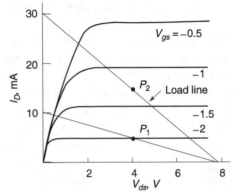

Fig. 10.48 I_D versus V_{ds} characteristics courtesy: R.E. Collin [12]

Microwave transistor amplifier design At microwave frequencies transistor amplifiers are designed using S-parameters of the transistors and using input and output matching networks as shown in Fig. 10.49. Assuming that the transistor amplifier is unilateral ($S_{12} = 0$) and the transistor is conjugate matched to the load and source, the maximum gain is given by

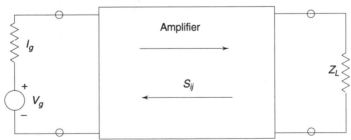

Fig. 10.49 *Microwave transistor amplifier*

$$G_{max} = \frac{|S_{21}|^2}{\left(1 - |S_{11}|^2\right)\left(1 - |S_{22}|^2\right)}$$

$$= \frac{1}{\left(1 - |S_{11}|^2\right)} \cdot |S_{21}|^2 \cdot \frac{1}{\left(1 - |S_{22}|^2\right)}$$

$$= G_i \cdot G_T \cdot G_o \tag{10.60}$$

where G_i and G_o are determined by the input and output matching networks, respectively, and G_T is the gain of the transistor. In general, the amplifier will be stable if $|S_{11}| < 1$ and $|S_{12}| < 1$.

Example 10.7 The S-parameters of a transistor at 5 GHz for a conjugate matched transistor amplifier are given by $S_{11} = 0.9 \angle -100°$, $\angle 90°$, $S_{21} = 2.4$, $S_{12} = 0$, $S_{22} = 0.8 \angle 40°$. Determine the maximum gain.
Solution
The maximum gain of the conjugate matched transistor amplifier is

$$G_{max} = \frac{(2.4)^2}{\left(1 - (0.9)^2 \left(1 - (0.8)^2\right)\right)}$$

$$= 84.21$$

$$= 19.3 \text{ dB}$$

10.8.3 Microwave Transistor Oscillator

Microwave transistor oscillators are designed by choosing the input and output port terminations in the unstable regions where both the input and output impedances of the transistor circuits will have a negative resistance. The oscillations will occur at a frequency at which the total reactance in the input and output circuits becomes zero. Even if the transistor is stable, it is made unstable by using feedback from the output to the input of the circuit. Any of the standard low frequency oscillation circuits such as the Hartley, Colpitts, or Clapp circuits, are used in which the frequency stability is achieved by using a resonator in either the input or output circuits as part of the feedback loop. Figure 10.50 shows a typical FET oscillator which is stabilized by using a resonator in the input circuit. The

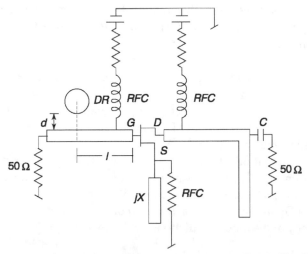

Fig. 10.50 *FET oscillator circuits (source: R.E. Collin [12])*

magnitude and the phase angle of the input reflection coefficient is controlled by adjusting the coupling spacing d between the resonator and the microstrip line, and by the length l of the input line, respectively. The output is matched by using a stub. Unstability of oscillations is met by using a series capacitance reactance in the common source lead. RF chokes are inserted in series with the dc biasing resistances and in shunt with feedback reactance. Si bipolar transistors are used for oscillators at frequencies up to 5 GHz, MESFETs are used from 5 GHz to 40 GHz, and HEMTs are used from 30 GHz to 100 GHz.

EXERCISES

10.1 The dc current through a point contact diode is 1.5 mA at room temperature. Determine the slope conductance and slope resistance, assuming that $n = 1.4$ and $V_T = 26$ mV.

10.2 A sinusoidal microwave signal of peak voltage 50 mV is applied to a microwave diode having $n = 1.4$, $V_T = 26$ mV. $I_a = 1\mu A$. Calculate the resulting dc current.

10.3 A microwave point contact diode detector has a reverse saturation current of 1 μA. Find the detector current for a microwave signal amplitude of 1 V at room temperature. $n = 1.4$ and $V_T = 26$ mV.

10.4 (a) A 0.8 V LO signal at frequency 10 GHz and a 0.5 V input signal at frequency 9.9 GHz are applied to the above diode when used as mixer. Calculate the frequency and magnitude of the IF signal current.

   (b) In a diode mixer circuit the signal frequency is 4 GHz and the local oscillator frequency is 3.93 GHz. Determine the sum frequency, the image frequency, and the IF.

10.5 Define and explain the terms conversion loss and noise figure of a diode mixer circuit. The conversion loss of a mixer is 5 dB. Find the approximate value of noise figure.

10.6 Explain what is meant by the notation 1-2-3 tripler. When the frequency input applied to varactor is 3 GHz, determine the idler frequency and the output frequency.

10.7 For a PIN diode equivalent circuit with $C_p = 0.3$ pF, $C_j = 0.2$ pF, $R_s = 0.3$ ohm and $L_s = 0$, show that $(wC_jR_s)^2 \ll 1$, and calculate the admittance of the diode at 5 GHz.

10.8 A GaAs Gunn diode has an active region length of 10 μm. If the electron drift velocity is 10^5 m/s, find the natural frequency and the critical voltage. The critical electric field is 3 kV/cm.

10.9 A GaAs Gunn diode oscillator operates at 10 GHz with the drift velocity of electrons = 10^5 m/s. Determine the effective length of the active region. What is the required dc voltage for oscillation?

10.10 A GaAs Gunn diode has a drift length of 10 μm. Determine the intrinsic frequency of the diode oscillations.

10.11 A transferred electron diode operates in transit time mode at 10 GHz with the velocity of propagation of 10^7 cm/sec. Calculate the approximate thickness of the device.

10.12 The active impedance of a suitably biased Gunn diode is $-10 + j\,7.5$ ohms at 10 GHz. The diode is to be used in a coaxial line oscillator circuit. Calculate the cavity length required when the characteristic impedance of the coaxial line used is 50 ohms.

10.13 The equivalent circuit parameters for an IMPATT diode oscillator are $R_d = -2$ ohm, $C_j = 0.5$ pF at a breakdown voltage of 80 V with $L_p = 0.55$ nH, and $C_p = 0.3$ pF. The oscillator operates into a resistive load of 2.2 ohm connected across C_p under a bias current of 80 mA. Determine (a) the resonant frequency, (b) the average power output, (c) the average power input, and (d) the conversion efficiency.

10.14 A small-signal IMPATT diode amplifier has an R_d of -3 ohm and is operated into a load of 5 ohm. What is the power gain in dB at the resonant frequency.

10.15 A tunnel diode has $R_j = 60$ ohm, $R_s = 9$ ohm, $C_j = 0.6$ pF, and $L_s = 1$ nH. Find the resistive cut-off frequency f_r and the self-resonant frequency f_0.

10.16 A varactor diode has the junction capacitance 0.5 pF with $v = 0$, the barrier potential 1.1 V and $n = 0.3$. Calculate the junction capacitance and the cut-off frequency for a reverse voltage of 0.8 V, if the substrate resistance $R_s = 0.7$ ohm.

10.17 A parametric diode amplifier has $R_s = 4$ ohm, $R_L = R_1 = R_2 = 0$. The characteristic impedance of the amplifier is 50 ohm. Calculate (a) the reflection coefficient and (b) the power gain.

10.18 A transistor operates at $f = 8.0$ GHz, with $S_{11} = 0.9 \angle 110°$, $S_{21} = 1.1 \angle -30°$, $S_{22} = 0.55 \angle -120°$. For unilateral operation calculate source and load reflection coefficients for a conjugate matched amplifier.

10.19 Calculate the gain factors G_i, G_T, and G_o for the above amplifier in dB.

10.20 In a bipolar transistor, the emitter charging time is 40 per cent of the total transit time. Calculate the percentage change in the cut-off frequency when the emitter current is increased from 100 mA to 200 mA. All other parameters may be assumed to remain constant.

REFERENCES

1. Davis, WA, *Microwave Semiconductor Circuit Design*, Van Nostrand Reinhold, New York, 1984.

2. Liao, SY, *Solid-State Microwave Ampilifier Design*, John Wiley, New York, 1981.

3. Watson, HA, *Microwave Semiconductor Devices and Their Circuit Applications*, McGraw-Hill, New York, 1969.

4. Ha, TT, *Solid-State Microwave Amplifier Design*, John Wiley, New York, 1981.

5. Vendelin, GD, *Design of Amplifiers and Oscillators by the S-Parameter Method*, New York, John Wiley, 1982.

6. G.E. Application note TPD-6104, "Bulk effect diodes and solid state microwave circuit modules."

7. *Hewlett-Packard Application Note 957-1*, "Broadbanding the shunt PIN diode SPDT switch."

8. *Hewlett-Packard Application Note 962*, "Silicon double-drift IMPATTs for high power CW applications".

9. Pengelly, RS, and John Wiley and Sons: *Microwave Field-effect Transistor-theory, Design and Applications*, Research Studies Press, Hertfordshire, England, Inc., New York, 1986.

10. Gonzalez, G, *Microwave Transistor Amplifiers, Analysis and Design*, Prentice-Hall, Inc., Englewood Cliffs, N.J., 1984.

11. Gentile, C, *Microwave Amplifiers and Oscillators*, McGraw-Hill Book Company, New York, 1987.

12. Collin, RE, *Foundation for Microwave Engineering*, Second Edition, McGraw-Hill, Inc., 1992.

13. John, A Seeger, *Microwave Theory, Components, and Devices*, Prentice-Hall, NJ, 1986.

14. Dennis, R, *Microwave Technology*, Prentice-Hall, N.J., 1986.

15. Bahl, I J and P Bhatia, *Microwave Solid State Circuit Design*, John Wiley and Sons, 1988.

Chapter 11

Applications of Microwaves

11.1 INTRODUCTION

The microwave devices and circuits described in the previous chapters have widespread practical applications in the development of microwave antennas, radar systems and communication systems. During World War II and shortly afterwards microwave engineering played a very important role in the development of microwave antennas and high resolution radar systems capable of detecting and locating enemy planes and ships. Even today radar is used in many varied forms, such as, missile tracking radars, fire control radars, weather detecting radars, air traffic control radars, etc., represent a major use of microwave frequencies. In more recent years microwave frequencies have also come into widespread use in communication links, generally referred to as microwave links between two ground stations directly or via satellites for communication of voice, picture, data, and other information. In the modern age, microwaves are increasingly used in domestic, industrial, scientific, and medical applications. Some of these applications are described below.

11.2 MICROWAVE RADAR SYSTEMS

The Radio Detection and Ranging (RADAR) is an electromagnetic device for detecting the presence and location of objects. The presence of objects and their coordinates (range and direction) are determined by the transmission and return of electromagnetic energy. The advantages of the use of microwaves in radar are that very compact, high gain and highly directional antennas can be used. The radar is classified into two categories—primary radar and secondary radar. In primary radar a microwave signal is transmitted and is reflected by the target to be detected. The reflected signal is received by the radar. From the knowledge of the time between the transmission and reception and the velocity of light in space, the range of the target can be calculated. The application of a primary radar is the detection of position and flight path of aircrafts, ships and missiles.

In secondary radar, the target in space receives the radar signal and transmits a coded signal for its identification as a friend. The various frequency bands used in radars are given in Table 11.1.

Table 11.1 *Radar frequency bands*

Band name	Nominal frequency range		
HF	3	– 30	MHz
VHF	30	– 300	MHz
UHF	300	– 1000	MHz
L	1	– 2	GHz
S	2	– 4	GHz
C	4	– 8	GHz
X	8	– 12	GHz
Ku	12	– 18	GHz
K	18	– 27	GHz
Ka	27	– 40	GHz
mm	40	– 3000	GHz

11.2.1 The Radar Equation

It is assumed that the radar transmitter sends out a sinusoidal wave, which is reflected back by the target to the common antenna, where it is picked up by the receiver as shown in Fig.11.1(a). The power density at the target is given by

$$P_d = \frac{P_t\, G_t}{4\,\pi\, R^2}\ \text{W/m}^2 \tag{11.1}$$

where P_t is the transmitted power and G_t is the antenna power gain relative to an isotropic radiator, and R is the distance of the target. The amount of power that will be reflected back in the direction of the incident wave is

$$P_{\text{ref}} = P_d\, \sigma \tag{11.2}$$

where σ is the effective echoing area which converts the incident power density into reflected power. If A_{eff} represents the effective area of the receiving antenna, the received power from the reflected component is

$$P_r = \frac{P_d\, \sigma}{4\,\pi\, R^2}\ A_{\text{eff}} \tag{11.3}$$

Using the last two equations and the relationship $A_{\text{eff}} = \dfrac{G\,\lambda_0^2}{4\,\pi}$ for the common antenna used for transmitting and receiving, the equation for the received power becomes

$$P_r = \frac{P_t\, G^2\, \lambda_0^2\, \sigma}{(4\,\pi)^2\, R^4} \tag{11.4}$$

which means that the received power drops off very rapidly with increase in target range R. The receiver noise power

$$N = FkT_0\, B_n \tag{11.5}$$

where F is the noise factor of the receiver, k is Boltzmann's constant, T_0 is the room temperature in Kelvin, and B_n is the noise bandwidth of the receiver. The signal-to-noise ratio at the input to the radar receiver is therefore

$$S/N = \frac{P_t G^2 \lambda_0^2 \sigma}{(4\pi)^3 R^4 FkT_0 B_n} \tag{11.6}$$

In this equation, no account is taken of the various losses, such as atmospheric attenuation and scattering of the radio waves by rain or precipitation. This equation is the basic radar equation describing all the parameters of the radar system which affect performance. The maximum range R_0 of the radar corresponds to the minimum detectable signal power S_{min} of the receiver, i.e.,

$$R_0 = \left[\frac{P_t G^2 \lambda_0^2 \sigma}{(4\pi)^3 FkT_0 B_n S_{min}} \right]^{1/4}$$

or, $R_0(dB) = 1/4 \, [P_t + 2G + 2\lambda_0 + \sigma + 171 - F - B_n - (S_{min} - N)]$ (11.7)

where $(4\pi)^3 kT_0 = -171$ dB. Navigational distance R_0 is often measured in international nautical miles where 1 nautical mile = 1852 m.

11.2.2 Duplexer

Duplexer is a device which couples the transmitter and receiver to the antenna while producing isolation between the transmitter and receiver. There are two types of duplexers possible, one using PIN switches and the other using circulators as shown in Figs 11.1 (b) and (c).

In Fig. 11.1(b) the duplexer uses transmit-receive (TR) and anti-TR PIN diode switches. Under transmit condition, the TR and ATR switches are closed or short circuited. The $\lambda/4$ sections transform these short circuits to open circuits across the main line to allow power flow from the transmitter to the antenna, while the receiver is protected from high power output of the transmitter by the TR short circuit. In the receive mode both the switches are open circuited. In the ATR branch this is transformed to short circuit across the main line. This prevents any received signal from entering the transmitter. The short circuit on the main line at the ATR connection point is transformed to a open circuit on the main line at the TR connection point and a receive line appears through the connection to the receiver to receive the signal. Since the operations are controlled by $\lambda/4$ lines, this arrangement results in a narrow band system. In Fig. 11.1(c) two ferrite circulators do the function of the duplexer over a broader bandwidth by a circulating action.

11.2.3 Pulsed Radar

When a target is stationary, CW sinusoidal signal returned from the target cannot provide any radar information since the time of return is difficult to measure with respect to time of transmission. Moreover, there would not be any frequency changes to give some information. Hence short RF pulses are used for the determination of the range of a target as explained in Fig.11.2.

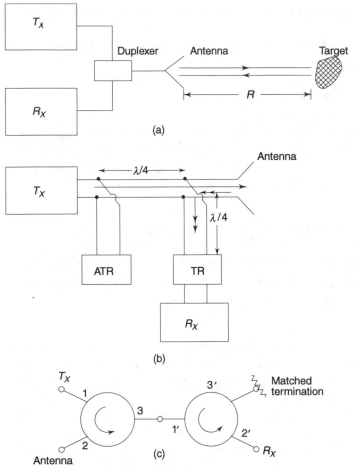

Fig. 11.1 (a) Basic radar system with duplexer (b) PIN switch duplexer (c) Ferrite circulator duplexer

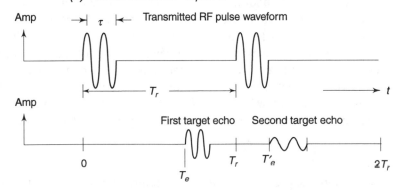

Fig. 11.2 Pulsed radar waveform and detection principle

Here a RF carrier at frequency f_0 is pulsed amplitude modulated with modulating pulse repetition frequency $f_r = 1/T_r$ before transmission. For unambiguous detection, the echo time T_e must be less than the pulse repetition period T_r otherwise an echo would be confused with the reference transmitted pulse number and the range will be ambiguous.

Range concept　The maximum unambiguous range R_{max} is determined by maximum echo time $(T_e = T_r)$ for which

$$R_{max} = \frac{cT_r}{2} \tag{11.8}$$

where c is the velocity of light. Again the minimum range R_{min} is determined by the pulse width τ because if an echo arrived within time $t < \tau$, it cannot be compared with the transmitted pulse. Hence

$$R_{min} = \frac{c\tau}{2} \tag{11.9}$$

When two targets are present, they can be resolved only if the second echo arrived after time τ from the arrival first time of the each. Therefore, for target resolution, the separation between the pulses should be greater than or equal to τ.

Power concept　For the RF pulse of width τ, with period T and current maximum I_m, the peak power $P_m = I_m^2 R_r$, where R_r is the radiation resistance of the antenna. The peak average power over a pulse width τ is

$$P = \frac{P_m}{2} = \frac{I_m^2 R_r}{2} \tag{11.10}$$

The pulse radar wave and the instantaneous power wave are shown in Fig.11.3. Here the spectrum of RF pulses contains an infinite side frequencies $f_0 \pm nf_r$, $n = 1, 2, 3, \ldots$ If f_r is low, most of the side frequencies will be close to f_0 and the complete spectrum will be received. The peak average power P will be used in the radar equation. If f_r is high, the receiver may receive only the carrier and closer side frequencies. The average power in the carrier component of the modulated wave, is

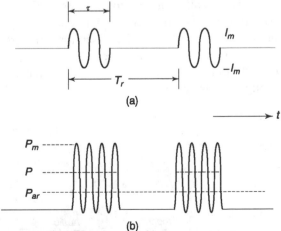

(a)

(b)

Fig. 11.3　*(a) Pulsed radar waveform (b) Corresponding instantaneous power waveform*

$$P_0 = P_{av}^\tau /T \tag{11.11}$$

where true average power $P_{av} = P\tau/T$. Here τ/T is called the duty factor. This average power P_0 must be used in the radar equation when prf is f_r and is high.

11.2.4 CW Doppler Radar

When a moving target is detected by a continuous wave radar signal, frequency shift occurs in the return signal given by doppler frequency shift

$$f_d = \frac{2 f_0}{c} \frac{dR}{dt} = \frac{2\dfrac{dR}{dt}}{\lambda_0} = \frac{2v\cos\theta}{\lambda_0} \tag{11.12}$$

where v is the relative velocity of the target making an angle θ with the position vector R of the target with respect to the radar antenna as shown in Fig.11.4 and dR/dt represents the range rate. When the target is moving towards the radar ($0 \le \theta < \pi/2$) receiver, the signal frequency f_0 increases by f_d and when it is moving away from the receiver ($\pi/2 < \theta \le \pi$) the frequency decreases by f_d. If the velocity is perpendicular ($\theta = \pi/2$) to **R**, there is no change in the frequency, or the doppler frequency is zero. By measuring the doppler frequency shift and noting the polarity of f_d the range rate for the target can be determined.

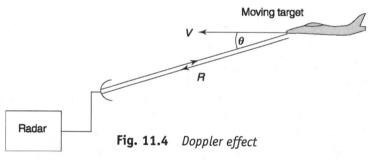

Fig. 11.4 *Doppler effect*

The doppler frequency shift can be utilised to design low cost radar for intruder alarms, speed detectors, and traffic signal control. Figure 11.5 shows a basic block diagram of a CW Doppler radar. A circulator is used for the duplexer with a single antenna for transmission and reception. A Doppler mixer produces the signal output at the doppler frequency by mixing the received signal with the signal leaked through the circulator.

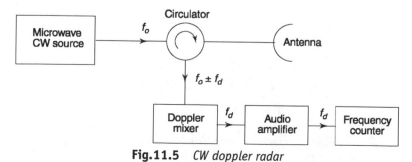

Fig.11.5 *CW doppler radar*

11.2.5 FMCW Radar

FM Radar In a FMCW radar, the carrier signal at a frequency f_0 is frequency modulated by the frequency $f_m(t)$ such that the instantaneous FM signal frequency of transmitted signal is given by

$$f_i = f_0 + f_m(t) \tag{11.13}$$

A basic block diagram of the FMCW radar is shown in Fig.11.6.

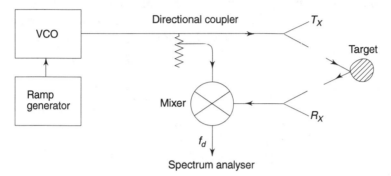

Fig. 11.6 *Basic block diagram of FMCW radar*

The received signal frequency is represented by $f_{ir} = f_0 + f_m(t - T_e)$, where T_e is the time of the echo signal after transmission. The frequency difference between the transmitted and the received signal is $f_d = f_m(t) - f_m(t - T_e)$. The transmitted and received waveform of a FMCW radar is shown in Fig. 11.7.

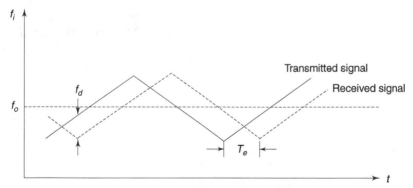

Fig. 11.7 *FMCW radar waveforms*

The commonly used frequency modulation wave shape is triangular. The slope of the line in any given time interval is given by

$$a = f_d/T_e \tag{11.14}$$

The range of the target is, therefore, expressed by

$$R = cf_d/2a \tag{11.15}$$

The advantages of the FMCW radar are
1. It operates at a continuous power level that is relatively low as compared to a pulse radar.
2. It is used mainly for short range because its minimum range is not limited.

FM Doppler Radar FMCW radar is used for the determination of the velocity and range of a moving target. When a triangular frequency modulation waveform returns from a moving target with a velocity v, the frequency difference between the transmitted and received signals contains the doppler frequency shift f_{dop} $= 2\,v\cos\theta/\lambda_0$ to result in waveforms as shown in Fig. 11.8, where θ is the angle between the direction of v and the radial vector of the target with respect to the radar as the origin.

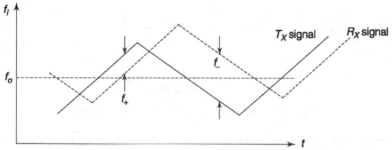

Fig. 11.8 *FMCW doppler shift waveforms*

If the target moves towards the radar, the frequency difference between the transmitted and received signals over the positive slopes decreases and is given by $f_+ = (f_d - f_{dop})$ and over the negative slope it increases ($f_- = f_d + f_{dop}$) where f_d is the frequency difference between T_x and R_x signals for stationary target. Hence

$$f_d = (f_+ + f_-)/2 \tag{11.16}$$

$$f_{dop} = (f_- - f_+)/2 \tag{11.17}$$

The range is given by $R = cf_d/2a$

$$= c(f_+ + f_-)/4a \tag{11.18}$$

where a is the slope of the line forming the triangular waves.
Now

$$f_{dop} = \frac{2(dR/dt)}{\lambda_0} \tag{11.19}$$

or,

$$v_r = \frac{dR}{dt} = \frac{\lambda_0 f_{dop}}{2}$$

$$= \frac{\lambda_0}{4}\,(f_- - f_+) \tag{11.20}$$

Thus by measuring the frequency shifts over positive and negative slopes, the range and range rate can be found.

FMCW doppler radar is used to find the altitude of aircraft, speed of vehicles, to detect road vehicles ahead and give a warning if the closing speed or distance is potentially dangerous by using a transmitter power of a few watts.

Example 11.2 A 1 kW, 3 GHz radar uses single antenna with a gain of 30 dB. The receiver has noise bandwidth of 1 kHz and noise factor 5 dB. A target of echoing area of 10 m^2 at a range of 10 nautical miles is to be detected. Calculate the minimum *S/N*.
Solution

$$\lambda_0 = 30/3 = 10 \text{ cm} = 0.1 \text{ m}$$
$$G = 30 \text{ dB} = 10^3 = 1000$$
$$P_t = 1000 \text{ W}, F = 5 \text{ dB} \rightarrow 10^{5/10} = 3.16$$
$$(4\pi)^3 kT_0 = 7.9433 \times 10^{-18}$$
$$R = 10 \text{ n mile} = 18520 \text{ m}$$
$$B_n = 1 \text{ kHz} = 1000 \text{ Hz}$$

$$S/N = \frac{P_T G^2 \lambda_0^2 \sigma}{(4\pi)^3 R^4 FkT_0 B_n}$$

$$= \frac{10^3 \times 10^6 \times 0.01 \times 10}{7.9433 \times 10^{-18} \times (18520)^4 \times 3.16 \times 10^3}$$

$$= \frac{10^5 \times 10^{18}}{7.9433 \times (18520)^4 \times 3.16}$$

$$= \frac{10^5}{2.953}$$

$$= 45.3 \text{ dB}$$

Example 11.3 A radar system operates at 10 GHz with a common antenna with a gain of 30 dB. The receiver has a bandwidth of 1 kHz and the noise factor is 5 dB. The transmitted power is 1 kW and the target echoing area is 10 m^2. Calculate its range for *S/N* = 10.
Solution

$$\lambda_0 = c/f = 30/10 = 3 \text{ cm} = 0.03 \text{ m} = 10 \log(0.03) = -15.23 \text{ dB meter}$$
$$B_n = 1 \text{ kHz} = 1000 \text{ Hz} = 10 \log 1000 = 30 \text{ dB Hz}$$
$$P_t = 1 \text{ kW} = 1000 \text{ W} = 10 \log 1000 = 30 \text{ dBW}$$
$$G(\text{dB}) = 30 \text{ dB}$$
$$\sigma = 10 \text{ m}^2 = 10 \log 10 = 10 \text{ dB Sq.m}$$
$$F = 5 \text{ dB}$$
$$S/N = 10 \log 10 = 10 \text{ dB}$$

Therefore,

$$R_0 \text{ (dB)} = \frac{1}{4} [30 + 2 \times 30 - 2 \times 15.23 + 10 + 171 - 5 - 30 - 10]$$

$$= \frac{1}{4} [271 - 75.46]$$

$$= 48.89 \text{ dB metre}$$

or, $$R_0 = 10^{(48.89/10)} \text{ m}$$

$$= 77.45 \text{ km}$$

11.2.6 Radar Cross-section

Radar cross-section of a target is defined as

$$\sigma = \frac{\text{Power re-radiated from the target}}{\text{Incident power density on the target}} \tag{11.21}$$

and is given by the conventional radar range equation

$$P_r = \frac{P_t \, G_t \, A_r \sigma}{(4\pi R^2)^2} = K \frac{\sigma}{R^4} \tag{11.22}$$

If P_t, G_t and R are kept fixed,

$$P_r \propto \sigma \tag{11.23}$$

Again σ is a function of aspect angle θ and hence P_r is a function of θ. By measuring P_r for different aspect angles θ the normalised radar cross-section $\sigma(\theta)/\sigma(0)$ of the target can be obtained for different aspects.

11.3 MICROWAVE COMMUNICATION SYSTEMS

Microwaves are used in wireless communication to an accommodate increased number of channels with a high signal-to-noise power ratio. Since the bandwidth can be a small fraction of the microwave carrier frequency f_c, by using a higher carrier frequency the overall system bandwidth can be increased, and more distinct channels can be transmitted at microwave frequencies than at much lower frequencies. Moreover, high resolution radiation beam (narrow) can be generated with relatively smaller size antennas and unlike UHF/VHF systems, atmospheric noise and man made interferences caused by electric transient, automobile ignition system, etc. are negligibly small. Since microwaves penetrate ionosphere and any possibility of ground reflected waves and surface waves suffer heavy attenuation, the microwave communication is line of sight (LOS) communication. In general, microwave communication system requires transmit-receive equipment at each end for the purpose of transmitting the baseband signal to the microwave carrier and receiving the baseband signal from the microwave carrier. The baseband signal is usually a multiplexed signal, carrying a number of individual telephone signals, video signals and data. Microwave communication systems can be broadly classified as terrestrial systems and satellite systems.

11.3.1 Terrestrial Systems

Since ground based LOS distance is limited due to the attenuation of microwaves caused by the geographical profile, trees, building, etc., long distance microwave communications are achieved through a multisection radio relay system. It consists of two terminal stations and a number of repeater stations spaced at intervals of about 32–80 km each of which amplifies the microwave signals to make up for transmission losses in space and retransmits again for the next stage. A basic block diagram of the terrestrial microwave communication system is shown in Fig.11.9.

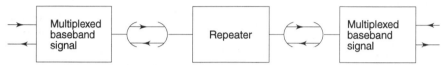

Fig. 11.9 *Basic block diagram of a terrestrial microwave communication system*

The systems may employ analog or digital modulation. In analog systems (FDM/FM) the information signals are frequency-division multiplexed (FDM) to form the baseband signal which is then frequency modulated (FM) onto the microwave carrier for transmission through an antenna. In digital systems (TDM/PSK) the information signals are time-division multiplexed (TDM) to form the baseband signal which is then phase modulated (PM) by phase shift keying (PSK) onto the microwave carrier.

(a) Terminal systems   FDM is more commonly used in that a number of telephone channels (300 Hz – 3.4 kHz) are transmitted on the same microwave carrier by FM. The telephone channels are arranged side by side in a baseband at intervals of 4 KHz as shown in Fig.11.10. The base band frequency limits are given in Table 11.2.

Table 11.2   *Baseband frequency limits*

Number of channels	Baseband frequency limit (kHz)
12/1 group	12–60 or 60–108
24/2 groups	12–108
60/1 super group	12–252 or 60–300
120/2 super group	12–552 or 60–552
240/4 super group	60–1052
600/10 super group	60–2540

In the terminal system each of the multichannel baseband inputs first modulates a 70 MHz IF carrier and then this IF signal is upconverted to a microwave carrier. All microwave carriers $f_1, f_2, \ldots$ are passed to a branching filter to form a FDM signal for transmission through a polarisation filter and antenna as shown in Fig.11.11.

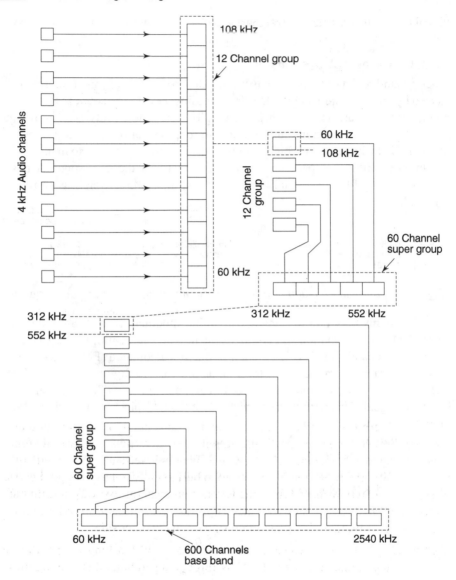

Fig. 11.10 *FDM multichannel telephone signals*

The multi-carrier received signal is received by the common antenna and passed through the polarisation filter to the receive branching filter. The output of the branching filter consists of a multichannel carrier $f_1', f_2', ...,$ which are down converted at each separate channel to 70 MHz IF and then demodulated to baseband signal.

(b) Repeaters The block diagram of a repeater system is shown in Fig.11.12. Incoming carriers $f_1, f_2, ...,$ polarised horizontally, are received from the left side by the antenna and converted to 70 MHz IF. The IF signals are upconverted to

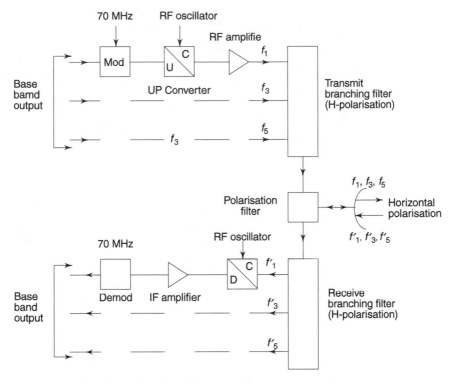

Fig. 11.11 *Block diagram of a terminal system*

different microwave carriers $f_1', f_2', \ldots$, and retransmitted on the next hop after passing through branching and polarisation filter as vertically polarised signals from the right side antenna. Similarly, the incoming vertically polarised signal $f_1, f_2, \ldots$, from the right side is converted to horizontally polarised signal carrier $f_1', f_2', \ldots$, and retransmitted from the left side antenna for two way communications. The signals at two sides are made oppositively polarised in order to minimise direct coupling of the transmitted signals on the other side.

Example 11.4 A microwave terrestrial link of 30 km long is operating at 4 GHz with radiated power of 100 W through a parabolic disk having maximum gain of 50 dB. The receiver uses similar antenna. Find the free space loss and the received power.

Solution

$$P_t(\text{dBm}) = 10 \log P_t(\text{mW}) = 10 \log 100 \times 10^3 = 50$$
$$L(\text{dB}) = 32.5 + 20 \log 30 + 20 \log 4 \times 10^3$$
$$= 134.1$$
$$P_r(\text{dBm}) = P_t(\text{dBm}) + G_t(\text{dB}) + G_r(\text{dB}) - L(\text{dB})$$
$$= 50 + 50 + 50 - 134.1 = 15.9$$

(c) Radio path planning In order to plan for obstacle free communication the first fresnel zone ellipsoid around the direct ray path must be determined. The following factors are considered for radio path planning.

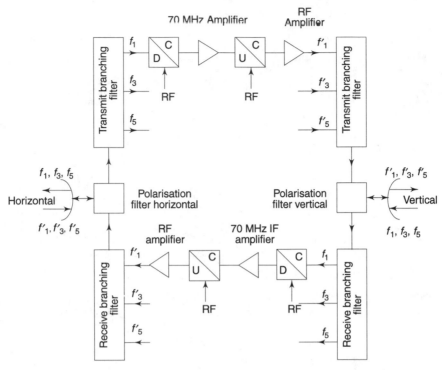

Fig. 11.12 *Block diagram of a heterodyne repeater*

Atmospheric refraction Since the dielectric constant of the atmosphere changes, the refractive index of the atmosphere decreases with increase in height above the earth surface. The propagation path of microwaves therefore bends towards the earth as shown in Fig. 11.13. Since the change in refractive index is very small, such a curved ray path can be modelled by an equivalent straight ray path over a spherical earth of radius a' greater than radius a of the original earth producing same result, for the received signal, where $a' \approx (4/3)a$. With this model, the maximum distance R_{max} between a transmitter and a receiver, is a function of antenna heights above ground when the ray path just grazes the equivalent earth's surface. This distance R_{max} is called *radio horizon*. From Fig. 11.13,

$$(a')^2 + R_1^2 = (a' + h_t)^2 \tag{11.24}$$

Therefore, $\qquad R_1^2 \approx 2a' \, h_t$, since $a' >> h_t$ $\qquad\qquad$ (11.25)

Similarly, $\qquad R_2^2 = 2a' \, h_r$ $\qquad\qquad\qquad\qquad\qquad$ (11.26)

The radio horizon

$$R_{max} \approx R_1 + R_2$$

$$= \sqrt{(2a')} \left(\sqrt{h_t} + \sqrt{h_r} \right) \tag{11.27}$$

Taking $a = 6376$ km,

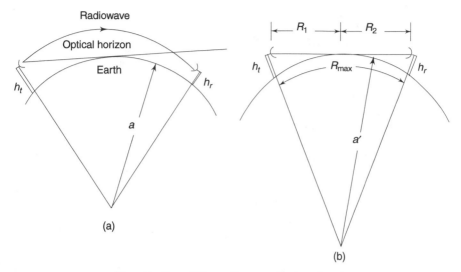

Fig. 11.13 *Effect of atmospheric refraction*

$$R_{max}(\text{km}) = \sqrt{17} \ (\sqrt{h_t} \ (\text{m}) + \sqrt{h_r} \ (\text{m})) \tag{11.28}$$

Contour maps If the distance $R = \sqrt{(17h)}$ of one antenna from the grazing point P is plotted with $R(\text{km})$ as abscissa and $h(\text{m})$ as ordinates, a contour is obtained as shown in Fig.11.14(a) where the vertical distance h of a point on the curve from the tangent drawn at any point P determines the distance R (km) by the equation

$$R(\text{km}) = \sqrt{[17h(\text{m})]} \tag{11.29}$$

When this curve is inverted, a baseline is formed by it as shown in Fig.11.14(b). Then the line-of-sight distance $R(\text{km})$ from the antenna to the horizon is obtained corresponding to the antenna height $h(\text{m})$ measured from the baseline to the tangent drawn on the base line at P. The profile of the terrain between the transmitter and receiver when plotted on this graph, is obtained known as the contour map. This map helps to achieve obstacle free line-of-sight transmission.

Fresnel ellipse The above line-of-sight concept is developed assuming that the propagating wavefront is plane. However, in practice the wavefront is not plane for a finite practical distance between T_x and R_x. If we model the transmitting source as a point source, the wavefront will be spherical. Each spherical wavefront can be assumed to consist of a large number of secondary source points P's which in turn generate secondary waves. These secondary waves reach the receiver with a phase lag in comparison to the direct path. Therefore, even an obstacle does not block direct ray path, it may fall on the secondary wavelet generation point P and reduces the signal at the receiver point due to interference with the direct ray path. It can be shown by the diffraction theory that the indirect path $TP + PR$ = direct path $TR + \lambda_0/2$ must be cleared for obstruction less trans-

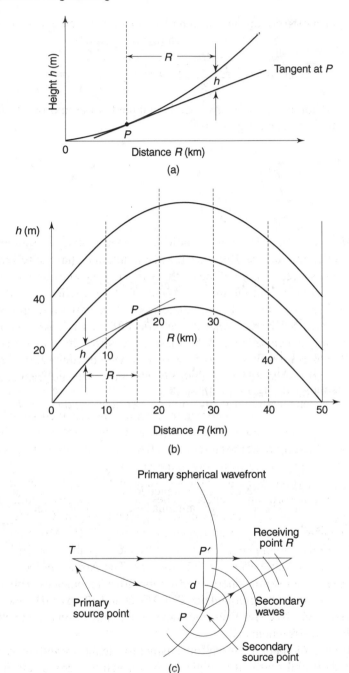

Fig. 11.14 *(a) Contour of h vs R (b) contour maps*
(c) Determination of Fresnal zone

mission. The locus of P to satisfy the equation $TP + PR = TR + \lambda_0/2$ describes an ellipsoid around the direct ray path TR. This ellipsoid surface encloses a volume called the first Fresnel-zone, which must be cleared of any obstacle to obtain free space propagation condition. The length of the perpendicular drawn from a point P on the ellipsoid to the direct ray path (axis) is given by

$$d = \sqrt{\left[\frac{TP' \times P'R}{TR} \lambda_0\right]} \tag{11.30}$$

The maximum clearance distance is obtained at the mid point ($TP' = P'R = TR/2$) of the range and is given by

$$d_{max} = \sqrt{\left[\frac{TP' \times TP'}{TR} \lambda_0\right]}$$

$$= \sqrt{\left(\frac{TR}{4} \lambda_0\right)} \tag{11.31}$$

Ground reflections The above radio horizon calculation assumes a perfect smooth reflecting surface. But since the magnitude and phase of reflection coefficient depends on the nature of the earth's surface, frequency and the polarisation of the waves, the contour map clearance will be applicable only when the Rayleigh's criterion for surface roughness is met. A reasonable approximation is that in the terrestrial links, microwave reflection coefficients have magnitudes less than unity but close to it and the phase angle is 180° for both horizontally and vertically polarised waves. According to the Rayleigh criterion, the height of surface irregularities should be less than $3.6 \lambda_0/\theta$, where θ is the grazing angle of the ray path with the earth surface in radians.

(d) Fading Fading is a phenomenon in which the received signal strength is reduced or lost due to atmospheric effects and ground reflections. Accordingly four types of fading-absorption fading, reflection multipath fading, atmospheric multipath fading and sub-refraction fading exist.

There are several elements in the atmosphere that cause attenuation of microwaves due to absorption of energy. These are (i) rain, (ii) cloud, (iii) fog and airgas molecules, such as water vapour (H_2O) and oxygen (O_2). Heavy rain seriously attenuates microwaves at frequencies above 10 GHz. Moderate rain, clouds and fog seriously attenuate microwaves at frequencies above 30 GHz. Snow does not have a remarkable effect on microwaves. The Airgas molecules attenuates microwaves by absorbing energy due to vibrational resonances. The peak attenuations occur for H_2O at frequencies 22.22 GHz and 176.47 GHz and for O_2 they are 60 GHz and 120 GHz. Thus microwave communications must be established excluding these frequencies.

After the radio path planning for satisfactory signal strength there is possibility of diurnal and seasonal changes in the refractive index of air. These cause changes in the relative path length between the direct and ground reflected rays as shown in Fig.11.15(a). The corresponding phase change may reduce the resultant signal strength or even cancellation of the signal. This is called *reflection multipath fading*. Again, due to decrease in refractive index with height, a second atmospheric ray path occurs in addition to the direct ray path as shown in

Fig.11.15(b). The corresponding phase difference between these two rays produces a cancellation of the signal, or *atmospheric multipath fading*. Due to another effect in the atmosphere, which reduces signal strength by bending the ray away from the receiving point as shown in Fig.11.15(c), is known as *subrefraction* fading.

The fading caused by change of refractive index occurs at frequencies below 8 GHz, whereas that caused by rain, cloud, fog and gas molecules (H_2O and O_2) occurs at frequencies above 10 GHz.

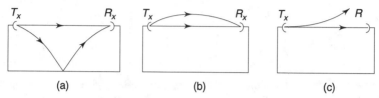

Fig. 11.15 *Fading mechanism (a) Reflection multipath fading (b) Atmospheric multipath fading (c) Subrefraction fading*

The fading is minimised by two techniques called *frequency diversity* and *space diversity* as explained in Fig. 11.16.

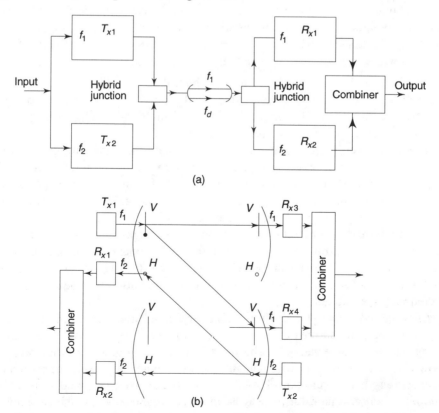

Fig. 11.16 *(a) Frequency diversity technique (b) Space diversity technique*

In *frequency diversity* technique, the signal is sent through the same path over two different frequencies f_1 and f_2, where $f_1 \sim f_2 \geq 100$ MHz so that statistically it is highly unlikely to simultaneously fade both the signal frequencies. Since this method requires two transmitters and two receivers for two frequencies, the system cost is very high. Moreover, licensing restriction on the allocation of frequencies in the microwave band prevents easy use of this technique.

In *space diversity* technique the signals are transmitted from left to right at least through two different paths in space at same frequency f_1 from a single antenna (vertically polarised) and received by two different antennas spaced vertically at a distance equal to a minimum of 50 wavelengths. Outputs of the two receivers are combined to give a single desired output. Similarly, right to left transmission is made at a different frequency f_2 with horizontal polarisation.

In case fading cannot be eliminated, a fading margin of 30 dB is given in the system design for a 99.99 % availability of the system reliably over the time during the worst fading month.

In communication link calculations, the free space propagation equation is no longer valid due to practical difficulties in the link as described above. Hence a modified equation is required to be formulated as follows.

Since the noise level in the receiver would degrade the S/N ratio in the analog system and produce errors in the digital system, output power P_0 of the final amplifier stage of the transmitter and minimum detectable receiving carrier power $S_{\min}$ should be considered in link calculations which results in the equation

$$P_0(\text{dBm}) \geq S_{\min}(\text{dBm}) + FM(\text{dB}) + L(\text{dB})$$
$$+ L_b(\text{dB}) + L_f(\text{dB}) - G_t(\text{dB}) - G_r(\text{dB}) \tag{11.32}$$

where, $FM(\text{dB})$ is the fading margin, $L_b(\text{dB})$ is the loss in the branching circuits of the transmitter and receiver, such as, filters and circulators, $L_f(\text{dB})$ is the loss in the transmitter and receiver antenna feeder lines. The overall system gain is defined as

$$G_s(\text{dB}) = P_0(\text{dBm}) - S_{\min}(\text{dBm}) \tag{11.33}$$

A typical value of G_s is 100 dB for a well designed system.

11.3.2 Satellite Communication Systems

In satellite communication systems the information is transmitted at microwave frequency through a highly directional antenna from a ground station to the geosynchronous satellite, which receives the signal through an on-board antenna, shifts the frequency, and amplifies it by means of a low-noise wide-band amplifier (transponder). Then this received signal is retransmitted towards the earth at frequency 2 GHz lower than the uplink frequency. The usual uplink frequency is 6 GHz and downlink frequency is 4 GHz covering the C-band. This frequency conversion is necessary to avoid interference between uplink and downlink. Satellites are used to handle the long distance telephone traffic, to relay TV signals across oceans and to provide national TV cables directly to the home.

The satellite frequencies are chosen such that the effect of the ionosphere is negligible and absorption by atmospheric gases and water vapour is very small. These satellites orbit around the earth in a circle at an approximate speed of 11,000 km/hour but are geostationary at approximately 35,800 km above the earth surface. Thus no tracking is required and a fixed antenna in the ground station establishes the link at all times.

The stage between the reception of the uplink signal and the transmission of the downlink signal is called a *transponder*. A satellite commonly contains 12 transponders having bandwidth of 36 MHz each. The whole satellite has a bandwidth of 500 MHz. There are satellites that operate at other frequency bands such as 14 GHz/11 GHz and 17 GHz/12 GHz. Figure 11.17 shows the block diagram of a typical transponder. The uplink transmitter power is nearly 25–110 watt, while the downlink transmission power is nearly 7 watt. Table 11.3 shows the satellite bands and the corresponding frequency ranges.

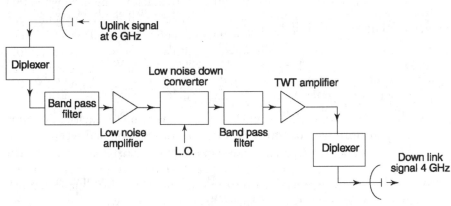

Fig. 11.17   *Block diagram of a transponder*

Table 11.3   *Satellite bands and frequencies*

Satellite band	Frequency range
6 GHz/4 GHz	5.93 – 6.42 GHz/3.705 – 4.195 GHz
14 GHz/11 GHz	14.0 – 14.5 GHz/11.45 – 11.7 GHz

Modulation of link signal   For TV transmission, the video broadcast signal bandwidth is 4.5 MHz including the synchronizing, blanking and equalizing pulses, which frequency modulates the microwave carrier at 6 GHz band. The sound signal has a bandwidth of 15 kHz which frequency modulates a 6.8 MHz sub-carrier. The sound information is frequency division multiplexed with the video information and relayed by a single transponder. The frequency deviation of the video signal and sound signal are 10.5 MHz and 2 MHz, respectively. Therefore, the bandwidth of 2 (10.5 + 2 + 6.8) = 36.5 MHz ≈ 36 MHz should be assigned to the transponders. The total bandwidth for 12 transponder plus guard bands is accommodated within 500 MHz.

For data and telephone signal relay, the time division multiplexing (TDM) or frequency division multiplexing (FDM) is used. In TDM, the data is converted to TDM signals which modulate a microwave carrier digitally for transmission. In FDM, modems are used to convert the data into signals which are compatible with voice frequency bandwidth. The microwave carrier is analog modulated by these signals. The voice frequency signal is also pulse code modulated (PCM) and then time division multiplexed and finally merged with the digital signal.

Satellite antennas   Spacecraft antennas are mounted on the body of the satellite to provide coverage of a certain zone on the earth's surface. Monopole, dipole, horn antennas, reflector antennas, and microstrip array antennas are used. Earth station antennas are large parabolic reflector antennas having high gain to provide a narrow (pencil) beam pointing towards the satellite. The monopole and dipole are used primarily at VHF/UHF frequencies for omni-directional coverage in T.T.C systems where the aperture antenna would have a very large size to increase the gain at very large wavelengths. Horn antennas are used when relatively wide beams are required for global coverage. It is difficult to obtain a gain much greater than 23 dB. The reflector antenna with feed placed at its focus produces plane waves giving maximum gain and narrow beam for coverage of a particular zone on the earth. In order to reduce the complexity in the mechanical design for the reflector, an array of printed antennas is the alternative to paraboloid antennas for high gain or narrow beamwidth.

In an ideal spacecraft, there would be one antenna beam for each earth station, completely isolated from all other beams, for transmission and reception. For multiple earth stations, a separate beam should be provided for each station. This would also require one antenna feed per earth station with a single reflector. INTELSAT V spacecraft has four reflector antennas. Each is illuminated by a complex feed that provides the required beam shape to cover all earth stations within a given coverage zone.

The largest reflector should be used to transmit at 4 GHz for the zone beams (peanut) to concentrate onto densely populated areas where much telecommunications traffic is generated. The smaller antennas are used to provide hemisphere transmit and receive beams, and the 14/11 GHz spot beams.

Most domestic satellites do not have complex antenna systems, but use orthogonal polarisation frequency reuse to double the effective bandwidth at 6/4 GHz for separate channels with cross polar isolation of 25 dB. Table 11.4 shows various type of antennas used in Indian satellites.

Table 11.3 *Indian satellite antennas*

Satellite	Antenna type	Freq. in MHz	Use
INSAT-1D	Horn/Dish	C Band (6000/4000)	T.T.C
	Patch (circular)	UHF (up link)	Metrological data
	Dish	(6000/4000)	Communication

INSAT-2A	Horn/Dish	C Band	T.T.C
INSAT-2B	Horn/Dish		
	Patch (circular)	UHF (up link)	Metrological data
	Dish	(6000/4000)	Communication
Ground applications			
INSAT-1D	Helix	UHF (up link)	Metrological data
INSAT-2A	Dish	C Band	Communication
INSAT-2B	Dish	C Band	Communication

Example 11.5 Let an illumination zone subtend an angle of $6° \times 3°$ when viewed from a geostationary orbit. What dimension must a reflector antenna have to illuminate half this area with a circular beam $3°$ in diameter at 11 GHz? Can a reflector be used to produce $6° \times 3°$ beam ? What gain would the antenna have ?

Solution

(i)   For a $3°$ circular beam

$$\theta_3 = \frac{75\,\lambda_0}{D}; \text{ at 11 GHz}, \rightarrow \lambda_0 = 0.0272 \text{ m}$$

Therefore, $D/\lambda_0 = 75/3 = 25$

or, $D = 25\lambda_0 = 0.68 \text{ m} \approx 2 \text{ ft.}$

(ii) To generate a beam of $6° \times 3°$ we need aperture with dimensions in two orthogonal planes as

$$D_{1,\,2} = \frac{74\,\lambda_0}{6}, \frac{75\,\lambda_0}{3} = 12.5\,\lambda_0, 25\,\lambda_0$$

Therefore,

$$G = \frac{30,000}{6° \times 3°} \approx 32 \text{ dB}$$

A feed horn of unequal beam width is required to illuminate such a reflector with elliptical aperture. But this results in poor polarisation characteristics.

Thus when orthogonal polarisation is to be transmitted or received, it is better to use a circular reflector.

(iii) $G = 30,000/3^2 \approx 35$ dB

Ground stations   The ground station has a need to achieve a low system noise temperature in the receiving mode or during down link, due to a very weak signal being received. The antenna characteristics of the earth stations are very important in their design. In a large station, a 30 m dia dish antenna is used, particularly in Intelsat network. There are small earth stations which use a 0.7 m dia dish antenna for reception of Direct Broadcast Satellite (DBS) TV.

The carrier-to-noise power ratio in a ground station is proportional to G/T (dB/K), where G is the gain of the antenna >25 dB to satisfy link requirements, T is the system noise temperature in Kelvin. Typically

$$G/T = G(\text{dB}) - 10 \log T$$
$$= -10 \text{ to } + 46 \text{ dB/K} \tag{11.34}$$

$G/T = 46$ dB/K (max.) corresponds to $G = 65$ dB, $T = 70$ K. The optimum G/T for a given application is a compromise between the cost of large antenna to increase G and the cost of lower system noise to decrease T. Since the beam width is very small, most large antennas are equipped with automatic tracking facilities so that the motion of the satellite can be followed.

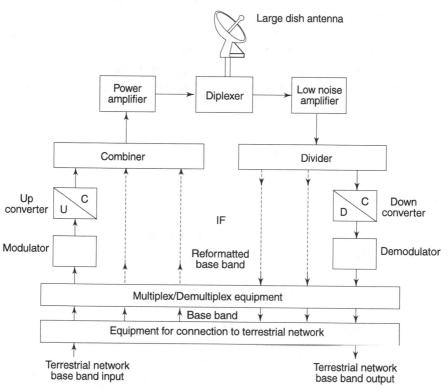

Fig. 11.18 *Ground stations*

Important parameters for satellite antennas
EIRP : EIRP is the most useful parameter of an antenna which is called the effective isotropically radiated power of the satellite antenna.

$$\text{EIRP} = \text{Antenna gain} * \text{Transmitter power output} = G_t P_t.$$
$$= 10 \log_{10} [P_t G_t] \text{ dBW} \tag{11.35}$$

Satellite received power It is expressed in terms of EIRP

$$P_r = [P_t G_t] \, G_r \left(\frac{\lambda_0}{4\pi R}\right)^2$$

$$= \text{EIRP} * G_r/L_p \tag{11.36}$$

$$\text{Path loss} = L_p = \left[\frac{4\pi R}{\lambda_0}\right]^2 \tag{11.37}$$

assuming no other losses in the antenna.

G/T Ratio for earth station In satellite communication, the noise in R_x is reduced as far as possible because the signals are weak due to large distances involved. The carrier-to-noise ratio at the demodulator is given by

$$C/N = \frac{P_r G}{[kT_S B]G} = \frac{P_r}{kT_s B} \tag{11.38}$$

Here T_s = system noise temperature = noise temperature of a noise source located at the input of a noise less R_x which gives the same noise power as the original R_x at the output. $T_s = T_1$ (thermal noise) + T_2 (sky noise). Therefore,

$$C/N = \frac{P_t G_t G_r}{kT_s B} \left[\frac{\lambda_0}{4\pi R}\right]^2 \alpha \frac{G_r}{T_s} \tag{11.39}$$

for a given satellite system. Therefore, G/T should be large to increase C/N.

Example 11.6 In a satellite communication link the earth station G/T ratio is 10 dBk, the satellite EIRP is 50 dBW and the propagation loss is 200 dB. If the sum of the fading margin, antenna pointing margin and equipment margin is 3 dB, calculate the received C/N ratio.

Solution

$$k = 1.38 \times 10^{-23} \text{ J/K}$$

$$C/N = \text{EIRP} - \text{margin} - \text{propagation loss} + G/T - 10 \log K$$

$$= 50 - 3 - 200 + 10 + 228.6$$

$$= 288.6 - 203$$

$$= 85.6 \text{ dB Hz}$$

11.4 INDUSTRIAL APPLICATION OF MICROWAVES

Microwaves have been used for industrial, scientific, and medical applications at frequencies 896, 950, 2450, 3300, 5800, and 10,525 MHz. Most industrial applications of microwaves utilised the heating effect of microwaves in the industrial processes: like cooking, baking, puffing, drying, curing, evaporating, sterilising, moulding, etc. Many industries such as food, chemicals, rubber, textiles, plastics, paper, ceramic, cosmetics, etc. use microwave energy.

11.4.1 Microwave Heating

In 1946, Dr. Parcy Spencer, an engineer with the Raytheon Corporation, USA, while operating a magnetron discovered that the candy bar in his pocket melted and also the popcorn placed in front of the magnetron sputtered, crackled, and popped all over the lab. He also observed that the egg placed in front of magnetron exploded and splattered hot yoke all over. In this process, Spencer discovered that microwaves can be used for heating and cooking.

Mechanism of microwave heating Any dielectric material exhibits loss due to non-zero conductivity as well as permittivity. Microwaves penetrate into such materials and dissipate as heat due to ohmic losses. The normal conductive effects are made up of free electron conduction, rotational and vibrational losses in the molecules (Fig.11.19a). As a result, an equivalent conductivity $\sigma\,(S/m)$ is defined such that the dielectric constant becomes complex

$$\varepsilon_r = \varepsilon_r' - j\varepsilon_r'' \tag{11.40}$$

where ε_r' is the relative permittivity and ε_r'' is the loss factor given by

$$\varepsilon_r'' = \sigma/\omega\varepsilon_0 \tag{11.41}$$

The losses in the dielectric are measured in terms of loss tangent

$$\tan\delta = \varepsilon_r''/\varepsilon_r' = \sigma/\omega\varepsilon_r'\varepsilon_0 \tag{11.42}$$

The amount of microwave radiation is measured in terms of power density. For incident microwaves, the power density inside the dielectric material is given by

$$P_d = \sigma|\mathbf{E}_i|^2$$
$$= \omega\varepsilon_r'\varepsilon_0 \tan\delta|\mathbf{E}_i|^2 \text{ W/m}^3 \tag{11.43}$$

where $\mathbf{E}_i$ is the internal electric field in V/m. $\mathbf{E}_i$ will not necessarily be uniform and this produces non uniform heating, resulting in 'hot spots'. The 'hot spots' are generated from the focusing action of high dielectric constant materials and also from the slope of any cavity formed inside the materials.

Microwave ovens A microwave oven is a metallic cooking chamber or cavity excited in multiple modes at frequency 2450 MHz by means of a magnetron oscillator as shown in Fig.11.19. The frequency of 2450 MHz is chosen to achieve a desired balance between the degree of heat generated within the food and the degree of penetration of the energy.

The microwave energy is generated by magnetron oscillator at a frequency of 2450 MHz. The energy is extracted from the resonant cavities by a magnetic loop formed by the centre conductor of output coaxial line. The other end of the co-axial line forms an **E**-field probe which feeds a rectangular waveguide of dimensions $a = 6.83$ cm and $b = 3.81$ cm. This waveguide is terminated with a short, at a distance of quarter guide wavelength from the **E**-field probe. Thus the waveguide is excited in the dominant TE_{10} mode. The other end of the waveguide is physically open to launch the microwave energy into the stainless steel cavity in which the food is placed. A rotating metallic stirrer (blade) reflects the microwave energy and produces multimode fields in the cavity. Due to the rotation of stirrer, the boundary conditions of the oven change with time and a statistically uniform field is produced throughout the cavity resulting in uniform heating of the food material. A plastic ceiling shield is placed between the stirrer chamber and the oven chamber for environmental protection.

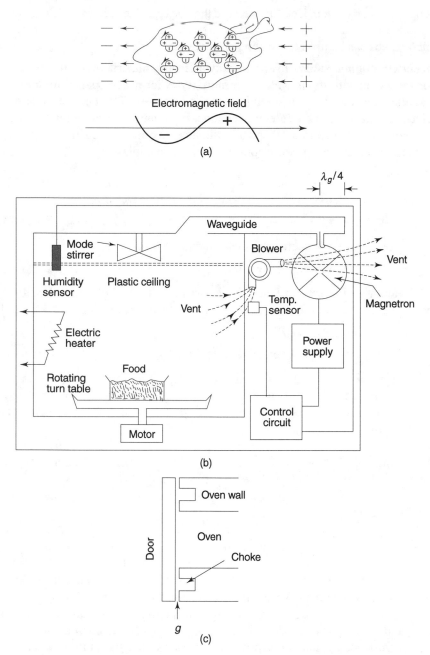

Fig. 11.19 *(a) Rotational and vibrational losses due to polarisation change in food molecules in microwaves (b) Microwave oven (c) Choke in oven door*

The typical input power to the oven is normally 1 to 1.5 kW, which in turn produces a maximum microwave power of 600–700 W at 2450 MHz. Since a microwave can cook from the inside to the outside of the material, an electric heater is being used to brown the food to give conventional cooked appearance on the outside. The molecules associated with the moisture contained in the food, are set in vibration at microwave frequency and the resulting friction produces heat necessary for the cooking process. Usually, microwaves can penetrate the food materials up to a depth of 2–3 cm. Therefore, large pieces are still cooked by the conduction of heat to the centre of the food. Since the cooking time is less and the process is even, the food retains its natural flavour and nutritional value.

The cooking time and temperature are controlled by using a microprocessor based controller circuit. The temperature is sensed from the exhaust air outside the oven cavity. The time of cooking is predetermined for certain ranges of food. The time and degree of heating can be set in the control circuit and once the required parameter is detected, the magnetron may be automatically switched off.

The microwave oven must be designed such that the door maintains shielding integrity with the oven cavity walls, to prevent leakage of microwave energy above the maximum permissible limit, as per the international regulation. The shielding integrity cannot be high for direct mechanical contact between door and oven cavity. A microwave quarter wave choke is designed in series with the door to oven gap, so that the gap g sees an open circuit towards the oven cavity which prevents leakage. Moreover, the choke is filled with a special ferrite material which readily absorbs microwaves and prevents the entry of any foreign matter inside. The quarter wave section must be cut accordingly. The door must have an additional safety lock which is automatically switched off the microwave power once the door is opened.

Medical applications   Microwaves are now extensively used in medical applications due to its less harmful effects than radioactive or X-ray exposure and improved focussing of power to heat specific tissue regions.

Two therapeutic applicators—diathermy and hyperthermia are very common. Since microwaves can penetrate the body to reach bones and deep muscles, localized heating can be achieved. Microwave energy is launched into a patient using a focussing device such as dielectric lens. In diathermy, microwave heating is used to relieve pain and to treat various inflammatory diseases and ailments. Microwave heating is also combined with radioactive exposure to destroy cancerous cells.

11.6.2   Industrial Control and Measurements

This application includes precision thickness measurement of metal sheets in rolling machines and determination of moisture content in a substance.

Thickness measurements   Sheet thickness measurements involve determination of complex reflection or transmission coefficient at microwave frequencies.

The metal sheet in the rolling process is passed between two horns placed face to face as shown in Fig.11.20. The circulator, connecting waveguides, phase shifter and the horns constitute a resonant cavity whose resonant frequency is made equal to that of the signal source by adjusting the phase shifter. For a given thickness the reflection coefficient at plane PP' is made a small value. Any change in thickness will cause changes in reflection coefficient which can be calibrated by using known thickness of a given metal.

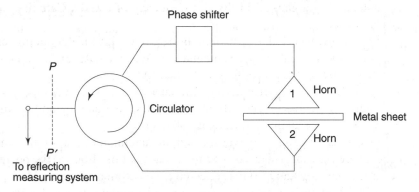

Fig. 11.20 *Thickness measurements*

Moisture content measurements   The moisture content in paper and textile is measured using microwave bridges. In this the attenuation and phase shift of the electromagnetic wave passing through the material or the complex reflection co-efficient of the reflected wave from the material sheet are measured. The principles of measurement set up are explained in Fig.11.21. The sensitivity of these methods can be increased by suitable selection of spacing between the horn radiator and the material sheet.

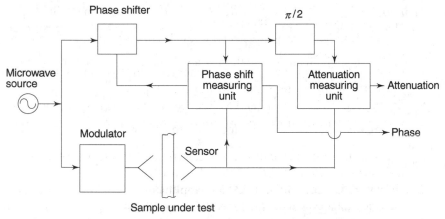

Fig. 11.21 *Moisture content measurement in solid material*

EXERCISES

11.1   A rectangular aperture ($a = 3$ and $b = 2$) with a constant field distribution is mounted on an infinite ground plane. Compute the half power beam width in the E-plane and the directivity.

11.2 A 10 m diameter parabolic reflector having $f/d = 0.5$ is operating at 3 GHz with aperture efficiency 75%. Find (a) overall directivity (b) half of the subtended angle.

11.3 An antenna has a gain of 55 dB at a frequency of 6 GHz. Calculate its effective area.

11.4 A 73 ohm antenna is fed by a 50 ohm lossless line with 5 kW input power. The total power radiated by the antenna is 4 kW. Calculate (a) the antenna efficiency and (b) the power loss in the antenna.

11.5 The complex relative permittivity of a substance at 3 GHz is $60 - j20$. Calculate (a) the loss tangent, (b) the equivalent conductivity and (c) the power density for an internal field of 100 mV/m.

11.6 A radar transmits 100 W power through an antenna of gain 40 dB. The echoing area of a target placed at 5 km distance is 1 m^2. Find (a) the power density at the target and (b) the reflected power from the target.

11.7 An FMCW radar operates at 10 GHz and employs triangular-wave modulation for which the rate of change of frequency is 1 GHz/S. The beat frequency from a target return is 5 kHz over the negative slope and 4.8 kHz over the positive slope. Find the target range, range rate, and the relative direction of motion of the target with respect to radar.

REFERENCES

1. Balanis, C A, *Antenna Theory-Analysis and Design*, Harper & Row, Publishers, New York, 1982.

2. Veley, V F, *Modern Microwave Technology*, Prentice-Hall, Inc., Englewood Cliffs, New Jersey, 1987.

3. Dennis Roddy, *Microwave Technology*, Prentice-Hall, Inc., Englewood Cliffs, New Jersey, 1986.

4. Gupta, K C, *Microwaves*, New Age International (P) Limited, Publishers, Formerly Wiley Eastern Ltd., New Delhi, 1983.

5. Stuchly, S, A Kraszowski and M Rzepecka, *Microwaves for Continuous Control of Industrial Proces*, *Microwave Jour.*, 12, No.9, Aug. 1969, p. 51.

6. "Special issue on the industrial, scientific and medical applications of microwaves", *Proc. IEEE*, 62, No. 1. Jan.1974.

7. Gupta, K C, and Bhal, I J, "RF applicators find jobs-on farms, in factories", *Microwaves, 17*, June 1978, pp. 52–64.

8. Brown, W C, "Satellite Power Stations: A New Source of Energy?" *IEEE spectrum*, March 1973. pp. 38–47.

9. Freedman, G, "The future of microwave heating equipment in the food industries", *K. Microwave Power*, Vol. 7, No. 4.

10. Gupta, C, "Microwaves in medicine", *Electron. Power*, Vol.27, No.5, pp. 403-406, 1981.

11. Kraszewski, A, and Stanislaw Kulinski: "An Improved Microwave Method of Moisture Content Measurement and Control", *IEEE Trans.* on Ind. Electron. and Control Inst. Vol. IECI-23, No. 4 Nov. 1976.

12. Kraszewski, A, "Microwave Aquametry—A Review", *J. Microwave Power*, Vol. 15, No. 4, 1980.

13. Stuchly, M A and Stuchly, S S, "Dielectric Properties of Biological Substances ", *J. Microwave Power*, Vol.15, No.1, 1980.

Chapter 12

Microwave Radiation Hazards

12.1 INTRODUCTION

Microwaves are used in the scientific and industrial applications as well as military and civilian world. But there are some adverse effects of high power microwave radiations. These are Hazards of Electromagnetic Radiation to Personnel (HERP), Hazards of Electromagnetic Radiation to Ordnance (HERO) and Hazards of Electromagnetic Radiation to Fuel (HERF). HERP is the potential of electromagnetic radiation to produce harmful biological effects in humans, HERO is the potential of electro explosive devices to be adversely affected by electromagnetic radiation, and HERF is the potential of electromagnetic radiation to cause spark ignition of volatile combustibles such as vehicle fuels.

12.2 HAZARDS OF ELECTROMAGNETIC RADIATION TO PERSONNEL (HERP)

Hazards of microwave radiation to personnel range from leukemia to sterility. There is also the possibility that weak electric and magnetic fields from high power transmission lines may affect biological cellular processes at the cell nucleus. HERP is caused by the thermal effect of radiated energy. Since biological substances, such as blood, brain, bone, muscle, and fat, behave as conductive dielectrics, the microwave energy directed onto the body may be scattered, reflected and absorbed depending on the field strength, the frequency, the dimension of the body and the electrical properties of the tissue. The absorbed microwave energy produces molecular vibration and converts this energy into heat. If the organism cannot dissipate this heat energy as fast as heat is produced, the internal temperature of the body will rise. This heat may damage these biological substances permanently. For example, if the lens of the eye were exposed to microwaves, its circulatory system would be unable to provide sufficient flow of blood for cooling and could cause cataracts. Similarly, the stomach, intestines and bladder are especially sensitive to thermal damage from high power microwaves.

Microwave frequencies for which the wavelengths are of the same order of magnitude as the dimensions of the human body produce close coupling between the body and the microwave field and a large amount of heat can be generated to cause severe damage in the body. Significant energy absorption will occur even when the body size is at least 1/10 of a wavelength. Although the biological damage occurs mostly due to electric field coupling, low frequency magnetic field coupling can also produce damage when exposure time is large.

12.3 HAZARDS OF ELECTROMAGNETIC RADIATION TO ORDNANCE (HERO)

Microwave energy is also dangerous to ordnance like weapon systems, safety and emergency devices and other equipment containing sensitive electro explosive devices (EEDs), in addition to the attending personnel and associated equipment. Radiated fields can cause unintentional triggering of EEDs.

Ordnance is more sensitive than humans partially because it does not have a circulatory system to dissipate internal heat. Ordnance reacts to peak power whereas, humans react to average power over some time. However, EEDs can more easily be protected from the effects of RF energy than humans by enclosing them with metallic enclosures which reflect back the incident microwave energy.

12.4 HAZARDS OF ELECTROMAGNETIC RADIATION TO FUEL (HERF)

HERF occurs due to the possibility of accidentally igniting fuel vapours by RF-induced arcs during fuel handling operations in proximity to high level RF fields. The probability of ignition may be significant for more than 50 volt-ampere arcs.

12.5 RADIATION HAZARD LEVELS FOR PERSONNEL

The most widely used parameter for the measure of microwave radiation level is average power density for a plane wave in free space

$$P_d = EH = E^2/377 = 377H^2 \qquad (12.1)$$

Unfortunately, the majority of hazardous fields in practice are not simple plane waves but have complicated amplitude, phase and polarisation distributions due to their standing wave, or near field or modulation characteristics. Therefore, some standards consider specific absorption rate (SAR) for specifying the limits. SAR is the rate of energy absorption for unit mass of substance measured in W/kg. SAR depends on the density of the biological substance, the substance conductivity, and the magnitude of the field. Mathematically,

$$\text{SAR} = \sigma E^2/m_d \text{ W/kg} \qquad (12.2)$$

where E(V/m) is the rms electric field within the material, σ(S/m) is the conductivity of the material and m_d(kg/m^3) is the mass density of the material.

Energy absorption takes place due to penetration of the microwaves inside the body. The microwave penetration depth inside a body is a function of frequency. Since both dielectric constant and conductivity of fat are less than those for other

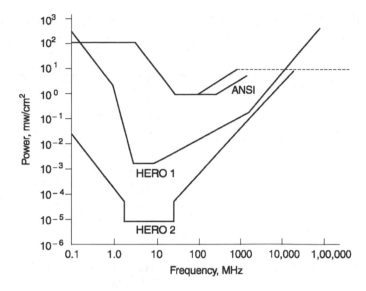

Fig. 12.1 *A comparison of HERP and HERO limits*

substances in the body, the penetration in fat is much greater than that for either the muscle or the blood over most of the microwave frequency range. Since the conductivities of the muscle and blood are higher, microwaves dissipate more in these substances resulting in greater rate of heating.

Within the body wherever the dielectric constant is higher, microwave penetration generates hot spots due to focussing action and shape of the cavities formed by bones. It is theoretically predicted that hot spots occur inside the human skull in the frequency range of 918–2450 MHz, whereas within the eyeball it occurs at frequencies nearly 1500 MHz. It is also found that the microwave absorption depends on the variation in body size and orientation with respect to wavelength. The typical average SAR value for a human being is about 0.03 W/kg for an incident power density of 1 mW/cm^2 at 700 MHz. This value peaks at 0.25 W/kg at 70 MHz, when the average height of a person is approximately half wavelength.

The rate of temperature rise due to microwave heating can be expressed by

$$\frac{dT}{dt} = \frac{Q}{S_p} \ (°C/s) \tag{12.3}$$

where Q (= SAR + metabolic rate of heat production per unit mass), the rate of heat loss per unit mass in W/kg and S_p is the specific heat of the substance in kcal°C/kg. Due to microwave irradiation, the temperature initially increases rapidly for a few minutes and then the thermo regulatory system of the body tends to stabilize the temperature. But the body temperature will start to rise again if the system cannot remove the excess heat at the same rate and therefore a hazard occurs.

Table 12.1 *USSR, US, Canada, and Sweden exposure standards*

Standard	Type	Frequency	Exposure limit	CW/ duration	Pulsed
USSR Govt. 1977	Occupational	10-30 MHz	20 V/m	Working day	both
		30-50 MHz	10 V/m	Working day	both
			0.3 A/m	Working day	both
		50-300 MHz	5 V/m	Working day	both
		0.3-300 GHz	$10\mu W/cm^2$	Working day	both
			$100\mu W/cm^2$	Working day	both
			$100\mu W/cm^2$	2 hours	both
			$1mW/cm^2$	2 hours	both
			$1mW/cm^2$	20 mins.	both
USSR Govt. 1970	General public	0.3-300 GHz	$1\mu W/cm^2$	24 hours	both
U.S. ANSI 1974	Occupational	10MHz-100GHz	$10mW/cm^2$	no limit	CW
			200 V/m		
			0.5 A/m		
			$1mWhr/cm^2$	0.1 hour	pulsed
U.S.Army and Air force 1965	Occupational	10MHz-300GHz	$10mW/cm^2$	no limit	both
U.S. Industrial Hygientist 1971	Occupational	100MHz-100GHz	$10mW/cm^2$	8 hours	both
			$25mW/cm^2$	10 min.	both
Canada Can. Stds Asso.1966	Occupational	10MHz-100GHz	$10mW/cm^2$	no limit	CW
			$1mWhr/cm^2$	0.1 hour	pulsed
Sweden worker protection authority	Occupational	0.3-300GHz	$1mW/cm^2$	8 hours	both
		10-300 MHz	$5mW/cm^2$	8 hours	both
		10MHz-300GHz	$25mW/cm^2$	any	CW, pulsed averaged over 1sec

Source: Health and Welfare, Canada, 1978.

12.6 RADIATION HAZARD LIMITS

International Radiation Protection Association (IRPA) sets the Permissible Exposure Levels (PEL) for the general public. IRPA guidelines are divided into two categories—Occupational Permissible Exposure Levels, which applies to personnel who work in the vicinity of RF for 8 hours a day, and General Public Permissible Exposure Levels, which applies to residents, who are exposed to the RF electromagnetic environment all throughout the year. The limits take into consideration the skin depth and Specific Absorption Rate (SAR). Table 12.1 gives the exposure standards for different countries.

The US Navy's Bureau of Medicine and Surgery established the biological hazard level that personnel should not be exposed to a power density above 10 mW/cm^2, when averaged over any 0.1 hour period, in the frequency range of 10 MHz to 100 GHz. This limit is 100 mW/cm^2 where the exposure is not continuous.

The American National Standards Institute Inc. (ANSI) developed ANSI C 95.1-1982 electromagnetic hazard limit of HERP and ordnance, as shown in Fig. 12.1. The ANSI standard uses the idea of a constant average SAR limit of 0.4 W/kg, which is 1/10 the value of the threshold for adverse effects.

The curve "HERO 2 Limit" represents the maximum safe fields for bare EEDs with lead wires arranged in optimum receiving orientation. The curve " HERO 1 limit " represents the safe field strength for fully assembled ordnance undergoing normal handling and loading operations. The limits for ordnance are generally lower than the limit for humans because ordnance is more sensitive than humans as they do not have natural systems to dissipate heat.

The modified military standard for military operations in ships at sea, is based on the ANSI limit but specified as 5W/cm^2 above 1 GHz. Restricted areas have reduced safety levels for children and pregnant women. The limit for the military in non-operations condition is the same as the ANSI limit.

The RF limits for HERP in the US is based on the body heating effects. However, the Soviet limit for HERP is 0.01 mW/cm^2, which is 30 dB lower than the US limit because they studied that there were biological effects other than thermal. The Chinese limit is 0.05 mW/cm^2.

12.7 RADIATION PROTECTION

Radiation protection can be practised by preventing radiation from entering into the beam of the transmit antenna or from coming close to any microwave generators or propagating medium. In areas where high power radars are used, the service and maintenance personnel must wear microwave absorptive suitmade out of stainless steel woven into a fire retardant synthetic fibre. The suit is lightweight, comfortable, and easy to put on. The attenuation produced by such a suit is above 20 dB at 2450 MHz, 20–35 dB from 650–1150 MHz, and 35–40 dB from 1–11 GHz.

EXERCISES

12.1 The microwave exposure limit for the frequency range 50–300 MHz is 5 V/m. If the biological substance has a conductivity of 150 S/m and a mass density of 1.3×10^3 kg/m^3, calculate the incident power density corresponding to this limit and the specific absorption rate.

12.2 A biological substance has conductivity of 140 S/m, mass density of 1.2×10^3 kg/m^3 and the specific heat 1.2 kcal.°C/kg. Find the rate of temperature rise in °C/s.

REFERENCES

1. "IRPA guidelines on limits of exposure to RF electromagnetic fields in the range from 100 MHz to 300 GHz", 1988.
2. "Radiation Hazards", *IEC Tech. Com.* TC 77.
3. "Standards for Safety Levels with Respect to Human Exposure to Radio Frequency Electromagnetic Fields", 3 KHz to 300 GHz. *IEEE C 95.1–1991/ANSI C95.1–1982*.
4. "Military Standard Definitions and Systems of Units, Electromagnetic Interference and Electromagnetic Compatibility Technology ", *MIL-STD-473A*, June 1977.
5. "American National Standard Safety Levels with Respect to Human Exposure to Radio Frequency Electromagnetic Fields", 300 KHz to 100 GHz, *ANSI C95.1-1982*, July 1982.
6. "Preclusion of Ordnance Hazards in Electromagnetic Fields", General Requirement for, *MIL-STD 1385 B*, August 1986.
7. "Electromagnetic Compatibility by Design McDonnel Aircraft Company", *McDonnel Douglas Corporation*, St. Louis, Missouri.
8. Bowman, R R, "Quantifying hazardous microwave fields", *Trans. Int. Microwave Power Inst.*, Vol. 8.
9. "Health Aspects of Radio Frequency and Microwave Exposure", *Health and Welfare*, Canada, 1978.
10. Lin, James, "Microwave Biophysics", *Trans. Int. Microwave Power Inst.*, Canada 1978.
11. Michaelson, S M, Thomson, RAE and Howland, RJ, "Biologic Effects of Microwave Exposure", *RADC-TR-67- 461*, 138, pp. 1967.
12. Michaelson, S M, "Biologic and pathophysiologic effects of exposure to microwaves", *Trans. Int. Microwave Power Inst.*, Vol. 8, 1978.
13. Repacholi, M H, "Control of Microwave Exposure in Canada", *Trans. Int. Microwave Power Inst.*, Vol. 8.
14. Stuchly, M A, and Repacholi, M H, "Microwave and Radio Frequency Protection Standards", *Trans. Int. Microwave Power Inst.*, Vol. 8, 1978.
15. Michael Barge, J, "Warning: Radhaz", *EMC Technology Magazine*, pp. 21–27, May, June 1989.
16. Michael Kachmar, "Radiation Protection, in Small to Extra-large", *Microwaves and RF*, pp. 41-42, July 1986.

Chapter 13

Microwave Measurements

13.1 INTRODUCTION

The basic measurement parameters in low frequency ac circuits containing lumped elements are voltage, current, frequency and true power. From these measurements, the values of the impedance, the power factor, and the phase angle can be calculated. At microwave frequencies, the amplitudes of the voltages and currents on a transmission line are functions of distance and are not easily measurable. However, in a lossless line, the power transmitted is independent of the location along the line. Therefore, it is more convenient to measure power instead of voltage and current. Much of the properties of devices and circuits at microwave frequencies are obtained from the measurement of S-parameters, power, frequency, phase shift, VSWR and the noise figure.

Due to the complications and high cost of direct microwave measuring devices and instrumentations, such as vector network analysers, spectrum analysers, power meters, etc., microwave measurements in the laboratory are often carried out using a 1 kHz square-wave modulating signal which modulates the microwave test signal. The transmitted and reflected signals are then demodulated and measured using low frequency instruments such as an oscilloscope and a low frequency (1kHz) tuned receiver, called a VSWR meter. The amplitude and phase information of the microwave test signals are available in the detected low frequency signal for calculating the desired parameters. These are described in the following sections. It is found appropriate to include the descriptions of some important measurement devices and instrumentations in brief, in this chapter.

13.2 TUNABLE DETECTOR

The low frequency square-wave modulated microwave signal is detected using a non-reciprocal detector diode mounted in the microwave transmission line. These diodes are specially designed point contact or metal-semi conductor Schottky barrier diodes. A detailed description of these diodes are given in Chapter 10. To match the detector to the transmission system a tunable stub is used as shown in Fig.13.1. Broad band detectors are also manufactured in coaxial form. In order to pick up propagating fields, a coaxial line tunable probe detector is used.

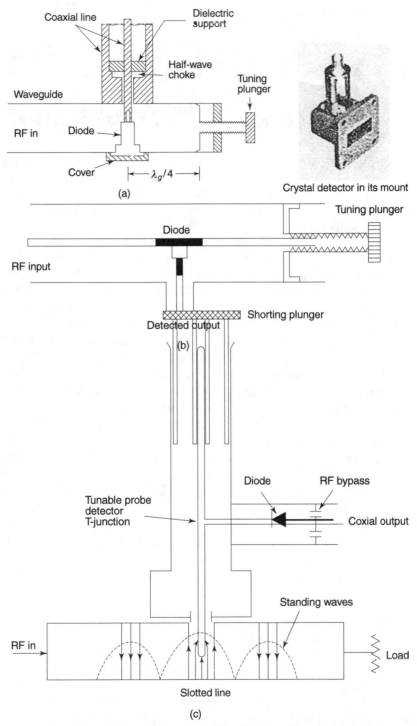

Fig. 13.1 *(a) Tunable waveguide detector (b) Coaxial detector (c) Tunable probe detector*

13.3 SLOTTED LINE CARRIAGE

A slotted line carriage (Fig.13.2) contains a coaxial E-field probe which penetrates inside a rectangular waveguide slotted section or a coaxial slotted line section from the outer wall and is able to traverse a longitudinal narrow slot. The longitudinal slot is cut along the centre of the waveguide's broad wall or along the outer conductor of the coaxial line over a length of 2-3 wavelengths where the electric current on the wall has no transverse component. The slot should be narrow enough to avoid any distortion in the original field inside the waveguide. The two ends of the slot are tapered to zero width for reducing the effect of discontinuity. The probe is made to move longitudinally at a constant small depth to achieve a uniform coupling coefficient between the electric field inside the line and the probe current at all positions. The probe samples the elec-

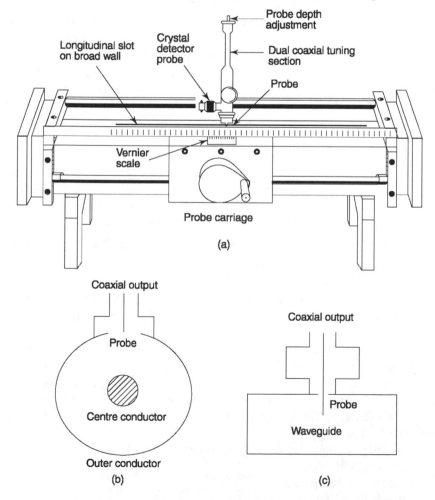

Fig.13.2 *(a) Slotted line carriage and schematic diagram(b) Cross-section of a coaxial slotted line(c) Cross-section of rectangular waveguide slotted line(d) Longitudional slot and electric wall currents*

tric field which is proportional to the probe voltage. This unit is primarily used for the determination of locations of voltage standing wave maxima and minima along the line. The probe carriage contains a stub tunable coaxial probe detector to obtain a low frequency modulating signal output to a scope or VSWR meter. The probe should be very thin compared to the wavelength and the depth also should be small enough to avoid any field distortion.

The slotted line with tunable probe detector is used to measure

1. VSWR and standing wave pattern
2. Wavelength
3. Impedance, reflection coefficient and return loss measurements by the minima shift method.

13.4 VSWR METER

A VSWR meter is a sensitive high gain, high Q, low noise voltage amplifier tuned normally at a fixed frequency of 1 KHz at which the microwave signal is modulated. The input to the VSWR meter is the detected signal output of the microwave detector and the output of the amplifier is measured with a square-law-calibrated voltmeter which directly gives the VSWR reading V_{max}/V_{min} for an input of V_{min}, after the meter is adjusted to unity VSWR for an input corresponding to V_{max} as shown in Fig.13.3. A gain control can be used to adjust the reading to the desired value. The overall gain is nearly 125 dB which can be altered in steps of 10 dB.

There are three scales on the VSWR meter. When the VSWR is between 1 and 4, reading can be taken from the top SWR NORMAL scale. For VSWR between 3.2 and 10, bottom of SWR NORMAL scale is used. When the VSWR is less than 1.3, a more accurate reading can be taken by selecting the EXPANDED scale, graduated from 1 to 1.3. The third scale at the bottom is graduated in dB.

13.5 SPECTRUM ANALYSER

A spectrum analyser is a broad band super het receiver which provides a plot of amplitude versus frequency of the received signal, i.e., the signal spectrum as explained in Fig.13.4.

The local oscillator is electronically swept back and forth between two frequency limits at a linear rate. The sweep voltage waveform is saw tooth type with zero flyback time to move the spot on the CRT horizontally in synchronism with the frequency sweep so that the horizontal position is a function of the frequency of the local oscillator. The amplitude of the input RF signal is obtained from the vertical deflection of the spot.

The basic design considerations for proper operation are:

1. Frequency sweep rate
2. Frequency sweep range
3. Bandwidth of IF amplifier
4. Centre frequency of IF amplifier.

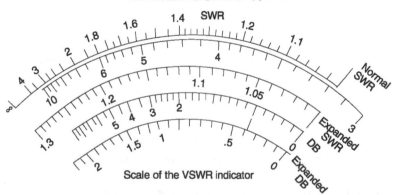

Fig. 13.3 *VSWR meter (Courtesy : Hewlett Packard)*

For highest resolution, the bandwidth should be kept minimum and consequently sweep speed should be very low in order to allow time to build up the voltage in the receiver circuit. The range of frequencies to be covered should be as small as possible. The IF frequency should be chosen high enough to avoid the image response. If f_i is image frequency, f_0 is local oscillator frequency, f_{if} is IF frequency and f_s is signal frequency, then

$$f_i = f_0 \pm f_{if} = f_s \pm 2f_{if} \qquad (13.1)$$

is the frequency that beats with the L.O. frequency and produces a frequency difference equal to the IF. Thus

$$f_{if} = f_s - f_0 \ ; f_s > f_0 \qquad (13.2)$$

$$= f_0 - f_s \ ; f_s < f_0 \qquad (13.3)$$

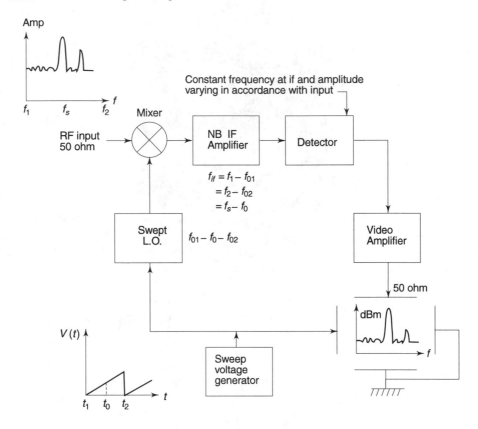

Fig. 13.4 *Basic block diagram of a spectrum analyser*

For example, when $f_{if} = 450$ kHz and $f_0 = 8000$ kHz, signal on either 8450 kHz or 7550 kHz will beat against f_0 and produce f_{if}. For $f_{if} = 450$ kHz, image frequency $f_i = 450 \times 2 = 900$ kHz off the signal frequency. For $f_{if} = 2000$ kHz, $f_i = 2000 \times 2 = 4$ MHz off the signal and can be tuned out easily.

The bandwidth and hence resolution of the spectrum analyser is determined by the bandwidth of IF amplifier.

13.6 NETWORK ANALYSER

The use of the slotted line for microwave measurements has the disadvantage that the amplitude and phase measurements are limited to single frequencies. Therefore, broadband testing is very time consuming and manpower cost is very high. A network analyser measures both amplitude and phase of a signal over a wide frequency range within a reasonable time. The basic measurements involve an accurate reference signal which must be generated with respect to which the test signal amplitude and phase are measured. A schematic block diagram of a complex network analyser is shown in Fig.13.5.

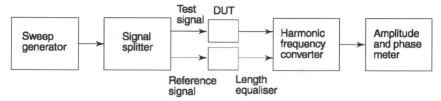

Fig. 13.5 *Schematic block diagram of a complex network analyser*

The microwave signal from a sweep oscillator is first divided by means of a power divider into test signal and a reference signal channel. The test signal is transmitted through the device under test, while the reference signal passes through a phase equalising length of line. Since processing of the microwave frequencies is not practical, both the test and reference signals are converted to a fixed intermediate frequency by means of a harmonic frequency converter. The output signals from the harmonic frequency converter are compared to determine the amplitude and phase of the test signal. The harmonic frequency converter uses a phase locked loop which helps the local oscillator to track the reference channel frequency as shown in Fig.13.6. This allows swept frequency measurements. The frequency conversion takes place in two steps. The first mixer converts RF to a fixed IF in the MHz range and then after amplification they are further converted to another fixed IF in the kHz range by means of a second mixer for the final amplitude and phase comparison.

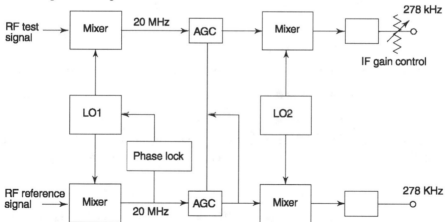

Fig. 13.6 *Schematic block diagram of the harmonic frequency converter*

The reflection and transmission measurements are carried out by using the reflection-transmission test unit as shown schematically in Fig. 13.7.

The reference line length can be balanced for transmission measurement, and the device under test is compared to the sliding short for reflection measurements. The direction couplers used in the bridge are accurately matched to ensure a good balance between the two channels.

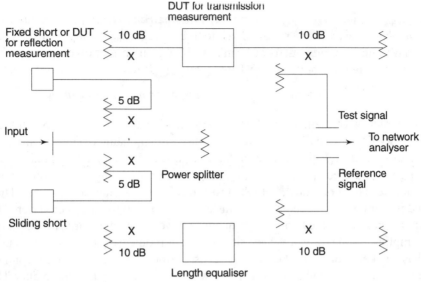

Fig. 13.7 *Reflection-transmission test unit*

For a two port-network, Fig.13.8 shows the test set up for S-parameters S_{11} and S_{21} measurements using a network analyser S_{ii} and S_{ij} are computed from the measured output of the dual directional couplers as follows:

$$S_{11} = V_2/V_1 \ (\phi_2 - \phi_1) \tag{13.4}$$

$$S_{21} = V_3/V_1 \ (\phi_3 - \phi_1) \tag{13.5}$$

For S_{22} and S_{12} the signal source and the load position are interchanged, so that

$$S_{22} = V_3/V_4 \ (\phi_3 - \phi_4) \tag{13.6}$$

$$S_{12} = V_2/V_4 \ (\phi_2 - \phi_4) \tag{13.7}$$

Therefore, from the measurements of amplitude and phase from the ports of the dual directional couplers, S-parameters of a two-port network can be determined.

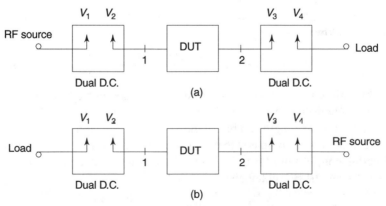

Fig. 13.8 *S-parameter test set (a) S_{11} and S_{21} (b) S_{22} and S_{12}*

13.7 POWER MEASUREMENTS

Power is defined as the quantity of energy dissipated or stored per unit time. The range of microwave power is divided into three categories—low power (less than 10 mW), medium power (from 10 mW to 10 W) and high power (greater than 10 W). The average power is measured while propagation in a transmission medium and is defined by.

$$P_{av} = \frac{1}{nT} \int_0^{nT} v(t)\, i(t)\, dt \; ; \tag{13.8}$$

where T is the time period of the lowest frequency involved in the signal and n cycles are considered. For a pulsed signal

$$P_{peak} = \frac{1}{\tau} \int_0^{\tau} v(t)\, i(t)\, dt , \tag{13.9}$$

$$P_{av} = P_{peak} * \text{Duty cycle} \tag{13.10}$$

$$\text{Duty cycle} = \text{pulse width} * p.r.f = \tau f_r = \tau/T < 1$$

where τ is the pulse width, T is the period and f_r is the pulse repetition frequency.

The most convenient unit of power at microwaves is dBm, where

$$P(\text{dBm}) = 10 \, \text{Log} \, \frac{p(mW)}{1\,mW} \tag{13.11}$$

viz.   30 dBm = 1W and –30 dBm = 1μW.

The microwave power meter consists of a power sensor, which converts the microwave power to heat energy. The corresponding temperature rise provides a change in the electrical parameters resulting in an output current in the low frequency circuitry and indicates the power. High power is often measured, especially for standards and calibration purposes, using microwave calorimeters in which the temperature rise of the load provides a direct measure of the power absorbed by the load.

The sensors used for power measurements are the Schottky barrier diode, bolometer and the thermocouple.

13.7.1 Schottky Barrier Diode Sensor

A zero-biased Schottky barrier diode is used as a square-law detector whose output is proportional to the input power. Since diode resistance is a strong function of temperature, the circuit is designed such that the input matching is not affected by diode resistance as shown by the equivalent circuit in Fig. 13.9. The diode detectors can be used to measure power levels as low as 70 dBm.

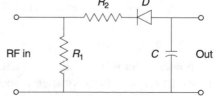

Fig. 13.9 *Schottky barrier diode sensor*

13.7.2 Bolometer Sensor

A bolometer is a power sensor whose resistance changes with temperature as it absorbs microwave power. The two most common types of bolometer are the *barretter* and the *thermistor*. The barretter is a short thin metallic (platinum) wire sensor which has a positive temperature coefficient of resistance. The thermistor is a semiconductor sensor which has a negative temperature coefficient of resistance and can be easily mounted in microwave lines as shown in Fig. 13.10 due to its smaller and more compact size. The impedances of these bolometers are in the range of 100–200 ohm. However, barretters are more delicate than thermistors, hence they are used only for very low power (< few mW). Medium and high power are measured with a low-power thermistor sensor, after precisely attenuating the signal. The sensitivity level of a thermistor is limited to about 20 dBm. The thermistor mount provides good impedance match, low loss, good isolation from thermal and physical shock and good shielding against energy leakage.

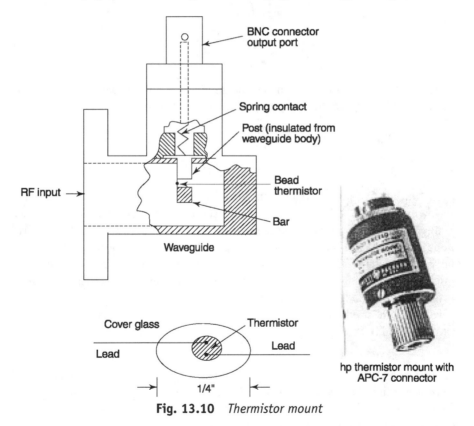

Fig. 13.10 *Thermistor mount*

13.7.3 Power Meter

The power meter is basically constructed from a balanced bridge circuit in which one of the arms is the bolometer as shown in Fig. 13.11. The microwave power applied to this arm will change the bolometer's resistance causing an unbalance

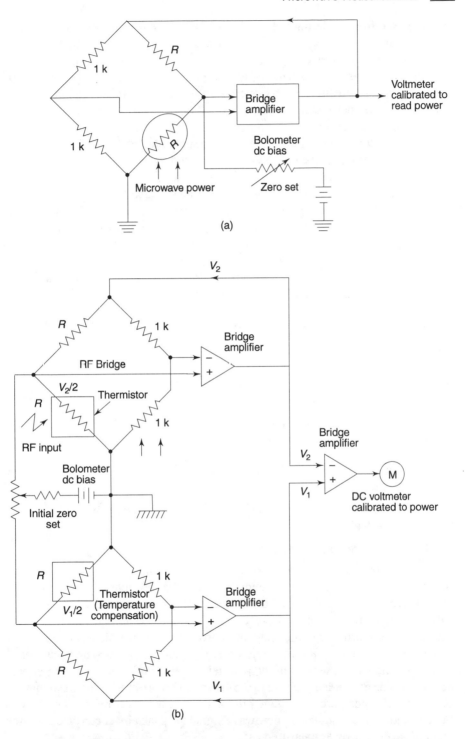

Fig. 13.11   *Principle of the power meter bridge circuit (a) Single bridge
(b) Double bridge for compensation*

in the bridge from its initial balance condition under zero incident power. The non-zero output is recorded on a voltmeter which is calibrated to read the level of the input microwave power.

The main disadvantages with using a single bridge are that (1) the change of resistance due to a mismatch at the microwave input port results in incorrect reading, and (2) the thermistor is sensitive to changes in the ambient temperature resulting in a false reading.

These problems are eliminated by using double identical bridges—the upper bridge circuit measures the microwave power, and the lower bridge circuit compensates the effect of ambient temperature variation ($V_1 = V_2$). The added microwave power due to mismatch is compensated automatically through a self-balancing circuit by decreasing the dc power V_2 carried by the RF sensing thermistor until bridge balance is restored or net change in the thermistor resistance is zero due to negative dc feedback.

The initial zero setting of the bridge is done by adjusting $V_2 = V_1 = V_0$ with no microwave input signal applied, when R is the resistance of the thermistor at balance. Without and with microwave present, the dc voltages across the sensor at balance are $\dfrac{V_1}{2}$ and $V_2/2$, respectively. The average input power P_{av} is equal to the change in dc power:

$$P_{av} = \frac{V_1^2}{4R} - \frac{V_2^2}{4R} = \frac{(V_1 - V_2)(V_1 + V_2)}{4R} \qquad (13.12)$$

For any change in temperature if the voltage changes by ΔV, the change in RF power is $P_{av} + \Delta P = (V_a + \Delta V)^2/4R - (V_2 + \Delta V)^2/4\ R$

or $\qquad P_{av} + \Delta P = \dfrac{(V_1 - V_2)(V_1 - V_2 + 2\Delta V)}{4R} \qquad (13.13)$

Since $V_1 + V_2 \gg \Delta V$ in practice, $\Delta P \approx o$. The meter responds to Eq. (13.13) to read microwave power P_{av}.

13.7.4 Thermocouple Sensor

A thermocouple is a junction of two dissimilar metals or semiconductors (n-type Si). It generates an emf when two ends are heated up differently by absorption of microwaves in a thin film tantalum-nitride resistive load deposited on a Si substrate which forms one electrode of the thermocouple as shown in Fig. 13.12. This emf is proportional to the incident microwave power to be measured.

Here C_2 is the RF by-pass capacitor and C_1 is the input coupling capacitor or dc block. The emf generated in the parallel thermocouples are added to appear across C_2. The output leads going to the dc voltmeter are at RF ground so that the output meter reads pure dc voltage proportional to the input microwave power. For a square-wave modulated microwave signal the peak power can be calculated from the average power measured as

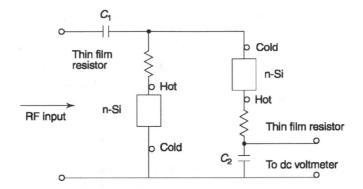

Fig. 13.12 *Thermocouple power sensor*

$$P_{peak} = \frac{P_{av} \times T}{\tau} \tag{13.14}$$

where T is the time period and τ is the pulse width.

13.7.5 High Power Measurements by the Calorimetric Method

High power microwave measurements can be conveniently done by the calorimetric method which involves conversion of the microwave energy into heat, absorbing this heat in a fluid (usually water) and then measuring the temperature rise of the fluid as shown in Fig.13.13. There are two types: one is the direct heating method and another is the indirect heating method. In the direct heating method, the rate of production of heat can be measured by observing the rise in the temperature of the dissipating medium. In indirect heating method, heat is transferred to another medium before measurement. In both the methods static calorimeter and circulating calorimeter are used.

Static calorimeters   It consists of a 50 ohm coaxial cable which is filled by a dielectric load with a high hysterosis loss. The load has sufficient thermal isolation from its surrounding. The microwave power is dissipated in the load. The average power input is given by

$$P = \frac{4.187 \, mC_p \, T}{t} \text{ watts} \tag{13.15}$$

where
    m = mass of the thermometric medium in gms
    C_p = its specific heat in cal /gms
    T = temperature rise in °C
    t = time in sec.

Circulating calorimeters   Here the calorimeter fluid (water) is constantly flowing through a water load. The heat introduced into the fluid makes exit temperature higher than the input temperature. Here average power

$$P = 4.187 \, v \, d \, C_p \, T \text{ Watts} \tag{13.16}$$

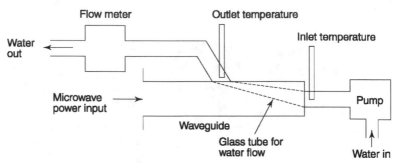

Fig. 13.13 *Microwave calorimeter*

where
   v = rate of flow of calorimeter fluid in cc/sec
   d = specific gravity of the fluid in gm/cc.
   T = temperature rise in °C
   C_p= specific heat in cal/gm
   A disadvantage of calorimeter measurements is the thermal inertia caused by the lag between the application of microwave power and the parameter readings.

13.8   INSERTION LOSS AND ATTENUATION MEASUREMENTS

When a device or network is inserted in the transmission line, part P_r of the input signal power P_i is reflected from the input terminal and the remaining part $P_i - P_r$ which actually enters the network is attenuated due to the non-zero loss of the network. The output signal power P_0 is therefore less than P_i. Therefore, insertion loss is defined by the difference in the power arriving at the terminating load with and without the network in the circuit.

Since,
$$\frac{P_0}{P_i} = \frac{P_i - P_r}{P_i} * \frac{P_0}{P_i - P_r} \qquad (13.17)$$

or,

$$10 \log \frac{P_0}{P_i} = 10 \log \left(1 - \frac{P_r}{P_i} \right) + 10 \log \left(\frac{P_0}{P_i - P_r} \right) \qquad (13.18)$$

Insertion loss = reflection loss + attenuation loss
where, by definition

Insertion loss (dB) $= 10 \log P_0/P_i$ $\qquad (13.19)$

Reflection loss (dB) $= 10 \log \left(1 - \frac{P_r}{P_i} \right)$

$= 10 \log \left(1 - |\Gamma|^2 \right)$

$= 10 \log \dfrac{4S}{(1+S)^2} \; ; S = \dfrac{1-|\Gamma|}{1+|\Gamma|} \qquad (13.20)$

$$\text{Attenuation loss (dB)} = 10 \log \left(\frac{P_0}{P_i - P_r} \right) \tag{13.21}$$

$$\text{Return loss (dB)} = 10 \log P_r/P_i = 20\log \mid \Gamma \mid \tag{13.22}$$

For perfect matching, $P_r = 0$, and the insertion loss and the attenuation loss become the same. The experimental set up for insertion and the attenuation measurements are shown Fig.13.14. The relative power levels are measured by using detectors and a VSWR meter. DC_1 and DC_2 are two identical directional couplers.

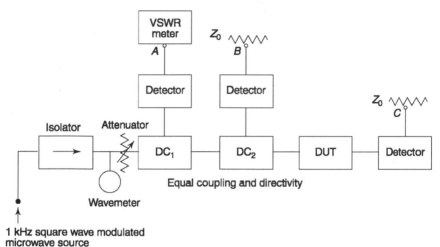

Fig. 13.14 *Insertion loss and attenuation measurements*

The following steps are involved for the insertion loss and attenuation measurements:

1. The microwave source is set to a suitable frequency and the 1 kHz square wave modulation level is adjusted for a peak reading on the VSWR meter at A with minimum input attenuation.

2. For a crystal detector to work in the square-law region the power level is adjusted to get a reading in the 30 dB range of the VSWR meter. The input power from port A is set to zero dB or 1.0 using gain control.

3. Frequency is read from the cavity frequency meter when a dip is observed in the VSWR meter.

4. Connecting matched load Z_0 to ports A and C and VSWR meter to port B, without disturbing any other set-up, the reading in the VSWR meter gives the ratio P_r/P_i, the return loss. The reflection loss $1 - (P_r/P_i)$ is calculated.

5. Now the input attenuator is adjusted to give an attenuation equal to the dB coupling of the directional coupler. The matched load is connected to ports A and B, and the VSWR meter to port C without disturbing any other set-up. The reading in the VSWR meter gives the ratio P_0/P_i, the insertion loss. Attenuation of the network under test can be determined by

subtracting the dB reflection loss from the dB insertion loss.

The main errors in this measurement are

1. P_i, P_0 and P_r may not all be capable of operating the crystal detector within its square-law region.
2. Both the directional couplers may not have the same characteristics.
3. There is some degree of mismatch between the various components in the set-up.

13.9 VSWR MEASUREMENTS

VSWR and the magnitude of voltage reflection coefficient Γ are very important parameters which determine the degree of impedance matching. These parameters are also used for the measurement of load impedance by the slotted line

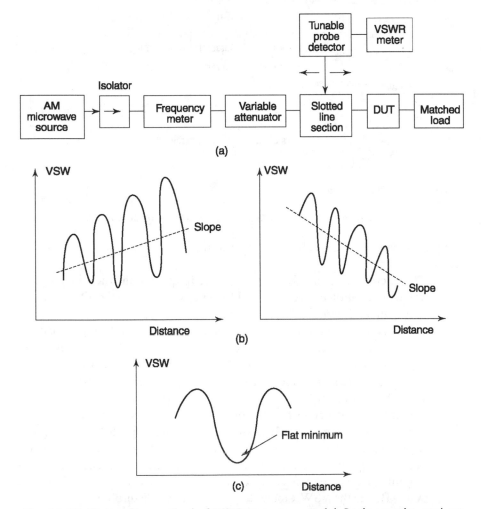

Fig. 13.15 *Slotted line method of VSWR measurement (a) Basic experimental set-up(b) Slant pattern due to mechanical slope error (c) Flat minimum position due to imperfect square wave modulating signal*

method as shown in Fig.13.15. When a load $Z_L \neq Z_0$ is connected to a transmission line, standing waves are produced. By inserting a slotted line system in the line, standing waves can be traced by moving the carriage with a tunable probe detector along the line. VSWR can be measured by detecting V_{max} and V_{min} in the VSWR meter : $S = V_{max}/V_{min}$.

13.9.1 Low VSWR (S < 20)

Low values of VSWR can be measured directly from the VSWR meter using the experimental set-up shown in Fig.13.15 as follows.

1. The variable attenuator is adjusted to 10 dB. The microwave source is set to the required frequency. The 1 kHz modulation is adjusted for maximum reading on the VSWR meter in a 30 dB scale. The probe carriage stub is tuned for maximum detected signal in VSWR meter.

2. The probe carriage is slid along the non-radiating slot from the load end until a peak reading is obtained in VSWR meter. The meter's gain control is adjusted to get the meter reading at 1.0 or 0 dB corresponding to the position of voltage maximum.

3. The probe is moved towards the generator to an adjacent voltage minimum. The corresponding reading in VSWR meter directly gives the VSWR = V_{max}/V_{min} on the top of SWR NORMAL scale for $1 \leq S \leq 4$ or on the EXPANDED scale for $1 \leq S \leq 1.33$.

4. The experiment is repeated for other frequencies as required to obtain a set of values of S vs f.

5. For VSWR between 3.2 and 10, a 10 dB lower RANGE should be selected and reading corresponding to V_{min} position should be taken from the second VSWR NORMAL scale from the top.

6. For VSWR between 10 and 40, a 20 dB RANGE sensitivity increase is required and reading is taken from the top of VSWR NORMAL scale (1 to 4) at the voltage minimum and should be multiplied by 10 to obtain actual VSWR.

7. For VSWR between 32 and 100, a 30 dB lower RANGE must be selected and reading is taken from the second VSWR NORMAL scale (3.2 to 10) from the top at the voltage minimum. The reading should be multiplied by 10 to obtain actual VSWR.

The possible sources of error in this measurements are

1. V_{max} and V_{min} may not be measured in the square-law region of the crystal detector.

2. The probe thickness and depth of penetration may produce reflections in the line and also distortion in the field to be measured. Depth of penetration should be kept as small as possible otherwise values of VSWR measured would be lower than actual.

3. Mechanical slope between the slot geometry and probe movement may cause different values of VSWR for measurement at different locations along the slot (Fig.13.15 b).

4. When VSWR < 1.05, the associated VSWR of connector produces significant error in VSWR measurement. Very good low VSWR (< 1.01) connectors should be used for very low VSWR measurements.

5. If the modulating 1 kHz signal is not a perfect square-wave, the microwaves will be frequency modulated and at each frequency there will be a different set of standing waves. This causes reduction in the sharpness of voltage minima and there may be error in the reading of minimum position as shown in Fig.13.15 (c).

6. Any harmonics and spurious signals from the source may be tuned by the probe to cause measurement error.

7. A residual VSWR of slotted line arises due to mismatch impedance between the slotted line and the main line as explained in Fig.13.16. Let

ρ_L = Actual load reflection coefficient
ρ_s = Slotted line reflection coefficient on main line
E_i = Incident electric field at any point on the main line
E_L = Reflected electric field from the load
E_s = Reflected electric field on the main line because of slotted line

Then, the total reflected field at a point $= \left| E_s \pm E_L \right|$. The maximum and minimum VSWR and reflection coefficients on the main line are

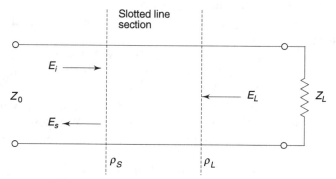

Fig. 13.16 *Residual VSWR of slotted line*

$$S_{\max} = \frac{E_i + (E_s + E_L)}{E_i - (E_s + E_L)} \tag{13.23}$$

$$S_{\min} = \frac{E_i + (E_s - E_L)}{E_i - (E_s - E_L)} \tag{13.24}$$

$$\rho_{\max} = \frac{S_{\max} - 1}{S_{\max} + 1} = |\rho_L| + |\rho_s| \tag{13.25}$$

$$\rho_{\min} = \frac{S_{\min} - 1}{S_{\min} + 1} = |\rho_L| - |\rho_s| \tag{13.26}$$

The above equations can be solved for ρ_L and ρ_s from the measurements of S_{max} and S_{min} on the line. Then the residual VSWR.

$$S_s = \frac{1 + |\rho_s|}{1 - |\rho_s|} \tag{13.27}$$

13.9.2 High VSWR (S > 20)

For high VSWR, the difference of power at voltage maximum and voltage minimum is large, so it would be difficult to remain on the detector's square-law region at maximum positions when the diode current may exceed 20 μ A. Therefore, VSWR measurement with a VSWR meter calibrated on a square-law basis ($I = kV^2$) will be inaccurate. Hence double minimum method as shown in Fig.13.17 is used where measurements are carried out at two positions around a voltage minimum point. The theory of this method can be established as follows. Let the ratio of line voltage near a minimum and the voltage at the minimum be

$$r_n = \frac{|V(x)|}{|V_{min}|} \tag{13.28}$$

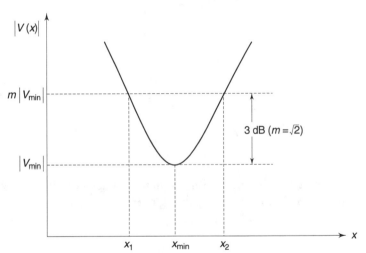

Fig. 13.17 *Double minima method*

For a load reflection coefficient $\Gamma = \rho e^{j\phi}$. The line voltage at a distance x from the load end is

$$|V(x)| = |V_{inc}|\,|1 + \rho e^{j(\phi - 2\beta x)}|$$

or,

$$|V(x)| = |V_{inc}|\,[1 + 2\rho \cos(\phi - 2\beta x) + \rho^2]^{1/2} \tag{13.29}$$

The voltage minimum

$$|V_{min}| = |V_{inc}|(1 - \rho)\ \text{at}\ x = x_{min} \tag{13.30}$$

If x_1 and x_2 are two points around x_{min} where $|V(x_1)| - |V(x_2)| = m|V_{\text{min}}|$,

$$m = \frac{|V(x_1)|}{|V_{\text{min}}|} = \frac{\left[1 + 2\rho\cos(\phi - 2\beta x_1) + \rho^2\right]^{1/2}}{1 - \rho} \tag{13.31}$$

By substituting $\rho = (S - 1)/(S + 1)$, VSWR can be expressed as

$$S = \frac{\left[m^2 - \cos^2\left(\dfrac{2\pi(x_1 - x_{\text{min}})}{\lambda g}\right)\right]^{1/2}}{\sin\dfrac{[2\pi(x_1 - x_{\text{min}})]}{\lambda g}} \tag{13.32}$$

where $\beta = 2\pi/\lambda_g$ and λ_g is the guide wavelength.
If x_1 is the point in the vicinity of x_{min},

$$\Delta x = 2(x_1 - x_{\text{min}}) \tag{13.33}$$

and

$$S = \sqrt{\left[\frac{m^2 - \cos^2\left(\dfrac{\pi\Delta x}{\lambda_g}\right)}{\sin^2\left(\dfrac{\pi\Delta x}{\lambda_g}\right)}\right]}$$

$$= \left[\frac{m^2 - 1}{\sin^2\left(\dfrac{\pi\Delta x}{\lambda_g}\right)} + 1\right]^{1/2} \tag{13.34}$$

For convenience of measurement with a square-law detector, if $m = \sqrt{2}$ is selected, where x_1 is 3 dB above the x_{min} point, then

$$S = \sqrt{\left[\frac{2 - 1}{\sin^2\left(\dfrac{\pi\Delta x}{\lambda_g}\right)} + 1\right]}$$

$$= \sqrt{\left[1 + \csc^2\left(\dfrac{\pi\Delta x}{\lambda_g}\right)\right]} \tag{13.35}$$

If $\pi\Delta x \ll \lambda_g$,

$$S \approx \csc\left(\frac{\pi\Delta x}{\lambda_g}\right)$$

$$= \frac{1}{\sin\left(\dfrac{\pi \Delta x}{\lambda_g}\right)}$$

$$\cong \frac{\lambda_g}{\pi \Delta x} \tag{13.36}$$

where $\Delta x = x_2 - x_1$. Thus high VSWR can be measured by observing the distance between two successive minima to find λ_g and distance Δx between two 3 dB points on both sides of $V_{\min}$.

The method follows the steps given below.

1. The probe is moved to a voltage minimum and the probe depth and gain control is adjusted to read 3 dB in the VSWR meter.
2. The probe is moved slightly on either side of the minimum to read 0 dB in the meter. This position x_1 is noted. The probe is then moved to the other side of the minimum to read 0 dB again at x_2.
3. By moving the probe between two successive minima a distance equal to $\lambda_g/2$ is found to determine the guide wavelength λ_g,
4. High VSWR is calculated from

$$S = \frac{\lambda_g}{\pi\,(x_1 \sim x_2)} \tag{13.37}$$

13.10 RETURN LOSS MEASUREMENT BY A REFLECTOMETER

The return loss and VSWR of a load can be determined by measuring the magnitude of the reflection coefficient with a reflectometer, a set-up in which two identical directional couplers are connected opposite to each other as shown in Fig.13.18. One coupler couples to the forward wave and the other to the reverse wave.

Let us assume that the directional couplers have infinite directivity, a voltage coupling coefficient C, main line VSWR 1 and the detectors have constant impedance and perfect matching to the line. When a unit input amplitude is fed to port 1, voltages at ports 4 and 2 are, respectively,

$$b_4 = C \tag{13.38}$$
$$b_2 = (1 - C^2)^{1/2} \tag{13.39}$$

Incident voltage at port 2 is reflected by the load under test. If Γ_L is the reflection coefficient, the reflected wave amplitude at port 2 is

$$a_2 = (1 - C^2)^{1/2}\,|\,\Gamma_L\,| \tag{13.40}$$

This will be coupled to port 3 to produce a voltage of

$$b_3 = (1 - C^2)^{1/2}\,C\,|\,\Gamma_L\,| \tag{13.41}$$

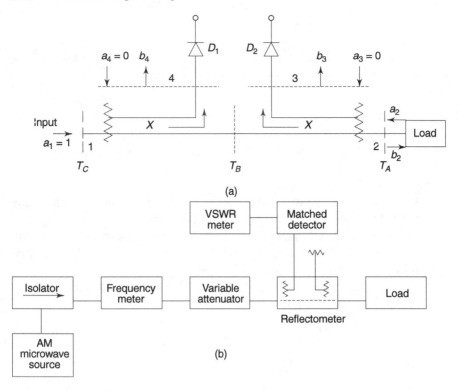

Fig. 13.18 *(a) Reflectometer (b) Experimental set-up for a reflectometer*

Then
$$|b_3/b_4| = (1 - C^2)^{1/2} |\Gamma_L| = K |\Gamma_L| \tag{13.42}$$

If coupling is extremely small i.e. $C \ll 1$, $K \approx 1$. Therefore,

$$|b_3/b_4| = |\Gamma_L|. \tag{13.43}$$

Thus, knowing the voltage ratio between ports 3 and 4 , the reflection coefficient and hence VSWR and return loss can be determined from the following relations.

$$\text{VSWR} = \frac{1 + |\Gamma_L|}{1 - |\Gamma_L|} \tag{13.44}$$

$$\text{Return loss} = - 20 \log |\Gamma_L| \tag{13.45}$$

The experiment is conducted first by terminating port 2 with a short and adjusting the output of the detector D_1 at port 4 to unity in VSWR meter while detector D_2 at port 3 is matched terminated. VSWR meter and match load at D_1 and D_2 are now interchanged. The output of the port 3 is noted which should ideally be equal to the output from port 4. Without disturbing the VSWR meter adjustment, the unknown load is connected at port 2 by replacing the short and the output at port 3 is noted to obtain $1/|\Gamma_L|$ directly from the VSWR meter.

This method is well suited for loads having low VSWR. The major source of error in this method is that the instability of the signal source causes a change of signal power level during measurement of input and reflected signal levels at different instants of time. Non-ideal directional couplers and detectors are also sources of error.

13.11 IMPEDANCE MEASUREMENT

Since impedance is a complex quantity, both amplitude and phase of the test signal are required to be measured. The following techniques are commonly employed for such measurements.

13.11.1 Slotted Line Method

The complex impedance Z_L of a load can be measured by measuring the phase angle ϕ_L of the complex reflection coefficient Γ_L from the distance of first voltage standing wave minimum d_{min} and the magnitude of the same from the VSWR, S. The following relations are important for the computation of Z_L.

$$Z_L = Z_0 \frac{1 + \Gamma_L}{1 - \Gamma_L} \tag{13.46}$$

$$\Gamma_L = \rho_L\, e^{j\phi L} \tag{13.47}$$

$$S = (1 + \rho_L)/(1 - \rho_L) \tag{13.48}$$

$$\phi_L = 2\beta d_{min} - \pi \tag{13.49}$$

$$\beta = 2\pi/\lambda_g \tag{13.50}$$

$\lambda_g = 2 \times$ distance between two successive minima.

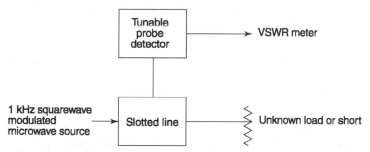

Fig. 13.19 *Determination of load impedance using slotted line*

The method of using slotted line to determine an unknown impedance is explained in Fig.13.19. The steps for measurement are summarized below.

1. Measure the load VSWR to find ρ_L from Eq. (13.48).
2. Measure the distance d between two successive voltage minima to find $g = 2\,d$ and $\beta = 2\pi/\lambda_g$.
3. Measure the distance d_{min} of the first voltage minimum from the load plane towards generator in the following manner.

Since it may not be possible to reach the first d_{min} by the probe close to the load directly using slotted line, an equivalent load reference plane on the slotted line is established by means of a short circuit at the load reference plane where a voltage minimum now occurs. Since a series of minima are produced on the slotted line at intervals of $\lambda_g/2$, the load reference plane can be shifted to a convenient minimum position near the centre of the slotted line as shown in Fig.13.20. The d_{min} can then be measured by observing the first minimum from this shifted reference plane when the load replaces the reference short.

4. Phase angle ϕ_L of the load is calculated from Eq. (13.49) and hence $\Gamma_L = p_L \, e^{j\,\phi_L}$ is found.
5. The unknown impedance Z_L is then calculated from Eq. (13.46).

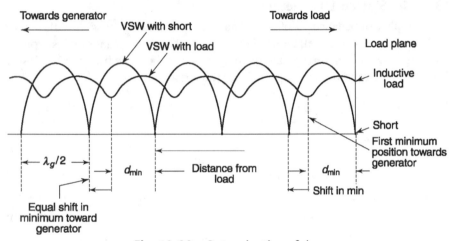

Fig. 13.20 *Determination of d_{min}*

To ease the calculation, Smith chart (Fig. 13.21) can be used to determine Z_L from the measurements of S and d_{min} as follows, where load VSWR $S = 2$, and $d_{min}/\lambda_g = 0.2$, say.

1. Draw the VSWR circle centred at 0 ($r = 1$) with radius cutting the r-axis at $S = 2$.
2. Move from the short circuit load point A on the chart along the wavelengths toward load scale by distance d_{min}/λ_g to B and join OB.
3. The point of intersection between the line OB and the VSWR circle gives the normalised load $z_L = Z_L/Z_0$ and hence the complex load $Z_L = Z_0(1.0 + 10.7)$.

13.11.2 Impedance Measurement of Reactive Discontinuity

The impedance of a shunt reactive discontinuity, such as a post or windows or a step in a microwave transmission line, can be measured using the slotted line method from the measurement of line VSWR and the distance of first voltage minimum from the discontinuity plane as follows.

Let jX be the reactance of the discontinuity at load distance point $d = 0$. The line is terminated by a matched load R_0 at $d = 0$. The total impedance of the combination is

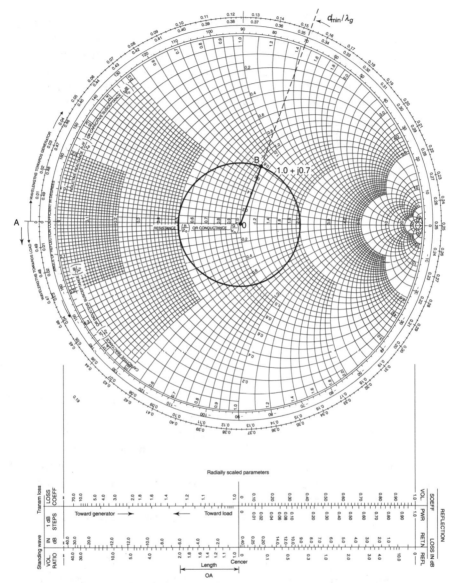

Fig. 13.21 *Smith chart for Z_L measurement*

$$Z_L = \frac{R_0 \cdot jX}{R_0 + jX} \tag{13.51}$$

The normalised value

$$\frac{Z_L}{R_0} = \frac{jX}{R_0 + jX} = \frac{X^2}{R_0^2 + X^2} + j\frac{XR_0}{R_0^2 + X^2}$$

$$= x + jy, \text{ say} \tag{13.52}$$

where,
$$x = \frac{X^2}{R_0^2 + X^2}, \quad y = j\frac{X R_0}{R_0^2 + X^2} \qquad (13.53)$$

with X unknown. Here we see that

$$X/R_0 = x/y \qquad (13.54)$$

or,
$$B/G_0 = y/x \qquad (13.55)$$

where $B = 1/X$ and $G_0 = 1/R_0$. The values of x and y are obtained from the Smith chart to find the value of the normalised susceptance of the discontinuity.

The procedure of measurement follows the following steps with reference to the experimental set-up shown in Fig.13.22.

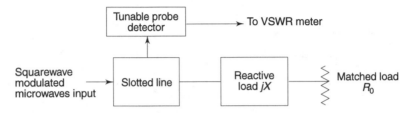

Fig. 13.22 *Experimental set up for the measurement of impedance of a discontinuity*

Discontinuity such as a tuning screw or windows in the line is connected to the slotted line and the output port is terminated by a matched load R_0. By adjusting the tuning stub, a voltage minimum position, x_1 is noted along the slotted line and the corresponding VSWR and λ_g are measured.

Next the discontinuity and matched load are replaced by a fixed short at the point of the discontinuity position. The first voltage minimum position x_2 is noted towards the load from x_1. Thus x_2 translates the plane of discontinuity on the slotted line scale so that the distance of first voltage minimum with discontinuity and matched load termination is $d_{min} = (x_1 \sim x_2)$.

By locating d_{min}/λ_g, S on the Smith chart, $Z_L/R_0 = x + jy$ can be read and $B/G_0 = y/x$ is calculated.

13.11.3 Impedance Measurement by Reflectometer

The reflectometer arrangement shown in Fig.13.18a cannot have ideal conditions of infinite directivity, constant impedance detectors and perfect impedance matching. When the unknown impedance is connected to the output port, the ratio of the signal amplitudes at ports 3 and 4 is, in general

$$\frac{b_3}{b_4} - \frac{A\Gamma_L + B}{C\Gamma_L + D} \qquad (13.56)$$

where A, B, C and D are functions of the S-parameters of the four ports formed by the reflectometer. By using two tuners T_A and T_B, a movable short, and a sliding load of low VSWR (<1.02), the above ideal conditions could be achieved as shown in Fig.13.23.

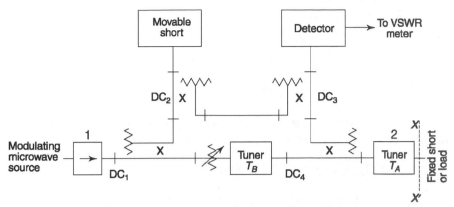

Fig. 13.23 *Reflectometer with tuners for amplitude and phase measurements*

Tuner T_A is adjusted to make $|b_3/b_4|$ constant while the phase of Γ_L is varied by changing the position of a sliding load at port 2, This makes $B = 0$. The tuner T_B is adjusted to make $|b_3/b_4|$ constant as the phase of Γ_L is varied by changing the position of a sliding short at port 2, then $C = 0$. Hence

$$\left|\frac{b_3}{b_4}\right| = \left|\frac{A}{D}\right| |\Gamma_L| = K|\Gamma_L| \tag{13.57}$$

or,

$$|\Gamma_L| = \frac{1}{K}\left|\frac{b_3}{b_4}\right| \tag{13.58}$$

For a given bridge, K is determined by noting $|b_3/b_4|$ using a fixed short of known reflection coefficient of -1 at port 4 and, therefore, by measuring $|b_3/b_4|$, the magnitude of reflection coefficient of any load at port 4 can be determined.

In order to measure the phase of the load reflection coefficient, four identical directional coupler reflectometer can be used as shown in Fig.13.23. The procedure of phase measurements is as follows.

1. First a fixed short is placed at XX' plane of the precession waveguide section at port 2. The movable short at port 4 and the attenuator are adjusted to obtain null in the detector output at port 3.
2. The fixed short at port 2 is replaced by the test load and a shift x of the movable short at port 4 is measured to obtain null in the detector output at port 3.

The phase of Γ_L is then given by

$$\phi_1 = \frac{2\pi\Delta x}{\lambda_g} \tag{13.59}$$

Thus

$$\Gamma_L = K \left| \frac{b_3}{b_4} \right| e^{j\phi_L} \tag{13.60}$$

and

$$Z_L = Z_0 \frac{1 - \Gamma_L}{1 + \Gamma_L} \tag{13.61}$$

The accuracy of phase measurement depends on the sensitivity of the detector for null reading and the vernier scale reading of the movable short.

13.12   FREQUENCY MEASUREMENT

Microwave frequency is measured using a commercially available frequency counter and cavity wavemeter. The frequency also can be computed from measured guide wavelength in a voltage standing wave pattern along a short circuited line by using a slotted line.

13.12.1   Wavemeter Method

A typical wavemeter is a cylindrical cavity with a variable short circuit termination which changes the resonance frequency of the cavity by changing the cavity length. As discussed in Section 7.5.5 TE_{011} mode is most suitable for wave meter because of its higher Q and absence of axial current. Since this is higher order mode, possibility of generation of lower order modes exits. Hence for practical purposes dominant TM_{010} mode is used in wavemeter applications. Wavemeter axis is placed perpendicular to the broad wall of the waveguide and coupled by means of a hole in the narrow wall as shown in Fig.13.24. This excites TM_{010} mode in the cavity due to the magnetic field coupling. A block of absorbing material (Polytron) placed at the back of the tuning plunger prevents oscillation on top of it. Thus the cavity resonates at different frequencies for different plunger positions. The tuning can be calibrated in terms of frequency by known frequency

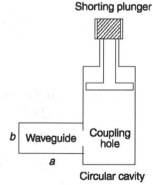

Fig. 13.24   *Wavemeter method of frequency measurements*

input signals and observing the dip in the display unit (power meter) connected at the output side of waveguide. The accuracy of such a wavemeter is in the range of 1 % to 0.005 % for available Q of 1,000–50,000, respectively. Since the power is absorbed in the wavemeter at resonance this is called *absorption type wavemeter*.

13.12.2   Slotted Line Method

Since the distance d_{min} between two successive minima of voltage standing wave pattern in a short circuited line is half wavelength $\lambda_g/2$, frequency can be determined from the relations

$$f(GH_Z) = 30/\lambda_{0\ (cm)} \tag{13.62}$$

$$\lambda_g = 2d_{min} \tag{13.63}$$

$$= \frac{\lambda_0}{\left[1 - (\lambda_0/2a)^2\right]^{1/2}} \text{ for waveguide} \tag{13.64}$$

$$= \frac{\lambda_0}{\sqrt{\varepsilon_r}} \text{ for coaxial line} \tag{13.65}$$

and measuring the d_{min} by the slotted line probe carriage.

13.12.3 Down Conversion Method

An accurate measurement of microwave frequency can be done by means of a heterodyne converter. A heterodyne converter (Fig.13.25) down converts the unknown frequency f_x by mixing with an accurately known frequency f_a, such that the difference $f_x - f_a = f_{IF}$ is amplified and measured by the counter. The frequency f_a is selected by first multiplying a local oscillator frequency (known) to a convenient frequency f_1 and then passing it through a harmonic generator that produces a series of harmonics of f_1. The appropriate harmonic $nf_1 = f_a$ is selected by the tuning cavity such that f_a can be added with f_{IF} and display f_x (counter reading $+ f_a$), the unknown frequency. In practice, the system starts with $n = 1$ and the filter frequency is selected by a feedback mechanism from IF stage until an IF frequency in the proper range is present. Typically, $f_1 = 100$ to 500 MHz for a range of f_x up to 20 GHz. For better accuracy a low noise oscillator and noiseless multiplier are to be selected.

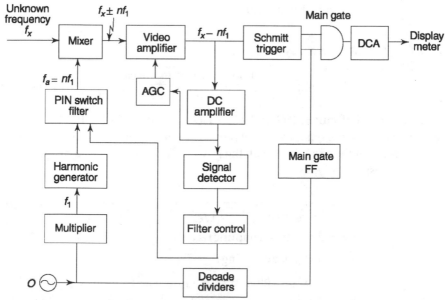

Fig. 13.25 *Down conversion method*

13.13 MEASUREMENT OF CAVITY Q

A difficult measurement at microwave frequencies is the accurate measurement of Q of a high Q cavity. This is due to the fact that the 3dB bandwidth of the cavity response curve is a very small fraction of the resonance frequency. More-over, the cavity has to be loaded during such measurements and Q becomes lower. There are three definitions of Q, s connected to the associated circuit which are summarized below.

Unloaded Q_0

$$Q_0 = 2\pi \frac{\text{Energy stored in the cavity}}{\text{Energy lost per cycle in the cavity}} \tag{13.66}$$

Q_0 is selectivity factor of the cavity, dependent on the geometrical portion of the cavity. The expressions for Q_0 of several cavities are given in Section 7.5 in terms of their dimensions and mode numbers.

Loaded Q_L

$$Q_L = 2\pi \frac{\text{Energy stored in the cavity}}{\text{Energy lost per cycle in the cavity + Energy lost per}} \tag{13.67}$$
$$\text{cycle in the external system}$$

Q_L is the Q of the entire system, including all sources of energy loss. Q_L can be determined by the formula

$$Q_L = f_0/\Delta f \tag{13.68}$$

where f_0 = resonance frequency

Δf = 3 dB bandwidth

External Q_E

Q_E is the Q of the external system.

$$Q_E = 2\pi \frac{\text{Energy stored in the cavity}}{\text{Energy lost per cycle in the external system}} \tag{13.69}$$

From above definition, $1/Q_L = 1/Q_0 + 1/Q_E$ and thus $Q_L < Q_o$. For an aperture coupled transmission type cavity, the input and output coupling factors β_1 and β_2 are a measure of the extent to which the power is coupled to the cavity and from the cavity, respectively, where

$$\beta_1 = \frac{4}{4 S_0 - (S_0 + 1)^2 T(f_0)} \tag{13.70}$$

$$= 1 \text{ (Critical coupling)}$$
$$< 1 \text{ (Under coupling)}$$
$$> 1 \text{ (Over coupling)}$$

and

$$\beta_2 = \beta_1 S_0 - 1 \tag{13.71}$$

$$Q_0 = Q_L(1 + \beta_1 + \beta_2) \tag{13.72}$$

Here S_0 = VSWR at the resonance frequency f_0.

$$T(f_0) = P_{out}/P_{in} \tag{13.73}$$

$$= \text{Transmission loss at the resonance frequency } f_0.$$

Measurement of both the transmission loss $T(f_0)$ and VSWR S_0 at resonance gives the data needed for calculating β_1 and β_2 and determining Q_0. A brief description of several methods of measurement of Q is given below.

13.13.1 Slotted Line Measurement of Q

A slotted line may be used to measure the Q of a reflection type cavity which is normally used in a microwave tube, through pure VSWR measurements or through measurement of the shift in position of a standing wave minimum as the generator frequency is varied. Here the VSWR in the line that feeds the cavity is uniquely related to the variation in amplitude of the cavity input reflection coefficient and the shift of minimum is related to the variation of phase angle of the complex voltage reflection coefficient. The measurement set-up is shown in Fig.13.26. The half-power frequency is found directly from the VSWR measurement, where the equivalent resonator reactance is assumed to be equal in magnitude to the equivalent resonator resistance. If $Z_{in} = R + jX$ is the input impedance in the vicinity of resonance of the cavity, VSWR

$$S = \frac{|Z_{in} + Z_0| + |Z_{in} - Z_0|}{|Z_{in} + Z_0| - |Z_{in} - Z_0|} \tag{13.74}$$

At resonance frequency f_0, $X = 0$, so that minimum VSWR S_0 is

$$S_0 = R/Z_0, \text{ if } R > Z_0$$

$$= Z_0/R, \text{ if } R < Z_0 \tag{13.75}$$

At half-power frequencies f_1 and f_2 of the unloaded cavity, $X = R$, so that

$$S_1 = \frac{\sqrt{\left[(R + Z_0)^2 + R^2\right]} + \sqrt{\left[(R - Z_0)^2 + R^2\right]}}{\sqrt{\left[(R + Z_0)^2 + R^2\right]} - \sqrt{\left[(R - Z_0)^2 + R^2\right]}} \tag{13.76}$$

or,

$$S_1 = S_0 + \frac{1}{2 S_0} + \sqrt{\left(S_0^2 + \frac{1}{4 S_0^2}\right)} \; ; R > Z_0$$

$$= 1/S_0 + S_0/2 + \sqrt{\left(1/S_0^2 + S_0^2/4\right)} \; ; R < Z_0 \tag{13.77}$$

The unloaded $Q_0 = f_0/(f_1 \sim f_2)$ can be determined from the above measurements. For a loaded cavity, minimum value S_0 as well as $f = f_1 \sim f_2$ increase and this results in a lower value of Q. The accuracy of measurement lies on the half-power VSWR and half-power bandwidth. In this method the measurement errors include the departure from square-law behaviours of the probe detector, frequency instability of the source, generator mismatch, probe and generator interaction at high VSWR.

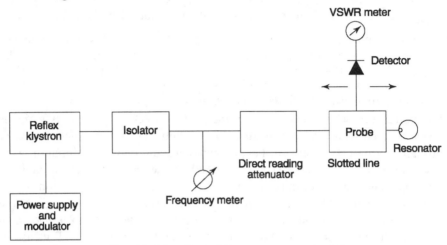

Fig. 13.26 *Slotted line measurement of Q*

13.13.2 Reflectometer Method of Measurement of Q

This method is suited for a reflection cavity and provides oscilloscope presentation by a swept microwave source. The method determines the magnitude of voltage reflection coefficient Γ at resonance, at half-power points and at a point far away from resonance $\Gamma \approx 1$. A schematic of the experimental set-up is shown in Fig. 13.27. The total errors in the measurement depend essentially on the accuracy of measuring the bandwidth and in setting half-power level for the reflection coefficient. The errors in half-power point for the reflection coefficient depend on the imperfect directivity of directional couplers and instability of the source frequency.

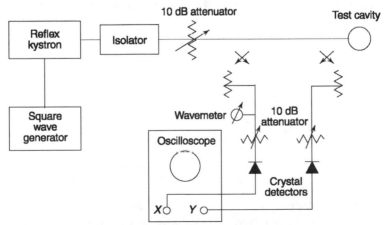

Fig. 13.27 *Reflectometer method of measurement of Q*

13.13.3 Q From Transmitted Power Measurement

This method uses the power transmitted through a cavity as a function of frequency, for measuring loaded Q. This method can be used for both transmission and the reflection type cavities. The transmission method has the advantage that the loaded Q can be measured directly regardless of the existence of coupling losses. However, this method cannot yield the unloaded Q without a number of additional measurements. The basic procedure can be carried out in several different ways as discussed below.

There are three main sources of error in the transmission method of Q measurement. The first relates to a possible mismatch of the generator and load between which the cavity is inserted. The second relates to the most important error that arises in measuring the bandwidth of the cavity response curve. Third is that due to the inaccuracies in relative power measurements, caused either by imperfect calibration of attenuators or power meters, or imperfect square-law response of detectors, errors in the readings of meters and attenuator dials, or by both calibration and reading errors.

(i)   CW measurement

The CW measurement set-up is illustrated in Fig.13.28. The transmitted CW power can be monitored with a power meter. Alternatively, the RF signal may be square-wave modulated and a tuned amplifier or a VSWR meter may be used at the output of the square-law crystal detector to indicate the transmitted powers at resonance frequency f_0 and half power points $f_0 \pm \Delta f/2$. Q is calculated from $f_0/\Delta f$. To avoid errors due to non square-law response from the crystal at different power levels, a calibrated attenuator may be used for determining the half-power frequencies by keeping input power level the same in all measurements.

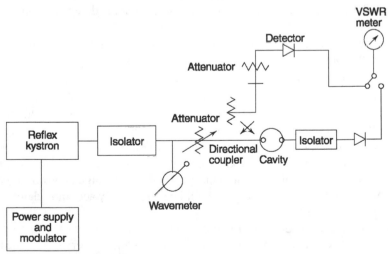

Fig. 13.28   *Q from CW transmitted power measurement*

(ii) Swept frequency measurement of Q A swept frequency technique (Fig.13.29) requiring less frequency stability of the RF source than the above

procedures, provides simultaneous display of two oscilloscope traces, one proportional to the incident power, the other to the power transmitted through the cavity. The method uses a pair of matched crystal detectors of the same response law over the power and frequency range of the measurements. The wave meter measures the resonance frequency f_0 and the half-power frequency from the dip in the input response curve at the output peak and half-power points, respectively, to yield loaded Q_L.

The procedure of measurement is as follows.

1. A linear sweep (saw-tooth wave) voltage is applied to the repeller of the reflex klystron so that a FM microwave signal is produced. Since the expected cavity bandwidth is much smaller than the frequency swing of FM signal, we can assume constant input voltage to the cavity at different frequencies around the resonance frequency of the cavity. Applying voltage to the triggering input of a double beam CRO from the same sweep generator, a response of the cavity is obtained in Y_2 beam.

    To determine the resonant frequency and 3 dB bandwidth a marker trace is generated by applying a detected signal from the auxiliary arm of the directional coupler to Y_1 beam on the CRO. This represents the klystron power output mode characteristics.

    By tuning the klystron frequency, cavity response is maximised and cavity peak response and peak of the klystron mode is made coincident. The frequency meter is adjusted to obtain a dip at the peak of cavity response. This indicates resonant frequency of the cavity. The flat top of the klystron mode curve is now adjusted to 3 dB point of the cavity response and the frequency bandwidth is noted by observing frequency meter readings corresponding to the dip in the response at 3 dB points.

2. VSWR of the cavity, terminated by match load, is determined at f_0 with a slotted line and VSWR meter.

3. Transmission loss at f_0 is determined by observing input and output powers of the cavity.

The α_0 and Q_L of the cavity are calculated from the above measured data and using Eqns. 13.68 – 13.73. This method is a rapid method of Q measurement. Obviously the Q_L of the frequency meter used should be considerably higher than that of the cavity under test to set the marker positions accurately. Reflex-klystrons are widely used as sources of swept frequency microwave power. The electronic tuning range of reflex klystron within which frequency modulation is possible, is comparatively narrow. Thus at 10 GHz this range is ordinarily about 10–15 MHz and the resonance curve should be completely accommodated within this frequency range.

A further limitation of measurement is that the input power to the cavity must be maintained constant within the cavity band. This limits very low Q measurement by this technique.

Accuracy of the measurement depends on the measurement of very low bandwidth. Therefore, resolution of the frequency meters should be very high. Eye estimation in measuring bandwidth also introduces some error. Errors can be reduced by taking the average of several observations.

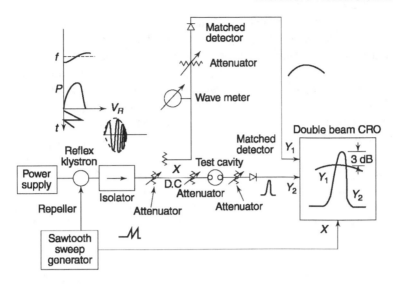

Fig. 13.29 *Swept frequency measurement of Q*

(iii) Swept frequency measurement of Q using electronic frequency marker
Bandwidth measurement accuracy is the most important factor in high Q measurement. Ordinarily when the test cavity Q is not very high, a high Q cavity wavemeter may be used to measure half-power frequencies. A cavity operating at 10 GHz with a high Q of 10,000 has a bandwidth of 1 MHz. It is difficult to adjust the microwave oscillator for such a small frequency difference. Moreover, it is difficult to measure this frequency difference since standard cavity wavemeters have smallest scale division of 1 to 5 MHz.

For very accurate measurement of high Q in a laboratory set-up swept frequency method using electronic frequency marker can be employed as shown in Fig.13.30.

The swept source is a klystron and is shown to be saw-tooth modulated. The same saw-tooth sweep is simultaneously applied to the horizontal input of a dual trace oscilloscope to give displays of incident and transmitted power versus frequency on the oscilloscope screen synchronously. Saw-tooth generates FM signal from the klystron with envelope variation in accordance with the klystron mode curve and the cavity response curve. Both the input and output signal of the cavity are detected by means of crystal detectors before being fed to the CRO. The markers are generated with the aid of an auxiliary frequency stable low frequency oscillator and the klystron in the main line which is frequency swept through the cavity resonance, varying in frequency from $f_0 - f_s$ to f_0 to $f_0 + f_s$. A portion of the swept signal from the directional coupler DC-1 and the output of the fixed frequency oscillator ($\sim$800 MHz) are combined in a harmonic crystal mixer which is connected to the input of a superheterodyne receiver. Harmonics of low frequency signal are mixed with the swept frequencics. Sum and difference frequencies are produced in the mixer. The superhet receiver can respond only to the time varying difference frequency component, $f_s(t)$ and produces a sharp audio output pip whenever the frequency of the input signal to the receiver equals f_r, the frequency to which the receiver is tuned. This occurs twice during a saw-tooth sweep, once below resonance, when $[f_0 - f_s(t) - f_0] = f_r$, and again above resonance, when $f_0 + f_s(t) - f_0 = f_r$, that is, whenever the swept signal frequency applied to the cavity differs from the resonant frequency by $\pm f_r$. If the output pip is applied to the Z-axis (intensity grid) of the oscilloscope, two bright spots will appear on the resonance curve, as shown, marking these two frequencies. The frequency separation of the markers is obviously $2f_r$, or twice the frequency reading of the tuning dial of the receiver.

The crystal detectors D_1 and D_2 must be well matched to the line and selected to give identical responses over the required frequency and power range and at the power level used. It is desirable that the RF power levels at both crystals be

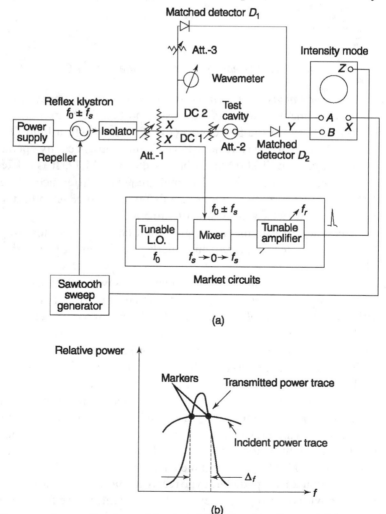

Fig. 13.30 (a) Swept frequency measurement of Q using electronic frequency markers(b) Input and output power traces

nearly the same. This makes it necessary that the sum of the attenuation in Att. 2 and the cavity insertion loss (in dB) in the transmitted power channel be approximately equal to the (dB) coupling loss and sum of the attenuations in the directional coupler arm.

The directional coupler should have high directivity in the frequency range of interest. The isolator should have at least 20 dB of isolation and VSWR better than 1.2. All the attenuators should have a good VSWR (<1.10). Broadband matching of the crystal detectors can be obtained by using well matched 10 dB fixed pads in tandem with the crystal mounts. The calibrated variable attenuator Att-3 should be of the precision type. For good matching, isolators may be used before the cavity and the detector.

Checking crystals

1. The klystron is tuned to f_0 by observing maximum transmitted signal through the cavity.
2. The cavity transmission trace B from D_2 should be centred with respect to the incident power trace A from D_1.
3. The dual trace oscilloscope vertical centering controls are adjusted so that the base lines of the two traces A and B coincide. During this adjustment, the RF power is completely removed.
4. Attenuator Att-3 is set to zero. Att-1 and 2 are adjusted so that the peak of the trace from D_2 just touches the trace from D_1.
5. The cavity is removed and Att-2 is adjusted so that the traces of the two crystals coincide.
6. The coincident traces are checked by increasing the attenuation of Att-1 by approximately 6 dB. The two traces should remain coincident over this range. Otherwise, discrepancies between the two crystals are noted and later applied as a correction. The amount of the discrepancy can be determined by adjusting Att-3 starting with some initially inserted attenuation and observing the change in setting required to align both traces exactly.
7. Steps 1–6 are repeated for a few frequencies in the neighborhood of f_0.

Measurement procedure

1. When the cavity is removed from the test position, the RF frequency sweep is adjusted so as to cover a sufficiently large portion of the cavity resonance curve. Calibrated attenuators 2 and 3 are adjusted so that the incident and transmitted power traces coincide. Coincidency of baselines of the two traces for zero RF power signal is done previously.
2. The test cavity is then inserted and the klystron repeller voltage is adjusted and it is tuned at resonance f_0 so that both the transmitted and incident power traces on the scope are centred. Att.-2 is adjusted so that the incident power trace touches the maximum power point of the transmitted power trace, corresponding to f_0 as shown in Fig.13.31(b). The change in attenuation of Att.-2 is the cavity transmission loss at resonance $T(f_0)$. Att.-3 is kept at 0 dB. Resonance frequency f_0 is observed by using a reaction wavemeter.

3. The incident power trace is lowered by adding 3 dB attenuation with Att.-3.
4. The markers are placed at the points of intersection of the incident and transmitted power traces at the above setting of Att.–3 by tuning the superhet receiver. The tuning frequency f_r is noted to determine half-power bandwidth $\Delta f = 2 f_r$.
5. Q_L is calculated from $f_0/2f_r$.
6. For the VSWR measurement, the klystron is square-wave modulated. A slotted line with a tunable probe detector is inserted at the input of the cavity and output is matched terminated. Detected signal from the slotted line is fed to a VSWR meter. VSWR S_1 are measured at frequencies around f_0. Also S_0 at f_0 is determined.

The Q_0 is calculated from equation 13.68 – 13.73.

Accuracy of measurement

Advantages of swept frequency measurements of Q with frequency markers are
- Requires less frequency stability of the RF source
- Accurate measurement of bandwidth
- Measurement time is small.
- *Q measurement by phase shift of modulation envelope of transmitted signal [8,9]*

When a microwave signal of frequency f_0 is sine wave amplitude modulated at a frequency f_1 and applied to a transmission cavity having a resonant frequency f_0, the modulation envelope of the transmitted signal will be delayed in phase with respect to that of the impressed signal by an angle

$$\theta = \tan^{-1}(2Q_L \, f_1/f_0) \tag{13.78}$$

If f_1 is made equal to $\Delta f/2$, where $\Delta f =$ half-power bandwidth, and $\theta = 45°$, then $Q_L = f_0/2f_1$. Since f_1 can be accurately known from the frequency calibration of the modulating generator, Δf and hence Q_L can be determined. When a sine wave amplitude modulated carrier wave is applied to a resonant circuit at its resonant frequency, its sidebands are symmetrically shifted in phase on transmission through the resonator, one sideband being advanced, the other delayed with respect to the input wave, as shown in Fig.13.31. This results in the phase shift in the modulation envelope shown in Fig.13.31(c), which may be measured by means of a phase meter. The advantages of this method are that it is capable of bandwidth measurements and the measurement is much less affected by frequency stability of the microwave generator. This follows from the fact that the phase shift of the modulation envelope of the transmitted wave is a maximum when the carrier is at the resonant frequency of the cavity.

The accuracy of measurement depends ultimately on the accuracy of the precision phase measuring instrument.

(v) Q Measurement from inflection points of the resonance curve

The tuned circuit behaves most nearly like a linear network in the neighbourhood of the two frequencies on the resonance curve and therefore, the second derivative, is zero at the inflection points of the curve. To a very good approximation, it can be shown that $Q_L = f_0/(\sqrt{3} \, \Delta f)$ for the response curve, where Δf is frequency

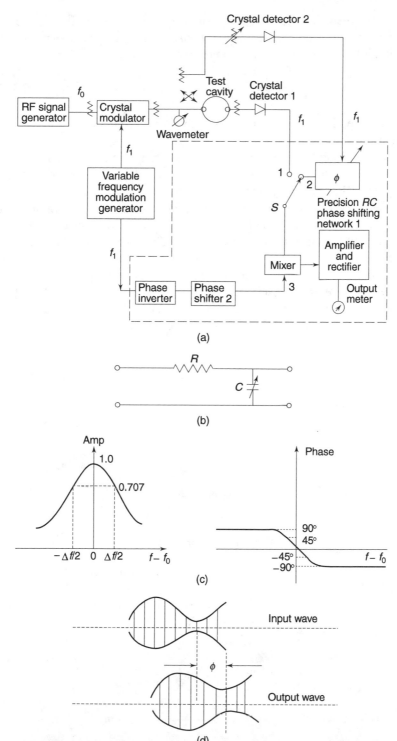

Fig. 13.31 *(a) Phase-shift method of measurement of Q (b) Precision RC phase shifting network (c) Transmission characteristics of single tuned resonator (d) Phase change of transmitted signal*

difference between the inflection points of the curve concerned. If the inflection point frequencies are denoted by f_1 and f_2, then f_0 is given very closely by $(f_1 + f_2)/2$, so that the above equation can be expressed by

$$Q_L = \frac{f_1 + f_2}{2\sqrt{(3|f_1 - f_2|)}} \tag{13.79}$$

The equation applies when a square-law detector is used in monitoring the resonator frequency response.

A general method to determine the two inflection point frequencies is illustrated in the power transmission curve shown in Fig.13.32(b), for a cavity resonant at f_0. The RF signal frequency is modulated about some centre value f at an audio frequency f_1 rate. The envelope of the RF (or detected) output of the cavity will be a distorted reproduction of the modulating waveform. The distortion will be a minimum if f is one of the inflection points. The linearity of behaviour of the tuned circuit at the inflection point frequencies provides the basis to determine these frequencies with fair precision. If a pure sinusoid of frequency f_1 is used to frequency modulate the RF oscillator, the second harmonic content should be a minimum if the applied RF frequency to the cavity corresponds to one of the inflection points. Figure 13.32(a) shows a block diagram for this method.

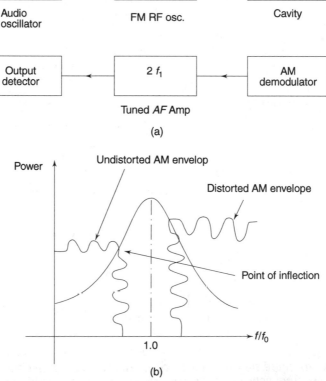

(b)

Fig. 13.32 *(a) Inflection point method of measurement of Q (b) Inflection point of resonance curve of cavity*

13.13.4   Decrement Method of Measurement of Q

Since it is difficult to measure VSWR or transmission in a very high Q system, a method is adopted based on measuring the time rate of decay of microwave energy stored in the very high Q resonator:

$$U = U_0 \, e^{-(\omega_0 t/Q)} \tag{13.80}$$

This method uses a pulsed excitation for the cavity system and observation of the rate of decay of the oscillations from U_1 to U_2 between exciting pulses at times t_1 and t_2.

Thus

$$10 \log_{10}(U_1/U_2) = 4.343 \, \omega_0(t_2 - t_1)/Q \tag{13.81}$$

or,

$$Q = 8.686 \pi f_0 \, \Delta t/A = 27.3 \, f_0 \, \Delta t/A \tag{13.82}$$

where

$$10 \log_{10}(U_1/U_2) = A, \; t_2 - t_1 = \Delta t \tag{13.83}$$

This set up is shown in Fig.13.33. An oscilloscope technique is used in conjunction with a pick-up probe and power indicating equipment. Since the fluctuations of the source frequency do not affect the rate of energy decay in the resonator, the decrement method does not require a very high degree of frequency stability of the generator. However, frequency variations do affect the level of energy build up in the cavity resulting in corresponding fluctuations in the decay curve trace on the oscilloscope. This problem can be overcome by photographing a trace due to a single pulse and then measuring the decay time on the photograph.

13.14   DIELECTRIC CONSTANT MEASUREMENT OF A SOLID

The dielectric constant ε_r is defined by the permittivity ε of the material with respect to that ε_0 of air or free space

$$\varepsilon_r = \varepsilon/\varepsilon_0, \; \varepsilon_0 = (10^{-9}/36\pi) \text{ farad/m} \tag{13.84}$$

Due to presence of non-zero conductivity, dielectric material exhibits loss, resulting in complex value represented by

$$\varepsilon_r = \varepsilon_r' + j \, \varepsilon_r'' \tag{13.85}$$

The loss tangent

$$\tan \delta = \varepsilon_r''/\varepsilon_r' \tag{13.86}$$

The measurement of the complex dielectric constant is required not only in scientific application but also for industrial applications such as microwave heating or ovens and to study the biological effects of microwaves.

The dielectric constant is not independent of frequency. Generally, the variation of ε_r with frequency is sufficiently gradual so that it can be considered to be

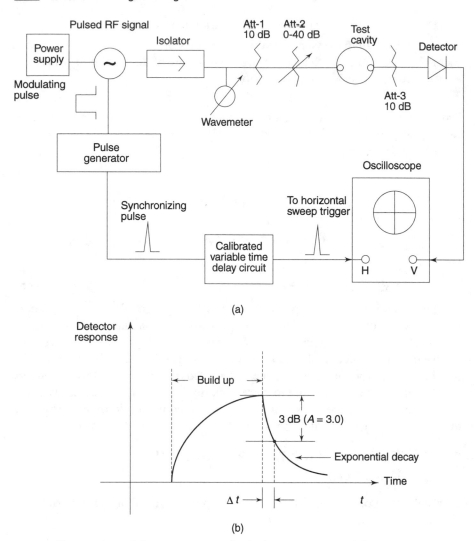

Fig. 13.33 *(a) Decrement method of measuring Q (b) Transient response of a cavity*

constant over a fairly wide frequency band for most common microwave applications. On the other hand, the per cent variation in ε_r'' is almost always greater than that of ε_r', so that ε_r'' should be measured near the frequency or frequencies of interest.

There are several methods available for dielectric constant measurement. The following sections describe two commonly used methods: the waveguide method and cavity perturbation method.

13.14.1 Waveguide Method

In this method it is assumed that the material is lossless. A dielectric sample AB completely fills a length of the waveguide and the end is terminated in a short as shown in Fig. 13.34. A voltage standing wave minimum is observed in the slotted line at C (say).

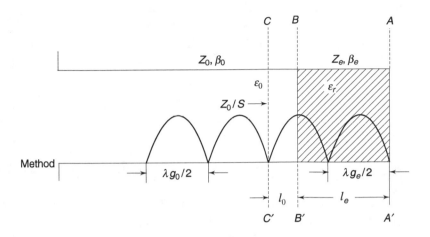

Fig. 13.34 *Waveguide method*

Let

$$l_e = AB = \text{the dielectric sample length} \qquad (13.87)$$

$$l_0 = BC \qquad (13.88)$$

Then the distance of V_{min} from short circuit $= l_e + l_0 = AC$. For a dielectric filled guide of characteristic impedance Z_e, input impedance at B is purely reactive,

$$Z_{in}' = jZ_e \tan \beta_e l_e. \qquad (13.89)$$

where β_e is the propagation constant.
Using this Z_{in}' as termination at B, input impedance at C for the empty guide is

$$Z_{in\,c} = \frac{Z_{in}' + jZ_0 \tan \beta_0 l_0}{Z_0 + jZ_{in}' \tan \beta_0 l_0} = 0, \text{ at } V_{min} \text{ point} \qquad (13.90)$$

Therefore,

$$Z_{in}' + jZ_0 \tan \beta_0 l_0 = 0 \qquad (13.91)$$

or,

$$jZ_e \tan \beta_e l_e + jZ_0 \tan \beta_0 l_0 = 0 \qquad (13.92)$$

or,

$$Z_0 \tan \beta_0 l_0 = -Z_e \tan \beta_e l_e \qquad (13.93)$$

Assuming nonmagnetic dielectric in the waveguide,

$$\frac{Z_0}{Z_e} = \frac{\beta_e}{\beta_0} \qquad (13.94)$$

or

$$Z_0 - \frac{\beta_e Z_e}{\beta_0} \tag{13.95}$$

Substituting this value in Eq. 13.93

$$\frac{\beta_e Z_e}{\beta_0} \tan \beta_0 l_0 = - Z_e \tan \beta_e l_e \tag{13.96}$$

or,

$$\frac{l_0 \tan \beta_0 l_0}{\beta_0 l_0} = - \frac{l_e \tan \beta_e l_e}{\beta_e l_e} \tag{13.97}$$

or,

$$\frac{l_0 \tan Y}{l_e Y} = - \frac{\tan X}{X} \; ; \text{where } X = \beta_e l_e, \; Y = \beta_0 l_0; \tag{13.98}$$

For dominant mode, $\beta_0 = 2\pi / \lambda_{g0}$ and $\lambda_{g0} = 2$ * distance between two successive V_{min} which can be measured by the slotted line. l_0 and l_e are also measured in the slotted line. Therefore, the left hand side (α say) of the above transcendental Eq. 13.98 is known and it can be written as

$$(\tan X)/X = -\alpha \tag{13.99}$$

The above equation can be solved for determining $X = \beta_e l_e$. Now

$$\beta_e = \frac{2\pi}{\lambda_{ge}} = \frac{2\pi}{\lambda_0} \sqrt{\left[\varepsilon_r - \left(\frac{\lambda_0}{\lambda_c} \right)^2 \right]}, \; \lambda_c = 2a \tag{13.100}$$

where a is the waveguide broadwall dimension.

Since β_e is known, ε_r can be determined from above equation. This equation has an infinite number of solutions for ε_r. Hence two different lengths of sample are taken for two sets of solutions.

For length l_e : $X = X_1$, X_2 , ...

$\varepsilon_r = \varepsilon_{r1}$, ε_{r2} , ...

For length l_e' : $X = X_1'$, X_2', ...

$\varepsilon_r' = \varepsilon_{r1}'$, ε_{r2}', ...

The desired value of ε_r is the solution which is common for the two samples.

13.14.2 Cavity Perturbation Method

Cavity perturbation techniques are highly sensitive and accurate and therefore, particularly advantageous in the determination of the dielectric constant and small loss tangents.

In the cavity perturbation technique, a small volume of the test sample is introduced in a cavity resonator at the position of maximum **E**-field for the measurement of ε (and at maximum **H**-field for the measurement of μ). Since the volume of the sample is small, the field in the cavity is assumed to remain undisturbed.

But introduction of dielectric sample changes the resonance frequency and the RF loss. These changes are reflected in the change of Q of the cavity from its unperturbed value. All these changes are related to the dielectric constant and the loss tangent.

Let the original cavity containing air having ε_0, μ_0 and resonance frequency f_0 is perturbed by some material ε, μ_0 placed inside the cavity at E_{max} and $H = 0$ position. For a sufficiently small sample the fields in the cavity outside the dielectric material are not greatly perturbed from those for the empty cavity case. Then the well established perturbation formulae for change in frequency and Q are given by

$$\omega - \omega_0 = \frac{\omega(\varepsilon'_e - 1)\int\limits_{V_s}|E_0|^2\,dv}{\int\limits_{V_c}|E_0|^2\,dv} \qquad (13.101)$$

$$\left(\frac{1}{Q}\right) = \frac{\varepsilon''_r|E_0|^2\,V_s}{\int\limits_{V_c}|E_0|^2\,dv} \qquad (13.102)$$

where
E_0, ω_0 = the field and resonant frequency of the original cavity
E, ω = the corresponding quantities of the perturbed cavity
V_c, V_s = volumes of the empty cavity and the sample filling the cavity, respectively

Assuming small perturbation, ω can be replaced by ω_0 and the simplified relations are

$$\frac{\Delta\omega_0}{\omega_0} = \frac{(\varepsilon'_r - 1)|E_0|^2_{V_s}\,V_s}{2\int\limits_{V_c}|E_0|^2\,dv} \qquad (13.103)$$

$$\left(\frac{1}{Q}\right) = \frac{\varepsilon''_r|E_0|^2_{V_s}.\,V_s}{\int\limits_{V_c}|E_0|^2\,dv} \qquad (13.104)$$

where $|E_0|_{Vs}$ has been assumed constant over the region of integration throughout the small sample and

$$\left(\frac{1}{Q}\right) = \left(\frac{1}{Q_c}\right) \sim \left(\frac{1}{Q_s}\right) \qquad (13.105)$$

Here $\qquad Q_c = Q$ of the unperturbed cavity
$\qquad\qquad Q_s = Q$ of the perturbed cavity

In principle, any type of cavity either rectangular or circular can be used with excitation by suitable mode. The following criteria are important.

(a) The cavity Q should be as high as possible to enhance the accuracy of the measurement and this is very stringent, particularly in low loss materials.

(b) The mode of operation should be such that the dielectric sample is conveniently placed at a uniform field region E_{max} where $H = 0$ for the measurement of ε' and ε''. Otherwise, if $H \neq 0$ in the perturbed region, complex permeability of the magnetically lossy material will add to frequency shift and RF loss in addition to the contribution offered by complex ε. It will be extremely troublesome to mathematically relate the individual contributions (reverse is the case for the measurement of μ).

(c) The accuracy of the method depends on the smoothness of the sample, and fitness of the sample in the cavity. The smaller the sample, the more critically the accuracy will depend upon the sample irregularities. It is, therefore, necessary to machine the sample very carefully for smoothness and size.

In view of the above, rectangular cavity in TE_{103} mode and circular cavity in TM_{010} mode is found suitable in this technique. Both the cavities thus produce E_{max} and $H = 0$ at the centre where the dielectric samples in the form of a thin rod are placed inside the cavity transversely as post parallel to **E**-field as shown in Fig.13.35. With this combination, the perturbation relations are described as follows.

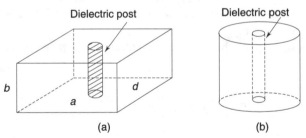

Dielectric post   Dielectric post

(a)   (b)

Fig. 13.35 *Cavity perturbation method (a) Rectangular cavity TE_{103} (b) Circular cavity TM_{010}*

TE_{103} *rectangular cavity* The non-zero field components for TE_{103} mode in the rectangular cavity $a \times b \times d$ are E_y, H_x and H_z only. Since the sample length is parallel to y, only E_y is required for finding ε_r.

$$E_y = \frac{\lambda}{2a} \sin (\pi x/a) \sin (3\pi z/d) \tag{13.106}$$

If the sample is placed at $x = a/2$, $z = d/2$

$$|E_0|^2 = |E_y|^2 = \left(\frac{\lambda}{2a}\right)^2 \tag{13.107}$$

$$V_s x = a/2$$

$$z = d/2$$

and

$$\int_{V_c} |E_0|^2 \, dv = \left(\frac{\lambda}{2a}\right)^2 \int_0^a \int_0^b \int_0^d \sin^2\left(\frac{\pi x}{a}\right) \sin^2\left(\frac{3\pi x}{d}\right) dx \, dy \, dz$$

$$= \left(\frac{\lambda}{2a}\right)^2 \frac{abd}{4} \tag{13.108}$$

Substituting Eq. 13.108 and Eq. 13.109 in Eqs. 13.104 and 13.105,

$$\frac{\Delta f_0}{f_0} = -2\,(\varepsilon_r' - 1)\,\frac{V_s}{V_c} \tag{13.109}$$

$$\left(\frac{1}{Q}\right) = 4\,\varepsilon_r''\frac{V_s}{V_c} \tag{13.110}$$

or,

$$\varepsilon_r' = 1 + 0.5\left(\frac{V_c}{V_s}\right)\frac{f_0 - f}{f_0} \tag{13.111}$$

$$\varepsilon_r'' = 0.25\left(\frac{V_c}{V_s}\right)\left(\frac{1}{Q_c} \sim \frac{1}{Q_s}\right) \tag{13.112}$$

TM$_{010}$ circular cavity   For the circular cavity in *TM*$_{010}$ mode $E_\rho = 0 = E_\phi$ and E_z is parallel to the length of the sample rod which is introduced axially. It can be shown in a similar way

$$\varepsilon_r' = 0.539\left(\frac{V_c}{V_s}\right)\left(\frac{\Delta f_0}{f_0}\right) + 1 \tag{13.113}$$

$$\varepsilon_r'' = 0.269\left(\frac{V_c}{V_s}\right)\Delta\left(\frac{1}{Q}\right) \tag{13.114}$$

The unique feature of the *TM*$_{010}$ mode is that its resonant frequency is independent of the length and therefore design is very simple.

13.15   MEASUREMENTS OF SCATTERING PARAMETERS OF A NETWORK

S-parameters can be conveniently measured following the Deschamps method which utilises measured values of complex input reflection coefficients under a number of reactive terminations. In this section measurements of *S*-parameters of a general two-port network and a four-port network called the magic-*T* are described.

13.15.1   *S*-Parameters of a Two-Port Network

The output end of the two-port network is terminated with a short circuit plunger to vary the reactive termination by moving the plunger in steps of at least 1/8th of the guide wavelength and the corresponding input reflection coefficients are measured. When the attenuation coefficient α in the line is small and the change of total length of the line is less than one guided wavelength, the points P's of the corresponding measured reflection coefficients describe an average circle on the polar chart with centre OC and radius CP as shown in Fig.13.36. However, for a lossless line $\alpha = 0$ and the reflection coefficients will describe a perfect circle. The reflection coefficients can be measured using a network analyser and reflection bridge set-up. Each pair of reflection coefficient points P's, diametrically opposite, are joined and the lines of intersection O' are marked. CO' is joined. Two perpendicular lines at C and O' are drawn to intersect the circle at Q and P respectively. PQ is joined to intersect CO'' at S_{11}. A line $S_{11}E$ parallel to $O'P$ is drawn to meet the circle at E. It can be shown that

$$|S_{11}| = |OS_{11}|$$ (13.115)

and

$$|S_{12}| = \frac{|S_{11}E|}{\sqrt{\rho \cdot R}}$$ (13.116)

where R is the radius of the circle and $\rho = \exp(-2\alpha l)$, the input reflection coefficient with output end of the line short circuited. The attenuation coefficient can be determined from the following equation when experiments are performed using two widely different lengths l and l' of the line

$$\alpha = \frac{1}{2(l - l')} \ln \left(\frac{R|S_{11}E|^2}{R'|S_{11}E'|^2} \right)$$ (13.117)

For $|S_{22}|$ and $|S_{21}|$, measurements are carried out by interchanging the ports.

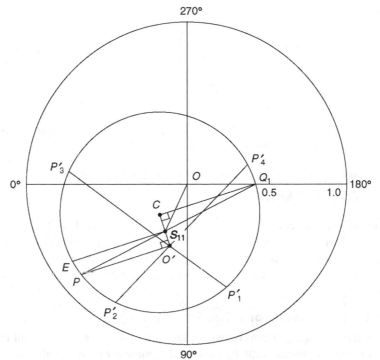

Fig. 13.36 *Deschamps method*

13.15.2 S-Parameters of a Magic-T

The S-parameters of a matched magic-T can be determined using the Deschamp's method by measuring the reflection coefficients at different ports under specific termination of other ports as explained below (Fig.13.37).

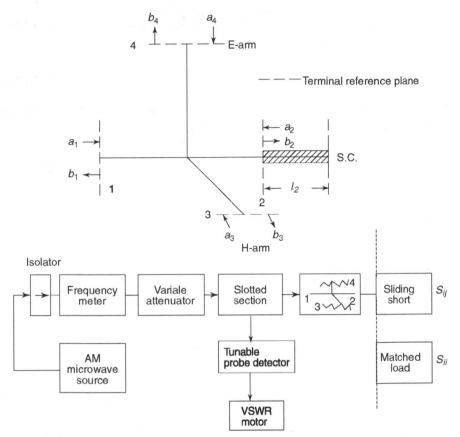

Fig. 13.37 *Experimental set-up for S-parameters measurement of Magic-T*

(a) Measurement of S_{ii} The diagonal elements are determined from the slotted line measurement of the VSWR S_i at the corresponding port with all other ports matched terminated.

$$| S_{ii} | = (S_i - 1) /(S_i + 1) \tag{13.118}$$

(b) Measurement of S_{ij} ($i \neq j$) To measure $S_{12,}$ ports 3 and 4 are match terminated and port 2 is terminated in a short circuit plunger when the input is fed at port 1. At the reference plane of port 2, the reflection coefficient will always be equal to 1 in magnitude but its phase will vary continuously with the distance of the short position from the same reference plane, leading to a reflection coefficient $\Gamma_2 = - e^{j2\theta_2}$. Since $S_{12} = S_{21}$ for an isotropic medium,

$$\Gamma_1 = S_{11} - \frac{S_{12}^2 \, e^{j2\theta_2}}{1 + S_{22} \, e^{j2\theta_2}} \tag{13.119}$$

or,

$$\Gamma_1 = S_{11} + \frac{S_{12}^2 \, S_{22}^*}{1 - |S_{22}|^2} + \frac{S_{12}^2 \, e^{j(2\phi - \phi_{22})}}{1 - |S_{22}|^2} \tag{13.120}$$

where $\qquad S_{22}^* = |S_{22}| e^{-j\phi_{22}} \tag{13.121}$

Here 2ϕ is a function of $2\theta_2 + \theta_{22}$ and $|S_{22}|$. As the length of the short circuit line increases, ϕ increases and Γ_1 describes a circle (Fig. 13.38) with

$$\text{Centre } OC = S_{11} + \frac{S_{12}^2 \, S_{22}^*}{1 - |S_{22}|^2} \tag{13.122}$$

radius $\qquad CP = \dfrac{S_{12}^2 \, e^{j(2\phi - \theta_{22})}}{1 - |S_{22}|^2} \tag{13.123}$

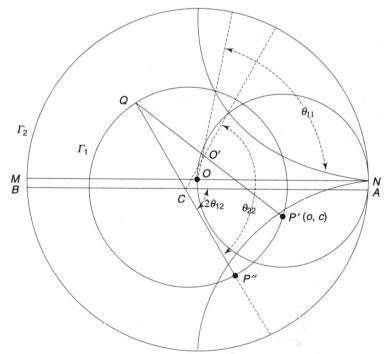

Fig. 13.38 *Deschamps's circle for magic-T*

The experimental measurement procedure is as follows:

1. Ports 2, 3 and 4 are matched terminated and S_{11} is measured directly at

port 1. $S_{11} = OO'$ is located on the Smith chart where O' is the 'iconocentre'.

2. Placing a sliding short at port 2, $\theta_2 = \beta l_2$ is varied to obtain Γ_1 circle on the Smith chart. The centre C is determined on the diagram.

3. Placing open circuit at the reference plane of port 2 ($\theta_2 = \pi/2, l_2 = \lambda_g/4$) the corresponding $\Gamma_1 = \Gamma_1'$ is noted and indicated on Γ_1 circle as P'.

Construction of Deschamp's circle is done in the following manner A straight line is drawn from P' to a point Q on the circle through O' and another line from Q to a point P'' on the circle through C. The angle of the phase or CP'' is equal to $2\theta_{12}$, where $\theta_{12} =$ angle of S_{12}.

Now

$$O'C = OC - OO' = S_{12}^2 S_{22}^* / (1 - |S_{22}|^2) \qquad (13.124)$$

Therefore, angle of $O'C = 2\theta_{12} - \theta_{22}$. Since angle of $CP'' = 2\theta_{12} =$ Angle BCQ, therefore, angle of CP''-angle of $O'C = \theta_{22} =$ Angle $P''CO'$.

Therefore,

$$|S_{11}| = OO', \ \theta_{11} = \text{angle of } NOO' \qquad (13.125)$$

and

$$|S_{22}^*| = O'C/CP'', \ \theta_{22} = \text{angle } P''CO' \qquad (13.126)$$

$$|S_{12}| = \sqrt{\left[\text{Radius of the circle} \times \left(1 - |S_{22}|^2\right)\right]} \qquad (13.127)$$

$$\theta_{12} = \text{angle } BCQ/2 \qquad (13.128)$$

A similar procedure is followed for other ports.

Ordinarily magic-T is not a matched one because of discontinuity present at the junction. Hence H-arm is matched by a tuning screw from the bottom wall and E-arm is matched by an inductive iris by trial and error method. With this arrangement the other two arms are automatically matched.

13.16 MICROWAVE ANTENNA MEASUREMENTS

The most important parameters required to be measured to determine the performance characteristics of microwave antennas are radiation amplitude patterns, radiation phase patterns, absolute gain, directivity, radiation efficiency, beamwidth, input impedance, bandwidth and polarisations. The accurate measurement methods for these parameters require standard antenna test ranges.

13.16.1 Antenna Test Ranges

There are two basic antenna test ranges used for antenna measurements. These are indoor and outdoor test ranges. Usual indoor test range is an anechoic chamber which consists of a rectangular volume enclosed by microwave absorber walls. These walls reduce reflections from the boundary walls and increase the measurement accuracy. Microwave absorbers are carbon impregnated polyurathene foam in the shape of pyramids. The materials are expensive for lower frequency ranges because the typical size of a pyramid is nearly 5'–6' for 100 MHz. Most of the antenna parameters to be measured require uniform plane wave field incidence on the test antenna placed at a far field distance from a transmitting antenna. Consequently the requirement of a large space limits the use of the costly indoor facility. Special indoor ranges such as compact range and near field range could be used where the former produces a plane wave field in a smaller distance by means of an offset fed reflector antenna having a special edge geometry. The latter one uses mathematical computations of the near field measurement data to obtain the far field information. Both these methods are very costly and have several limitations. In this section, far field outdoor test range is discussed for antenna parameter measurements.

Outdoor antenna test range The most popular microwave antenna test range is the free space outdoor range in which the antennas are mounted on tall towers as shown in Fig.13.39. The reflections from the surrounding environment are reduced by

1. Selecting the directivity and side lobe level of the transmitting antenna.
2. Making line-of-sight between the antennas obstacle free.
3. Absorbing or redirecting the energy that is reflected from the range surface or from any obstacle.

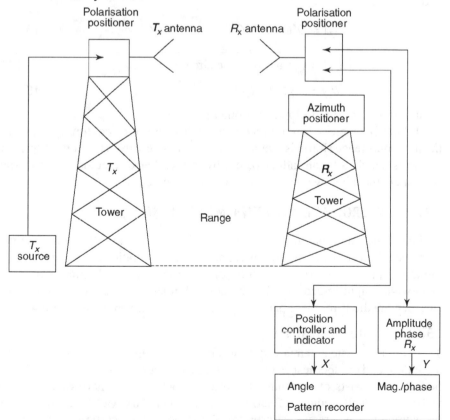

Fig. 13.39 *Outdoor antenna test range*

13.16.2   Radiation Pattern Measurements

The radiation pattern is a representation of the radiation characteristics of the antenna as a function of elevation angle θ and azimuthal angle ϕ for a constant radial distance and frequency. The three-dimensional pattern is decomposed into two orthogonal two-dimensional patterns in **E** and **H** field planes where the Z-axis is the line joining the transmitting and receiving antennas and perpendicular to the radiating apertures. Due to the reciprocal characteristics of antennas, the measurements are performed with the test antenna placed in the receiving mode. The source antenna is fed by a stable source and the received signal is measured using a receiver. The output of the receiver is fed to Y-axis input of an *XY* recorder. The receiving antenna positioner controller plane and the angle information is fed to X-axis input of the *XY* recorder. Thus the amplitude vs angle plot is obtained from the recorder output.

Initially two antennas are aligned in the line of their maximum radiation direction by adjusting the angle and height by the controller and antenna mast. Effects of all surroundings are removed or suppressed through increased directivity and low side lobes of the source antenna, clearance of LOS, and absorption of energy reaching the range surface.

The following precautions are taken for better accuracy in the measurements:

1. Effects of coupling between antennas—inductive or capacitive—causes error in measurement. The former exists at lower microwave frequencies and negligible if range $R \geq 10\ \lambda$. Mutual coupling due to scattering and reradiation of energy by test and source antenna causes error in measurement.

2. Effect of curvature of the incident phase front produces phase variation over the aperture of test antenna and this restricts the range R. For a phase deviation at the edge $\leq \pi/8$ radians, $R \leq 2D^2/\lambda$, where D is the maximum size of the aperture.

3. Effect of amplitude taper over the test aperture will give deviation of the measured pattern from the actual. This occurs if the illuminating field is not constant over the region of the test aperture. Tolerable limit of amplitude taper is 0.25 dB, for which decrease in gain is 0.1 dB.

4. Interference from spurious radiating sources should be avoided.

13.16.3   Phase Measurement

The phase of the radiated field is a relative quantity and is measured with respect to a reference as shown in Fig.13.40. This reference is provided either by coupling a fraction of the transmitted signal to the reference channel of the receiver or by receiving the transmitted signal with a fixed antenna placed near the test antenna. The fixed antenna output is fed to the reference channel of the receiver and the phase pattern is recorded as the antenna under test is rotated in the horizontal plane.

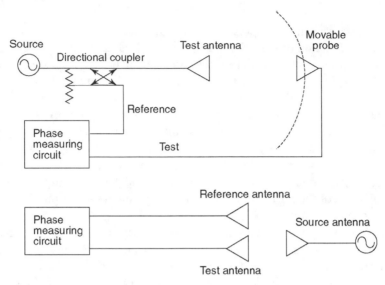

Fig. 13.40 *Phase pattern measurement set-up*

13.16.4 Phase Centre Measurement

When an antenna radiates, there is an equivalent point in the antenna geometry which represents the radiation centre. At the far field region the phase pattern of this antenna remains constant with angle when measured with respect to this point. Therefore, the phase centre of the test antenna is determined by positioning the rotational axis of the test antenna mast such that the phase pattern within the main beam remains constant.

13.16.5 Beamwidth

The beamwidth of the antenna is calculated from the angle subtended by the 3 dB or 10 dB points on both sides of the radiation maximum in the main beam.

13.16.6 Gain Measurements

The gain is the most important parameter to be measured for microwave antennas because it is used directly in the link calculations. There are three basic methods that can be used to measure the gain: standard antenna method, two antenna method and three antenna method.

Standard antenna method　This method uses two sets of measurements with the test and standard gain antennas. Using the test antenna of gain G_r in receiving mode, the received power P_r is recorded in a matched recorder. The test antenna is then replaced by a standard gain antenna of gain G_s and the received power P_s is again recorded without changing the transmitted power and geometrical configuration. Then

$$\frac{P_r}{P_s} = \frac{G_r}{G_s}$$

or,

$$G_r(\text{dB}) = G_s(\text{dB}) + 10 \log\left(\frac{P_r}{P_s}\right) \qquad (13.129)$$

Thus by measuring the received power with test and standard gain antennas and knowing gain G_s of the standard gain antenna, the gain of the test antenna can be found.

Two antenna method   In this method the signal is transmitted from a transmitting antenna of gain G_t and the signal is received by the test antenna of gain G_r placed at far-field distance R. The received power is expressed by

$$P_r = \frac{P_t\, G_t\, G_r\, \lambda^2}{(4\,\pi\,R)^2}$$

or,

$$G_r(\text{dB}) + G_t(\text{dB}) = 20 \log\left(\frac{4\,\pi\,R}{\lambda}\right) + 10 \log\left(\frac{P_r}{P_t}\right); \qquad (13.130)$$

where P_r is the received power and P_t is the transmitted power. When the two antennas are selected identical, $G_t = G_r$, so that

$$G_r(\text{dB}) = G_t(\text{dB}) = 10 \log\left(\frac{4\,\pi\,R}{\lambda}\right) + 5 \log\left(\frac{P_r}{P_t}\right); \qquad (13.131)$$

By measuring R, λ and P_r/P_t, the gain G_r can be determined.

Three antenna method   In the two antenna method if the measuring systems are not exactly identical, error will be introduced. Hence the three antenna method is the most general method to find gain of all the three antennas. Any two antennas are used at a time i.e. 1 and 2, 2 and 3, and 3 and 1, respectively. The following equations can be developed for the received and transmitted powers

$$G_1(\text{dB}) + G_2(\text{dB}) = 20 \log\left(\frac{4\,\pi\,R}{\lambda}\right) + 10 \log\left(\frac{P_{r2}}{P_{t1}}\right) \qquad (13.132)$$

$$G_2(\text{dB}) + G_3(\text{dB}) = 20 \log\left(\frac{4\,\pi\,R}{\lambda}\right) + 10 \log\left(\frac{P_{r3}}{P_{t2}}\right) \qquad (13.133)$$

$$G_3(\text{dB}) + G_1(\text{dB}) = 20 \log\left(\frac{4\,\pi\,R}{\lambda}\right) + 10 \log\left(\frac{P_{r1}}{P_{t3}}\right) \qquad (13.134)$$

Since R and λ are known and (P_r/P_t)'s are measured, the right hand side of the above equations are known. Then three unknown quantities G_1, G_2 and G_3 can be determined from the three equations.

A typical block diagram of the measurement set-up for two and three antenna methods are shown in Fig.13.41.

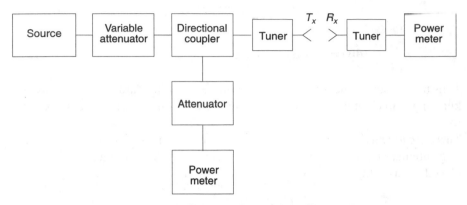

Fig. 13.41 *Block diagram of antenna gain measurements*

For accuracy of the measurements, care must be taken so that
1. All antennas meet the far field criteria : $R \geq 2D^2/\lambda$.
2. The antennas are aligned for bore-sight radiation face-to-face.
3. The measuring system is frequency stable.
4. Impedance mismatched in the system components is minimum.
5. Polarisation mismatch is minimum.
6. Reflection from various background and support structure is minimum.

13.16.7 Directivity Measurements

The directivity of an antenna can be determined from the measurements of its radiation pattern in two principal planes, E and H planes and finding the half-power beamwidths θ_E and θ_H degree in these planes, respectively.

$$D_0 = \frac{41,253}{\theta_E \, \theta_H} \quad \text{or,} \quad \frac{72,815}{\theta_E^2 + \theta_H^2} \tag{13.135}$$

This method is accurate for the antennas having negligible side lobes.

13.16.8 Radiation Efficiency

$$\text{The radiation efficiency} = \frac{\text{Total power radiated, } P_{\text{rad}}}{\text{Total power accepted at its input}}$$

$$= \frac{P_{\text{rad}}}{P_{\text{in}} - P_{\text{ref}}} = \frac{\text{Gain}}{\text{Directivity}} \tag{13.136}$$

where P_{in} is the input power and P_{ref} is the reflected power at the input. Therefore, the radiation efficiency can be determined from the measurement of gain and directivity.

13.16.9 Polarisation Measurements

The polarisation of an antenna is conveniently measured by using it in the transmitting mode and probing the polarisation by a dipole antenna in the plane that contains the direction of the electric field as shown in Fig.13.42. The dipole is rotated in the plane of polarisation and the received voltage pattern is recorded and analysed as follows.

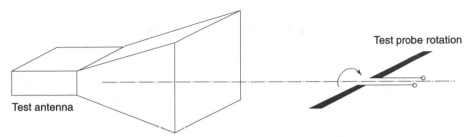

Fig. 13.42 *Polarisation measurements*

Linear polarisation   For linear polarisation, the output voltage pattern will be a figure of eight.

Circular polarisation   For circular polarisation, the output voltage pattern will be a circle.

Elliptical polarisation   For an elliptical polarisation, the nulls of the figure of eight are filled and a dumb-bell polarisation curve is obtained which will be tilted and a polarisation ellipse can be drawn as shown by dashed curve in Fig.13.43. The sense of rotation of the circular and elliptical polarisations can be determined by comparing the responses of two circularly polarised antennae, one left and the other rightwise rotation. The polarisation of the test antenna will be the same as that of one of these two directions for which the response is larger.

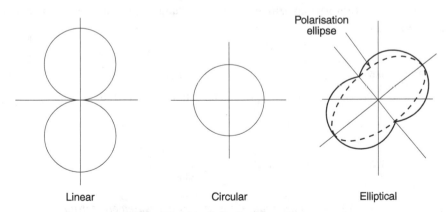

Fig. 13.43 *Polarisation patterns*

13.17   RADAR CROSS-SECTION (RCS) MEASUREMENTS

The radar cross-section of a target is defined by

$$\sigma = \frac{4\pi \times \text{Power re-radiated per unit solid angle}}{\text{Incident power density}} \quad (13.137)$$

and is expressed in terms of received power

$$P_r(\theta) = \frac{P_t\, G_t\, A_e\, \sigma(\theta)}{\left(4\pi R^2\right)^2} \qquad (13.138)$$

where

P_t = Transmitted power

G_t = Gain of transmit antenna relative to an isotropic radiator

A_e = Effective area of the receiving antenna

R = Distance of the target

θ = Aspect angle of the target which is the angle of direction of re-radiated power with respect to the line joining the T_x and the centre of the target

When all the factors remain constant in above equation

$$\sigma(\theta) = KP_r(\theta) \qquad (13.139)$$

where K is a constant. Normalising with respect to the angle $\theta = 0$,

$$\frac{\sigma(\theta)}{\sigma(0)} = \frac{P_r(\theta)}{P_r(0)} \qquad (13.140)$$

Thus by measuring the received power, the radar cross-section of a target placed at far field location, can be determined.

There are two basic radar cross-section terms i.e. monostatic or back scattering cross-section and bistatic cross-section. The former is defined for the reflected signal received by the common transmit/receive antenna. The other one is defined when the scattered signal is measured in the forward direction other than towards the source end. We will be interested in the back scattered radar cross-section in this chapter.

The basic laboratory experimental set up for radar cross-section measurement is shown in Fig.13.44.

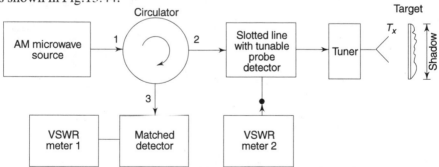

Fig. 13.44 *Basic radar cross-section measurement set-up*

Since signals reflected from the background cause error in the measurement, the tuner is adjusted to obtain minimum reading in VSWR meter-1 ($S < 1.05$) without the target present. Without disturbing the tuner, the target is now positioned and received signal strength is adjusted to read $S = 1$ in the VSWR meter-2 when $\theta = 0°$. The target is now rotated in azimuthal angular directions and corresponding readings in the VSWR meter-2 are noted to obtain normalised RCS Vs angle information.

The possible sources of error in the above method are

1. When background cancellation was done by using the tuner, the target shadow region at its back contributed to background echo which is not present when the target was positioned. This produces imperfect background cancellation.
2. Finite isolations from isolator port 1 to 3 and from port 2 to 1 produce error in the measurement.

In order to improve the isolation between transmit and receive signal, the circulator can be replaced by arranging a separate receiving antenna system by the side of the transmitting antenna. Under these circumstances the RCS is called quasi-monostatic. However, transmit and receive antennas are considered to be located at almost same origin when the target is far away from the antennas.

For better cancellation of actual background echo, RCS of a target is often defined in terms of an equivalent sphere whose RCS = σ_{sphere} is precisely known and is independent of aspect angle. Following measurements are recommended inside an anechoic chamber of low reflection coefficient as shown in Fig.13.45.

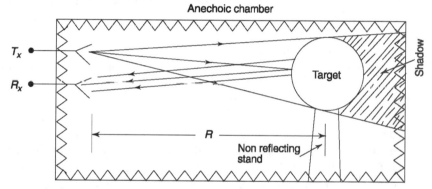

Fig. 13.45 *Quasi-monostatic radar cross-section measurements*

1. With an equivalent sphere of almost same vertical cross-section of the target the return signal strength is measured as P_{rs}.
2. Without disturbing anything in the chamber the sphere is replaced by the target under test and the received power P_{rt} is measured.

   Then the RCS of the target is obtained as

$$\sigma_{\text{target}}(\theta) = \frac{P_{rt}(\theta)}{P_{rs}} \times \sigma_{\text{sphere}} \tag{13.141}$$

EXERCISES

13.1 The signal power at the input of a device is 10 mW. The signal power at the output of the same device is 0.20 mW. Calculate the insertion loss in dB of this component.

13.2 A crystal detector generates a signal of 10 mV for an incident microwave power of -25 dBm. What is the detector sensitivity in mV/mW? Why is the microwave signal amplitude modulated by a 1 kHz square wave before it is applied to a crystal detector?

13.3 Calculate the VSWR in dB in a waveguide when the load is a 3 dB attenuator terminated by a short circuit.

13.4 The reflection coefficient of a load is $0.5/\angle -30°$. Using the Smith chart determine the normalised admittance of the load.

13.5 A waveguide with a load is matched by a tuning screw located at position 1. What impedance will be presented by the waveguide to the generator if the screw is moved by half guide wavelength towards the load from position 1 ?

13.6 A microwave signal is modulated by a rectangular pulse of width 1 μs. If the average power is 200 W and the pulse repetition rate is 500 pulses per second, calculate the value of the peak power.

13.7 In attenuation measurement of an attenuator the microwave source is modulated by 1 kHz square wave. The VSWR meter is peak to 0 dB with the 30 dB range. When the attenuator is inserted, the VSWR meter reads a value of 2 on the VSWR scale in 40 dB range. Find the attenuation provided by the attenuator.

13.8 A slotted line is used to measure VSWR of the load at 2 GHZ by double minima method. If the distance between the positions of twice minimum power is 0.5 cm, find the value of VSWR on the line and the magnitude of the voltage reflection coefficient.

13.9 In a reflectometer set-up two identical directional couplers are used to measure the incident and the reflected power. If the power level of the reverse coupler is 12 dB down from the level of the forward coupler, what is the VSWR on the line?

13.10 A rectangular waveguide is terminated with an unknown load. By using a slotted line, load VSWR is found to be equal to 2 and distance between two successive minima is 1.5 cm. When the load is replaced by a short, the position of a given minimum shifts 0.3 cm towards the generator. Find the normalised value of the load.

13.11 A slotted line with a short circuit termination measures two successive minima at 25.3 cm apart. When an unknown load is connected, the VSWR is 3.1 and the minimum occurs at 16.5 cm position. Find (a) voltage reflection coefficient at the load, (b) the load impedance when the characteristic impedance of the line is 50 ohm.

REFERENCES

1. Montgomery, C G, "Technique of microwave measurements", *RLS*, Vol.11.
2. Sucher, M and J Fox, *Handbook of microwave measurement,* 3rd. Edn. Vol. II, pp. 431.
3. Lebedev, I, *Microwave Engineering,* MIR Publishers, pp. 313.

4. Lerner, D S, and H A Wheeler, "Measurement of bandwidth of microwave resonator by phase shift of signals-modulation", *IRE, Trans. on MIT*, Vol. MTT-8. pp 343, May, 1960.

5. El-Ibiary, M Y, "Q of resonant cavities, measurement by phase shift method", *Electronic Tech.*, pp 284 July, 1960.

6. Ragon, G L, "Microwave Transmission Circuits", *RLS* ; Vol. 9.

7. Collin, R E, *Foundation of Microwave Engineering* , McGraw-Hill, 1966.

8. Harrington, J F, *Time-Harmonic Electromagnetic Fields*, McGraw-Hill, 1961.

9. Reich, Skolnik, Ordung and Krauss, *Microwave Principles*, East-West Press, 1972.

10. Singh, A, "An improved method for the determination of Q of cavity resonators", *IRE Trans*, on Microwave Theory and Techniques, Vol. MTT-6, pp 155-160, April, 1958.

11. Ginzton, E L, "Microwave Q Measurement in the Presence of Coupling Losses", *IRE Trans*, on Microwave Theory and Techniques, Vol. MTT-6, pp 383-389, October, 1958.

12. Barrington, A E, and J R, Rees, "A simple 3 cm Q-meter", *Proc. I.E.E.* Vol.105 B, pp 511-512, November,1958.

13. Boverly, "Inflection Point Method of Measuring Q at Very High Frequencies", *Sperry Engineering Rev.*, Vol. 4, No. 3, pp 16-19 (May-June, 1951). Sperry Gyroscope Co., Great Neck, New York.

14. LeCaine, H, "The Q of Microwave Cavity by Comparison with a Calibrated High Frequency Circuit", *Proc. IRE*, Vol. 40, pp 155-157, February 1952.

15. *Proc IEEE*, 1966, February, 1966 pp 311.

16. "Electronics Letters", Vol.10, No.8. 18th April 1966.

17. The Narda Microwave Corporation Catalog No.12.

18. Wilson, I G, Schramm, C W, and Kinzer, JP, "High Q Resonant Cavities for Microwave Testing", *BSTJ*,Vol. 25, pp 408, July, 1946.

19. Kinzer, J P and Wilson, I G, "Some Results on Cylindrical Cavity Resonators", BSTJ, Vol. 26, pp 410, July 1847.

20. Shaw T M and Windle, J J, "Microwave Techniques for the Measurement of the Dielectric Constant of Fibers and Films of High Polymers", *J. Appl. Physics*, Vol. 21, No.10, Oct. 1950.

21. Surber, Jr. W H, and Crouch, G E, Jr., "Dielectric Measurement Methods for Solid at Microwave Frequencies", *J. Appl. Physics*, Vol.19, No.12, December 1948.

22. Birnbaum, G and J Franeau, "Measurement of the Dielectric Constant and Loss of Solids and Liquids by a Cavity Perturbation Method", *J. Appl. Physics*, Vol. 20, No. 8, August 1949.

23. Dakin T W, and Works, C, "Microwave Dielectrics Measurements", *J. Appl. Physics*, Vol.18, No. 9, September 1947.

24. *Proceeding IEEE Trans*. Vol. 62, No.1, January 1974.

25. Deschamp, G A, "Determination of reflection coefficient and insertion loss of a waveguide junction", *A. Appl. Physics,* 1953, 28, pp. 1046-1050.

26. Storer, J E, et. al., "A simple graphical analysis of a two-port junction", *Proc. I.R.E.*, 1953, 41, pp.1004-1013.

27. Wenger, N C, "The launching of surface waves on an axial-cylindrical reactive surface", *IEEE Trans*. Antennas Propagate, Jan. 1965, pp. 126-134.

28. Collin, R E, *Field theory of guided waves*, McGraw-Hill Book Company, pp. 482, 1960.

29. Balanis, C A, *Antenna Theory Analysis and Design*, Harper and Row, Publishers, New York,1982.

INDEX